DEVELOPMENTS IN
EARTH SURFACE PROCESSES, 15

VOLUME FIFTEEN

GEOMORPHOLOGICAL MAPPING

DEVELOPMENTS IN EARTH SURFACE PROCESSES, 15

1. PALEOKARST: A SYSTEMATIC STUDY AND REGIONAL REVIEW
   P. BOSÁK, D. FORD, J. GLAZEK and I. HORÁCEK (Editors)
   [OUT OF PRINT]
2. WEATHERING, SOILS & PALEOSOLS
   I.P. MARTINI and W. CHESWORTH (Editors)
3. GEOMORPHOLOGICAL RECORD OF THE QUATERNARYOROGENY IN THE HIMALAYA AND THE KARAKORAM
   JAN KALVODA (Editor)
   [OUT OF PRINT]
4. ENVIRONMENTAL GEOMORPHOLOGY
   M. PANIZZA
5. GEOMORPHOLOGICAL HAZARDS OF EUROPE
   C. EMBLETON and C. EMBLETON-HAMANN (Editors)
6. ROCK COATINGS
   R.I. DORN
7. CATCHMENT DYNAMICS AND RIVER PROCESSES
   C. GARCIA and R.J. BATALLA (Editors)
8. CLIMATIC GEOMORPHOLOGY
   M. GUTIÉRREZ
9. PEATLANDS: EVOLUTION AND RECORDS OF ENVIRONMENTAL AND CLIMATE CHANGES
   MARTINI, A. MARTINEZ CORTIZAS and CHESWORTH (Editors)
10. MOUNTAINS WITNESSES OF GLOBAL CHANGES RESEARCH IN THE HIMALAYA AND KARAKORAM: SHARE-ASIA PROJECT
    RENATO BAUDO, GIANNI TARTARI and ELISA VUILLERMOZ (Editors)
11. GRAVEL-BED RIVERS VI: FROM PROCESS UNDERSTANDING TO RIVER RESTORATION
    HELMUT HABERSACK, HERVÉ PIÉGAY and MASSIMO RINALDI (Editors)
12. THE CHANGING ALPINE TREELINE: THE EXAMPLE OF GLACIER NATIONAL PARK, MT, USA
    DAVID R. BUTLER, GEORGE P. MALANSON, STEPHEN J. WALSH and DANIEL B. FAGRE (Editors)
13. NATURAL HAZARDS AND HUMAN-EXACERBATED DISASTERS IN LATIN AMERICA: SPECIAL VOLUMES OF GEOMORPHOLOGY
    EDGARDO M. LATRUBESSE (Editor)
14. THE WESTERN ALPS, FROM RIFT TO PASSIVE MARGIN TO OROGENIC BELT: AN INTEGRATED GEOSCIENCE OVERVIEW
    PIERRE-CHARLES DE GRACIANSKY, DAVID G. ROBERTS and PIERRE TRICART (Editors)

DEVELOPMENTS IN
EARTH SURFACE PROCESSES

VOLUME FIFTEEN

GEOMORPHOLOGICAL MAPPING

METHODS AND APPLICATIONS

MIKE J. SMITH
School of Geography, Geology and the Environment, Kingston University

PAOLO PARON
School of Geography and the Environment, Oxford University, United Kingdom & UNESCO-IHE, Institute for Water Education, Delft, The Netherlands

JAMES S. GRIFFITHS
School of Earth, Ocean & Environmental Sciences University of Plymouth, United Kingdom

ELSEVIER

Amsterdam • Boston • Heidelberg • London • New York • Oxford
Paris • San Diego • San Francisco • Singapore • Sydney • Tokyo

Elsevier
The Boulevard, Langford Lane, Kidlington, Oxford OX5 1GB, UK
Radarweg 29, PO Box 211, 1000 AE Amsterdam, The Netherlands

First edition 2011

British Library Cataloguing in Publication Data
A catalogue record for this book is available from the British Library

Library of Congress Cataloging-in-Publication Data
A catalog record for this book is available from the Library of Congress

ISBN: 978-0-444-53446-0
ISSN: 0928-2025

Printed and bound in Great Britain
11 12 13 14 15 10 9 8 7 6 5 4 3 2 1

CONTENTS

FOREWORD

When Paolo Paron first suggested the idea of this book on geomorphological mapping to me several years ago, I immediately recognised the potential of the concept for several reasons. First of all, my own interest in the topic began more than four decades ago when I originally discovered the gaps in my own rather Davisian education. I well remember the excitement I felt back in the 1960s when on my first trip to Europe, I was introduced to detailed geomorphologic mapping where many new symbols and detailed geomorphological maps were being introduced by Demek (1967, 1972), Verstappen (1970, 1983) and many others. At about the same time when I was returning to the United States, St. Onge (1968) first introduced the concept of large-scale geomorphological mapping in North America when he wrote a paper on the topic in Fairbridge's (1968) seminal *Encyclopedia of Geomorphology.* This was the volume that first taught me geomorphology beyond Thornbury's (1954/1968) *Principles of Geomorphology* of my undergraduate and graduate education. Thornbury had little to say about geomorphologic mapping even though the book had a whole chapter devoted to the 'Tools of the Geomorphologist'. Instead, only standard topographic maps and aerial photographs were mentioned by Thornbury, probably because in those days the techniques of geomorphologic mapping had not yet been refined to the fine art and science they later became.

Another factor that Paolo Paron and I discussed when we planned this volume was our mutual desire to try to make the newer ideas and techniques of geomorphological mapping more available to the poorer nations of the lesser developed world. We did not necessarily achieve this because the costs and uncertainties of the modern publishing world necessitated a fairly high price for this publication, but the increasing availability of these materials in electronic domains that can be more easily accessed over the Internet means that we will have succeeded at least in part in our original objectives. This book, *Geomorphological Mapping: Methods and Application*, is thus an attempt to explain and give examples of how this highly technical methodology can be applied and utilised to solve complex problems in land use and provide some of the more benign answers to development in the developing world.

As one might expect, the origins of geomorphological mapping are diverse and the resultant techniques are replete with differences, as this volume shows. The history of the development of geomorphological mapping grew from a need for a more analytical approach to interpretation of landforms more than half a century ago. As Hayden (1986) has noted, the study of landforms in the nineteenth and early years of the twentieth century was marked by rather static descriptive physiographies in which landscapes were discussed largely in writing. These older papers and books were generally accompanied by artistic block diagram drawings to illustrate the author's conclusions about what could be some wholly imagined geologic history.

An approximate dividing line between these older notions of geomorphology and newer thinking was World War II (Klimaszewski, 1982), after which the science of geomorphology took on a more modern and useful flavour. A more pragmatic geomorphology emerged, particularly in Europe, where geomorphologists became interested in comprehensive regional analyses of landforms that considered all the features and aspects of the landscape together. The natures and relationships of past landform processes in an area were compared and contrasted to the active processes of the present day. The significance and influences of landforms and relief on vegetation, hydrology and human cultural development were investigated. The interpretation of the complexity of such landscapes necessitated objective scientific methods of graphic portrayal of these landform factors. The detailed geomorphological map thus became, in many countries, the main research method in geomorphology (Demek, 1982). Extensive, complex and highly colourful graphic symbologies were developed, commonly different for different countries or for different applications. Some subjectivity unfortunately crept into some of the maps, particularly where loosely documented geologic histories were allowed to control some age assignments. The overall result, however, was the fairly accurate geomorphologic mapping of much of Europe, particularly where maps in planning, engineering and management were desired. In North America, however, a certain distaste existed for many aspects of central planning, with the result that the mapping techniques were not applied much there. As the editors note herein (Smith et al., 2011), the continued fragmentation of legend systems between the different users and other problems led to the relegation of much geomorphologic mapping throughout the remainder of the twentieth and into the present century as a rather adjunct activity. The new viewpoints and methods expressed herein, however, seem to have signalled an end to such lack of recognition of the true usefulness of this methodology. In fact, a bit of a renaissance in geomorphological mapping seems to be underway at the present time, as is attested to by Pavlopoulos et al. (2009), and especially by the recent 41st International Binghamton Geomorphological Symposium that was held on *Geospatial Technologies and Geomorphological Mapping* (Bishop et al., 2011).

This current volume, *Geomorphological Mapping: Methods and Applications*, is the fifteenth in our series on *Developments in Earth Surface Processes*. It is a professional handbook of techniques and applications targeted at academics and practitioners who wish to use geomorphological mapping in their work. This volume synthesises an historical perspective to the use of field-based geomorphological mapping in which new digital tools and techniques are now being used effectively in the process. Material is brought together for digital mapping from remote sensing into environments of cartography, geographic information systems and digital terrain analysis. Extensive case studies with plentiful use of diagrams and colour plates in the volume show the diverse nature of geomorphological mapping as it is practiced in the twenty-first century. Accompanying electronic resources can add to the usefulness of the work for geomorphologists who are interested in mapping in the field. Those active in geomorphology, engineering geology, the insurance industry, assessors of environmental impacts and allied areas should find the text of considerable value in their work. The authors and editors are convinced that the integrative methodology displayed in this volume has much to offer the practitioners and others who may wish to learn more about this increasingly specialised but highly useful, analytical methodology.

John F. Shroder Jr.
Editor-in-Chief
Developments in Earth Surface Processes

REFERENCES

Bishop, M.P., James, A., Walsh, S.J., Shroder Jr., J.F., 2011. Geospatial technologies and geomorphological mapping: concepts, issues and research directions. In: Bishop, M.P., James, A., Walsh, S.J. (Eds.), Geospatial Technologies and Geomorphological Mapping: 41st International Binghamton Geomorphological Symposium. Elsevier.

Demek, J., 1967. Generalization of geomorphological maps. In: Progress Made in Geomorphological Mapping, vol. 9. Geografický ústv ČSAV, Brno, pp. 36–72.

Demek, J. (Ed.), 1972. Manual for Detailed Geomorphological Mapping. IGU Commission on Geomorphic Survey and Mapping, Academia, Prague.

Fairbridge, R.W., 1968. Encyclopedia of Geomorphology. Rheinhold Book Co., New York, NY.

Hayden, R.S., 1986. Geomorphological mapping. In: Short, N.M., Short, R.W. (Eds.), Geomorphology from Space: A Global Overview of Regional Landforms. NASA, Washington, DC, pp. 637–656.
Klimaszewski, M., 1982. Detailed geomorphological maps. ITC J. 3, 265–271.
Pavlopoulos, K., Evelpidou, N., Vassilopoulos, A., 2009. Mapping Geomorphological Environments. Springer-Verlag, Berlin.
Smith, M.J., Griffiths, J., Paron, P., 2011. Future developments of geomorphological mapping. In: Smith, M.J., Paron, P., Griffiths, J. (Eds.), Geomorphological Mapping: Methods and Applications. Elsevier, London, pp. 589–593.
St. Onge, D., 1968. Geomorphological maps. In: Fairbridge, R.W. (Ed.), Encyclopedia of Geomorphology. Rheinhold Book Co., New York, NY, pp. 388–403.
Thornbury, W.D., 1954. Principles of Geomorphology. John Wiley & Sons, New York, NY.
Verstappen, H.Th., 1970. Introduction to the ITC-system of geomorphological survey. Geogr. Tijdschr. 4 (1), 85–91.
Verstappen, H.Th., 1983. Applied Geomorphology: Geomorphological Surveys for Environmental Development. Elsevier, Amsterdam, pp. 255–275.

FOREWORD

Between the International Association of Geomorphologists' (IAG) International Conferences in Zaragoza (2005) and Melbourne (2009), the IAG decided to establish a series of new working groups concerned with important issues in the discipline which would benefit from international collaboration. One of these issues was that of Applied Geomorphological Mapping.

As Professor Verstappen points out in Chapter 2, geomorphological mapping is not in itself new, and when the British Geomorphological Research Group, a parent of the IAG, was established five decades ago, one of its first roles was to try and establish a certain uniformity and logic of approach and annotation. Pioneering symbol-based geomorphological mapping was also carried out in many European countries but sometimes suffered from the fact that the aims, scale and purpose of the mapping were not always clearly identified. Such maps were all too frequently seen to be the *object* of research rather than a *tool* of research. Moreover, it was not always clear to whom they were aimed. This approach was sometimes derided and geomorphological mapping ceased to be as fashionable as it once had been. A strong regional and descriptive bent in geographical geomorphology, into which geomorphological mapping fitted, was replaced by a move towards reductionist process studies. Having said that, some first-class mapping work was undertaken that proved to be of great value in resource mapping and hazard evaluation, not least by organisations such as ITC in the Netherlands, CSIRO in Australia, and various UK geomorphologists working in the Middle East and elsewhere.

Recent years have seen a resurgence in the use of geomorphological maps for applied research. This is reflected in this volume, which contains an impressive range of studies and a list of authors who are impressive for their internationalism. Maps remain a very powerful tool for transmitting information to clients, but their value has been hugely magnified of late because of the availability and use of new techniques, including remote sensing, computation, digitisation, geostatistics, modelling, GPS, GIS, etc. This is all made clear by Dramis et al. in Chapter 3, who draw attention to the value of a multi-scale approach. The use of geomorphological maps has now been extended to new environments, and planetary and submarine geomorphological mapping are particularly exciting areas of research. Maps tell us where things are, what they are like, how their properties and distributions have changed through time and how phenomena correlate spatially. These functions are fundamental for locating and understanding resources and risks, the twin foundations of applied geomorphology.

I believe that this timely volume will highlight the fact that skills in applied geomorphological mapping are a very necessary and basic part of the training for all geomorphologists.

Andrew Goudie
IAG President 2005–2009

LIST OF CONTRIBUTORS

Niels Steven Anders
Institute for Biodiversity and Ecosystem Dynamics, University of Amsterdam, Nieuwe Achtergracht 166, WV Amsterdam, The Netherlands
n.s.anders@uva.nl

Ian Anderson
Halcrow Group Ltd, Martlett House, Chichester, UK

Sara Benetti
School of Environmental Sciences, University of Ulster, Coleraine BT52 1SA, Northern Ireland

James Brasington
Department of Geography, University of Canterbury, Private Bag 4800, Christchurch 8140
james.brasington@canterbury.ac.uk

Denys Brunsden
Vine Cottage, Sea Lane, Chideock near Bridport, Dorset DT6 6LD, UK

Antonello Cestari
C.U.G.R.I., Great Risks Interuniversity Consortium, University of Salerno, via Ponte Don Melillo, Fisciano, SA 84084, Italy
acestari@unisa.it

Lieven Claessens
Land Dynamics Group, Wageningen University and Research Centre, Wageningen, The Netherlands
International Potato Center (CIP), Sub-Saharan Africa Regional Office, Nairobi, Kenya

Jonathan D.A. Clarke
Geoscience Australia, P.O. Box 378, Canberra, ACT 2601, Australia

Francesco Dramis
Department of Geological Sciences, Roma Tre University, Largo San Leonardo Murialdo 1, Rome, Lazio 00146, Italy
dramis@uniroma3.it

Paul Dunlop
School of Environmental Sciences, University of Ulster, Coleraine BT52 1SA, Northern Ireland
p.dunlop@ulster.ac.uk

Martin Geilhausen
Department of Geography and Geology, University of Salzburg, Hellbrunnerstr. 34, A-5020 Salzburg, Austria
martin.geilhausen@sbg.ac.at

James S. Griffiths
SoGEES, University of Plymouth, Drake Circus, Plymouth, Devon PL4 8AA, UK
jim.griffiths@plymouth.ac.uk

Domenico Guida
Department of Civil Engineering, University of Salerno, Via Ponte Don Melillo, Fisciano, SA 84084, Italy
dguida@unisa.it

Marcus Gustavsson
Helsingforsgatan 71, S-75264 Uppsala, Sweden

Peter L. Guth
Department of Oceanography, United States Naval Academy, 572C Holloway R, Annapolis, MD 24102, USA
pguth@usna.edu

Andrew B. Hart
Geo-Hazard, Scott Wilson Ltd, Scott House, Alencon Link, Basingstoke, UK

Yuichi Hayakawa
Center for Spatial Information Science, University of Tokyo, 5-1-5 Kashiwanoha, Kashiwa 277-8568, Japan
hayakawa@csis.u-tokyo.ac.jp

Gareth J. Hearn
Geo-Hazard, Scott Wilson Ltd, Scott House, Alencon Link, Basingstoke, UK
gareth.hearn@scottwilson.com

Tomislav Hengl
Institute for Biodiversity and Ecosystem Dynamics, University of Amsterdam, Nieuwe Achtergracht 166, 1018 WV Amsterdam, The Netherlands
t.hengl@uva.nl

Vanessa M.A. Heyvaert
Geological Survey of Belgium, Royal Belgian Institute for Natural Sciences, Jennerstraat 13, B-1000 Brussels, Belgium

Murray Hicks
National Institute for Water and Atmosphere, New Zealand
m.hicks@niwa.co.nz

John K. Hillier
Department of Geography, Loughborough University, Leics, UK, LE11 3TU
j.hillier@lboro.ac.uk

Bernhard Höfle
Department of Geography, University of Heidelberg, Heidelberg, Germany

Graham Hunter
3D Laser Mapping Ltd, 1a Church Street, Bingham, Nottingham, UK

David K.C. Jones
Horsepen, Main Street, Beckley near Rye, Sussex TN31 6RS, UK

Jasper Knight
School of Geography, Archaeology and Environmental Studies, University of the Witwatersrand, Private Bag 3, Wits 2050, Johannesburg, South Africa
jasper.knight@wits.ac.za

Fred Labrosse
Department of Computer Science, Aberystwyth University, UK
ffl@aber.ac.uk

E.Mark Lee
15 Whernside Avenue, York YO31 0QB, UK

Aaron Micallef
IOI-Malta Operational Centre, University of Malta, Level 3, Chemistry Building, MSD 2080, Malta
micallefaaron@gmail.com

Wishart A. Mitchell
Department of Geography, Durham University,
Durham DH1 3LE, UK
w.a.mitchell@durham.ac.uk

Mark Neal
Department of Computer Science, Aberystwyth University, UK
mjn@aber.ac.uk

Colm Ó Cofaigh
Department of Geography, Durham University, Durham DH1 3LE, UK

Takashi Oguchi
Center for Spatial Information Science, University of Tokyo, 5-1-5 Kashiwanoha, Kashiwa 277-8568, Japan
oguchi@csis.u-tokyo.ac.jp

Jan-Christoph Otto
Department of Geography and Geology, University of Salzburg, Hellbrunnerstr. 34, A-5020 Salzburg, Austria
jan-christoph.otto@sbg.ac.at

Colin F. Pain
Geoscience Australia, P.O. Box 378, Canberra, ACT 2601, Australia
colin.pain@ga.gov.au

Paolo Paron
UNESCO-IHE, Institute for Water Education, Delft, The Netherlands
School of Geography and the Environment, Oxford University,
Oxford, UK
P.Paron@unesco-ihe.org

Steve Parry
GeoRisk Solutions Ltd, Suite 1502, Hollywood Centre, 233 Hollywood Road, Sheung Wan, Hong Kong, China
parrysteve@gmail.com

Norbert Pfeifer
Institute of Photogrammetry and Remote Sensing, Vienna University of Technology, Vienna, Austria

Emmanuel Reynard
Institute of Geography, University of Lausanne, Anthropole, CH-1015 Lausanne, Switzerland

James Rose
Department of Geography, Royal Holloway, University of London, Egham, Surrey TW20 0EX, UK
British Geological Survey, Keyworth, Nottingham, UK
j.rose@rhul.ac.uk

Martin Rutzinger
ITC-Faculty of Geo-Information Science and Earth Observation, University of Twente, Enschede, The Netherlands
rutzinger@itc.nl

Fabio Sacchetti
School of Environmental Sciences, University of Ulster, Coleraine BT52 1SA, Northern Ireland

Arie Christoffel Seijmonsbergen
Institute for Biodiversity and Ecosystem Dynamics, University of Amsterdam, Nieuwe Achtergracht 166, 1018 WV Amsterdam, The Netherlands
a.c.seijmonsbergen@uva.nl

Mike J. Smith
School of Geography, Geology and the Environment, Kingston University, Penrhyn Road, Kingston upon Thames, Surrey KT1 2EE, UK
michael.smith@kingston.ac.uk

David Theler
Institute of Geography, University of Lausanne, Anthropole, CH-1015 Lausanne, Switzerland
dtheler@hotmail.com

Damià Vericat
Forest Technology Centre of Catalonia, Spain
damia.vericat@ctfc.cat

Peter Verkinderen
Department of Languages and Cultures of the Near East and North Africa, Ghent University, Sint-Pietersplein 6, B-9000 Ghent, Belgium

Herman Theodore Verstappen
International Institute of Geo-Information Science and Earth Observation (ITC), University of Twente, Mozartlaan 188, Enschede 7522HS, The Netherlands
hergraverstappen@planet.nl

Michael Vetter
Institute of Photogrammetry and Remote Sensing, Centre of Water Resources, Vienna University of Technology, Vienna, Austria

Jan Walstra
Department of Languages and Cultures of the Near East and North Africa, Ghent University, Sint-Pietersplein 6, B-9000 Ghent, Belgium
jan.walstra@ugent.be

Thad Wasklewicz
Department of Geography, East Carolina University, A-227 Brewster Building, Greenville, NC 27858, USA
wasklewiczt@ecu.edu

Malcolm Whitworth
School of Earth and Environmental Sciences, University of Portsmouth, Drake Circus, Portsmouth, Devon PL4 8AA, UK
malcolm.whitworth@port.ac.uk

Richard Williams
Institute of Geography & Earth Sciences, Aberystwyth University, Llandinam Building, Penglais Campus, Aberystwyth, SY23 3DB
rvw@aber.ac.uk

Vanessa N.L. Wong
Geoscience Australia, P.O. Box 378, Canberra, ACT 2601, Australia
Present address: Southern Cross GeoScience, Southern Cross University, P.O. Box 157, Lismore, NSW 2480, Australia

SECTION 1

Geomorphological Mapping

CHAPTER ONE

Introduction to Applied Geomorphological Mapping

James S. Griffiths[a], Mike J. Smith[b] and Paolo Paron[c]
[a]SoGEES, University of Plymouth, Plymouth, UK
[b]School of Geography, Geology and the Environment, Kingston University, Surrey, UK
[c]UNESCO-IHE, Institute for Water Education, Delft, NL & School of Geography and the Environment, Oxford University, UK

Contents

The survival of humans is heavily dependent on a very narrow zone within the Earth's crust, from the water on the surface, to the few metres depth of agricultural soil, to the couple of hundred metres from which we extract potable groundwater. Although we do extract mineral resources and some groundwater from greater depths, the vast majority of human activities take place on the land, in rivers and lakes, or in the coastal and nearshore zone and predominantly within 100 m of the ground surface. The main exceptions to this depth limit are as follows: hard rock tunnels; deep sea drilling and production rigs; hydrocarbon exploration and exploitation; cross-ocean cables and deep mining. However, it is realistic to conclude that the overwhelming majority of human activities interact with the landforms that make up the surface and near surface of terrestrial, nearshore and offshore 'landscapes'. The scientific investigation of these landscapes, the processes that have formed them over time, the materials which they are composed of, the individual elements that combine to create them and the way they will evolve through time is the discipline of geomorphology. Understanding geomorphology, therefore, can be seen as fundamental to the safe, economic and sustainable development of the planet Earth.

Geomorphology is part of the broad range of disciplines that fall under the general heading of earth sciences, which includes both geology and

Developments in Earth Surface Processes, Volume 15
ISSN: 0928-2025, DOI: 10.1016/B978-0-444-53446-0.00001-X

geography. In Europe geomorphology has traditionally been associated with 'physical geography', whereas in North America it has usually been regarded as part of 'physical geology'. Until the 1960s, a significant part of geomorphological research was engaged in the history of landscape development, but during the 1960s and 1970s there was a shift in Anglo-American geomorphology into smaller scale process studies (Smith et al., 2002). However, landscape development continued as a core component of physical geology studies in the United States (Costa and Graf, 1984). The original geomorphological emphasis on the study of landscapes meant that, in common with the rest of the earth sciences, there has always been the need to compile spatial data and then to present these data in plan form as maps. Although the methodology and representation of much spatial data has a long history (e.g. accurate geological maps date back to William Smith's first map of the United Kingdom in 1815; Winchester, 2001), the presentation of geomorphology in map form has not reached the same level of standardisation, despite some attempts in Europe to produce comprehensive legends (Demek and Embleton, 1978).

The combination of a lack of a standard methodology or commonly accepted legend, plus the move of academic geomorphologists away from spatially extensive studies of landscape development, led to geomorphological mapping being regarded as a somewhat sterile area of study (Gustavsson et al., 2006). The indications are that it was seen as being of limited value in mainstream geomorphological research. However, at the time that the process–response geomorphologists were beginning to concentrate on small-scale studies, the compilation of geomorphological information was found to be fundamental to many applied studies of the Earth's surface, including coastal zone management; route alignment work for roads, railways and pipelines; soil erosion studies; military work using terrain classification for trafficability and tactical analysis; river catchment management; geohazard assessments, notably for civil engineering projects and, increasingly, in offshore studies particularly when seeking resources and identifying the potential hazards to their exploitation (e.g. gas hydrates and submarine landslides). Thus, since the 1980s we have seen the creation and use of applied geomorphological maps from many terrestrial and marine environments, and these have been produced by practitioners of applied geomorphology rather than academic geomorphology. In order to understand this development, it is necessary to define what the technique of geomorphological mapping entails. Lee (2001) described geomorphological mapping as one of the group of techniques under the general category of 'terrain evaluation' employed to systematically record the shape or morphology of the ground, landforms, landscape-forming processes and materials that constitute the surface of the Earth. Lee (2001) identified three forms of geomorphological map:

1. Regional surveys of terrain conditions, for general geomorphological investigations, land use planning or in baseline studies for environmental impact assessment (e.g. the 1:25,000 scale maps of Torbay; Doornkamp et al., 1988),
2. General assessments of resources or geohazards at scales between 1:50,000 and 1:10,000 (e.g. Bahrain Surface Materials Resources Survey; Doornkamp et al., 1980; ground problems in the Suez City area, Egypt; Jones, 2001),
3. Specific-purpose large-scale surveys to delineate and characterise particular landforms (e.g. the 1:500 scale investigations around the Channel Tunnel portal, Folkestone; Griffiths et al., 1995).

Given this background, and the widening interest in the role and importance of geomorphological mapping, the International Association of Geomorphologists commissioned this volume to provide a state-of-the-art review of the development of the technique and see the way it is now being employed both in the academic and in the professional world. The intention of this book is not to produce a standardised mapping methodology or to provide a detailed geomorphological legend, but it is an attempt to bring together leading exponents in the preparation and use of geomorphological maps and illustrate how they are being used to investigate a wide range of environmental issues. The book is divided into three sections:

1. *Geomorphological mapping*: It details the history of geomorphological mapping, focusing upon the development of methods and their evolution within different national 'schools'; outlines the aims and objectives of mapping and looks at quantitative risk assessment,
2. *Techniques of applied geomorphological mapping*: It reviews the techniques of mapping, including traditional field mapping and recognises the increasing use of digital data gathering techniques for mapping,
3. *Case studies in applied geomorphological mapping*: It presents examples of different industrial applications of geomorphological maps from a variety of environmental settings to demonstrate the wide range and application of mapping in both academic and professional arenas.

The actual content of each of these sections is described in more detail below. A final conclusion looks at the future development of geomorphological mapping. What became apparent to the editors during the compilation of this volume is that the techniques employed to create geomorphological maps are becoming increasingly sophisticated and the range of applications of the maps is becoming ever wider. This volume demonstrates that geomorphological mapping is a technique that all geomorphologists should be familiar with and be able to utilise in the collection and presentation of geomorphological data. The technique is also one that can provide a firm basis for the investigation of many environmental issues, notably in the field of geohazards and risk assessment.

1. GEOMORPHOLOGICAL MAPPING

Geomorphological mapping flourished in different countries and schools all with different aims, especially during the 1960s–1980s. The first contribution in this section is by Verstappen (2011) who illustrates the early development of geomorphological and landform mapping in Europe (western and eastern) and Australia with several examples of legend types and cartographic development. Dramis et al. (2011) focuses on the types, purposes and content of geomorphological maps, spanning from the 'traditional' symbol-based maps to the most modern digital techniques. This chapter presents some of the state-of-the-art techniques in object-oriented geomorphological mapping, with examples from Italy. It highlights the importance of moving towards an objective and multi-scalar method for the representation of the landscape that can be of great benefit to a wider community of users, including environmental analysts and planners. Paron and Claessens (2011) focuses on the need to integrate geomorphological mapping in national mapping programmes, natural hazard zonations and emergency programmes and landscape planning. The chapter shows how the new digital mapping and web-mapping reality can aid in disseminating the importance of geomorphological investigation and mapping. The last chapter of this section by Hearn and Hart (2011) looks at the practical issues involved in quantitative risk assessment/analysis of landslides, showing how some of the conceptual models are quite theoretical. This contribution attempts to bridge the gap between the diffuse hazard susceptibility maps and the more useful hazard and risk maps required for quantitative risk assessment of landslides. The examples in this chapter from less economically developed countries illustrate well the need for practical but sound hazard mapping in data poor environments where vulnerability may be increasing as new developments take place.

2. TECHNIQUES OF APPLIED GEOMORPHOLOGICAL MAPPING

The systematic recording of landform morphology requires some kind of geodetic framework and a methodology through which this is performed. Early geomorphological mapping required physical site visits in order to record plan-form position and, in some instances, composition on to a topographic base map. Knight et al. (2011) detail the techniques used for field-based geomorphological mapping and whereas, by volume, it has largely been replaced as a technique, it remains a common, and important, aspect of large-scale surveys.

One of the drivers of the resurgence in geomorphological mapping is technology: the availability of new data sources has allowed new insights and rapid mapping to be performed, organised within the framework of a geographic information system (GIS). The addition of new sources of digital spatial data has opened up vast regions of the Earth's surface (and indeed other planets) for study that would have otherwise been uneconomic or impossible to achieve. Oguchi et al. (2011) detail the vast range of data sets that are currently available and outline their potential application areas.

More mundanely, but of no less significance, is the organisation of spatial data into a digital data framework. The ability to use a 'layers' paradigm to organise input data and produce layers of thematic, mapped, output is of great significance. Smith (2011) outlines this 'layered' approach and introduces methods for the visualisation and digital recording of landforms. This remains an entirely manual process, limited by the skill and experience of the operator. Accurate automated and semi-automated landform extraction techniques remain a current research focus, and Seijmonsbergen et al. (2011) introduce the main techniques and their applications. Finally, no technical section would be complete without discussion of the *presentation* of mapped landforms. Otto et al. (2011) provide a brief synopsis of cartographic techniques and their applications to geomorphological mapping. There is specific focus upon the review and selection of an appropriate legend system. The chapter concludes with the digital dissemination of geomorphological information and, in particular, web servers, virtual globes and static maps.

3. CASE STUDIES IN APPLIED GEOMORPHOLOGICAL MAPPING

Thirteen case studies have been compiled, including three from the marine environment. The three marine examples illustrate how the use of modern marine geophysical techniques has revolutionised our ability to interpret marine geomorphology. Dunlop et al. (2011) used publically available multi-beam swath bathymetry data to construct a glacial geomorphology map of the continental margins north and northwest of Ireland. Hillier (2011) has created digital elevation models of the Hawaiian volcanoes to establish their height and volume, data that are critical to understanding the volcanic hazard. Micallef (2011) demonstrates how the range of marine geophysical techniques can be used to produce geomorphological maps in order to assess submarine landslides, presenting a case study of the Storegga slide in the North Sea between Norway and Scotland.

The value of geomorphological mapping in mass movement investigations is a theme that emerges from a number of the terrestrial case studies. Griffiths et al. (2011) use traditional field mapping and remote sensing interpretation to produce an engineering geomorphological map of a landslide that potentially could have affected the Channel Tunnel Terminal in Kent, United Kingdom. Parry (2011) looks at mapping as a technique for assessing landslide risk in Hong Kong. In a more academic investigation, Theler and Reynard (2011) use mapping as a tool for assessing sediment transfer processes in small catchments in Switzerland prone to debris flows. Whitworth et al. (2011) make use of terrestrial laser scanning to produce geomorphological assessments of complex landslide systems in the Cotswolds area of the United Kingdom. As a useful adjunct to this, the value of airborne laser scanning for compiling a range of geomorphological data is illustrated by Rutzinger et al. (2011) for three different test sites in the Austrian Alps.

Pain et al. (2011), also use airborne laser scanning alongside airborne electromagnetic surveys and satellite imagery to evaluate the hydro-geomorphology of the River Murray area in southeast Australia. Williams et al. (2011) have embraced the new geomatics technology as well, using terrestrial laser scanning coupled with high-resolution digital elevation models in the investigation of sediment transport rates in a braided river system in New Zealand. By way of contrast, Knight (2011) provides an example of more traditional field mapping approach, linked to remote sensing interpretation, to compile maps of a lowland glaciated landscape in north-central Ireland.

The final two case studies illustrate the role geomorphological mapping can have in anthropological investigations. Walstra et al. (2011) map the late Holocene evolution and human impact in the Mesopotamian region (southwest Iran) using remote sensing and a GIS. As a contrast, Guth (2011) provides a case study from the D-Day landings in Normandy (June 1944) of the way geomorphological maps allow military commanders to see the way the landscape will influence military operations.

The range of case studies presented is only illustrative of the potential applications of geomorphological mapping. They do illustrate the move away from traditional field mapping through increasing use of digital data capture systems. However, what does emerge from the studies is that interpretation of the data requires extensive and detailed understanding of geomorphological processes and landforms; this is a knowledge and skills base that still requires widespread fieldwork experience. It is also apparent that geomorphological maps are complex tools and to be of value beyond the academic community may often require careful explanation and presentation. The critical importance of communicating geomorphological data effectively remains a challenge that new multimedia tools are helping us to address.

REFERENCES

Costa, J.E., Graf, W.I., 1984. The geography of geomorphologists in the United States. Prof. Geogr. 36, 82–89.

Demek, J., Embleton, C. (Eds.), 1978. Guide to Medium-Scale Geomorphological Mapping. International Geographical Union, Stuttgart.

Doornkamp, J.C., Brunsden, D., Jones, D.K.C., Cooke, R.U., 1980. Geology, Geomorphology and Pedology of Bahrain. GeoBooks, Norwich.

Doornkamp, J.C., Griffiths, J.S., Lee, E.M., Tragheim, D., Charman, J.H., 1988. Applied Earth Science Mapping of the Torbay Region. 2 vols.+11 maps. Open File Research Report for the Department of the Environment and Torbay Borough Council.

Dramis, F., Guida, D., Cestari, A., in press. Nature and aims of geomorphological mapping. In: Smith, M.J., Paron, P., Griffiths, J. (Eds.), Geomorphological Mapping: A Handbook of Techniques and Applications. Elsevier, Amsterdam.

Dunlop, P., Sacchetti, F., Benetti, S., Ó Cofaigh, C., in press. Mapping Ireland's glaciated continental margin using marine geophysical data. In: Smith, M.J., Paron, P., Griffiths, J. (Eds.), Geomorphological Mapping: A Handbook of Techniques and Applications. Elsevier, Amsterdam.

Griffiths, J.S., Brunsden, D., Lee, E.M., Jones, D.K.C., 1995. Geomorphological investigation for the Channel Tunnel and Portal. Geogr. J. 161, 257−284.

Griffiths, J.S., Lee, E.M., Brunsden, D., Jones, D.K.C., in press. The Cherry Garden landslide, Etchinghill escarpment, South-east England. In: Smith, M.J., Paron, P., Griffiths, J. (Eds.), Geomorphological Mapping: A Handbook of Techniques and Applications. Elsevier, Amsterdam.

Gustavsson, M., Kolstrup, E., Seijmonsbergen, A.C., 2006. A new symbol-and-GIS based detailed geomorphological mapping system: renewal of a scientific discipline for understanding landscape development. Geomorphology 77, 90−111.

Guth, P.L., in press. Military applied geomorphological mapping: Normandy case study. In: Smith, M.J., Paron, P., Griffiths, J. (Eds.), Geomorphological Mapping: A Handbook of Techniques and Applications. Elsevier, Amsterdam.

Hearn, G., Hart, A., in press. Goemorphological contributions to landslide risk assessment: theory and practice. In: Smith, M.J., Paron, P., Griffiths, J. (Eds.), Geomorphological Mapping: A Handbook of Techniques and Applications. Elsevier, Amsterdam.

Hillier, J.K., in press. Submarine geomorphology: quantitative methods illustrated with the Hawaiian volcanoes. In: Smith, M.J., Paron, P., Griffiths, J. (Eds.), Geomorphological Mapping: A Handbook of Techniques and Applications. Elsevier, Amsterdam.

Jones, D.K.C., 2001. Ground conditions and hazards: Suez City development, Egypt. In: Griffiths, J.S. (Ed.), Land Surface Evaluation for Engineering Practice, 18. Geological Society Engineering Geology Special Publication, pp. 159−170.

Knight, J., in press. Uses and limitations of field-mapping of lowland glaciated landscapes. In: Smith, M.J., Paron, P., Griffiths, J. (Eds.), Geomorphological Mapping: A Handbook of Techniques and Applications. Elsevier, Amsterdam.

Knight, J., Mitchell, W., Rose, J., in press. Geomorphological field mapping. In: Smith, M.J., Paron, P., Griffiths, J. (Eds.), Geomorphological Mapping: A Handbook of Techniques and Applications. Elsevier, Amsterdam.

Lee, E.M., 2001. Geomorphological mapping. In: Griffiths, J.S. (Ed.), Land Surface Evaluation for Engineering Practice, vol. 18. Geological Society Engineering Geology Special Publication, pp. 53−56

Micallef, A., in press. Marine geomorphology: geomorphological mapping and the study of submarine landslides. In: Smith, M.J., Paron, P., Griffiths, J. (Eds.), Geomorphological Mapping: A Handbook of Techniques and Applications. Elsevier, Amsterdam.

Oguchi, T., Hayakawa, Y., Wasklewicz, T., in press. Data sources. In: Smith, M.J., Paron, P., Griffiths, J. (Eds.), Geomorphological Mapping: A Handbook of Techniques and Applications. Elsevier, Amsterdam.

Otto, J.-C., Gustavsson, M., Geilhausen, M., 2011. Cartography: design, symbolisation and visualisation of geomorphological maps. In: Smith, M.J., Paron, P., Griffiths, J. (Eds.), Geomorphological Mapping: A Handbook of Techniques and Applications. Elsevier, Amsterdam.

Pain, C.F., Clarke, J.D.A., Wong, V.N.L., in press. Applied geomorphic mapping for land management in the River Murray corridor, SE Australia. In: Smith, M.J., Paron, P.,

Griffiths, J. (Eds.), Geomorphological Mapping: A Handbook of Techniques and Applications. Elsevier, Amsterdam.

Paron, P., Claessens, L., in press. Makers and users of geomorphological maps. In: Smith, M.J., Paron, P., Griffiths, J. (Eds.), Geomorphological Mapping: A Handbook of Techniques and Applications. Elsevier, Amsterdam.

Parry, S., in press. The application of geomorphological mapping in the assessment of landslide hazard in Hong Kong. In: Smith, M.J., Paron, P., Griffiths, J. (Eds.), Geomorphological Mapping: A Handbook of Techniques and Applications. Elsevier, Amsterdam.

Rutzinger, M., Höfle, B., Vetter, M., Pfeifer, N., in press. Digital terrain models from airborne laser scanning for the automatic extraction of natural and anthropogenic linear structures. In: Smith, M.J., Paron, P., Griffiths, J. (Eds.), Geomorphological Mapping: A Handbook of Techniques and Applications. Elsevier, Amsterdam.

Seijmonsbergen, A.C., Hengl, T., Anders, N.S., in press. Semi-automated identification and extraction of geomorphological features using digital elevation data. In: Smith, M.J., Paron, P., Griffiths, J. (Eds.), Geomorphological Mapping: A Handbook of Techniques and Applications. Elsevier, Amsterdam.

Smith, B.J., Warke, P.A., Whalley, W.B., 2002. Landscape development, collective amnesia and the need for integration in geomorphological research. Area 33 (4), 409–418.

Smith, M.J., in press. Digital mapping: visualisation, interpretation and quantification of landforms. In: Smith, M.J., Paron, P., Griffiths, J. (Eds.), Geomorphological Mapping: A Handbook of Techniques and Applications. Elsevier, Amsterdam.

Theler, D., Reynard, E., in press. A geomorphological map as a tool for assessing sediment transfer processes in small catchments prone to debris-flows occurrence: a case study in the Bruchi torrent (Swiss Alps). In: Smith, M.J., Paron, P., Griffiths, J. (Eds.), Geomorphological Mapping: A Handbook of Techniques and Applications. Elsevier, Amsterdam.

Verstappen, H.T., in press. Old and new trends in geomorphological and landform mapping. In: Smith, M.J., Paron, P., Griffiths, J. (Eds.), Geomorphological Mapping: A Handbook of Techniques and Applications. Elsevier, Amsterdam.

Walstra, J., Heyvaert, V.M.A., Verkinderen, P., in press. Mapping late Holocene landscape evolution and human impact – a case-study from Lower Khuzestan (SW Iran). In: Smith, M.J., Paron, P., Griffiths, J. (Eds.), Geomorphological Mapping: A Handbook of Techniques and Applications. Elsevier, Amsterdam.

Whitworth, M., Anderson, I., Hunter, G., in press. Geomorphological assessment of complex landslide systems using field reconnaissance and terrestrial laser scanning. In: Smith, M.J., Paron, P., Griffiths, J. (Eds.), Geomorphological Mapping: A Handbook of Techniques and Applications. Elsevier, Amsterdam.

Williams, R., Brasington, J., Vericat, D., Hicks, M., Labrosse, F., Neal, M., in press. Monitoring braided river change using terrestrial laser scanning and optical bathymetric mapping. In: Smith, M.J., Paron, P., Griffiths, J. (Eds.), Geomorphological Mapping: A Handbook of Techniques and Applications. Elsevier, Amsterdam.

Winchester, S., 2001. The Map that Changed the World. Penguin Books, London, 338 pp.

CHAPTER TWO

Old and New Trends in Geomorphological and Landform Mapping

Herman Theodoor Verstappen
International Institute of Geo-Information Science and Earth Observation (ITC), University of Twente, Enschede, The Netherlands

Contents

1. THE ADVENT OF GEOMORPHOLOGICAL MAPPING

Geomorphological mapping began about a century after the advent of geological mapping and the standardisation of a legend system (Finkle, 1988). The earliest geomorphological map was probably made by Passarge (1914) on the Stadtremba 1:25,000 topographic map sheet in Germany. The legend of Passarge's map differed from modern examples because it did not encompass all aspects of geomorphology, and it emphasised mainly descriptive morphographic features and metrical elements. It did not receive much attention at that time, and geomorphologists tended to produce 'sketch' maps at small scales that were either largely structural or physiographic-pictorial (Raisz, 1931). Most of the maps from this early period dealt with only one phenomenon (e.g. river terraces) and left large portions of the map sheet blank.

The development of modern concepts of geomorphological mapping started in the early 1950s. In Switzerland, Helbling (1952) included a

Developments in Earth Surface Processes, Volume 15
ISSN: 0928-2025, DOI: 10.1016/B978-0-444-53446-0.00002-1

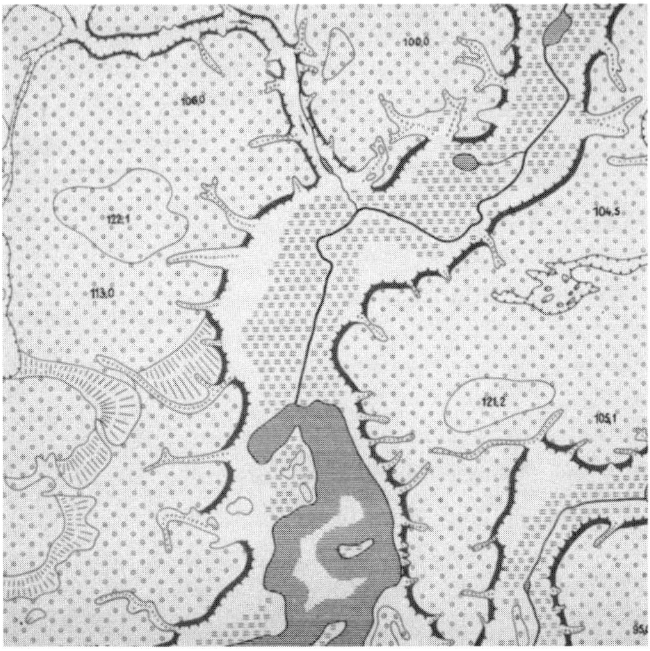

Figure 2.1 Detailed geomorphological map of an outwash landscape of the Poznan stage, Weichselian glaciation, NW Poland. Scale 1:50,000 (Galon, 1962). The legend of the detailed geomorphological map of northern Poland includes 15 landform categories. This example shows the outwash plain (IV.10 – screen of small circles) at places surrounded by periglacial foot slopes (VII.29) and dissected by small V-shaped valleys (IX.35). It is separated by an escarpment of a height of 10–20 m (XV.57) or more than 20 m (XV.58) from a peat-filled valley (XIII.5 – screen with dashes), where a river and some lakes (XV.60/61) occur. Contour lines and spot heights (XV.62) complete the map.

geomorphological map in his Ph.D. thesis on the Sern valley and thereafter Annaheim (1956) took an interest in the subject in that country. The greatest developments came from Poland where Klimaszewski (1956, 1963) launched a countrywide 1:50,000 scale geomorphological survey together with Galon (1962) who specialised particularly on the Polish lowlands (Figure 2.1). Then, other European countries such as France, Germany, and Switzerland developed similar maps. At the beginning, emphasis was mainly focused upon detailed mapping at scales ranging from 1:10,000 to 1:100,000, and methods and legends for medium- and small-scale mapping followed soon after.

Since the essence of geomorphological mapping is the representation of the terrain configuration, landforms are the cartographic units to be distinguished, regardless of the mapping scale. However, while in detailed mapping geomorphological processes are emphasised, the underlying structural factors in landform development became important particularly in medium- and small-scale mapping. This explains why morphostructure was advocated as the highest category of landform classification especially by geomorphologists of the former Soviet Union where small-scale geomorphological mapping was the trend (Simonov et al., 1960, in St-Onge, 1964; Bashenina, 1972). It is evident that making a small-scale geomorphological map from detailed maps does not only amount to data reduction and generalisation but also requires a different approach. Most analytical geomorphological maps are complex as a result of the diversity of the data to be included, such as morphography, morphogenesis, morphodynamics, morphometry, chronology, and lithology. This has led to the development of a great variety of legends in different countries on the basis of which some common concepts gradually emerged.

2. THE DIVERSITY OF LEGENDS

When comparing early geomorphological maps and their legends, one is struck by the diversity of the then prevailing concepts and cartographic conventions. This results in part from the terrain configuration of the surveyed areas. For example, legends for lowland areas tend to be simpler than those required in hilly and mountainous terrain and the need for coloured symbols is thus limited (Figure 2.1). That simple legends may, to a certain extent, also be feasible in areas of relief is shown in Figure 2.2 which depicts marine terraces in southern Italy (Verstappen, 1983).

However, the use of colours is inevitable for the cartographic representation of all types of geomorphological information. The coloured area symbols are at present commonly used for indicating morphogenesis, as proposed first by Klimaszewski (1956) in Poland (Map 2.1) and Joly in France. Gellert and Scholz (1960) produced maps of the lowlands of the former Deutsche Demokratische Republik (DDR) with coloured area symbols indicating the chronological sequence of landform development

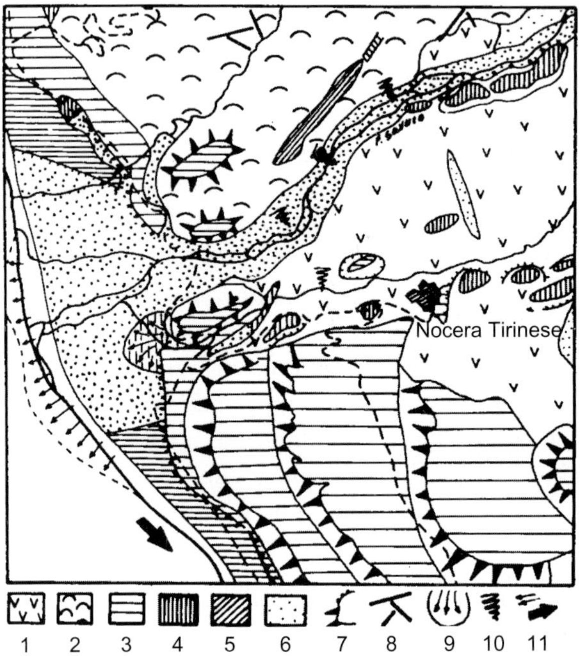

Black and white geomorphological map of the lower Savuto Valley and adjacent areas of Calabria, Italy at a scale of 1:100,000.

Key:

1. Denudational mountains in soft metamorphic/igneous rocks
2. Denudational hills in unconsolidated Quaternary sandstones/conglomerates
3. Marine terraces
4. Fluvial terraces
5. Erosion glacis
6. Floodplain and delta
7. Major and minor scarps
8. Faults
9. Fan
10. Sheet erosion
11. Coastal accretion and beach driting

Figure 2.2 Example of a black and white geomorphological map in Savuto Valley, Italy. Scale 1:100,000 (Verstappen, 1983).

(Map 2.2). This was a logical option because most of the landforms occurring dated from several Pleistocene glacial periods and were easily distinguishable. The advantage of this method was that it concurred with geological maps. However, this legend was not universally applicable.

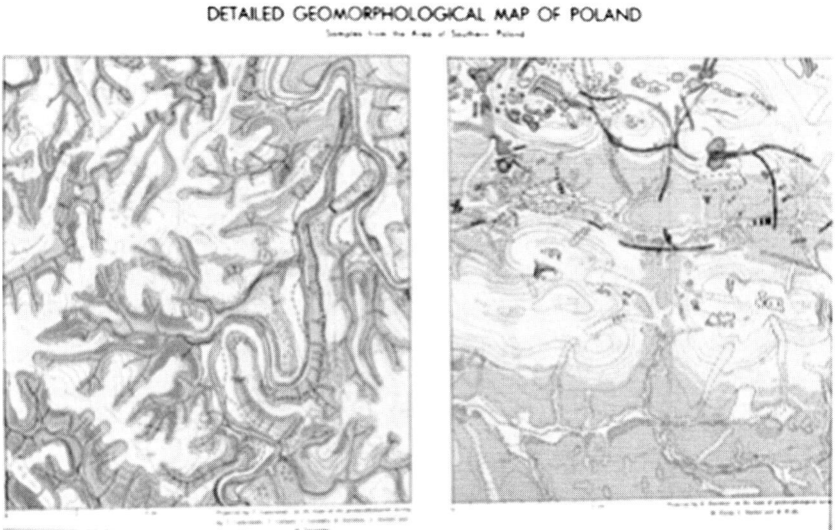

Map 2.1 Detailed geomorphological map of Poland. Two examples from the area of Southern Poland (1:25,000–1:50,000) (Klimaszewski, 1956). The electronic version is available at http://www.appgema.net/.

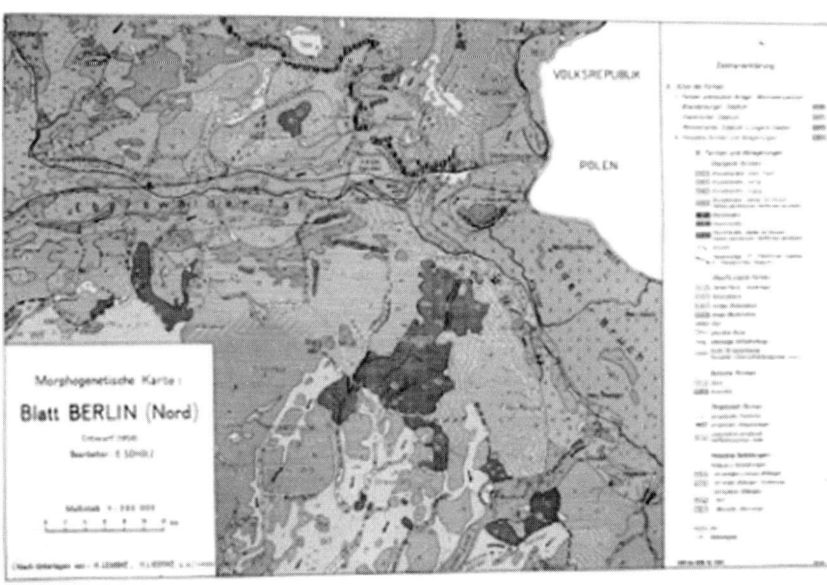

Map 2.2 Morphogenetic map of the former DDR. Sheet Berlin (Nord) 1:200,000 (Gellert and Scholz, 1960). The electronic version is available at http://www.appgema.net/.

Another option is to relate coloured area symbols to lithology as applied by Tricart (1955, 1969, 1972) in France.

The legend developed by Klimaszewski (1956) was first used in the denudational hills and mountains of southern Poland. It was therefore considerably more complex than the one used by Galon (1962) farther North. Klimaszewski introduced landforms as the highest category of his legend and gave further information on geomorphological processes and chronology. The legend of the geomorphological map of western Germany [then Federal Republic of Germany (FRG)], introduced by the German Research Council, is of comparable complexity (Leser, 1974). At the Centre for Applied Geomorphology (CGA) in Strasbourg (France), emphasis was put on lithology, for which the coloured area symbols were used. This was justified because of the obvious relations that exist between rock types, landforms and geomorphological processes. Morphochronology was also included (Tricart, 1955, 1969; Bourdiec et al., 1963). Tricart produced many maps of this type, placing the emphasis on the granulometry and chemical/mechanical characteristics of the rocks and superficial deposits. He used line symbols in various colours for indicating chronology. Joly (1963) proposed a slightly different approach for the compilation of the detailed geomorphological map of France at the scale of 1:50,000 (Tricart, 1970), with a legend depicting superficial deposits, their thickness and granulometry. Colours were used for the various morphogenetic systems with different hues showing successive generations of landforms. A decimal system was devised for the taxonomic classification of these landforms. In the legend used in Hungary (Pecsi et al., 1962; Pecsi, 1964), the coloured area symbols were used for distinguishing the major morphogenetic landform types. However, lithological influences were also stressed, while the processes were indicated by screens. An interesting contribution is that the chronology is indicated by ciphers. This is a very flexible solution because the ciphers can simply be omitted where the age of the landforms is not exactly known. The legends used in the former Soviet Union were essentially morphogenetic (Bashenina, 1972). The coloured mapping units thus represent major landform types and complexes called mesoforms. The chronology was also emphasised and was indicated by density of colours. Lithology was a minor component as well as morphometric information. Morphostructures were emphasised in medium- and small-scale maps. A map of Quaternary surface deposits was added where their thickness exceeded 10 m. Morphometric data rank high in a number of

Belgian geomorphological maps next to the morphogenetic information (Macar et al., 1960). The same principles were applied to the legend of the geomorphological map of the Netherlands at the scale of 1:50,000 (Maarleveld et al., 1974; Van Noord, 1993). This map was complementary to the earlier existing geological and soil map series at the same scale and this explains the strong emphasis on morphometry. The coloured area symbols were allocated to eight major relief classes, defined by slope gradient and length and subdivided into 18 relief types according to relief amplitude. Morphogenesis and past and present processes were listed in the legend where they are represented by screens.

3. THE NEEDS FOR STANDARDISATION AND FLEXIBILITY

Considerable efforts have been made to unify the legends used in various countries or at least to make them more easily comparable. From 1960 onwards, the Sub-commission for Geomorphological Mapping of the International Geographical Union (IGU) Commission on Applied Geomorphology (since 1968, the IGU Commission of Geomorphological Survey and Mapping) has been diligent in this field. It produced a manual for detailed and medium-scale geomorphological mapping (Demek, 1972). An attempt was even made at compiling a unified key for worldwide use (Gellert, 1968; Bashenina, 1972; Gellert and Scholz, 1974).

Although a universal legend designed to the smallest detail is unlikely to be practical, the efforts served well for achieving agreement on essential issues such as the use of coloured area symbols for large morphogenetic (groups of) landforms. The legends of analytical geomorphological maps tend to be very complex as a result of the diversity of data to be included on the morphometric, morphographic, morphogenetic, morphochronologic characteristics of the terrain and, in addition to morphostructures, lithology and surficial deposits. The geomorphological maps produced generally contained a great amount of information and are documents of high scientific value. This mapping methodology enabled geomorphologists to study and depict all aspects of every part of the terrain. As a result, it gave impulse to the development of modern geomorphological research in the same manner as the introduction of modern approaches to field observations, laboratory investigations and modelling. Overcrowding of maps should be avoided, however, because this hampers their efficient use particularly when only a part of the map content is relevant and when interdisciplinary research projects are concerned in which non-specialists participate.

Full standardisation is only required in the case of the production of a map series at a national or international level. Otherwise, it is more appropriate to apply general concepts with some flexibility as to optimally suit the purpose of the survey and specific characteristics of the mapped area. One may even contemplate the production of one or several special-purpose maps, on the basis of an analytical geomorphological survey. Another consideration is that in many parts of the world there is insufficient topographic and geomorphologic information for producing a geomorphological map on the basis of the systems developed in Europe and other parts of the world. The first attempt at overcoming these problems is the ITC (International Institute for Geo-information Science and Earth Observation; formerly International Training Centre for Aerial Survey) System of Geomorphological Survey (Verstappen and Van Zuidam, 1968/1975; revised edition 1991; Verstappen, 1970). It is not

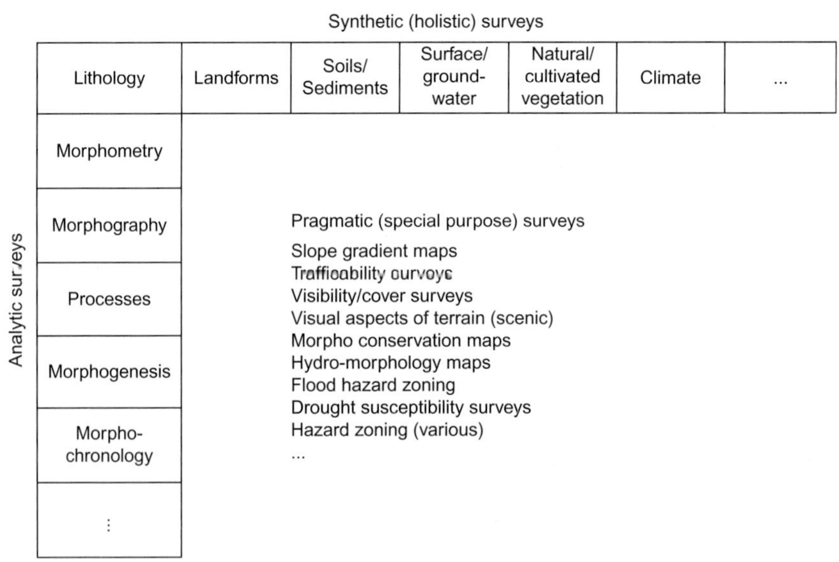

Figure 2.3 Contents and relationships of various types of geomorphological maps (Verstappen and Van Zuidam, 1991).

surprising that this flexible system for geomorphological survey and mapping was developed at the International Institute for Aerospace Survey and Earth Sciences in the Netherlands: the large international student body and worldwide research activities of this organisation simply made it necessary. The survey system encompasses analytic, synthetic (holistic) and pragmatic (special-purpose) surveys at various scales (Figure 2.3).

The method of generalisation, by simplification and omission, is shown in Figure 2.4.

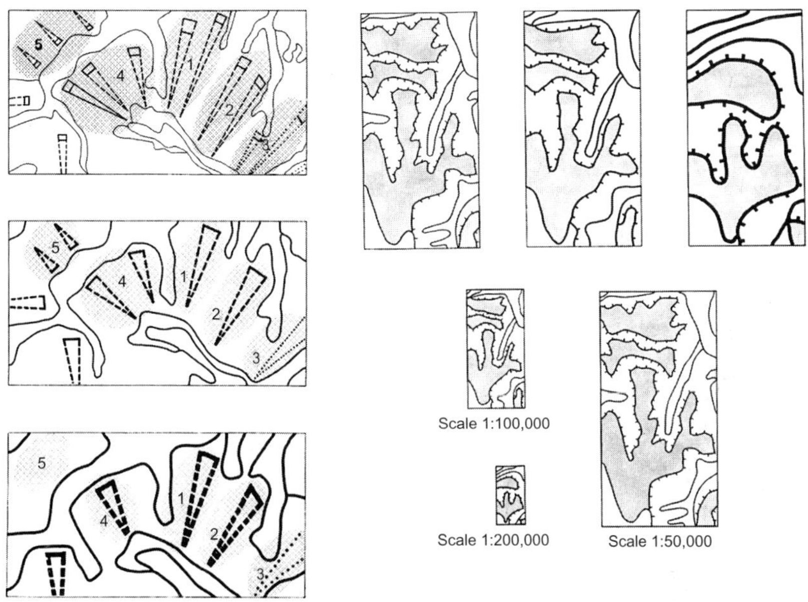

Figure 2.4 Generalisation of the map contents for scale reduction (Verstappen and Van Zuidam, 1991). Left: Generalisation of line symbols. Glacis symbols shown at the mapping scale 1:50,000 (top) are reduced in number, by using one symbol instead of two and two instead of three, to produce a map at the scale of 1:100,000 (centre). Further combination of symbols (4) and omission (5) is needed for producing a map at the scale of 1:200,000 (bottom). Top-right: Generalisation of geomorphological units. All parts of the structurally controlled plateau mapped at the scale of 1:50,000 (left) can be shown at the scale of 1:100,000 (centre), by simplification of boundaries, smoothing of irregularities and combining small forms. Further reduction to the scale of 1:200,000 (right) requires combination of two areas into one while maintaining the relative proportion of the unit to the surrounding units. Lines are further smoothed as well. Lower-right: The resulting outline of the structural plateau at the three map scales.

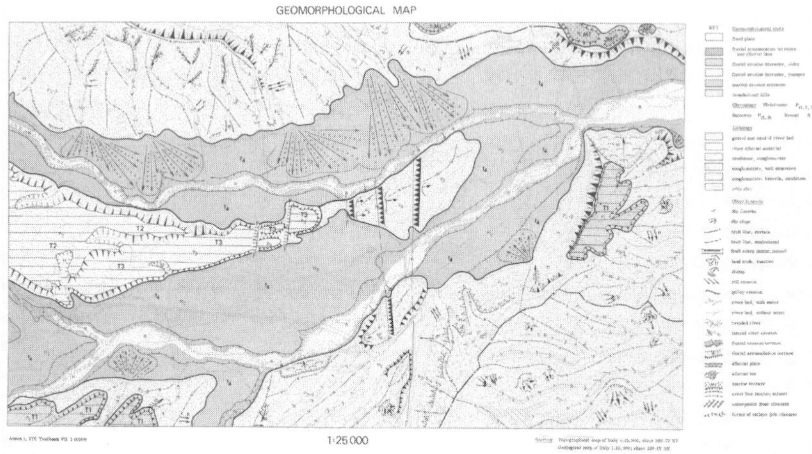

Map 2.3 Geomorphological map of part of the Crati Basin, southern Italy, 1:25,000 ITC System of Geomorphological Survey (Verstappen and Van Zuidam, 1968/1975; Verstappen, 1970). The electronic version is available at http://www.appgema.net/.

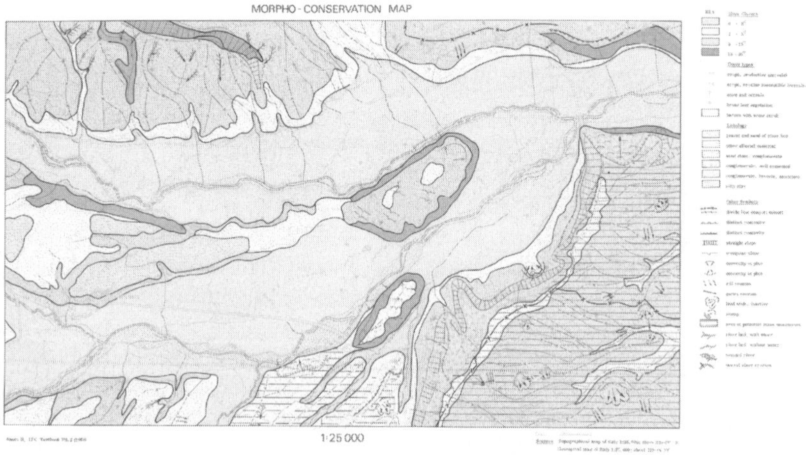

Map 2.4 Morpho-conservation map of the area shown in Map 2.3. The electronic version is available at http://www.appgema.net/.

A map example of an area in southern Italy (Verstappen, 1977b) is also given in Map 2.3, the mode of generalisation is shown in Map 2.4 and two types of applied maps are shown in Maps 2.5 and 2.6, respectively.

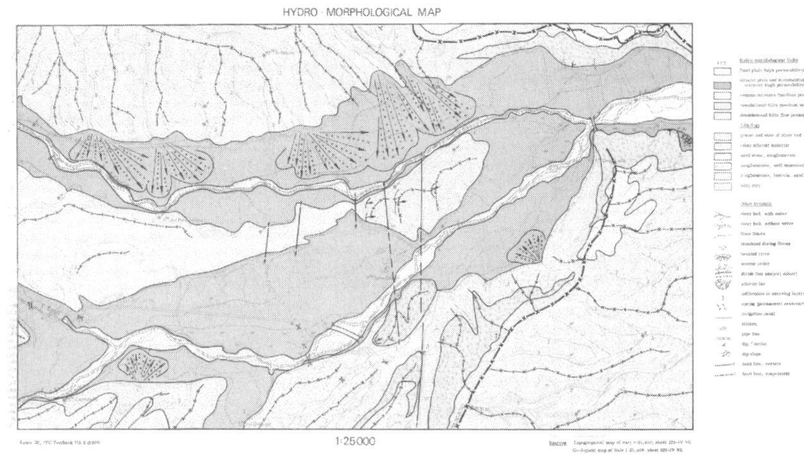

Map 2.5 Hydro-morphological map of the area shown in Map 2.3. The electronic version is available at http://www.appgema.net/.

Two more recent map examples, from southern Italy and northern Spain (Verstappen and Van Zuidam, 1991), respectively, are given in Maps 2.7 and 2.8. The survey starts with the compilation of a topographic base maps with drainage lines and contour lines or spot heights, on which subsequently the various morphogenetic units and sub-units are plotted, using coloured area symbols. Geomorphological processes are indicated by black line symbols, lithology by grey screens or hachures and chronology by a numbering system. The same base map is also used for compiling applied, special-purpose maps, as required. The possibility of storing the collected data in a database instead of making maps has been included in the revised edition. Geographical Information Systems (GIS), such as the Integrated Land and Water Information System (ILWIS) developed at ITC, serve this purpose (Meijerink, 1988). The methodology of this procedure is shown in Figure 2.5.

4. THE USE OF AERIAL PHOTOGRAPHS AND SATELLITE DATA

In some countries, particularly in Eastern Europe, access to existing aerial photographs was difficult or even completely impossible for scientific purposes. This situation affected the precision of the maps and the

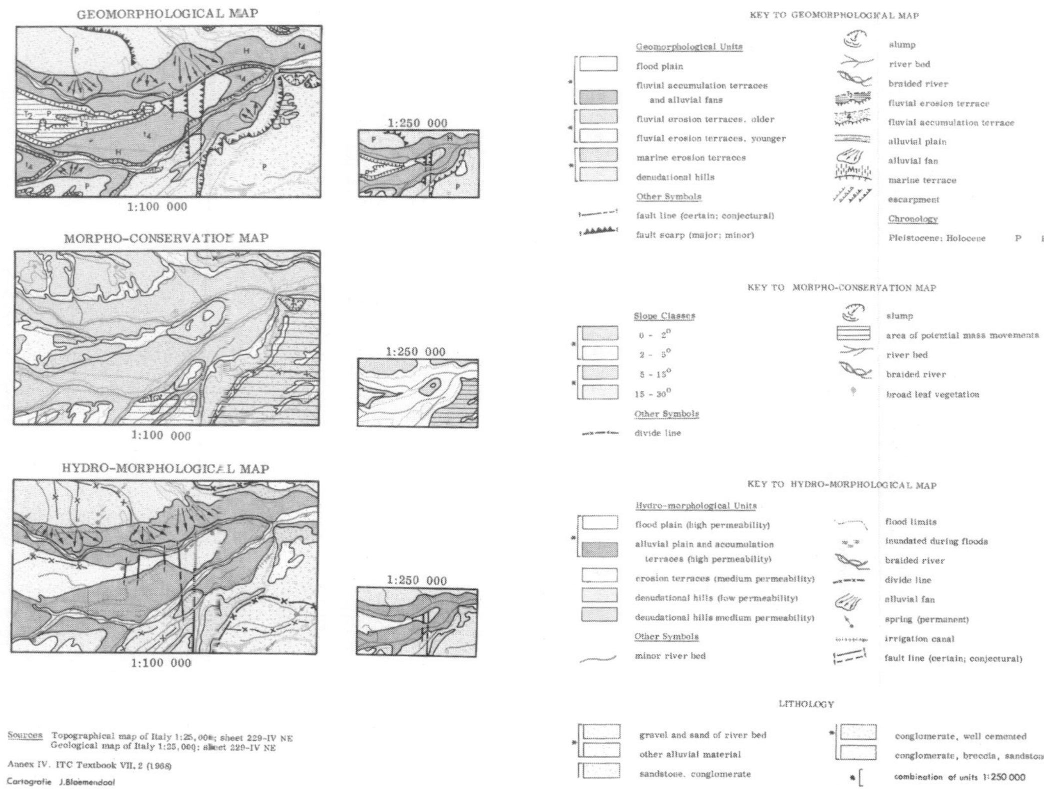

Map 2.6 Scale reduction and genera isation of the area shown in Map 2.3: 1:100,000 and 1: 250,000. The electronic version is available at http://www.appgema.net/.

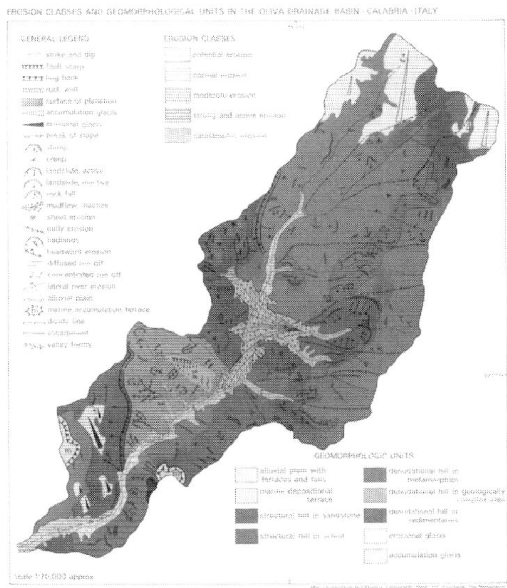

Map 2.7 Applied Geomorphological map of the Oliva basin, Italy, scale 1:70,000, from a 1:10,000 survey, using the ITC System of Geomorphological Survey, second ed. Geomorphological units are shown by colours; erosion classes by hachures and geomorphological details by line symbols (RAO, 1975; followed by Verstappen and Van Zuidam, 1991). The electronic version is available at http://www.appgema.net/.

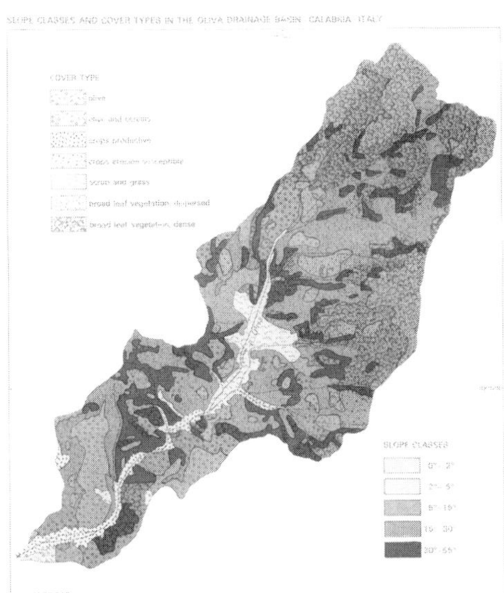

Map 2.8 Slope classification (colours) and cover types (screens) of the area shown in Map 2.7. The electronic version is available at http://www.appgema.net/.

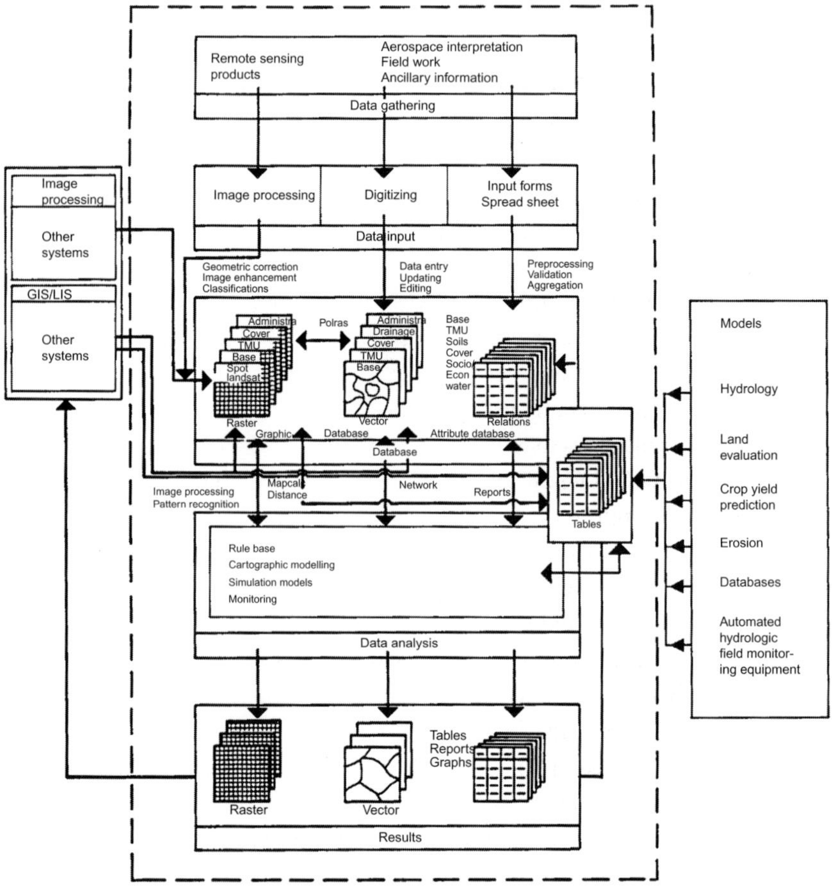

Figure 2.5 Structure of the GIS ILWIS used in the revised second edition of the ITC System of Geomorphological Survey (Meijerink, 1988; Verstappen and Van Zuidam, 1991).

progress of the surveys. However, the availability and utilisation of aerospace data gradually became general practice all over the world (Verstappen, 1977a,b). Their application to geomorphological survey and soon mapping was recognised, and the methods for their integration in the survey procedures were investigated. Where adequate sequential aerial photographic coverage was available, airborne morphodynamic studies became feasible (Verstappen, 1977a,b). The free availability of satellite imagery and digital elevation models (see Oguchi, 2010) substantially contributed to these developments. Geomorphological mapping without the use of aerospace data is obsolete at present. A few decades ago, the spatial resolution of the existing remote sensing images limited their use to small- and medium-scale mapping. The spatial resolution and coverage of present Earth observation satellites (less than 0.5 m) are now adequate for detailed geomorphological surveys. In addition, their excellent metric qualities facilitate mapping procedures; distortions inherent to aerial photographs are absent. Radar data provide terrain information for commonly cloud-obscured parts of the world, and the short recurrence interval of satellite passes has opened new opportunities for studying the effects of actual geomorphological processes and other morphodynamical aspects. Some satellites also provide stereoscopic data and allow the derivation of elevation data. Most important of all, however, is the digital form in which the satellite data are gathered. Digital data handling is now complementary to visual observation and has resulted in the merging of satellite data with other sources of information in GIS. Digital terrain models are now a standard tool in geomorphological mapping (Van Asselen and Seijmonsbergen, 2006).

The interpretation of remotely sensed data can be the starting point of geomorphological survey at all mapping scales. This leads, in combination with the study of the existing literature and all further available relevant information, to the compilation of a preliminary geomorphological map prior to fieldwork. This serves to provide an initial idea about terrain configuration, types and distribution of geomorphological phenomena and problems likely to appear, as well as an aid in planning a field survey. The classification of the geomorphological units and features may, at this stage, still be rather descriptive, and it is common that the morphogenesis of landforms becomes clear only after field investigations. Field transects and sampling sites can be selected efficiently in all geomorphological units on the basis of accessibility and survey requirements. The geomorphological maps can only be completed after the fieldwork has been terminated, after samples taken have been investigated in the laboratory and after a thorough second interpretation of the imagery. Subsequently, the required applied maps can be prepared. Storing all data so gathered grid-wise or within landform units for further use in a GIS can also be considered.

5. LANDFORM MAPPING IN SYNTHETIC (HOLISTIC) SURVEYS OF TERRAIN

Analytical geomorphological surveys provide full information about the geomorphology of the area studied, including processes and morphogenesis. However, they do not normally include data about other environmental parameters concerning geology, soils, hydrology, vegetation and land use which may be required for purposes of regionalisation and land management. A synthetic, holistic, approach therefore may provide the additional information required to place the geomorphological information in an environmental context and make it operational for planned resource development and environmental management. In holistic land surveys, the breakdown of terrain into units of several levels is usually based on geomorphology, and particularly on landforms and processes; the integration of mono-disciplinary, analytical, surveys and multidisciplinary, synthetic, land surveys does not pose major problems. It is evident that the introduction of aerial photographs and, more recently, satellite imagery gave further impetus to holistic surveys. First, because they provide an exact and detailed picture of the landforms, and second, because they give insight into the ecological relationships existing in the region between the various landscape elements such as lithology, geomorphology, soils, hydrology, vegetation, and land use.

Early attempts at holistic surveys predate the advent of aerial photography. However, its development is, in various ways, distinct from that of analytical geomorphological survey and mapping. The emphasis from the outset was mainly on pragmatic issues and particularly with the exploration of extensive and insufficiently explored areas (e.g. in Siberia or in Kazakhstan; Blagovolin and Timofeev, 1993). Small- and medium-scale mapping thus became common practice. In contrast with the general trend in analytical geomorphological surveying and mapping, detailed holistic surveys are more recent. There is also a difference in the countries where the methodology for analytical geomorphological and holistic reconnaissance surveying was developed. The latter occurred in countries with large unexplored areas and in organisations engaged in regional development. The British Directorate of Overseas Surveys (DOS), for example, worked in former British territories in Africa, whilst the Commonwealth Scientific and Industrial Research Organisation (CSIRO) launched land system surveys in Australia. Also in Canada, the former USSR, and more recently Brazil, similar developments occurred.

The concept of subdividing a territory into a number of characteristic regions and sub-regions dates back to the early nineteenth century. However, its further development was slow and occurred simultaneously in various countries. As a result, a variety of classifications and

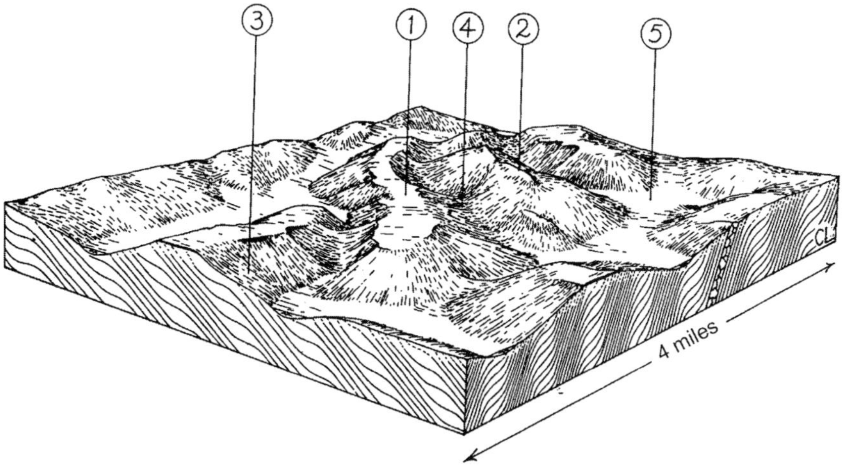

Figure 2.6 Block diagram of the Masaka land system, Uganda, illustrating a DOS resource survey: (1) plateau crest, (2) quartzite ridge, (3) convex interfluve and slope, (4) small valley and (5) main valley floor (Brunt, 1967).

terminologies developed. Bourne (1931) used the descriptive term 'site' that is still commonly used for indicating a small land surface having more or less uniform climatological, geomorphological, geological and pedological characteristics, for which also the term 'facet' is used. Troll (1939, 1966) introduced the equivalent 'ecotope' for emphasising the landscape ecological relations he observed on aerial photographs. An assemblage of sites is often referred to as a 'region', when surface area is concerned, or a 'catena', in the case of transects. It is generally recognised that geomorphology has a key position in holistic land surveys. As an example, Figure 2.6 relates to a DOS of the Masaka area in Uganda carried out in a holistic way. Christian (1958) emphasised the importance of landform characteristics in the Australian land system surveys. Solntsev (1962) engaged in holistic surveys in the former USSR, stating that the geological–geomorphological foundation was always the principal factor for discriminating landscapes as shown in Figure 2.7. Nevertheless, in several countries such as Canada (Gimbarzessky, 1966) and France (Rey, 1968) holistic surveys have been implemented focusing on botanical landscape elements. A justification for this is that cover types, including natural vegetation and land use patterns, are also visible on remotely sensed imagery as landforms.

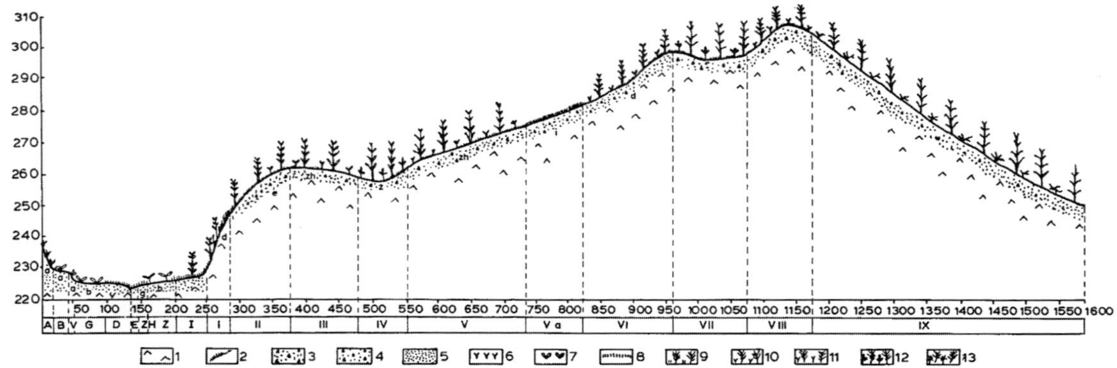

Facies of the Lower Patrushikha Valley Urochishche (Cyrillic capital letters):

A - Escarpment between first and second left-bank terrace; weeds on site of original pinewood.
B - First left-bank terrace; gardens on site of pinewood.
V - Escarpment between first terrace and upper left-bank floodplain; weeds on site of pinewood.
G - Upper left-bank floodplain; gardens on site of floodplain woods.
D - Lower left-bank floodplain; wet meadow on site of floodplain woods.
E - Stream channel.
ZH - Lower right-bank floodplain; wet meadow on site of floodplain woods.
Z - Upper right-bank floodplain; gardens on site of floodplain woods.
I - Transitional facies between upper floodplain and right-bank valley wall; dry mixed-grass-and-cereal-grass meadow on site of pinewood.

Facies of Uktus Urochishche (roman numerals):

I - Steep, rocky slope of southwest exposure; pinewood with steppe grasses.
II - Southwest-facing slope of medium gradient; pinewood with sparse grasses.
III - Gentle slope; pinewood with mixed grasses.
IV - Shallow depression; pinewood with tall grass.
V - Gentle slope; pinewood with mixed grasses.
Va - Dry meadow with cereal grasses and mixed grasses. (This is an anthropogenous part of Facies V.)
VI - Southwest-facing slope of medium gradient; pinewood with sparse grasses.
VII - Shallow depression; pinewood with tall grass.
VIII - Southwest-facing slope of medium gradient; pinewood with sparse grasses.
IX - Northeast-facing slope of medium gradient; pinewood with ferns.

Landscape cross-section through the valley of the Patrushikha River and the Uktus Mountains (horizontal scale 1:5000; vertical scale 1:1000).

Conventional symbols:

1 - Pyroxenites and dunites
2 - Rock outcrops on slopes
3 - Eluvium
4 - Drift
5 - Alluvium
6 - Weeds
7 - Vegetable gardens
8 - Meadow
9 - Pinewood with steppe grass
10 - Pinewood with sparse grass
11 - Pinewood with mixed grasses
12 - Pinewood with tall grass
13 - Pinewood with ferns

a - Cultivated, truncated
b - Humic gley
v - Alluvial meadow
g - Thin, stony humic gley
d - Undeveloped grey-brown forest
e - Thin, stony, moderately podzolised grey-brown forest
zh - Thin, stony, weakly podzolised grey-brown forest
z - Moderately thick grey forest
I - Moderately thick, stony, weakly podzolised grey-brown forest

Figure 2.7 Landscape cross section with facies description as used in synthetic mapping of terrain in the former USSR (Solntsev, 1962).

6. APPLIED GEOMORPHOLOGICAL SURVEYING AND MAPPING

The importance of geomorphological survey and mapping for a variety of practical purposes was gradually understood not only among geomorphologists but also by many scientists of neighbouring disciplines engaged in natural resource inventories and planned development. With time, decision-makers in governmental circles and in the private sector also became aware of the usefulness of geomorphological surveying and mapping for specific purposes. This methodology flourished to the benefit of society and also of our science. Among the early workers in applied geomorphological surveys, the names of Brunsden (1993), Pecsi (1964), Tricart (1955, 1969) and Verstappen (1970, 1983, 1991) and their collaborators should be mentioned.

Early applications were in the area of river floods and the related drainage basin development and in the area of accelerated erosion in agricultural areas and the required measures of slope stabilisation. Engineering applications soon followed. However, not all information contained in analytical geomorphological maps and in synthetic, holistic surveys is required for these specific types of survey. Thus, a careful selection from among all the available information has to be made to satisfy the requirements for such specific purposes. To do this properly, a full understanding of the problem at hand is essential, and this can be achieved only in cooperation with all those involved in a project. Flexibility of the legend and the map contents is thus essential. Where the focus is on natural disaster reduction, a risk assessment map quantifying the number of people affected and estimating the potential damage to dikes, buildings and infrastructure has become a common extension of the survey work. It serves to convince decision-makers that it makes sense to spend money on protective measures, formulating and legalising emergency scenarios and fostering other means of preparedness. These applied geomorphological surveys thus not only become an interdisciplinary issue but also involve a multi-sectorial component that combines several segments of society. These applied surveys are usually implemented project-wise by researchers at the request of, and in cooperation with, local or national government authorities and the private sector of international organisations.

The interdisciplinary context of applied geomorphological surveys needs some explanation. The beginning should always be a thorough geomorphological analysis of the area of study on the basis of which – in combination with information from other scientific disciplines – one or more applied maps can then be compiled. This procedure has already been briefly mentioned earlier in this chapter, using the ITC System of geomorphological survey as an example (see Maps 2.3–2.6). A more elaborate example is shown in the map of the Oliva Basin in Calabria, southern Italy (Rao, 1975; Map 2.7). The geomorphological map shows the geomorphological units by coloured area symbols, the erosion hazard classes by screens and specific erosion features by line symbols (Map 2.7). In addition, the slope gradient classes and the erosion-related land cover types are represented in Map 2.8. Other criteria have to be considered in the case of flood hazard surveys. It is then essential to establish a linkage between the hydrological regime and the landform characteristics of the drainage basin(s) concerned. In this way, it becomes possible to establish the areas that are susceptible to flooding and to determine the frequency

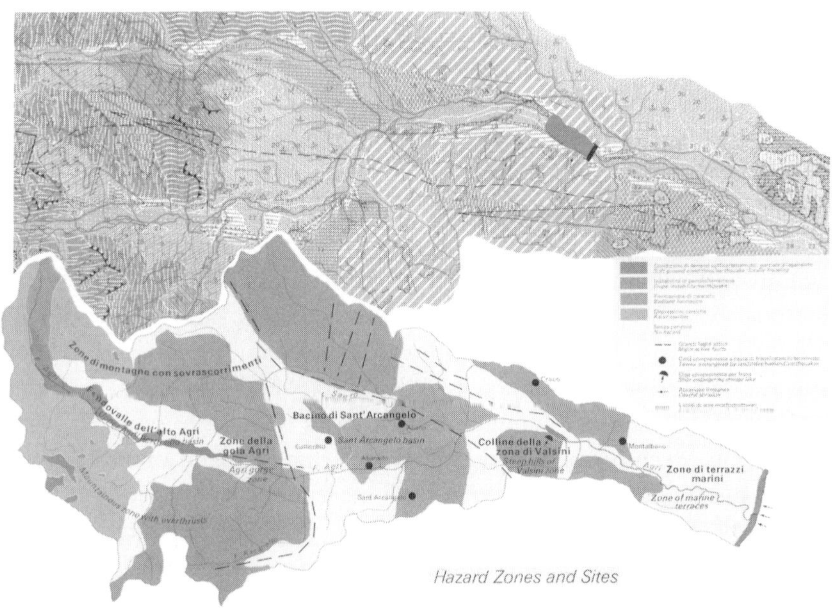

Map 2.9 Geomorphological map of part of the Agri Basin, Basilicata, southern Italy, 1:150,000. The inset map shows the hazard zones and hazard sites for the entire basin; 1:300,000. Based on a field survey at the scale of 1:50,000 using aerial photographs 1:35,000, Landsat images and existing topographic and geologic maps (Verstappen and Van Zuidam, 1991). The electronic version is available at http://www.appgema.net/.

of occurrence of such events. Volcanic hazard surveys require thorough study of the terrain configuration and good knowledge of the eruption types of the volcano concerned to produce a reliable hazard zoning. Finally, optimal modes of early warning should be considered.

In most cases, to reach the final synthetic product, for example a hazard zoning map, a considerable amount of investigation and a detailed geomorphological survey are required. This is because (i) only detailed geomorphological research can lead to reliable applications and (ii) practical results have to be presented in such a way that the map is also understandable by non-specialists. A survey of the Agri river basin in southern Italy (Verstappen, 1977a,b) provides an example (Map 2.9). The field survey, carried out at the scale of 1:50,000, resulted in a geomorphological map at the scale of 1:150,000 and in a map indicating hazard zones and sites at the scale of 1:300,000. Maps 2.10 and 2.11 show detailed geomorphological maps of the Cosenza Province, Calabria, Italy, at the scale of 1:100,000 (Verstappen and Van Zuidam, 1991) and of a part of the Huerva valley, Ebro basin, Spain, at the scale of 1:50,000 (Verstappen and

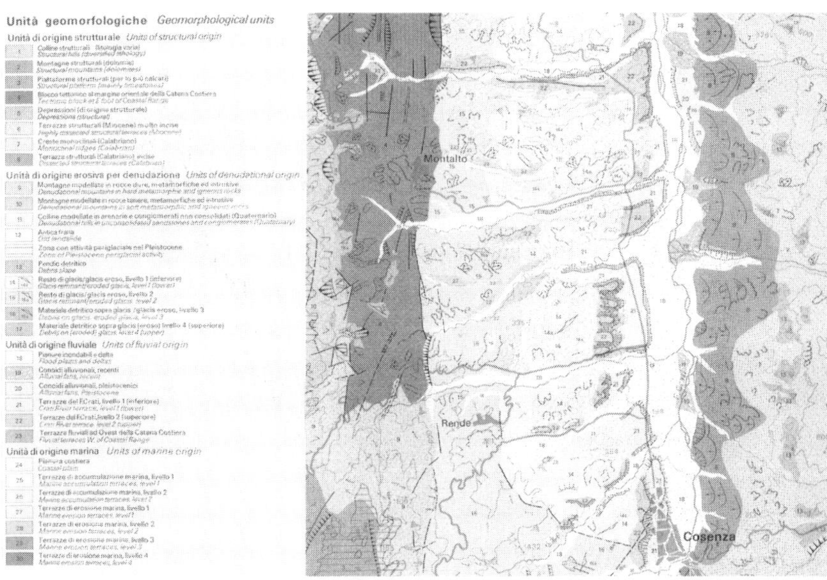

Map 2.10 Geomorphological map of part of the southwest Cosenza province, Calabria, Italy, 1:100,000 using the ITC System of Geomorphological Survey. The numbered geomorphological units are listed on the left. Based on field survey 1:25,000 and aerial photographs 1:30,000 (Verstappen and Van Zuidam, 1991). The electronic version is available at http://www.appgema.net/.

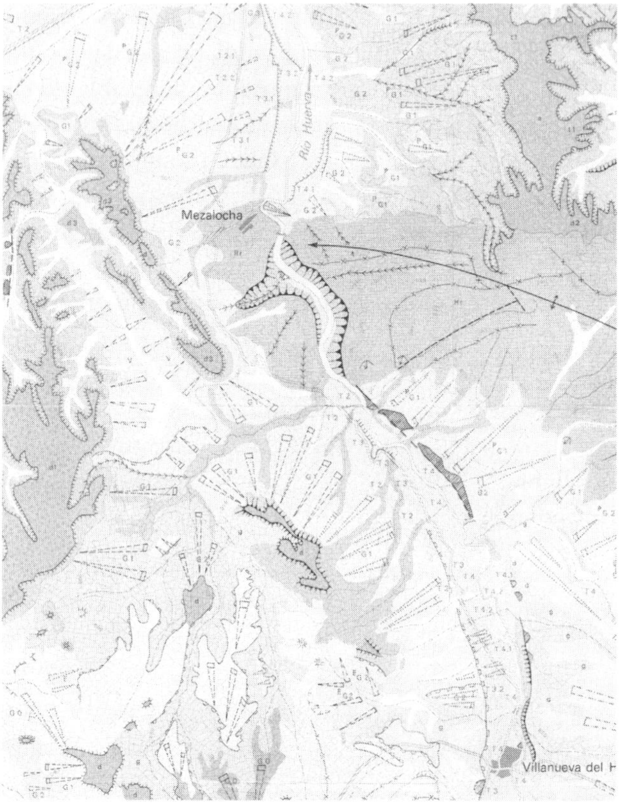

Map 2.11 Fragment of a geomorphological map of the Huerva Valley, northern Spain, 1:50,000. Based on field survey and aerial photographs 1:30,000 (Verstappen and Van Zuidam, 1991). The electronic version is available at http://www.appgema.net/.

Van Zuidam, 1991) compiled using the ITC System of geomorphological survey and mapping. Both served as a basis for further applied research. Integrating the geomorphological data obtained during the field survey and from remotely sensed imagery with information derived from other kinds of survey and with statistical data used to be a tedious and time-consuming procedure. At present, however, GIS are at our disposal to pursue this aim efficiently. Map 2.12 of the Komering basin in southern Sumatra (Meijerink, 1988) exemplifies this. It shows the geomorphological units by coloured area symbols and can be used for predicting their flood susceptibility given various management options, using the modelling facilities of the GIS ILWIS.

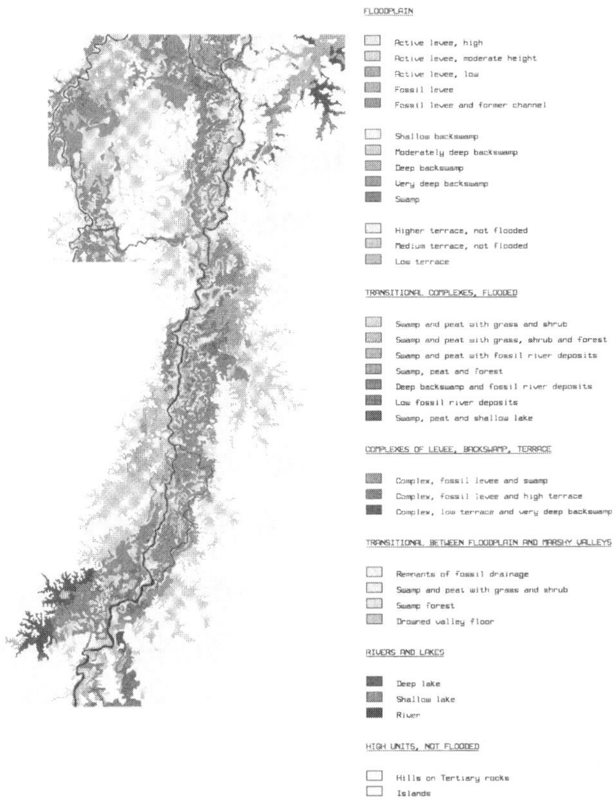

Map 2.12 Terrain mapping units map of the Komering Basin, southern Sumatra, Indonesia, produced using the ILWIS GIS. ILWIS modelling facilities were used to predict areas susceptible to flooding given various management options. Based on field survey by ITC students, aerial photographs 1:100,000 and Landsat images. Map compilation B. Maathuis (Meijerink, 1988; Verstappen and Van Zuidam, 1991). The electronic version is available at http://www.appgema.net/.

7. SUMMARY AND CONCLUSIONS

The chapter provides a concise review of the development of geomorphological mapping from the analytical surveys that developed mainly in continental Europe for academic purposes to the regional surveys that began with small-scale mapping for resource inventories in some Anglo-Saxon countries and the Soviet Union. These two approaches are complementary and are now in use at all scales. In addition, a variety of applied geomorphological mapping systems for specific purposes have been developed. The contents of these maps accentuate the geomorphological factors most relevant for the specific purpose and the legends are designed accordingly. A number of electronic maps, based on the ITC System and other methods of geomorphological survey, illustrate the survey methods.

It can be concluded that geomorphological maps have now become a generally recognised geoscientific document in much the same way as geological and soils maps. Furthermore, it is evident that systematic geomorphological surveying has contributed substantially to the advancement of geomorphology and to the development of a variety of applications. These results have become possible by the rapid development of remote sensing on one hand and by the increasing global need for reliable environmental information on the other.

Access to remotely sensed data, initially obtained from aerial photographs only, has been significantly increased with the advent of satellite remote sensing. High-resolution satellite data (less than 0.5 m) are valid for detailed mapping, while low-resolution satellite data can be used for reconnaissance mapping and global monitoring. The development of GIS and modelling techniques has also opened up new opportunities for applied geomorphological survey and mapping.

REFERENCES

Annaheim, H., 1956. Zur Frage der geomorphologischen Kartierung. Petermanns Geogr. Mitt. 103, 315–319.

Bashenina, N.V., 1972. Geomorphologische Kartierung des Gebirgsrelief im Maszstab 1:200,000 auf Grund einer Morphostruktur Analyse. Z. Geomorphol. NF 16/2, 125–128.

Blagovolin, N.S., Timofeev, D.A., 1993. Geomorphology in the former USSR. In: Walker, J.H., Grabau, W.E. (Eds.), The Evolution of Geomorphology. A Nation-by-Nation Summary of Development. John Wiley & Sons, pp. 483–499.

Bourdiec, F., 1963. Légende des cartes géomorphologiques détaillées, 1:20,000 et 1:25,000. Publ. C.G.A., Strasbourg, France.

Bourne, R., 1931. Regional survey and its relation to stock taking of the resources of the British Empire. Oxford Forestry Mem. 13, 16–18.

Brunsden, D., 1993. The nature of applied geomorphology. In: Panizza, M., Soldati, M., Barani, D. (Eds.), Proceedings of the First European Intensive Course on Applied Geomorphology, Modena-Cortina d'Ampezzo, 24 June–3 July 1992, pp. 3–11.

Brunt, M., 1967. The methods employed by the Directorate of Overseas Surveys in the assessment of land resources. Etudes de Synthèse 6, 3–10.

Christian, C.S., 1958. The concept of land units and land systems. Proceedings of Ninth Pacific Science Congress 20, pp. 75–81.

Demek, J. (Ed.), 1972. Manual of Detailed Geomorphological Mapping. IGU Commission for Geomorphological Mapping. Academia, Prague.

Finkle, Ch.F.J., 1988. The Encyclopedia of Field and General Geology. Van Nostrand Reinhold Company, New York, pp. 1−911

Galon, R., 1962. Instruction to the detailed geomorphological map of the Polish lowland. Geography and Geomorphology Department, Polish Academy of Sciences, Torun.

Gellert, J.F., 1968. Das System der Komplexgeomorphologischen Karten. Petermanns Geogr. Mitt. 112 (3), 185−190.

Gellert, J.F., Scholz, E., 1960. Konzeption und Methodik einer morphogenetischen Karte der DDR. Geogr. Berich. 14, 1−19.

Gellert, J.F., Scholz, E., 1974. Bemerkungen zur international vereinheitlichten Legende für mittelmassstäbliche Übersichtskarten von 1:200,000 zu 1:500,000. Stud. Geograf. Brno 41, 32−36.

Gimbarzessky, Ph., 1966. Land inventory interpretation. Photogramm. Eng. 32 (6), 967−976.

Helbling, E., 1952. Morphologie des Serntales. Ph.D Thesis. University of Bern, Bern.

Joly, F., 1963. Recherche d'une méthode de cartographie géomorphologique pour une carte des pays arides et semi-arides du monde à l'échelle du 1:1.000.000. B.S. Hellénique, Athène 4, 82−99.

Klimaszewski, M., 1956. The principles of the geomorphological map of Poland. Przeglad Geograficzny 28 (Suppl.), 32−40.

Klimaszewski, M., 1963. The principles of the geomorphological map of Poland. Geogr. Stud. 46, 69−70.

Leser, H., 1974. Geomorphologische Karten im Gebiete der BRD nach 1945. Bericht über die Aktivität des Arbeitskreises 'Geomorphologische Karte der BRD'. Catena 1, 297−326.

Maarleveld, C.G., ten Cate, J.A.M., de Lange, G.W., 1974. Die geomorphologische Karte der Niederlande. Z. Geomorphol. NF 18/4, 484−494.

Macar, P., de Béthume, P., Mammerickx, J., Seret, G., 1960. Travaux préperatoires à l'élaboration d'une carte géomorphologique détaillée de Belgique. Ann. Soc. Géol. Belge 84, 179−198.

Meijerink, A.M.J., 1988. Data acquisition and data capture through terrain mapping units. ILWIS, Integrated Land and Water Information System. ITC Publ. 7, Enschede, pp. 23−44.

Oguchi, T., Hayakawa, Y., 2011. Data sources. In: Smith, M.J., Paron, P., Grifith, J. (Eds.), Geomorphological Mapping: A Handbook of Techniques and Applications. Elsevier, London.

Passarge, S., 1914. Morphologischer Atlas. Erläuterungen zu Lief. 1, Morphologie des Messtischblattes Stadtremba (1:25,000). Mitt. Geogr. Gesell. Hamburg 28.

Pecsi, M., 1964. Geomorphological mapping in Hungary in the service of theory and practice. Applied Geography in Hungary. Akad. Kiadó, Budapest, pp. 1−18.

Pecsi, M., Ádám, L., Góczán, L., Hahn, Gy., Keresztesi, Z., Marosi, S., et al., 1962. Zeichenschlüssel zu der genetischen geomorphologische Übersichtskarte Ungarns. Hung. Acad. Sci. Publ. Budapest 1−85.

Raisz, E., 1931. The physiographic method of representing scenery on maps. Geogr. Rev. 21, 297−304.

Rao, D.P., 1975. Applied geomorphological mapping for erosion surveys: an example from the Oliva basin, Calabria. ITC J. 3, 341−350.

Rey, P., 1968. Photographie Aérienne et vegetation. Proceedings Toulouse Conference on Aerial Surveys and Integrated Studies. UNESCO, Paris, pp. 187−207.

Solntsev, N.A., 1962. Basic problems in Soviet landscape science. Sov. Geogr. 3, 3−15.

St-Onge, D., 1964. Geomorphological map legends, their problems and their value in optimum land utilization. Geogr. Bull. 22, 5–12.

Tricart, J., 1955. Un nouvel instrument au service de l'agronomie. Afr. Soils 4/1, 1–12.

Tricart, J., 1969. Cartographic aspects of geomorphological surveys in relation to development programmes. UN/ECOSOC 9, 75–83.

Tricart, J., 1970. Normes pour l'établissement de la carte géomorphologique détaillée de la France. Mém. Doc. CNRS 12, 1–267.

Tricart, J., 1972. Cartographie géomorphologique. Mém. Doc. CNRS 12, 1–267.

Troll, C., 1939. Luftbildplan und ökologische Bodemforschung. Z. Ges. Erdkunde Berlin 53, 241–298.

Troll, C., 1966. Landscape ecology. ITC Publ. S4 Delft, pp. 1–14.

Van Asselen, S., Seijmonsbergen, A.C., 2006. Expert-driven semi-automated geomorphological mapping for a mountainous area using a laser DTM. Geomorphology 78, 309–320.

Van Noord, H., 1993. A geomorphological mapping system at scale 1:10,000 and its application possibilities. In: Panizza, M., Soldati, M., Barani, D. (Eds.), Proceedings of the First European Intensive Course on Applied Geomorphology, Modena-Cortina d'Ampezzo, 24 June–3 July 1992, pp. 31–42.

Verstappen, H.Th., Van Zuidam, R.A., 1968/1975. ITC system of geomorphological survey (English, French and Spanish). Delft/Enschede, ITC-Textbook VII 2, 1–53.

Verstappen, H.Th., Van Zuidam, R.A. (with Meijerink, A.M.J. and Nossin, J.J.), 1991. The ITC System of Geomorphic Survey: A Basis for the Evaluation of Natural Resources and Hazards (English, French and Spanish). Revised ed. Enschede, ITC Publ. 10, 1–89.

Verstappen, H.Th., 1970. Introduction to the ITC System of geomorphological survey. Geograf. Tijd. 4 (1), 85–91.

Verstappen, H.Th., 1977a. Remote Sensing in Geomorphology. Elsevier, Amsterdam, pp. 1–214.

Verstappen, H.Th., 1977b. A geomorphological survey of the SW Cosenza Province, Calabria, Italy. ITC J. 1777 (4), 578–594 (map compilation: M.E. Hoschtitzky-Dantas).

Verstappen, H.Th., 1983. Applied Geomorphology. Geomorphological Surveys for Environmental Development. Elsevier, Amsterdam, pp. 1–437.

CHAPTER THREE

Nature and Aims of Geomorphological Mapping

Francesco Dramis[a], Domenico Guida[b] and Antonello Cestari[c]
[a]Department of Geological Sciences, Roma Tre University, Rome, Italy
[b]Department of Civil Engineering, University of Salerno, Fisciano, Italy
[c]C.U.G.R.I., Great Risks Interuniversity Consortium, University of Salerno, Fisciano, Italy

Contents

1. INTRODUCTION

Geomorphological maps are amongst the best tools for understanding the physical context of the Earth's surface. They provide a full objective description of landforms (*morphography*) identified with specific names and depicted with their correct shape or, where not allowed by the map scale, by appropriate symbols. Geomorphological maps should include

Developments in Earth Surface Processes, Volume 15
ISSN: 0928-2025, DOI: 10.1016/B978-0-444-53446-0.00003-3

information on the spatial properties (dimensions, slope, curvature, relief) of landforms (*morphometry*); their origin and evolution in relation to endogenous/exogenous genetic agents and processes (*morphogenesis*), also considering the effects of bedrock lithology/structure control; their relative or absolute age (*morphochronology*); their activity status and rate of genetic processes (*morphodynamics*) and the type of bedrock and near-surface deposits.

These data, collected at different scales in relation to the purposes of an investigation, from systematic field survey and the interpretation of aerial photographs and/or satellite imagery, are commonly reported on topographic sheets or on enlarged remotely sensed images (ortho-photomaps, ortho-photoplans, photo-mosaics and so on) in order to highlight their spatial distribution and mutual relationships.

Since the first published geomorphological map (Passarge, 1914), the importance of these documents has increased progressively, as testified to by the large number of scientific programmes of systematic survey and mapping promoted in different countries, even at a national level (Klimaszewski, 1956; Macar et al., 1961; Galon, 1962; Pecsi, 1963; Savigear, 1965; Tricart, 1965, 1972; Verstappen, 1970; Maarleveld et al., 1974; Barsch and Liedtke, 1980; Ten Cate, 1983; Barsch et al., 1987; Evans, 1990; Brancaccio et al., 1994; Buza, 1997; Kneisel et al., 1998; Wakamatsu et al., 2002; Baker, 2009; Gustavsson and Kolstrup, 2009).

Today, geomorphological mapping is present as a preliminary investigation method in practically all land management projects and geological risk assessment and zoning. Moreover, geomorphological baseline data are increasingly required by other sectors of environmental research such as land ecology, forestry and soil science (Tricart, 1969; Cooke and Doornkamp, 1974; Panizza, 1978; Guida et al., 1996; Brunsden, 2003).

The following sections are dedicated to 'traditional' symbol-oriented geomorphological maps distinguished in terms of purpose and scale. After a short description of the modern tools available for the acquisition, storage and display of geomorphic data, the efforts currently performed by geomorphologists in the transition process from traditional symbol-based mapping systems to full-coverage, multiscale, object-oriented geomorphological models will be discussed. The last part of the chapter will present the geographical information system (GIS)-based, object-oriented method of geomorphological mapping presently applied to landslide hazard assessment at Salerno University (Italy).

2. TYPES OF GEOMORPHOLOGICAL MAPS

Two main categories may be distinguished among geomorphological maps: *basic geomorphological maps* and *derivative geomorphological maps* (Dramis and Bisci, 1998).

Basic geomorphological maps (*analytical maps*; Verstappen, 1977) are produced by simple graphic transfer of data directly collected from field survey or aerial-photograph interpretation (Verstappen and van Zuidam, 1968; Klimaszewski, 1982; van Zuidam, 1985), from geological maps, soil maps, vegetation maps, land use maps and so on. A typical aspect of these maps is the ability to make interpretations not necessarily previewed by the practitioner.

Basic geomorphological maps may be made following two different perspectives: the first is concerned with the evolution of the landscape over geological timescales (*morpho-evolution maps*); the second takes into consideration the typology, and activity status, of geomorphological processes affecting the investigation area (*morphodynamic maps*).

Morpho-evolution maps represent Earth surface evolution in relation to endogenous agents (such as large-scale crustal vertical movements, surface tectonics and volcanism) and exogenous processes connected with past to modern climates, and, for more recent times, human activities. These maps are produced at scales that are not too large, in order to allow a general view of fairly large geomorphological features (such as planation surfaces, alluvial and marine terraces and fault scarps) that can be recognised more easily over a relatively wide area, even after being modified by subsequent geomorphological processes or tectonics.

Morphodynamic maps consider phenomena connected with present surface geodynamics including the effects of human activities. They are made at a more detailed scale, thus representing, with the necessary accuracy, all the landforms and near-surface deposits related to geomorphological processes affecting the investigated area. In this type of map, a detailed representation of bedrock lithology (possibly classified according to the mechanical behaviour of outcropping formations) and structural setting is important. According to the survey project purpose, some additional information could be provided concerning 'non-geomorphological' aspects such as paleoseismology, volcanic activity, soils, surface water, groundwater, vegetation cover and land use (*synthetic maps*; Verstappen, 1977).

From the analysis of morphodynamic maps it is possible to outline the overall framework of the recent/present morphogenesis of the investigated area as well as to formulate reasonable predictions of the future behaviour of recognised surface phenomena, also assessing scenarios of first-generation geomorphological events in previously unaffected areas. Therefore, regardless of their significant scientific value, morphodynamic maps may assume a primary role in land management projects (urbanisation, road construction, pipelines, parks and so on) and in projects aiming to mitigate geological risks.

Derivative geomorphological maps are obtained through selection, generalisation and reuse of data reported in basic maps with the purpose of zoning the spatial/temporal distribution of significant geomorphological processes such as landsliding, floods, co-seismic surface deformations, volcanic eruptions and tsunamis (*pragmatic geomorphological maps*; Verstappen, 1977; Ten Cate, 1990). Derivative maps are more easily readable than the original basic maps and may also be used by non-specialists, including engineers, land planners and decision-makers. A typical example is that of geomorphological stability maps (Panizza, 1973).

Geomorphological hazard maps are derivative maps that describe the 'nature of risk-causing surface phenomena, and their magnitude and frequency of occurrence' (Petley, 1998). They can be based either on the knowledge of an expert geomorphologist or on the application of statistical/deterministic models.

Computer-assisted procedures, mostly based on the analysis of geological—geomorphological, meteo-climatic and land use parameters, may be used to assess the susceptibility of land (i.e. the probability that a geomorphological event of given typology and magnitude may occur in a given area) to the occurrence (expanded, reactivated or newly generated) of potentially dangerous processes (Dikau, 1990; Parise, 2001; Cardinali et al., 2002; van Westen et al., 2008; Leoni et al., 2009). If the recurrence time interval of events triggering surface processes (extreme rainfall, high magnitude earthquakes) is considered, it is possible to assess, for the study area, different levels of geomorphological hazard (i.e. the probability that a geomorphological event of a given typology and magnitude may occur in a given area over a given time interval). Notwithstanding unavoidable assessment uncertainties and mistakes, hazard maps derived from large-scale geomorphological maps may be particularly useful (Petley, 1998).

3. GEOMORPHOLOGICAL MAP SCALE

Scale is one of main issues in geomorphological mapping. The spatial scales of geomorphological features span over a large range, from 10^7 km^2 (continents, ocean basins) to 10^{-8} km^2 (glacial striations, ripples) (Tricart, 1965). Moreover, the persistence time ranges from 10^8 years (for the largest features) to less than 10^2 years (for the smallest ones) in relation to their size (Table 3.1) as described by the following general equation (Baker, 1986):

$$S = aT^b$$

where S is the size of the feature, T is its duration time, a is constant indicating the intensity factor of the related geomorphic process (i.e.

Table 3.1 Spatial/Temporal Order of Magnitude of Earth Surface Features

Order	km^2	Corresponding Earth Surface Features	Approximate Persistence (years)
1	10^7	Continents, ocean basins	10^8–10^9
2	10^6	Shields	10^8
3	10^4	Medium-scale tectonic units (sedimentary basins, mountain massifs, domes)	10^7–10^8
4	10^2	Smaller tectonic units (fault blocks, volcanoes, sedimentary sub-basins)	10^7
5	10^2–10	Large-scale erosional/depositional units (deltas, major valleys, piedmonts)	10^6
6	10–10^{-2}	Medium-scale erosional/depositional units (floodplains, alluvial fans, moraines, smaller valleys)	10^5–10^6
7	10^{-2}	Small-scale erosional/depositional units (ridges, terraces, sand dunes)	10^4–10^5
8	10^{-4}	Larger geomorphic process units (hillslopes, sections of stream channels)	10^3
9	10^{-6}	Medium-scale geomorphic process units (pools and riffles, river bars, solution pits)	10^2
10	10^{-8}	Microscale geomorphic process units (fluvial and aeolian ripples, glacial striations)	

Source: Modified from Baker (1986).

rapidity of expenditure energy per unit area) and *b* is a scaling factor (equal to about 1.0).

Taking into account the timescale of geomorphological phenomena, Baker (1986) considers three main categories:

1. *macroscale*, over which major phases of erosion/deposition occur, controlled by regional warping, mountain building and crustal plate movement,
2. *mesoscale*, which treats major changes in landforms and landscapes over hundreds to thousands of years involving a complex interplay between tectonic and climatic controls on geomorphological processes (e.g. growth/recession of glaciers, aggradation/degradation of river bed and progradation/recession of shorelines),
3. *microscale*, over which the major variables of tectonism and climate are assumed to be constant (processes that characterise sand dunes, glaciers, rivers or beaches reflecting only the short-term events that dictate local flow physics).

Considering that genetic mechanisms, persistence times and, more generally, the nature of the geomorphological features change with changing landform dimensions (Schumm and Lichty, 1965; Cullingford, 1980; Brunsden, 1993, Evans, 2003, Slaymaker et al., 2009), it follows that maps with significantly different scales cannot address the same geomorphological contexts unless they have different objectives. Therefore, the choice of the map scale is strongly constrained by the project targets (Brunsden et al., 1975; Baker, 1986).

According to the level of cartographic detail, geomorphological maps were classified by Demek and Embleton (1978) into three groups:

- *large-scale geomorphological maps* (map scale >1:100,000),
- *medium-scale geomorphological maps* (map scale from 1:100,000 to 1:1,000,000),
- *small-scale geomorphological maps* (map scale <1:1,000,000).

However, considering the previous definition of geomorphological maps, it seems more appropriate to apply the scheme proposed by Dramis and Bisci (1998) (Table 3.2):

- *large-scale geomorphological maps* (map scale >1:25,000),
- *medium-scale geomorphological maps* (map scale from 1:25,000 to 1:250,000),
- *small-scale geomorphological maps* (map scale <1:250,000).

Table 3.2 Map Scale Classes, Ranges and Mappable Lengths

Scale	Scale Range	Maximum/Minimum Mappable Lengths (40 cm/2 mm on the map) (km)
Small	<1:1,000,000	>400/>2
	1:1,000,000–1:500,000	400/2–200/1
	1:500,000–1:250,000	200/1–100/0.5
Medium	1:250,000–1:100,000	100/0.5–40/0.2
	1:100,000–1:50,000	40/0.2–20/0.1
	1:50,000–1:25,000	20/0.1–10/0.05
Large	1:25,000–1:10,000	10/0.05–0.4/0.02
	1:10,000–1:5000	0.4/0.02–0.2/0.01
	>1:5000	<0.01

3.1 Large-Scale Geomorphological Maps

Large-scale geomorphological maps are made with enough detail to allow the correct representation of morphographic, morphometric, morphogenetic, morphochronologic and morphodynamic features of most landforms recognisable on slopes, valley floors, plains, coasts and so on.

Adequate information should be given on the main stratigraphic–sedimentologic characteristics and thickness of landform-related near-surface deposits, as well as on the outcropping bedrock lithology (possibly classified on the basis of lithotechnical characteristics) and structural setting (layering, foliations, faults, joints and so on).

To better understand the genesis of landforms and evaluate their possible future trends, the map contents should be enriched with data concerning surface/groundwater, vegetation cover, land use and so on.

The production of large-scale geomorphological maps is essentially based on systematic field survey. The interpretation of remotely sensed imagery (aerial photographs, satellite imagery) should only be used as a supporting tool during different project stages:

- to set up a preliminary geomorphological framework of the investigation area,
- to check the correct cartographic design of the surveyed field features,
- to perform the final revision of the field-based geomorphological map.

Where possible, in order to allow the easy and rapid transfer of field data, it is advisable to use aerial photographs with a scale close to that of the base topographic map sheet. Field observations should also be supported by laboratory analyses (sedimentological, paleontological, palinological, chronological) as well as by computer-assisted topographic analyses developed using digital elevation models (DEMs).

Field work should also include a detailed survey of bedrock lithology, possibly classified according to the main lithotechnical characteristics of the outcropping formations (Tricart, 1965; Panizza, 1972; Peña Monné, 1997; Dramis and Bisci, 1998). Data should also be collected on bedrock stratigraphy and structure (layering attitude, faults, jointing), as well as on the nature and thickness of near-surface deposits and weathering horizons, especially in the case of process-oriented (morphodynamic) maps (Evans, 1990; Dramis and Bisci, 1998).

Even if data concerning bedrock geology can be taken from pre-existing large-scale geological maps, it is best practice to inspect rock outcrops during the survey campaign (if necessary, with the help of an expert geologist). The same process should occur for near-surface deposits whose characteristics (lithology, texture, fabric, thickness, water content) play an important role in landscape evolution.

The analysis of the lithological composition of clasts may also be useful (Bridgland, 1986; Jones, 2000; McClanegan et al., 2001; Wanders et al., 2004):

- to reconstruct the extension and boundaries of ancient fluvial basins prior to the formation of contemporary systems,
- to quantify the individual contribution to moraine construction by glacial tongues originating from lithologically different valleys,
- to understand if debris deposits are fed by the upper slope or have been transported long distances.

Clast fabric may provide information on transportation/deposition mechanisms and transporting fluid direction. Particularly important in this context are the orientation of clast long axis (commonly perpendicular to flow lines in river channels) and clast imbrication, the best indicator of flow direction (Yagishita, 1989; Nichols, 2009).

The chronological reference of landforms is essentially based on the age of related deposits as provided by dating with different relative and absolute methods (^{14}C, Uranium series, ^{39}Ar/^{40}Ar, ^{40}K/^{40}Ar, ^{210}Pb, OSL – optically stimulated luminescence, TL – thermoluminescence and so on) of material included therein (Lowe and Walker, 1997). Some specific methods (cosmogenics, dendrochronology, lichenometry, weathering level) also allow the dating of surfaces (Darlymple, 1991; Winchester and Harrison, 2000;

Watchman and Twidale, 2002; Gosse, 2007). In any case, independently from the existence of absolute dates, both landforms and near-surface deposits should be placed within a temporal succession (on the base of their reciprocal spatial relationships). Indirect information regarding the landform/deposit age and paleoenvironmental genetic conditions may be obtained by paleomagnetic or thermo-chronological data.

In the case of morpho-evolution maps, it is convenient to organise near-surface deposits not in contact among each other according to morphostratigraphic sequences (North American Commission on Stratigraphic Nomenclature, 1983).

The activity status of surface features may be deduced by field observations (e.g. detailed stratigraphic observations, archaeological investigations, characteristics of vegetation cover, lichenometry) supported by the comparison of multitemporal aerial photographs and/or high-definition satellite images and the analysis of archive data (local history, periodicals, newspapers, minutes of governmental meetings, notarial acts, maps, paintings, photographs, scientific papers and reports and so on) (Dramis and Bisci, 1998). Significant data can be obtained from the examination of cracks and other disturbances affecting buildings (Coltorti et al., 1986). For more recent events, interviews with residents may provide useful information.

A possible field classification of landforms, in terms of activity, may consider three main categories (Dramis and Bisci, 1998):

1. *Active landforms* – landforms visibly evolving under the action of their genetic agents and related geomorphic processes,
2. *Quiescent landforms* – active landforms characterised by discontinuous, step-like evolution mapped in a dormant stage,
3. *Inactive landforms* – landforms produced in a geomorphological context definitely different from the present one and evolving under the action of agents (different from the genetic ones) that generally tend to destroy or bury them.

At scales above 1:5000, geomorphological maps are particularly suitable for outlining a detailed framework of the spatial–temporal evolution of landforms (and related deposits) such as shorelines, river beds, landslides and weathering features (Sauro, 1977; Fenti et al., 1979; Seijmonsbergen and van Westen, 1990; Faccini et al., 2008). Mapping activities may also include geophysical investigations, exploration boreholes, field/laboratory geotechnical data (regarding near-surface deposits and outcropping bedrock) and instrumental monitoring of landform activity status. Also information on surface/groundwaters may be included to better understand the morphodynamics of the investigated area. These cartographic documents, called *engineering geomorphological maps* (Griffiths and Marsh, 1986; Fookes, 1997; Griffiths, 2001), can play a significant role in land management activities such as stability analysis in built-up areas, preliminary investigations for engineering works, waste disposal areas and seismic microzoning.

3.2 Medium-Scale Geomorphological Maps

Medium-scale geomorphological maps provide a representation of large landscape units (volcanic hills, fault slopes, tectonic basins, mesas, cuestas, inselbergs, planation surfaces, alluvial/coastal terraces, alluvial plains, glacial valleys, dune fields and so on) which can be reproduced in full, or at least for a large part of their extension, thus allowing the depiction of mutual relationships and morphochronologic sequences.

Smaller landforms, such as those present on slopes and valley floors, are grouped together or reproduced by means of not-to-scale symbols. Also the subdivisions of landforms, near-surface deposits and genetic processes should be necessarily more generalised than in large-scale maps. As an example, slope processes connected with gravity (landslides, soil creep) and running water slope processes (slope wash, gullying) may be grouped in the single category of denudation processes. At smaller scales, it is more appropriate to use comprehensive terms such as fluvio-denudational slope and fluvial-depositional plain.

As far as bedrock geology is concerned, the relevant data are normally extracted from pre-existing cartographic documents. In some cases, bedrock geology is represented together with landforms as geological–geomorphological units (e.g. fluvio-denudational slope on limestones and planation surface on sandstone).

Where not derived by the generalisation of large-scale maps, medium-scale geomorphological maps are essentially produced by concurrent aerial-photograph interpretation and field work. Field observations are usually restricted to sample areas or representative transects with the aim of collecting interpretative keys from remote sensing analysis.

3.3 Small-Scale Geomorphological Maps

Small-scale maps can be classified into three groups:

1. Maps produced by a number of 'desk studies' such as the generalisation of previous larger scale maps, extrapolation of known situations

from comparable areas and bibliography data (e.g. the IGU Geomorphological Map of Europe on the scale of 1:2,500,000 by Bashenina et al., 1968, 1971),
2. Maps directly derived from satellite imagery interpretation (e.g. the 1:15,000,000 scale geomorphological map of the world directly constructed from space imagery by Bashenina and Talôskaya, 1981; the 1:1,000,000 scale map of Argentine Pampa by Canoba, 1982; the landforms map of part of New South Wales, Australia, by Pain, 1985),
3. Derivative maps, simply obtained by generalisation of larger scale geomorphological maps.

Small-scale geomorphological maps represent the structural framework of the land surface and the long-term geomorphological history of major depositional and erosional units, volcanic hills and effusive rocks and morphotectonic mega- and macro-structures. They are used in education to 'show the complex integration of the natural environment' (Embleton, 1985) as well as in land management at the country level, providing a 'first approach' land classification particularly useful for wide regions.

4. NEW TOOLS IN GEOMORPHOLOGICAL MAPPING

The recent advances in satellite technology and the ability of modern personal computers to manage large volumes of digital data have introduced radical changes in geomorphological mapping, providing a positive solution to some 'classical' problems of the 'traditional' cartographic approach.

Particularly relevant in this context is the role of the *global positioning system* (GPS), *satellite imagery data*, high-definition *DEMs* and *GIS*. Oguchi et al. (2011) provided a more detailed description of data sources, whereas Smith (2011) details manual mapping and Seijmonsbergen et al. (2011) detail automated and semi-automated mapping techniques.

4.1 Global Positioning System

The GPS can provide accurate measurements of the latitude, longitude and elevation of a survey/sampling point by means of geometric trilateration (a method for determining the intersections of three sphere surfaces given their centres and radii) of a constellation of geostationary satellites

(Leick, 1995). For this reason, GPS has become more and more widespread among field geomorphologists (Cornelius et al., 2006), particularly for active processes (Coe et al., 2003). Voženílek (2000) compared GPS-aided geomorphological mapping and conventional surveying techniques, highlighting the utility of the tool in terms of accuracy and data management.

4.2 Satellite Imagery

Data collected by satellite sensors, mostly in a digital form, offer the opportunity of observing the landscape at a regional scale (some even stereoscopically), permit identification of features not perceptible on site or on larger scales as well as landscape changes at regular intervals of time (Campbell, 1987; Drury, 1990; Smith and Pain, 2009). Satellite imagery cannot be substituted in full for those collected by field work and aerial-photograph analysis, however the use of high-resolution satellite imagery (up to 0.5 m with GeoEye-1) may provide valuable support to the geomorphologic interpretation of the landscape, especially in constructing medium/small-scale maps (Ulaby and McNaughton, 1975; Townshend, 1981; Hayden, 1986; Bocco et al., 2001; Etzelmüller et al., 2001; Rao, 2002).

Multispectral sensors (panchromatic, colour, and near infrared, short wave infrared and mid infrared bands), thermal radiation scanners and active microwave sensors (side-looking airborne radar or synthetic aperture radar) may provide detailed information on land surface features, highlighting small elevation differences and ground irregularities even in cloudy regions. Radar data can also provide information on land surface properties such as slope and dielectric behaviour of outcropping materials.

4.3 Digital Elevation Models

DEMs, that is digital imagery in which each matrix point has a value corresponding to its altitude above sea level, can be derived by digitising elevation data from topographic maps or, directly, from stereo imagery, interferometric synthetic aperture radar or light detection and ranging (LiDAR) (Dikau, 1989, 1992; Oguchi, et al., 2011). These models provide a 3D representation of the investigation area allowing observations from different viewpoints and with different vertical scales. These may also be rendered by draping over the DEM aerial photographs, topographical maps, geological maps and geomorphological maps (Teeuw, 2007; Aringoli et al., 2008). Moreover, morphometric data, such as slope gradients and breaks, slope aspect, altimetric belts, surface roughness or grain, as well as parameters concerning hydrographic networks can be automatically extracted from DEMs.

The availability of detailed DEMs allows analysis of landscape morphology and related processes in terms of topographic morphometry or *geomorphometry* (Wilson and Gallant, 2000; Hengl et al., 2008). This investigation method, in particular, provides a significant contribution to tectonic geomorphology, whose principal goal is to extract information regarding the rates and patterns of active deformation from landscape topography (Montgomery and Brandon, 2002). In this context, the study of bedrock channels plays an important role, especially in understanding the relationship between relief, elevation and denudation rates (Howard et al., 1994; Whipple et al., 1999). Indeed, the long profiles of bedrock rivers may yield valuable information about the distribution of recent deformation within the underlying region (Merritts and Vincent, 1989; Burbank et al., 1996; Lavé and Avouac, 2001; Montgomery and Brandon, 2002).

4.4 Geographical Information System

GIS packages are reference tools for the collection, storage, analysis and cartographic display of geospatial data (Burrough, 2000; Krönert et al., 2001), including topographic base data. Input land surface elements from geomorphological mapping may be selected and distributed into different georeferenced *layers*, which can be superposed and compared, enabling advanced spatial data analyses such as map overlay, adjacency, connectivity and containment to be performed. A GIS built on geomorphological data, criteria and rules is termed a *geomorphological information system* (GmIS) (Meijerink, 1988; Létal, 2005).

Although 'traditional' cartographic documents are '*static maps*' (that is not modifiable after their printing), those produced by means of GmIS may be considered '*dynamic maps*', whose printouts are simple reproductions taken at a given update stage (Eklundh, 2001). Moreover, a GmIS allows the simple and rapid processing of thematic layers and production of numerical analyses. Further advantages include the automatic extraction of data from topographical maps, such as calculating slope gradient and aspect, changing map scale, projections and coordinate systems (Bonham-Carter, 1994; Longley et al., 2001), joining two or more adjacent maps without loss of design quality or selecting geomorphological features from the database to produce special purpose maps. This last feature provides a positive solution to the 'classical' problem of geomorphological maps in fully representing, in a readable form, all the requested aspects of the land surface (Gustavsson, 2005). However, there remain graphic limitations in the reproduction of classical readable *general purpose* geomorphological maps, covering the whole scientific remit of land surface features.

In a GmIS database, land surface features can be stored on map layers as pixels (*raster data*) or points, open lines or polygons (*vector data*) which may be combined with *attribute data*, describing their characteristics (see Smith, 2011). These latter can be divided into *spatial data* (feature location, topology and geometry), *temporal data* (feature age or time of data collection) and *thematic data* (feature type).

In more detail, the GmIS structure should include the following data organised according to a cross-validation scheme with informative levels verifying each other (*congruence control*):

- *Vector data* representing land features (geomorphological database *sensu stricto*),
- *Raster data* representing images (output data from pixel/object-oriented analysis),
- *Triangulated irregular networks* (TINs) representing land surface by means of irregularly distributed nodes and lines with three-dimensional coordinates (x, y and z) that are arranged in a network of triangles (physical model of the investigated area),
- *Addresses and locators* defining geographical positions (depository of surveyed data).

Moreover, a GmIS database should include information about surface and sub-surface properties such as stratigraphy and lithology.

GmIS data can also be stored as objects and groups of objects, not separated into layers but gathered into hierarchically arranged classes. This latter approach reflects more accurately the 'real world' even if it has the disadvantage of time-consuming problems (Heywood et al., 2002).

Some limitations in application of input data may result from their accuracy and reliability (as an example, the data extracted from geological maps, such as layering or lithological boundaries, are sometimes uncertain, inhomogeneous and imprecise, combining original field mistakes with map drawing mistakes). Therefore, it would be necessary to review the survey methods, substituting generic descriptions with GPS-located numerical data and ordering field-surveyed land features in a pre-processed model based on remotely sensed data. The conceptual validity of the model should be verified by definition and cross-validation of numerical parameters obtained from pixel/object-oriented analysis. The field data should be recorded on bespoke forms (paper based) or transferred to a laptop and directly processed using mobile GIS software (e.g. ArcPad from ESRI, TerraSync from Trimble and Mobile GIS from Tensing).

5. PROBLEMS AND EFFORTS IN CURRENT GEOMORPHOLOGICAL MAPPING

As discussed earlier, a geomorphological map should contain substantial information regarding landform genesis, chronology and dynamics as well as near-surface and outcropping bedrock. However, this goal has proved hard to achieve (Gustavsson, 2005; Gustavsson et al., 2006). In fact, the huge amount of data to be mapped and the need to keep maps sufficiently readable has forced geomorphologists from different countries to adopt legends which, under the influence of local environmental conditions and academic schools, do not consider sufficiently, or even ignore, some of these fundamental landscape aspects (Gilewska, 1967; Demek et al., 1972; Demek and Embleton, 1978; Salomé et al., 1982; Gustavsson, 2005). For example, bedrock lithology is not present in Polish maps (Klimaszewski, 1982), whereas different outcropping rocks represent the fundamental landform units in French and Italian geomorphological maps (Joly and Tricart, 1970; Tricart, 1972; Panizza, 1988; Brancaccio et al., 1994; Dramis and Bisci, 1998); geometrically homogeneous land sectors divided by discontinuity lines are used as basic landform units in the British and Alpine Geomorphology Research Group (AGRG) legends (Savigear, 1965; Cooke and Doornkamp, 1974; Brunsden et al., 1975; De Graaff et al., 1987; Rose and Smith, 2008); in contrast, ITC legends (Verstappen and Van Zuidam, 1968; Verstappen, 1970, 1977; van Zuidam, 1982) show slope form as contour lines whereas the base map units generally are large genetically homogeneous areas. Some legends are extremely complicated and difficult to read (Barsch and Liedtke, 1980; Barsch et al., 1987; Kneisel et al., 1998), whereas others are extremely simple with limited information (Kienholz, 1978).

Summarising, the 'traditional' symbol-based mapping systems adopted in different countries, sometimes for national projects, are not comparable with each other and unable to provide a complete representation of landscape complexity (features and evolution processes) at the different scales and are therefore insufficient to fulfil all the scientific and practical needs of society (Klimaszewski, 1982, 1990; Barsch et al., 1987; Ten Cate, 1990; Gustavsson et al., 2006).

On the other hand, multiscale mapping models, coherently managed with a GIS (Mark and Smith, 2004) and easily readable and applicable to multidisciplinary landscape studies at the regional level, are increasingly required by land administrators and decision-makers in different sectors of land management (such as geo-hazard zoning for risk mitigation, land conservation, inventory of geo-sites, soil mapping, hydrology, landscape ecology, environmental engineering, forestry and agronomy).

To fulfil these requirements, geomorphological maps should represent, as precisely as possible, the spatial properties of landforms, reducing the use of symbols in favour of correctly bounded geometric elements (*full-coverage mapping*). In this regard, the extremely wide range of landform sizes implies the need for new mapping models, able to represent correctly the same area at different scales.

These models should:

- increase typology, quality, quantity and combinations of manageable and representable geomorphological data. In particular, the information associated with each 'object' should be flexible enough to allow the representation of all related attributes (e.g. the terrace edge of Figure 3.2, besides being a linear entity, is also part of the polygon which defines the terrace itself and the underlying river bed),
- interact with the analysis and data representation of other disciplinary sectors at different scales,
- conform with *spatial data transfer standard* (SDTS) in order to promote and facilitate the transfer of digital spatial data between dissimilar computer systems (Goodchild et al., 1999).

A positive response to these requirements is provided by the use of GIS-based mapping models rooted on the following principles:

- *Exhaustivity and mutual exclusivity*: All the geomorphological objects should be recognised, delimited by discrete or indeterminate boundaries according to the *fuzzy–set theory* (Borrough, 1996) and classified in only one distinct class (Fisher et al., 2000, 2004, 2007; MacMillan et al., 2000b; Arrel et al., 2007) or in fuzzy non-exclusive geomorphic types (Zhu, 1999; Borrough et al., 2000, 2001),

- *Understandability and applicability*: Terminology, classification schemes and procedures should be easily understandable and applicable,
- *Repeatability and independence*: The obtained products (in particular the object limits) should be reproducible and independent from any operator decision, possibly by automatic landform recognition,
- *Hierarchical multiscalar congruence*: The mapping process should cover adequately and congruently, areas with different geomorphological characters at scales of different detail,
- *Operational flexibility and structural coherence*: The GIS structure should be modifiable by the inclusion of further data and goals without implementing new classification schemes.

Problems and efforts in current geomorphological mapping may be synthesised in the following basic points: data interoperability, hierarchical data structure and full-coverage object-oriented data management.

5.1 Interoperability

In geomorphology, as in other earth sciences, specified land objects and their structural/functional interrelations 'have to be seen as a mental model, simplifying real world conditions' (Dikau et al., 1991). Therefore, the semantic rules supporting a GIS-based geomorphological mapping system can be defined as relationships between computer representations and the corresponding 'real world' features within a certain context (Bishr, 1998). Moreover, GIS-based mapping operators should be able to interact among them even if working at different sites and with different computer systems.

A possible way to achieve this state of interoperability may be provided by the development of a definitive and authoritative ontological nomenclature of the geospatial domain, grounded on the idea that a knowledge base can be defined through the development of a set of unique, domain-specific concepts for objects and processes describing geospatial information. In the 'concept space', a set of such concepts exists as an interlinked network of nodes between and within domains. Based on existing equivalency between concepts and category meanings, each node in the 'category space' can be linked to its corresponding node in the 'concept space' (Ng, 1998). By explicitly defining these links, a formal ontological data structure can be created.

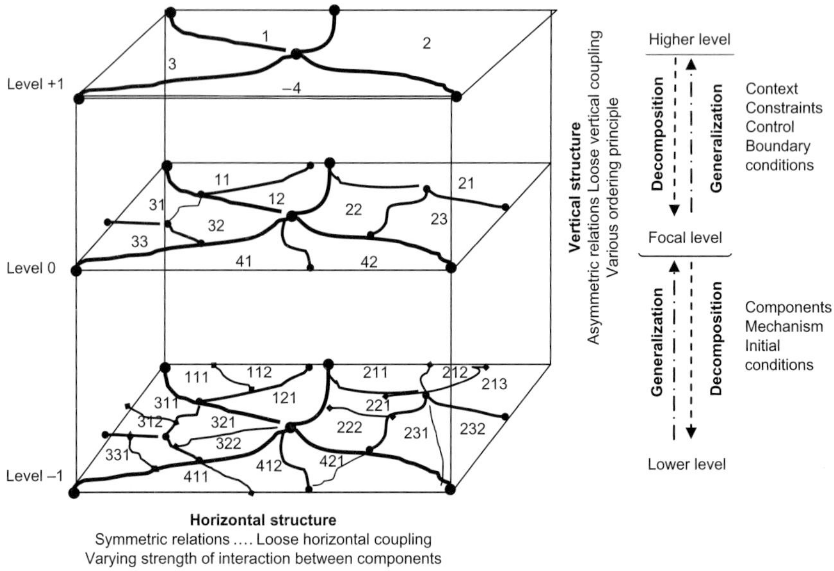

Figure 3.1 Illustration of hierarchical ordering/coding and horizontal/vertical relationship between the focal (initial) level and the higher/lower levels. In the focal to higher level transition, a set of generalisation algorithms allows the adaptation of time-spatial context, number and typology of control factors and boundary conditions. In the focal versus L-level transition, a set of decomposition algorithms are involved to extract basic components and mechanisms, modifying the previous initial conditions. *Modified from Wu (1999).*

5.2 Hierarchical Taxonomy and Multiscale Geomorphological Mapping

The problem of multiscale geomorphological mapping may be approached in the following manner: (1) the *principles of allometry* (Bull, 1975), that is the space–time relationships of landforms, including the energy rate involved in the genetic process and their persistence time (Huxley, 1972; Church, 1996; Small, 1996) and (2) the *hierarchy theory*, a set of principles to order structurally complex multilevel systems (Figure 3.1), with symmetrical, horizontal and asymmetrical upwards/downwards relationships (Koestler, 1967; Webster, 1977; O'Neil et al., 1986; Haigh, 1987; Seelbach et al., 1997; Wu, 1999; Krönert et al., 2001; Pereira, 2002).

A noteworthy aspect is the integration of 'traditional' symbol-based geomorphological legends with the hierarchically ordered land classification systems, largely applied in different sectors of the environmental sciences (Linton, 1951; Christian, 1958; MEXE, 1965; Christian and

Stewart, 1968; Ollier, 1977; Howard and Mitchell, 1980; Bailey, 1987; Speight, 1990; Dikau et al., 1991; Bisci and Dramis, 1992; Guida et al., 1996; Wielemaker et al., 2001; Pain and Kilgour 2003; McKenzie et al., 2005; Blasi et al., 2007; Pain et al., 2007; Paron et al., 2007). Through this approach, the land surface can be viewed as a mosaic of geomorphic objects that, by increasing observation detail, can be decomposed into smaller and smaller ones and vice versa. In this ordering system, called a *nested sequence*, each hierarchy level 'includes the cumulative effects of lower levels in addition to some new considerations (called emergent properties in the technical literature)' (Slaymaker et al., 2009).

5.3 Full-Coverage Object-Oriented Mapping

Full-coverage object-oriented mapping may be performed by expert judgement-based intercomparison between 'traditional' field mapping and pixel or object-oriented grid analysis for automatic landform recognition (Heil, 1980; Franklin, 1987; Molenaar, 1989; Hughes, 1991; Graff and Usery, 1993; Fels and Matson, 1996; Schmidt and Hewitt, 2004). The second procedure is based on *grid segmentation* techniques, allowing the partitioning of DEMs or remotely sensed imagery into non-overlapping regions (segments) representative of geomorphic entities (Baatz and Schäpe, 2000; MacMillan et al., 2000b; Blaschke and Strobl, 2001; Schiewe et al., 2001; Blaschke, 2003; Burnett and Blaschke, 2003; Drăguţ and Blaschke, 2006; Anders et al., 2009). With this technique, the geomorphic entities are designed with 'non-subjective' and repeatable boundaries better achieving quantitative landscape analysis and environmental design.

Two image objects are considered similar when they are near to each other in a certain 'feature space'; the semantic links between image objects are established on principles of object-oriented programming. An object is constituted by certain sub-objects; sub-objects are elements of super-objects. Sub-objects inherit certain characteristics from their respective super-objects and vice versa (Blaschke and Strobl, 2001). The decomposition of land features into smaller units, characterised by distinctive mechanisms, magnitude and evolution rates, may provide a positive contribution to a deeper understanding of their evolutionary trends, also in view of assessing related risk levels and setting up appropriate remedial measures.

Object-oriented geomorphological mapping is increasingly used in the automatic or semi-automatic definition of landforms, with particular reference to those connected with slope and fluvial processes. The capacity of overcoming the 'three-dimensional' limitations related to symbol-oriented methods and grid-based analysis (boundary/segment representation of geomorphic entities) will induce widespread diffusion of this system in the future. However, the transition to the full use of object-oriented geomorphological mapping will be not simple or immediate. In fact, before reaching the goal of a reliable automatic recognition of landforms from remote sensing imagery, the 'traditional' symbol-oriented mapping system will continue to be used at least as the first operative step of the object-oriented methodology.

6. EXPERIENCES OF GIS-BASED, OBJECT-ORIENTED MULTISCALE GEOMORPHOLOGICAL MAPPING

A new GIS-based, full-coverage, object-oriented geomorphological mapping system has been applied in Italy, in several national and regional projects on engineering geomorphology, landscape ecology and hydrology (Cascini et al., 2005; Rossi et al., 2006; Blasi et al., 2007). These activities constitute the 'core sector' of a GmIS at the Department of Civil Engineering and Great Risks interuniversity Consortium, Salerno University (Italy).

Intercomparison between 'traditional' mapping ('expert judgement-based') and automatic landform recognition allows a 'non-subjective' and repeatable delineation of the geomorphic entities in order to better pursue quantitative landscape analyses and environmental design. The hierarchical taxonomy shown in Table 3.3 is a modified version of the scheme applied in these projects (Guida et al., 1996, 2009 Cascini et al., 2005; Blasi et al., 2007; De Pippo et al., 2007). The informatic structure of the different taxonomic levels is organised in terms of 'nested topologic entities' (closed polygons, open lines and punctual symbols) supported by an attribute list. Moving upward towards smaller scales, polygons may change to lines or symbols. Moving downwards, symbols may change to lines or polygons, lines may change to polygons, whereas polygons may be decomposed into smaller features (Figure 3.2). In cartography this is termed a scale-dependent renderer.

Levels 1 (*physiographic domain*), 2 (*physiographic region*) and 3 (*physiographic province*) correspond to morphologically distinctive surface features significant at the continental, subcontinental and regional levels

Table 3.3 The Salerno University Hierarchical Multiscale Taxonomy

Level	Scale Range	Land Features Taxonomy	Corresponding Land Units in Other Classification Schemes	Persistence Time
1	<1:1,000,000	Physiographic domain	Physiographic domain (MacMillan et al., 2000a) Land region *p.p.* (Crofts, 1991) Land system *p.p.* (Linton, 1951)	10^8–10^9 years
2	1:1,000,000 1:500,000	Physiographic region	Physiographic region (MacMillan et al., 2000a) Land region *p.p.* (Crofts, 1991) Land system *p.p.* (Linton, 1951) Geotectonic region (Blasi et al., 2007)	10^8 years
3	1:500,000 1:250,000	Physiographic province	Physiographic province (MacMillan et al., 2000a) Land region (Crofts, 1991) Land system *p.p.* (Linton, 1951) Morphotectonic province (Guida et al., 1996; Blasi et al., 2007)	10^7–10^8 years
4	1:250,000 1:100,000	Landform system	Physiographic system *p.p.* (MacMillan et al., 2000a) Land region (Linton, 1951) Morphological system *p.p.* (Guida et al., 1996; Blasi et al., 2007)	10^7 years
5	1:100,000 1:50,000	Landform sub-system	Land system *p.p.* (Linton, 1951) Land system (Crofts, 1991) Morphological system *p.p.* (Guida et al., 1996; Blasi et al., 2007)	10^6 years
6	1:50,000 1:25,000	Landform pattern	Landform type *p.p.* (MacMillan et al., 2000a) Land facet (Crofts, 1991) Facet (Linton, 1951) Morphological unit (Guida et al., 1996; Blasi et al., 2007)	10^5–10^6 years
7	1:25,000 1:10,000	Landform complex	Landform type *p.p.* (MacMillan et al., 2000a) Land facet *p.p.* (Crofts, 1991) Facet *p.p.* (Linton, 1951)	10^4–10^5 years

(*continued*)

Table 3.3 (continued)

Level	Scale Range	Land Features Taxonomy	Corresponding Land Units in Other Classification Schemes	Persistence Time
8	1:10,000 1:5000	Landform unit	Landform element *p.p.* (MacMillan et al., 2000a) Land site *p.p.* (Crofts, 1991) Site *p.p.* (Linton, 1951)	10^3–10^2 years
9	>1:5000	Landform element	Landform element *p.p.* (MacMillan et al., 2000a) Land site *p.p.* (Crofts, 1991) Site *p.p.* (Linton, 1951)	10^2 years or less

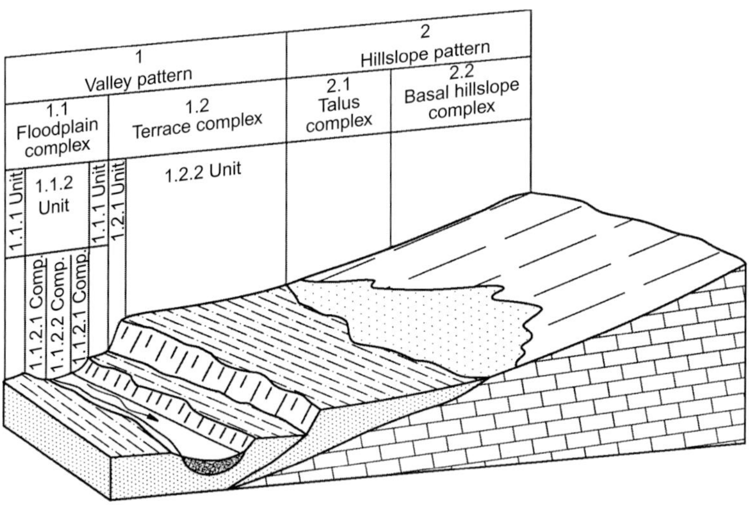

Figure 3.2 Nested hierarchic sequence of landforms.

(respectively), such as great mountain chains, sedimentary basins and forelands. The identification/delineation criteria of surface features are related to the physiographic expressions of long- to mid-term orogenetic/epeirogenetic activity over wide areas, primarily acquired from remotely sensed imagery and coarse resolution DEMs (~500 m × 500 m). The related maps are adequate to illustrate inter-regional/regional landscape features (Blasi et al., 2007), atmospheric circulation and neotectonics.

Level 4 (*morphological system*) includes prominent landscape components such as plateaus, valleys, plains and coastal belts. Their identification/delineation criteria imply the definition of coarse topo-position, polygenetic and polyphase consistency, acquired by automatic landform recognition from satellite imagery and coarse resolution DEMs (~100 m × 100 m). Additional data from previous studies and selected field surveys may be required. The resulting maps may be used for subregional landscape analysis (Guida et al., 1996; Blasi et al., 2007), environmental planning and hydro-geomorphology studies.

Level 5 (*morphological sub-system*) includes mid-size landscape components such as small ridges, hillslopes, large valley floors, piedmonts and moraine amphitheatres. The identification/delineation criteria imply the definition of detailed landform topo-position, morphometrics and morphogenetic consistency, acquired by automatic landform recognition from mid-resolution DEMs (~25 m × 25 m), and aerial-photograph interpretation with supplementary field work. The resulting maps may be used in local landscape analysis (Guida et al., 1996; Blasi et al., 2007), environmental planning and detailed hydro-geomorphology studies.

Level 6 (*morphological pattern*) includes large compound landforms (e.g. alluvial terraces, glacial cirques, coastal cliffs, talus belts). The identification/delineation criteria imply the definition of landform detailed topo-position, morphometrics and morphogenetic consistency, acquired by automatic landform recognition from mid- to mid-fine resolution DEMs (~25 m × 25 m to ~10 m × 10 m), aerial-photograph interpretation and field work. The resulting maps may be used in detailed landscape analysis (Guida et al., 1996; Blasi et al., 2007), local environmental planning and detailed hydro-geomorphology studies.

Levels 7–9 are essentially based on detailed field survey. The identification/delineation criteria imply the definition of landform detailed topo-position, morphometrics and morphogenetic consistency, acquired by automatic landform recognition from fine DEMs (~10 m × 10 m to ~5 m × 5 m), and the interpretation of large-scale aerial photographs. The resulting maps are commonly used as preliminary tools for programming further *in situ* investigations (Guida et al., 1996; Blasi et al., 2007). Level 7 (*landform complex*) includes mid-size landform produced by single or multiple geomorphic processes (e.g. large river channels, coastal arcs, large compound landslides).

Level 8 (*landform unit*) includes small landforms formed by single or multiple geomorphic processes (e.g. alluvial terrace scarp, moraine arcs and mid-size landslides) or landform components (e.g. terrace scarp slide, alluvial fan channel, coastal cliff notch, landslide scar and landslide accumulation zone). Level 9 (*landform element*) includes non-decomposable landforms with reference to the project purposes. Mapping at this level typically includes special investigation methods such as geotechnical tests, geophysical soundings, boreholes, laboratory tests and instrumental monitoring.

Level 8 usually represents the starting (focal) point for the production of lower level maps by nested landform composition. However, the focal scale level may change substantially in relation to the mapping project purposes.

The Salerno University mapping procedure (Guida et al., 2009) includes the following steps (Figure 3.3):

1. Production of a 'traditional' field-surveyed, symbol-based geomorphological map, normally at scales ranging between 1:5000 and 1:25,000, in relation to the mapping project purpose, focusing on morphography, morphometry and morphogenesis. The data source is a detailed field survey supported by aerial-photograph interpretation (1a); the legend is a symbol-oriented list of significant relief features (1b); the result is a 'traditional' field-surveyed, symbol-based geomorphological map (1c). The geological aspects of bedrock and near-surface deposits as well as other geomorphological/environmental relevant aspects of land units (such as dominant process and age) are digitally recorded as attributes and transferred into the database,
2. Aerial-photograph interpretation (2a), at a scale close to that of the survey base toposheet, to produce a full-coverage geomorphological map (2c) from expert judgement. At this stage, the geomorphological features are delimited and coded in a nested structure with boundary lines at different reliability levels (2b),
3. Primitive topological transformation (3a) of the mapped units supported by attribute list (3b),
4. Construction of an object-oriented, GIS-based geomorphological map,
5. DEM-based geomorphometrical analysis (5a), automatic multiscale landform recognition (5b) and object-oriented remotely sensed imagery processing (5c) (Baatz and Schäpe, 2000; Baatz et al., 2002; Arko and Stein, 2005; Minàr and Evans, 2008; Schneevoigt et al., 2008).

The main topics of the Salerno University geomorphological mapping model are presented in the annexes A and B. Annexe A illustrates the transition steps from a traditional symbol-oriented geomorphological map

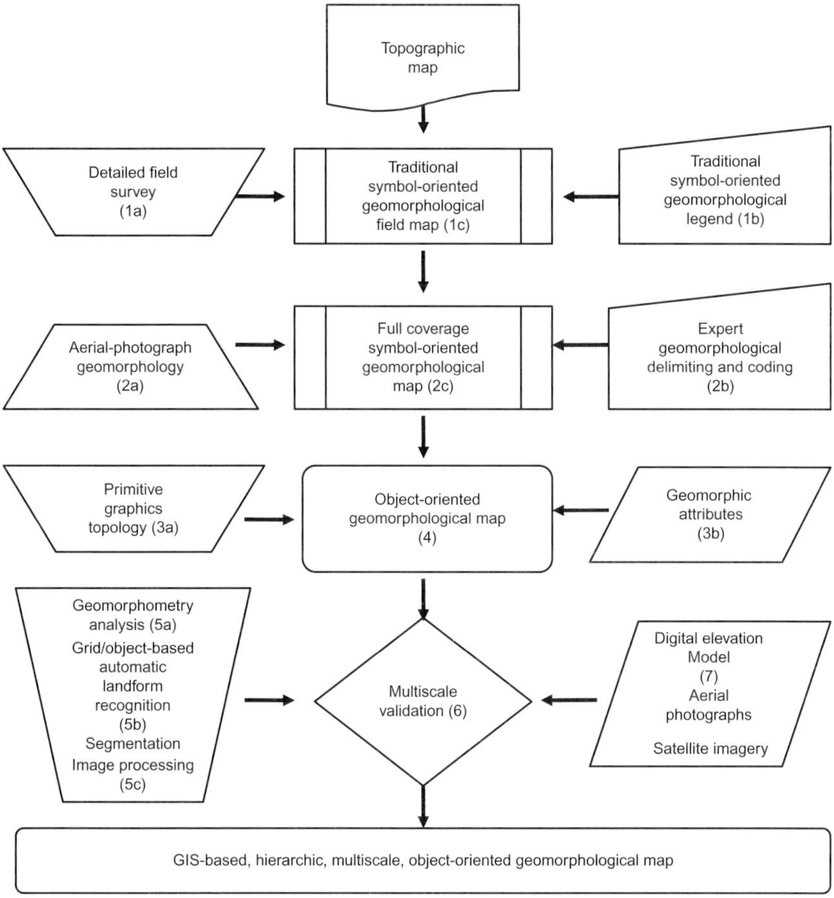

Figure 3.3 Flow diagram of the Salerno University geomorphological mapping system. The progressive numbers indicate the sequence of steps and sub-steps; the trapezoidic shapes indicate the field, laboratory and analytical data inputs; the rhomboid shapes indicate the graphical or code tools used to transfer inputs into preliminary (1c), intermediate (2c) and final (4) geomorphological map; the rhombus indicates the decision about the acceptance of the map into the GmIS.

to a full-coverage, object-oriented geomorphological map (landslide hazard map of Roveta Valley, Abruzzi, Italy). In Annexe B, the Salerno University GmIS (UNISA_GmIS) and the generalisation process from the 1:5000 (focal level) object-oriented geomorphological map of the Fisciano Campus area to 1:25,000 and 1:100,000 scales are presented. Annexe C shows some examples of 'traditional' symbol-oriented geomorphological maps.

7. CONCLUDING REMARKS

Over the last few decades, traditional symbol-oriented geomorphological mapping methods have been widely used for land management, especially in the field of geo-hazard evaluation and risk mitigation. However, these methods are not able to meet the current scientific and technical requirements of land management. In fact, despite major improvements introduced by new investigation tools and GIS-based procedures, symbol-oriented legends are unable to provide a suitable representation of the landscape complexity for a multipurpose, multidisciplinary and multiscalar approach to land management.

A proper representation of landscape complexity can be obtained through the characterisation of the spatial properties of landforms based on hierarchically arranged geometric elements (geomorphological objects), translatable from larger to smaller scales and vice versa by generalisation/decomposition. In this context, the GIS-based, object-oriented mapping system applied at the Salerno University may be considered as a milestone in a 'road map' towards a shared 'cartographic language', which preserves the previous experiences and provides, at the same time, appropriate support for present-day environmental projects.

REFERENCES

Anders, N.S., Seijmonsbergen, A.C., Bouten, W., 2011. Segmentation optimization and stratified object-based analysis for semi-automated geomorphological mapping. Rem. Sens. Environ. doi:10.1016/j.rse.2011.05.007.

Aringoli, D., Calista, M., Gentili, B., Pambianchi, G., Sciarra, N., 2008. Geomorphological features and 3D modelling of Montelparo mass movement (Central Italy). Eng. Geol. 99 (1–2), 70–84.

Arko, L., Stein, A., 2005. Texture-based landform segmentation of LiDAR imagery. Int. J. Appl. Earth Obs. Geoinformation 6 (3–4), 261–270.

Arrell, K., Fisher, P., Tate, N., Bastin, L., 2007. A fuzzy k-means classification of elevation derivatives to extract the natural landforms in Snowdonia, Wales. Comput. Geosci. 33 (10), 1366–1381.

Baatz, M., Benz, U., Dehghani, S., Heynen,M., Höltje, A., Hofmann, P., et al., 2002. Definiens imaging – eCognition user guide 3. München. <http://www.definiens-imaging.de/documents/userguide.htm> (accessed 10.07.04).

Baatz, M., Schäpe, A., 2000. Multiresolution segmentation – an optimization approach for high quality multi-scale image segmentation. In: Strobl, J., Blaschke, T., Griesebner, G. (Eds.), Angewandte Geographische Informationsverarbeitung, XII. Wichmann-Verlag, Heidelberg, pp. 12–23.

Bailey, R.G., 1987. Suggested hierarchy of criteria for multiscale ecosystem mapping. Landsc. Urban Plan. 14, 313–319.

Baker, C., Skene, D., Babu, D., 2009. A nationally consistent geomorphic classification of the Australian Coastal Zone. Abstracts of the Seventh International Conference on Geomorphology, July 2009, Melbourne, Australia (CD-ROM).

Baker, V.R., 1986. Introduction: regional landform analysis. In: Short, N.M., Blain Jr., R.W. (Eds.), Geomorphology From Space: A Global Overview of Regional Landforms. NASA, Scientific and Technical Information Branch, Washington, DC. Chapter 1 – GES DISC, Goddard Earth Sciences. <http://disc.sci.gsfc.nasa.gov/geomorphology/GEO_1/GEO_CHAPTER_1.shtml>.

Barsch, D., Fischer, K., Stäblein, G., 1987. Geomorphological mapping of high mountain relief, Federal Republic of Germany (with geomorphology map of Königsee, scale 1:25,000). Mt. Res. Dev. 7 (4), 361–374.

Barsch, D., Liedtke, H., 1980. Principles, scientific value and practical applicability of the geomorphological map of the Federal Republic of Germany at the scale of 1:25,000 (GMK 25) and 1:100,000 (GMK 100). Z. Geomorphol. N.F. Suppl. Band 36, 296–313.

Bashenina, N.V., Talôskaya, N.N., 1981. Space imagery analysis for a geomorphological map of the world. Sov. J. Remote Sens. 6, 861–871.

Bashenina, N.V., Blagovolin, N.S., Demek, J., Dumitrashko, N.V., Ganeshin, G.S., Gellert, J.F., et al., Legend to the International Geomorphological Map of Europe 1:2,500,000, Fifth version. Czechoslovak Academy of Sciences, Institute of Geography, Brno.

Bashenina, N.V., Gellert, J., Joly, F., Klimaszewski, M., Scholz, E., 1968. Project to the unified key to the detailed geomorphological map of the world. Folia Geogr. Ser. Geogr. Phys. 2, 1–40.

Bisci, C., Dramis, F., 1992. Geomorphologic-seismic zonation of the Marche Region (Central Italy) using computer aided techniques: preliminary considerations. ITC J. 1992-2, 196–201.

Bishr, Y., 1998. Overcoming the semantic and other barriers to GIS interoperability. Int. J. Geogr. Inf. Sci. 12 (4), 299–314.

Blaschke, T., 2003. Object-based contextual image classification built on image segmentation. IEEE Proceedings, Washington, DC (CD-ROM).

Blaschke, T., Strobl, J., 2001. What's wrong with pixels? Some recent developments interfacing remote sensing and GIS. GIS-Zeitschrift für Geoinformationssysteme 6, 12–17.

Blasi, C., Guida, D., Siervo, V., Paolanti, M., Michetti, L., Capotorti, C., et al., 2007. Defining and mapping the landscape of Italy. Advance and Application of Landscape Character Mapping, Proceedings of the 7th IALE Congress – part 1, pp. 572–573.

Bocco, G., Mendoza, M., Velázquez, A., 2001. Remote sensing and GIS-base regional geomorphological mapping – a tool for land use planning in developing countries. Geomorphology 39, 211–219.

Bonham-Carter, G.F., 1994. Geographic Information Systems for Geoscientists – Modelling with GIS. Pergamon Press, Oxford.

Brancaccio, L., Castiglioni, G.B., Chiarini, E., Cortemiglia, G., D'Orefice, M., Dramis, F., et al., 1994. Carta geomorfologica d'Italia – 1:50.000. Guida al Rilevamento. Quaderni del Servizio Geologico Nazionale, ser. III 4, 1–42.

Bridgland, D.R., 1986. Clast Lithological Analysis. Technical Guide 3. Quaternary Research Association, Cambridge, UK.

Brunsden, D., 1993. The persistence of landforms. Z. Geomorphol. N.F. Suppl. Band 93, 13–27.

Brunsden, D., 2003. Geomorphology, engineering and planning. Geogr. Pol. 76, 185–205.

Brunsden, D., Doornkamp, J.C., Fookes, P.G., Jones, D.K.C., Kelly, J.M.H., 1975. Large-scale geomorphological mapping and highway engineering design. Q. J. Eng. Geol. 8, 227–253.

Bull, W.B., 1975. Allometric change of landforms. GSA Bull. 86 (11), 1489–1498.
Burbank, D.W., Meigs, A., Brozovic, N., 1996. Interactions of growing folds and coeval depositional systems. Basin Res. 8, 199–223.
Burnett, C., Blaschke, T., 2003. A multi-scale segmentation/object relationship modelling methodology for landscape analysis. Ecol. Modell. 168, 233–249.
Burrough, P.A., 1996. Natural objects with indeterminate boundaries. In: Burrough, P.A., Frank, A.U. (Eds.), Geographic Objects with Indeterminate Boundaries. Taylor & Francis, London, pp. 3–28.
Burrough, P.A., 2000. Principles of Geographical Information Systems, Spatial Information Systems and Geostatistics. Clarendon Press, Oxford.
Burrough, P., van Gaans, P., Hansen, A., 2000. High resolution landform classification using fuzzy k-means. Fuzzy Sets Syst. 113 (1), 37–52.
Burrough, P., Wilson, J., van Gaans, P., Hansen, A., 2001. Fuzzy k-means classification of topoclimatic data as an aid to forest mapping in the Greater Yellowstone Area, USA. Landsc. Ecol. 16, 523–546.
Buza, M., 1997. A general geomorphological map of Romania on the scale of 1:25,000, Zlatna sheet. GeoJournal 41 (1), 85–91.
Campbell, B., 1987. Introduction to Remote Sensing. Guilford Press, New York.
Canoba, C.A., 1982. Geomorphological mapping using Landsat imagery: a case study in Argentina. ITC J. 1982–3, 324–329.
Cardinali, M., Reichenbach, P., Guzzetti, F., Ardizzone, F., Antonini, G., Galli, M., et al., 2002. A geomorphological approach to the estimation of landslide hazards and risks in Umbria, Central Italy. Nat. Hazards Earth Syst. Sci. 2, 57–72.
Cascini, L., Guida, D., Lanzara, R., Sorbino, G., 2005. Il Sistema Informativo del Presidio Territoriale. Rubbettino, Cosenza.
Christian, C.S., 1958. The concepts of land units and land systems. Proceedings of the Ninth Conference of the Pacific Science Association, Bangkok, Thailand, 1957, vol. 20, pp. 74–81.
Christian, C.S., Stewart, G.A., 1968. Methodology of integrated surveys. Proceedings of the Toulouse Conference, 1964, Natural Research Series, UNESCO, vol. 6, pp. 233–280.
Church, M., 1996. Space, time and the mountain: how do we order what we see? In: Rhoads, B.L., Thorn, C.E. (Eds.), The Scientific Nature of Geomorphology. John Wiley & Sons, Chichester, pp. 17–170.
Coe, J.A., Ellis, W.L., Godt, J.W., Savage, W.Z., Savage, J.E., Michael, J.A., et al., 2003. Seasonal movement of the Slumgullion landslide determined from global positioning system surveys and field instrumentation, July 1998–March 2002. Eng. Geol. 68, 67–101.
Coltorti, M., Dramis, F., Gentili, B., Pambianchi, G., Sorriso-Valvo, M., 1986. Aspetti geomorfologici. In: Crescenti, U., (Ed.), La grande frana di Ancona, Studi Geologici Camerti, vol. spec., pp. 29–39.
Cooke, R.U., Doornkamp, J.C., 1974. Geomorphology in Environmental Management. Clarendon Press, Oxford.
Cornelius, S.C., Sear, D.A., Carver, S.J., Heywood, D.I., 2006. GPS, GIS and geomorphological field work. Earth Surf. Process. Landforms 19 (9), 777–787.
Crofts, R.S., 1991. Mapping techniques in geomorphology. In: Goudie, A., Anderson, M., Burt, T., Lewin, J., Richards, K., Whalley, B., et al., Geomorphological Techniques, second ed. George Allen & Unwin, London, pp. 66–75.
Cullingford, R.A., Davidson, D.A., Lewin, J., 1980. Timescales in Geomorphology. John Wiley & Sons, Chichester.
Darlymple, G.B., 1991. The Age of the Earth. Stanford University Press, Stanford, CA.

De Graaff, L.W.S., De Jong, M.G.G., Rupke, J., Verhofstad, J., 1987. A geomorphological mapping system at scale 1:10,000 for mountainous areas. Z. Geomorphol. N.F. 13 (2), 229–242.
Demek, J., Embleton, C. (Eds.), 1978. Guide to Medium-Scale Geomorphological Mapping. E. Schweizerbart'sche Verlagsbuchhandlung (Nägele u. Obermiller), Stuttgart.
Demek, J., Embleton, C., Gellert, J.F., Verstappen, H.T. (Eds.), 1972. International Geographical Union Commission on Geomorphological Survey and Mapping. Academia, Prague.
De Pippo, T., Guida, D., Lanzara, V., Siervo, V., Valente, A., 2007. Criteri, metodi e procedure innovative per la redazione di cartografia geomorfologica gerarchica multiscalare: proposte operative in ambiente GIS. Convegno Nazionale AIGEO 'Ambiente Geomorfologico e Attività dell'Uomo: Risorse, Rischi, Impatti', Torino, 28–30 marzo 2007, pp. 230–234.
Dikau, R., 1989. The application of a digital relief model to landform analysis. In: Raper, J.F. (Ed.), Three Dimensional Applications in Geographical Information Systems. Taylor & Francis, London, pp. 51–77.
Dikau, R., 1990. Derivatives from detailed geoscientific maps using computer methods. Z. Geomorphol. N.F. 80, 45–55.
Dikau, R., 1992. Aspects of constructing a digital geomorphological base map. Geol. Jahrb. A 122, 357–370.
Dikau, R., Brabb, E.E., Mark, R.M., 1991. Landform classification of New Mexico by computer. US Department of the Interior, US Geological Survey. Open-file report, pp. 91–634.
Drăguţ, L., Blaschke, T., 2006. Automated classification of landform elements using object-based image analysis. Geomorphology 81, 330–344.
Dramis, F., Bisci, C., 1998. Cartografia Geomorfologica. Manuale di Introduzione al Rilevamento ed alla Rappresentazione Degli Aspetti Fisici del Territorio. Pitagora Editrice, Bologna.
Drury, S.A., 1990. A Guide to Remote Sensing. Interpreting Images of the Earth. Oxford Science Publications, Oxford.
Eklundh, L. (Ed.), 2001. Geografisk Informationsbehandling. second ed. Byggforskningsrådet, Stockholm.
Embleton, C., 1985. Techniques, problems and uses of mega-geomorphological mapping. In: Hayden, R.S. (Ed.), Global Mega-Geomorphology, NASA CP-2312, pp. 84–88.
Etzelmüller, B., Hoelze, M., Heggem, E.S.F., Isaksen, K., Mittaz, C., Vonder Mühll, D., et al., 2001. Mapping and modelling the occurrence and distribution of mountain permafrost. Nor. Geogr. Tidsskr. 55, 186–194.
Evans, I.S., 1990. Cartographic techniques in geomorphology. In: Goudie, A., Anderson, M., Burt, T., Lewin, J., Richards, K., Whalley, B., et al., Geomorphological Techniques, second ed. George Allen & Unwin, London, pp. 97–108.
Evans, I.S., 2003. Scale-specific landforms and aspects of the land surface. In: Evans, I.S., Dikau, R., Tokunaga, E., Ohmori, H., Hirano, M. (Eds.), Concepts and Modelling in Geomorphology: International Perspectives. TERRAPUB, Tokyo, pp. 61–84.
Faccini, F., Piccazzo, M., Robbiano, A., Roccati, A., 2008. Applied geomorphological map of the Portofino Municipal Territory (Italy). J. Maps 2008, 451–462.
Fels, J.E., Matson, K.C., 1996. A cognitively based approach for hydro-geomorphic land classification using digital terrain models. Third International Conference/Workshop on Integrating GIS and Environmental Modeling, Santa Fe, New Mexico, 21–25 January 1996, National Centre for Geographic Information and Analysis, Santa Barbara, CA (CD-ROM).

Fenti, V., Silvano, S., Spagna, V., 1979. Methodological proposal for an engineering geomorphological map. Forecasting rockfalls in the Alps. IAEG Symposium on Engineering Geological Mapping for Planning, Design and Construction in Civil Engineering, Newcastle Upon Tyne, UK, September 1979. Bull. Eng. Geol. Environ. 2007, 134–138.

Fisher, P.F., 2000. Sorites paradox and vague geographies. Fuzzy Sets Syst. 113 (1), 7–18.

Fisher, P.F, Wood, J., Cheng, T., 2004. Where is Helvellyn? Multiscale morphometry and the mountains of the English Lake District. Trans. Inst. Br. Geogr. 29, 106–128.

Fisher, P.F., Wood, J., Cheng, T., 2007. Higher order vagueness in a dynamic landscape: multi-resolution morphometric analysis of a coastal dunefield. J. Environ. Inform. 9 (1), 56–70.

Fookes, P.G., 1997. Geology for engineers: the geological model, prediction and performance. Q. J. Eng. Geol. 30, 293–424.

Franklin, S.E., 1987. Geomorphic processing of digital elevation models. Comput. Geosci. 13, 603–609.

Galon, R., 1962. Instruction to the Detailed Geomorphological Map of the Polish Lowland. Geography Institute, P.A.N., Torun.

Gilewska, S., 1967. Different methods of showing the relief on the detailed geomorphological maps. Z. Geomorphol. N.F. 11 (4), 481–490.

Goodchild, M.F., Egenhofer, M.J, Fegeas, R., Kottmann, C.A. (Eds.), 1999. Interoperating Geographic Information Systems. Kluwer, New York.

Gosse, J.C., 2007. Cosmogenic nuclide dating: overview. In: Elias, S.A. (Ed.), Encyclopedia of Quaternary Science. Elsevier, Amsterdam, pp. 409–411.

Graff, L.H., Usery, E.L., 1993. Automated classification of generic terrain features in digital elevation models. Photogramm. Eng. Remote Sens. 59, 1407–1409.

Griffiths, J.S. (Ed.), 2001. Land Surface Evaluation for Engineering Purposes. The Geological Society of London, London, , Special Publication.

Griffiths, J.S., Marsh, A., 1986. The role of geomorphological and geological techniques in a preliminary site investigation. In: Hawkins, A. (Ed.), Site Investigation Practice. Engineering Geology Special Publication, vol. 2., Geological Society of London, pp. 261–267.

Guida, D., Guida, M., Lanzara, R., Vallario, A., 1996. Unità territoriale di riferimento per la pianificazione ambientale: esempi a diversa scala nell'area di Monte Bulgheria (Cilento, Campania). Geologia Tecnica e Ambientale 3 (1996), 1–66.

Guida, D., De Pippo, T., Cestari, A., Siervo, V., Valente, A., 2009. Applications of the hierarchic GIS-based geomorphological mapping system. In: Marchetti, M., Soldati, M. (Eds.), The Role of Geomorphology in Land Management, Abstract Volume, Third AIGEO National Conference, 13–18 September, Modena, Italy, pp. 109–110.

Gustavsson, M., 2005. Development of a Detailed Geomorphological Mapping System and GIS Geodatabase in Sweden. Licentiate Thesis, May 2005, Uppsala University, Sweden.

Gustavsson, M., Kolstrup, E., 2009. New geomorphological mapping system used at different scales in a Swedish glaciated area. Geomorphology 110, 37–44.

Gustavsson, M., Kolstrup, E., Seijmonsbergen, A.C., 2006. A new symbol-and-GIS based detailed geomorphological mapping system: renewal of a scientific discipline for understanding landscape development. Geomorphology 77, 90–111.

Haigh, M.J., 1987. The holon hierarchy theory and landscape research. Catena 10, 181–192.

Hayden, R.S., 1986. Geomorphological mapping. In: Short, N.M., Blain Jr., R.W. (Eds.), Geomorphology From Space: A Global Overview of Regional Landforms. NASA, Scientific and Technical Information Branch, Washington, DC. Chapter 11 – GES DISC, Goddard Earth Sciences. <http://disc.sci.gsfc.nasa.gov/geomorphology/GEO_11/GEO_CHAPTER_11.shtml>.

Heil, R.J., 1980. The digital terrain model as a data base for hydrological and geomorphological analysis. Auto-Carto IV, vol. II. Proceedings of the International Symposium on Cartography and Computing: Applications in Health and Environment. Reston, VA, American Congress on Surveying and Mapping. American Society of Photogrammetry. Falls Church, VA, pp. 132–139.

Hengl, T., Hannes, I., Reuter, H.I., 2008. Geomorphometry: Concepts, Software, Applications. Elsevier, Amsterdam.

Heywood, I., Cornelius, S., Carver, S., 2002. An Introduction to Geographical Information Systems. second ed. Pearson Prentice Hall, Upper Saddle River, NJ.

Howard, A.D., Dietrich, W.E., Siedl, M.A., 1994. Modeling fluvial erosion on regional to continental scales. J. Geophys. Res. 99, 13971–13986.

Howard, J.A., Mitchell, C.W., 1980. Phyto-geomorphic classification of landscape. Geoforum 11, 85–106.

Hughes, J.H., 1991. Object-Oriented Databases. Prentice Hall, Englewood Cliffs, NJ.

Huxley, J.S., 1972. Problems of Relative Growth. second ed. Dover, New York.

Joly, F., Tricart, J., 1970. Légende pour la carte géomorphologique de la France au 1:50.000, vol. 77. Centre National de la Recherche Scientifique, Paris, 78 pp.

Jones, A.P., 2000. Late quaternary sediment sources, storage and transfers within mountain basins using clast lithological analysis: Pineta Basin, central Pyrenees, Spain. Geomorphology 34 (3–4), 145–161.

Kienholz, H., 1978. Maps of geomorphology and natural hazards of Grindelwald, Switzerland: scale 1:10,000. Arct. Antarct. Alp. Res. 10 (2), 169–184.

Klimaszewski, M., 1956. The principles of the geomorphological survey of Poland. Przegl. Geogr. 28, 32–40.

Klimaszewski, M., 1982. Detailed geomorphological maps. ITC J. 1982–3, 265–271.

Klimaszewski, M., 1990. Thirty years of geomorphological mapping. Geogr. Pol. 58, 11–18.

Kneisel, C., Lehmkuhl, F., Winkler, S., Tressel, E., Schröder, H., 1998. Legende für geomorphologische kartierungen in Hochgebirgen (GMK Hochgebirge). Trierer Geographische Studien 18, 7–24.

Koestler, A., 1967. The Ghost in the Machine. Macmillan, New York.

Krönert, R., Steinhardt, U., Volk, M. (Eds.), 2001. Landscape Balance and Landscape Assessment. Springer-Verlag, Berlin.

Lavé, J., Avouac, J.P., 2001. Fluvial incision and tectonic uplift across the Himalayas of Central Nepal. J. Geophys. Res. 106, 26561–26592.

Leick, A., 1995. GPS Satellite Surveying. second ed. John Wiley & Sons, New York.

Leoni, G., Barchiesi, F., Catallo, F., Dramis, F., Fubelli, G., Lucifora, S., et al., 2009. GIS methodology to assess landslide susceptibility: application to a river catchment of Central Italy. J. Maps 2009, 87–93.

Létal, A., 2005. Aplikace GIS v geomorfologické mapové tvorbě. Disertační práce, Přírodovědecká Fakulta, University Karlovy, Prague.

Linton, D.L., 1951. The delimitation of morphological regions. In: Stamp, L.D., Wooldridge, S.W. (Eds.), London Essays in Geography, Annals of the Association of American Geographers, 41(3), pp. 199–218.

Longley, P.A., Goodchild, M.F., Maguire, D.J., Rhind, D.W., 2001. Geographic Information Systems and Science. John Wiley & Sons, Chichester.

Lowe, J.J., Walker, M.J.C., 1997. Reconstructing Quaternary Environments. second ed. Longman, New York.

Maarleveld, G.C., Ten Cate, J.A.M., De Lange, G.W., 1974. Die geomorphologische karte der Niederlande. Z. Geomorphol. N.F. 18 (4), 484–494.

Macar, P., de Béthune, P., Mammerickx, J., Seret, G., 1961. Travaux préparatoires à l'èlaboration d'une carte géomorphologique de Belgique. Annales del la Societé Géologique de Belgique 84, 179–197.

MacMillan, R.A., McNabb, D.H., Jones, R.K., 2000a. Automated landform classification using DEMs: a conceptual framework for a multi-level, hierarchy of hydrologically and geomorphologically oriented physiographic mapping units. Fourth International Conference on Integrating GIS and Environmental Modeling – Problems, Prospects and Research Needs. GIS/EM4, Banff, Alberta, Canada, September 2000. <http://www.colorado.edu/research/cires/banff/pubpapers/198/>.

MacMillan, R.A., Pettapiece, W.W., Nolan, S.C., Goddard, T.W., 2000b. A generic procedure for automatically segmenting landforms into landform elements using DEMs, heuristic rules and fuzzy logic. Fuzzy Sets Syst. 113 (1), 81–109.

Mark, D.M., Smith, B., 2004. A science of topography: from qualitative ontology to digital representations. In: Bishop, M.P., Shroder, J.F. (Eds.), Geographic Information Science and Mountain Geomorphology. Springer-Verlag, Berlin, pp. 75–100.

McClenaghan, M.B., Bobrowshy, P.T., Hall, G.E.M., Cook, S.J., 2001. Drift Exploration in Glaciated Terrain. Geological Society of London, London, Special Publication 185.

McKenzie, N.J., Jacquier, D.W., Maschmedt, D.J., Griffin, E.A., Brough, D.M., 2005. The Australian Soil Resource Information System Technical Specifications, Version 1.5. Australian Collaborative Land Evaluation Program, National Committee on Soil and Terrain Information. <http://www.asris.csiro.au>.

Meijerink, A.M.J., 1988. Data acquisition and data capture through terrain mapping unit. ITC J. 1, 23–44.

Merritts, D., Vincent, K.R., 1989. Geomorphic response of coastal streams to low, intermediate and high rates of uplift. Mendocino triple junction region, northern California. Geol. Soc. Am. Bull. 101, 1373–1388.

MEXE – Military Engineering Experimental Establishment, 1965. The classification of terrain intelligence. Reports of the Combined Pool (AER), 1960–64. Report 915.

Minàr, J., Evans, I.S., 2008. Elementary forms for land surface segmentation: the theoretical basis of terrain analysis and geomorphological mapping. Geomorphology 95, 236–259.

Molenaar, M., 1989. Single valued vector maps. A concept in Geographic Information Systems. GIS 2 (1), 18–26.

Montgomery, D.R., Brandon, M.T., 2002. Topographic controls on erosion rates in tectonically active mountain ranges. Earth Planet. Sci. Lett. 201, 481–489.

Ng, T.D., 1998. Semantic interoperability for Geographic Information Systems. DLI 98 Berkeley, All Project Meeting, University of Arizona, Berkeley.

Nichols, G., 2009. Sedimentology and Stratigraphy, second ed. Wiley-Blackwell, Chichester.

North American Commission on Stratigraphic Nomenclature, 1983. North American stratigraphic code. Am. Assoc. Pet. Geol. Bull. 67 (5), 841–875.

Oguchi, T., Hayakawa, Y., Wasklewicz, T., 2011. Data sources. In: Smith, M.J., Paron, P., Griffiths, J. (Eds.), Geomorphological Mapping: A Handbook of Techniques and Applications. Elsevier, Amsterdam.

Ollier, C.D., 1977. Terrain classification: methods, applications and principles. In: Hails, J.R. (Ed.), Applied Geomorphology. Elsevier, Amsterdam, pp. 277–316.

O'Neil, R.V., De Angelis, R.L., Waide, J.B., Allen, T.F.H., 1986. A Hierarchical Concept of Ecosystems. Princeton University Press, Princeton, NJ.

Pain, C.F., 1985. Mapping of landforms from Landsat imagery: an example from New South Wales, Australia. Remote Sens. Environ. 17, 55–65.

Pain, C.F., Chan, R., Craig, M., Gibson, D., Kilgour, P., Wilford, J., 2007. RTMAP Regolith Database Field Book and Users Guide, second ed. CRC LEME report 231. <http://crcleme.org.au/>.

Pain, C., Kilgour, P., 2003. Regolith mapping – a discussion. In: Roach, I.C. (Ed.), Advances in Regolith. CRC LEME Geoscience Australia, Canberra, pp. 309–313.

Panizza, M., 1972. Schema di legenda per carte geomorfologiche di dettaglio. Bollettino della Società Geologica Italiana 91, 207–237.

Panizza, M., 1973. Proposta di legenda per carte di stabilità geomorfologica. Bollettino della Società Geologica Italiana 92, 303–306.

Panizza, M., 1978. Analysis and mapping of geomorphological processes in environmental management. Geoforum 9, 1–15.

Panizza, M., 1988. Geomorfologia Applicata. La Nuova Italia Scientifica, Rome.

Parise, M., 2001. Landslide mapping techniques and their use in the assessment of the landslide hazard. Phys. Chem. Earth C 26 (9), 697–703.

Paron, P., Vargas, R., 2007. Landform of selected study areas in Somaliland and Southern Somalia. Integrated landform mapping approach at semi-detailed scale using remote sensing and GIS techniques. FAO-SWALIM, project report. L-02, Nairobi, Kenya. <http://www.faoswalim.org/ftp/Land_Reports/Cleared/L-02%20Landform%20of%20Selected%20Study%20Areas%20in%20Somaliland%20and%20Southern%20Somalia.pdf>.

Passarge, S., 1914. Morphologischer Atlas. Lieferung I: Morphologie des Messtiscb lattes Stadtremda. Mittelungen der Geographischen Gesellschaft, Hamburg.

Pecsi, M., 1963. Legende der Dettailierten Geomorphologischen Karten Ungarns Budapest Geographische Institut. Bayerische Akademie der Wissenschaften, Munich.

Peña Monné, J.L., 1997. Cartografia Geomorfologica Basica y Aplicada. Geoforma Ediciones, Logroño.

Pereira, G.M., 2002. A typology of spatial and temporal relations. Geogr. Anal. 34 (1), 21–33.

Petley, D.N., 1998. Geomorphological mapping for hazard assessment in neotectonic terrain. Geogr. J. 164 (2), 183–201.

Rao, D.P., 2002. Remote sensing application in geomorphology. Trop. Ecol. 43 (1), 49–59.

Rose, J., Smith, M.J., 2008. Glacial geomorphological maps of the Glasgow region, western central Scotland. J. Maps 2008, 399–416.

Rossi, G., Cancelliere, A., Giuliano, G., 2006. Role of decision support system and multicriteria methods for the assessment of drought mitigation measures. In: Andreu, J., Rossi, G., Vagliasindi, F., Vela, A. (Eds.), Drought Management and Planning for Water Resources. Taylor & Francis, Boca Raton, FL, pp. 204–240.

Salomé, A.I., Van Dorsser, H.J., Rief, Ph.L., 1982. A comparison of geomorphological mapping systems. ITC J. 1982–3, 272–274.

Sauro, U., 1977. Propositions pour une cartographie morphologique à grande échelle des champs de lapiés. Studi Trentini di Scienze Naturali 54, 163–176.

Savigear, R.A.G., 1965. A technique of morphological mapping. Ann. Am. Assoc. Geogr. 55 (3), 514–538.

Schiewe, J., Tufte, L., Ehlers, M., 2001. Potential and problems of multi-scale segmentation methods in remote sensing. GIS – Geo-Informationssysteme 6, 34–39.

Schmidt, J., Hewitt, A., 2004. Fuzzy land element classification from DTMs based on geometry and terrain position. Geoderma 121 (3–4), 243–256.

Schneevoigt, N.J., van der Linden, S., Thamm, H., Schrott, L., 2008. Detecting Alpine landforms from remotely sensed imagery. A pilot study in the Bavarian Alps.. Geomorphology 93, 104–119.

Schumm, S.A., Lichty, R.W., 1965. Time, space and casuality in geomorphology. Am. J. Sci. 263, 110–119.

Seelbach, P.W., Wiley, M.J., Kotanchik, J.C., Baker, M.E., 1997. A landscape-based ecological classification system for river valley segments in lower Michigan (MI-VSEC Version 1). Fisheries Division Research Report 2036, Department of Natural Resources, State of Michigan, Lansing, USA.

Seijmonsbergen, A.C., van Westen, C.J., 1990. Geomorphological, geotechnical, and natural hazard maps of the Hintere Bregenzerwald area (Voralberg, Austria). Alpine Geomorphology Research Group Laboratory for Physical Geography and Soil Science University of Amsterdam, The Netherlands.

Seijmonsbergen, A.C., Hengl, T., Anders, N.S., in press. Automated mapping. In: Smith, M.J., Paron, P., Griffiths, J. (Eds.), Geomorphological Mapping: A Handbook of Techniques and Applications. Elsevier, Amsterdam.

Slaymaker, O., Spencer, T., Dadson, S, Slaymaker, O., Spencer, T., Embleton-Hamann, C. (Eds.), 2009. Geomorphology and Global Environment Change. Cambridge University Press, Cambridge.

Small, C.G., 1996. The Statistical Theory of Shape. New York. Springer-Verlag.

Smith, M.J., in press. Digital mapping. In: Smith, M.J., Paron, P., Griffiths, J. (Eds.), Geomorphological Mapping: A Handbook of Techniques and Applications. Elsevier, Amsterdam.

Smith, M.J., Pain, C., 2009. Applications of remote sensing in geomorphology. Prog. Phys. Geogr. 33 (4), 568–582.

Speight, J.G., 1990. Landform. In: McDonald, R.C., Isbell, R.F., Speight, J.G., Walker, J., Hopkins, M.S. (Eds.), Australian Soil and Land Survey Field Handbook, second ed. Inkarta Press, Melbourne, pp. 9–57.

Teeuw, R.M., 2007. Mapping Hazardous Terrain Using Remote Sensing. Geological Society, London, Special Publication 283.

Ten Cate, J.A.M., 1983. Detailed systematic geomorphological mapping in the Netherlands and its applications. Geologie en Mijnbouw 62, 611–620.

Ten Cate, J.A.M., 1990. Sea-level rise and geomorphological mapping. Geogr. Pol. 58, 19–39.

Townshend, J.R.G., 1981. Regionalization of terrain and remotely sensed data. In: Townshend, J.R.G. (Ed.), Terrain Analysis and Remote Sensing. George Allen and Unwin, London, pp. 109–132.

Tricart, J., 1965. Principes et Méthodes de la Géomorphologie. Masson et Cie, Paris.

Tricart, J., 1969. Cartographic aspects of geomorphological surveys in relation to development programs. World Cartography, vol. 9, U.N. Department of Economic and Social Affairs, New York, pp. 75–83.

Tricart, J., 1972. Normes pour l'établissement de la carte geomorphologique detaillée de la France: (1:20.000, 1:25.000, 1:50.000). Memoires et Documents, année 1971, n.s. 12, Paris, France, 105 pp.

Ulaby, F.T., McNaughton, J., 1975. Classification of physiography from ERTS imagery. Photogramm. Eng. Remote Sens. 41, 1019–1027.

van Westen, C.J., Castellanos Abella, E.A., Sekhar, L.K., 2008. Spatial data for landslide susceptibility, hazards and vulnerability assessment: an overview. Eng. Geol. 102 (3–4), 112–131.

van Zuidam, R.A., 1982. Consideration on systematic medium-scale geomorphological mapping. Z. Geomorphol. N.F. 26 (4), 473–480.

van Zuidam, R.A., 1985. Aerial Photo-Interpretation in Terrain Analysis and Geomorphologic Mapping. Smits Publishers, The Hague.

Verstappen, H.Th, 1970. Introduction to the ITC system of geomorphological survey. Koninlijk Nederlands Aardrijkkundig Genootschap. Geografisch Nieuwe Reeks 4 (1), 85–91.

Verstappen, H.Th, 1977. Remote Sensing in Geomorphology. Elsevier, Amsterdam.

Verstappen, H.Th., Van Zuidam, R.A, 1968. ITC Textbook of Photo-Interpretation, VII: 2 – ITC System of Geomorphological Survey. ITC, Delft.
Voženílek, V., 2000. Spatial database for geomorphological mapping by GPS techniques. Geographica 36, 97–105.
Wakamatsu, K., Matsuoka, M., Kubo, S., Hasegawa, K., Sugiura, M., 2002. A nationwide gis-based engineering geomorphological map and site characteristics of K-net and Kik-net stations. Proceedings of the Japan Earthquake Engineering Symposium, September 2002, Tokyo, Japan, vol. 11, pp. 47–52.
Wandres, A.M., Bradshaw, J.D., Weaver, S., Maas, R., Ireland, T., Eby, N., 2004. Provenance analysis using conglomerate clast lithologies: a case study from the Pahau terrane of New Zealand. Sediment. Geol. 167 (1–2), 57–89.
Watchman, A.L., Twidale, C.R., 2002. Relative and 'absolute' dating of land surfaces. Earth Sci. Rev. 58 (1), 1–49.
Webster, R., 1977. Quantitative and Numerical Methods in Soil Classification and Survey. Clarendon, Oxford.
Whipple, K., Kirby, E., Brocklehurst, S., 1999. Geomorphic limits to climate-induced increases in topographic relief. Nature 401, 39–43.
Wielemaker, W.G., de Bruin, S., Epema, G.F., Veldkamp, A., 2001. Significance and application of the multi-hierarchical landsystem in soil mapping. Catena 43, 15–34.
Wilson, J.P., Gallant, J.C., 2000. Terrain Analysis. John Wiley & Sons, Chichester.
Winchester, V., Harrison, S., 2000. Dendrochronology and lichenometry: colonization, growth rates and dating of geomorphological events on the east side of the North Patagonian Icefield, Chile. Geomorphology 34 (3–4), 181–194.
Wu, J., 1999. Hierarchy and scaling: extrapolating information along a scaling ladder. Can. J. Remote Sens. 5 (4), 367–380.
Yagishita, K., 1989. Gravel fabric of clast-supported resedimented conglomerate. In: Taira, A., Masuda, F. (Eds.), Sedimentary Facies in the Active Plate Margin. Terra Scientific Company, Tokyo, pp. 33–42.
Zhu, A., 1999. A personal construct-based knowledge acquisition process for natural resource mapping. Int. J. Geogr. Inf. Sci. 13 (2), 119–141.

CHAPTER FOUR

Makers and Users of Geomorphological Maps

Paolo Paron[a,b] and Lieven Claessens[c,d]
[a]UNESCO-IHE, Institute for Water Education, Delft, The Netherlands
[b]School of Geography and the Environment, Oxford University, Oxford, UK
[c]Land Dynamics Group, Wageningen University and Research Centre, Wageningen, The Netherlands
[d]International Potato Center (CIP), Sub-Saharan Africa Regional Office, Nairobi, Kenya

Contents

1. INTRODUCTION

Geomorphology is a young discipline compared to geology and soil science emerging from earth science and geography, and systematic geomorphological mapping is even younger, becoming widely used as a tool of investigation mainly in Europe, during the 1950s and 1960s (St-Onge, 1968, 1981; Verstappen, 1983; Hayden, 1986; Griffiths, 2002). As in other disciplines, mapping in geomorphology has developed as a means of

Developments in Earth Surface Processes, Volume 15
ISSN: 0928-2025, DOI: 10.1016/B978-0-444-53446-0.00004-5

theoretical and applied research, beyond simply the graphical representation of data (Hayden, 1986). Over the last 60 years, mapping has contributed significantly to the understanding of landscape evolution in the geologically recent past and has adapted techniques from other fields for the study of the landforms (see Gregory, 2010, pp. 17–39 for a concise history of geomorphology). After initial development of geomorphological mapping programmes, interest in the technique declined. However, the relatively recent availability of remotely sensed data is seeing this trend reversed (Lee, 2001; NAP, 2010). Software developments have stimulated further work, allowing the storage, analysis and output of complex data sets. This chapter aims to review who makes and who uses geomorphological maps outside the academic world. It goes through a partial review of the national technical geological surveys, applied research institutions, humanitarian organisations and the private sector to show different aims, scales and foci of national geomorphological mapping programmes. It also provides an example of a specific type of geomorphologic mapping applied to natural hazards in a developing country. Finally, it provides suggestions for better integration between makers and external users of geomorphological information.

2. GEOMORPHOLOGICAL MAPPING CHARACTERISTICS

Some characteristics of geomorphology are quite distinct as the technique lies at the interface of several bio-geosciences, and this is both a barrier to its use and application as well as a potentially rich source of innovation. Having such a complex, broad, base poses significant challenges in, for instance, understanding surface processes and their interrelations with the lithosphere, biosphere, atmosphere, hydrosphere and anthrosphere. At the same time, this broad base requires that geomorphologists necessarily have a holistic view of landforms and their evolution. Nevertheless, this complexity can become an obstacle when there is a need for displaying the landscape, its forms, materials and processes on a single map. Even the advent and utilisation of geographical information systems (GIS), which allow the grouping, layering and 2.5/3D visualisation of landforms (Smith and Clark, 2005), have not ameliorated the significant graphical density of geomorphologic maps and this can present a barrier for users who are not trained as geomorphologists. This poses a major challenge for the dissemination and widening of the user community.

Geomorphological maps share many similarities with geological and pedological maps. Yet the economic importance of geomorphology has rarely been exploited by government and commercial users, in comparison to the development of geology and pedology (AGI, 2004). This is also reflected by the number of now almost universally accepted unifying theories and underlying paradigms that have been developed in geological and soil sciences (e.g. plate tectonics and catena concepts). In contrast, whilst geomorphology has its own theories, such as cascading process systems (Chorley and Kennedy, 1971), morphological evolutionary systems (Thornes and Brunsden, 1977) and climatic geomorphology (Bűdel, 1980), these have not received universal acceptance in the geomorphological community.

Geological mapping programmes initially developed due to the need to identify economic resources in rocks, including metalliferous minerals, oil, gas and groundwater. Arising from these mapping programmes has been a greater understanding of geological processes. In the field of geological mapping, national geological surveys are present in almost every country in the world. One of the main mandates of all geological surveys is to provide a national baseline in terms of geological mapping of the rocks, minerals and resources of the country. Furthermore, in an effort for the global standardisation of geological information, a number of international programmes have been developed, e.g. OneGeology (http://www.onegeology.org/).

A similar path has been followed by the soil science communities with global efforts focused upon standardising mapping units and symbology; for example, soil standards and mapping (e.g. http://www.isric.org/, http://soils.usda.gov/use/worldsoils/mapindex/ and http://eusoils.jrc.ec.europa.eu/) and land cover mapping (e.g. http://www.glcn.org/index_en.jsp and http://edc2.usgs.gov/glcc/glcc.php).

The process of standardisation of mapping criteria has not been fully developed by geomorphologists, though efforts have been made (Demek, 1972; Gustavvson et al., 2006).

The semiotic and semantic complexity of geomorphological maps imposes a graphical simplification when used in applied contexts and when the end-users are not geomorphologists. This simplification follows two directions: (1) the representation of just one of several layers on a map (Savigear, 1965) and (2) the creation of compound terrain units that group together a variety of information (e.g. the Land System units of developed by CSIRO in the '60s in Australia and conterminous countries, http://www.publish.csiro.au/nid/289.htm last access 7th Oct 2011). More recent examples of the application of geomorphological maps in engineering studies have also followed the first form of simplification thus targeting specific user needs (Verstappen, 1970; Brunsden et al., 1975, 1979; Schmitz, 1980; Griffiths et al., 1995).

Table 4.1 summarises the mapping approaches of a range of disciplines and specifically whether they are used predominantly by non-specialists or experts. It is clear that most thematic maps are primarily used by a restricted community of expert users and this makes their dissemination relatively limited. This is currently the case for geomorphological mapping.

3. MAKERS AND USERS

It is possible to identify three major classes of geomorphological map's makers and users and are as follows:

1. *national technical departments*, which generally produce omni-comprehensive geomorphological maps, commonly at scales ranging from 1:100,000 to 1:25,000,
2. *private companies*, generally in the field of engineering, environment, insurance and so on, usually with scales of 1:10,000 or bigger,
3. *research and development institutions*, primarily some United Nations specialised agencies (e.g. FAO, UNEP, CGIAR, UNESCO), usually with scales of 1:250,000 or smaller.

The main difference between the first group of *national technical departments* and the second two groups of companies and institutions is that the departments are guided by the broad need to explain and display morphology, morphogenesis, evolution, activity, and the related deposits, of all landforms of a given area. The companies and institutions tend to be query driven and therefore present simpler maps answering specific questions such as 'which areas are flood prone?', 'where is the highest probability of landslide activity?', 'which landform information can be used for digital soil mapping?', 'where are sand and gravel deposits?', 'what will happen to river morphology after the opening of a bed mine?', 'where is the safest location for the next refugee camp?' and so on. Another difference between the three groups is that they generally focus upon different geographical extents and scales: the *private companies* are concerned with relatively small areas for operational purposes, whereas the research and development agencies characteristically have supranational scope with purposes varying from identifying mapping standards for specific topics to creating global data sets. The national surveys have generally national or sub-national coverage, with the

Table 4.1 Users and Makers of Different Types of Maps

Type of Maps	Makers	Users: Non-Specialised	Users: Specialised
Topographic maps	Ordnance survey	Lay people Trekkers . . .	Topographers Planners Civil protection agencies Geologists Geomorphologists Soil scientists Military . . .
Street maps	Ordnance survey Commercial navigation companies Geographical software developers Open source Geospatial community	Lay people Taxi and truck drivers Trekkers GPS users	Planners Express courier and taxi drivers . . .
Weather maps	Meteorological surveys	Lay people Trekkers Holiday makers	Meteorologists Civil protection agencies Military . . .
Nautical charts	Navy and ordnance survey	Sailors Holiday makers . . .	Commercial maritime traders Marine conservationists Military
Geological maps	Geological surveys Geologists	Amateur geologists	Geologists Engineers Architects Land planners Environmentalists
Soil maps	Soil survey Soil scientists	Farmers	Pedologists Planners Agronomists Archaeologists
Geomorphological maps	Geomorphologists		Geomorphologists Pedologists

(*continued*)

Table 4.1 (continued)

Type of Maps	Makers	Users	
		Non-Specialised	**Specialised**
			Archaeologists Ecologists Environmentalists ...
Landscape maps	Landscape ecologists Ecologists Architects and urban planners		Planners Ecologists Architects
Emergency	UNOSAT Respond consortium Ushahidi (http://www.ushahidi.com)	Civil population	UN Civil protection agencies NGO

production of several sheets covering the entire territory of a country. In this chapter, we identify the national survey as producers of geomorphological maps, given their broad-based approach, and the other two actors as users given their question-driven approach. The next sections of this chapter follow this double classification. We do not consider the purely academic development of geomorphological maps.

4. EXAMPLES OF NATIONWIDE MAP MAKERS

There has been a long tradition of geomorphological mapping in European universities (Klimaszewski, 1956, 1982; Galon, 1962; Tricart, 1965; Verstappen and van Zuidam, 1968) and international organisations such as the International Geographical Union (Demek, 1972; Demek and Embleton, 1978; see also Chapter 2). However, it is not common for *national technical departments* (e.g. geological surveys, forestry departments, soil departments) outside Europe to produce geomorphological maps. Most 'industrialised' nations have completed their national geological mapping programmes and for many these are regularly updated. Fewer countries have nationwide geomorphological mapping programmes. In

Europe geomorphological mapping is moving forward while in the United States, for instance, there is no such programme.

In Europe there are many examples of extensive national geomorphological mapping programmes, for instance, Austria, the Netherlands, France, Germany, Hungary, Italy, Poland, Romania, Spain and Switzerland. Outside Europe, Russia developed an extensive geomorphological mapping programme, Australia has a regolith mapping programme that uses extensively geomorphological mapping principles, and China has recently developed a geomorphological atlas of China. India has also developed nationwide geomorphic mapping, and in South America, Brazil has completed extensive mapping of the geomorphology of both the Amazon rainforest and parts of its arid lands. In the following sections, examples from some of the above-mentioned countries are presented, in order to briefly illustrate the variety of methods, scales and maps that have been produced. A new UK two-map sheet at 1:1 million has just been published by the British Geological Survey that presents the engineering geology of both bedrock and superficial deposits. This latter one contains a great deal of geomorphological information.

4.1 Germany

In Germany a priority research programme on Digital Geomorphological Mapping of the Federal Republic of Germany ended in 1986. It produced two sets of geomorphological maps: 8 sheets and accompanying booklets at a scale of 1:100,000 and 27 sheets (with 24 booklets) at a scale of 1:25:000 (http://gidimap.giub.uni-bonn.de/gmk.digital/home_en.htm). All these products are available for interrogation and viewing using a WebGIS (http://gidimap.giub.uni-bonn.de/gmk.digital/webgis_en.htm), with static raster versions available for download (http://gidimap.giub.uni-bonn.de/gmk.digital/downloads_en.htm). In addition, there is access to a detailed bibliography on geomorphological mapping and a set of geomorphological mapping symbols.

At both scales these maps present a complex legend that accounts for a detailed surface and sub-surface lithology description, geomorphological processes, morphometry, hydrology and the age of the substratum. The base of the geomorphological maps is given by the related topographic sheet. An example of a German geomorphological map at the scale of 1:25,000 is shown in Figure 4.1. Figure 4.2 shows the detailed legend for the same sheet.

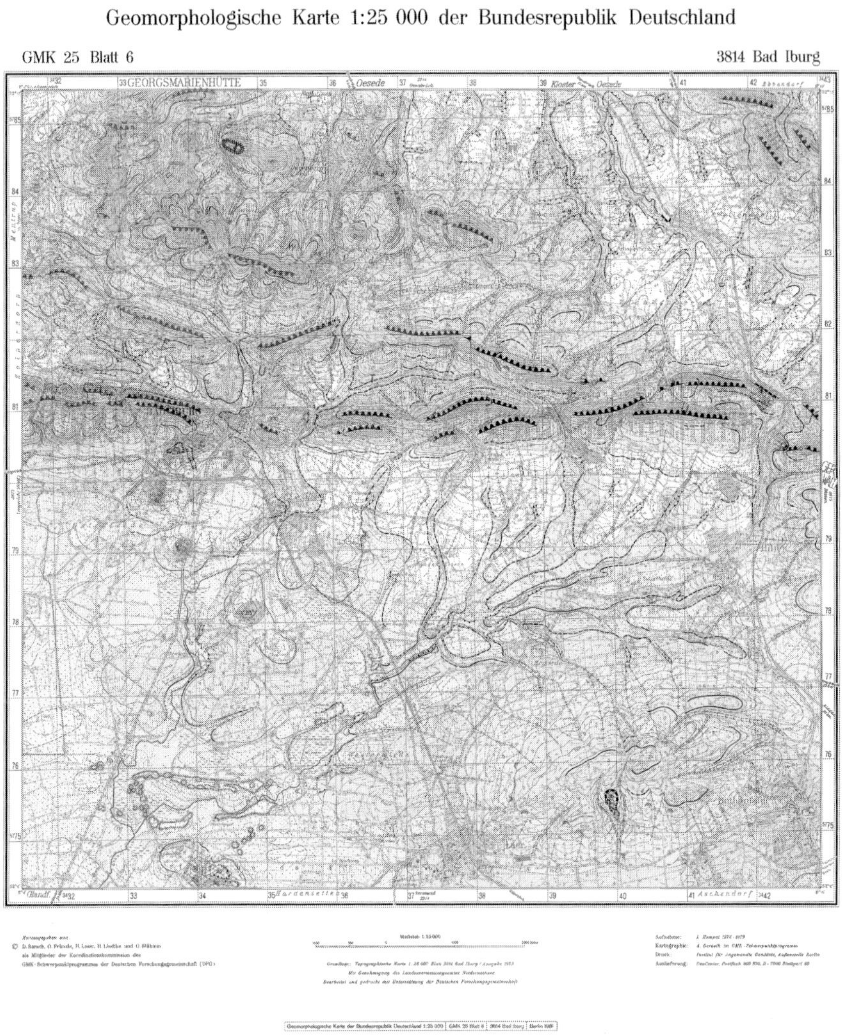

Figure 4.1 Example of German geomorphological map (Bad Iburg) at a scale of 1:25,000. *Downloaded from http://gidimap.giub.uni-bonn.de/gmk.digital/downloads_en.htm on 13 August 2010.*

4.2 Spain

In Spain, nationally consistent geomorphological mapping of the entire nation started in 1986, as part of Project MAGNA (http://www.igme.es/internet/default.asp). The programme has produced a series of geomorphological maps at a scale of 1:50,000 and accompanying booklets are

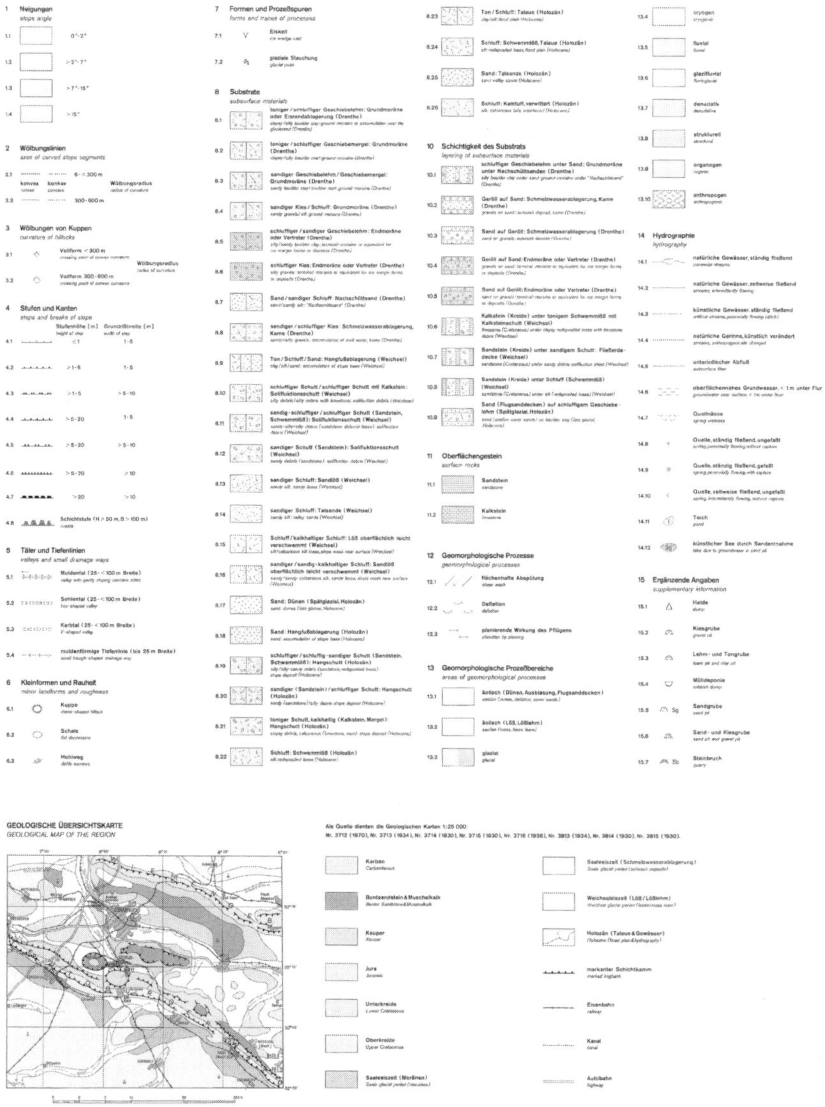

Figure 4.2 Legend of the Bad Iburg geomorphological map of Figure 4.1.

available from the Instituto Tecnológico GeoMinero de España (ITGE). In some cases, the maps have a scale of 1:25,000. The maps and accompanying booklets contain information about landforms (grouped by morphogenetic agent), superficial deposits (grouped by type of deposit and genetic origin) and the substratum. A good proportion of the national

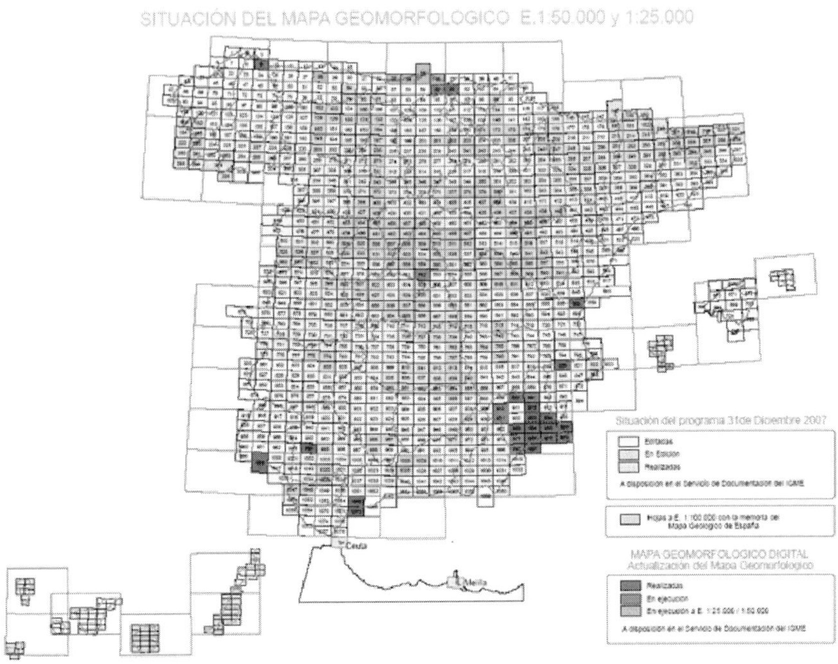

Figure 4.3 National coverage of Spanish geomorphological maps up to December 2007.

territory is complete (Figure 4.3) and collated digitally within a GIS (Rodriguez Garcia and Perez Cerdan, 2006).

An example of a Spanish geomorphological map is shown in Figure 4.4. These maps focus on two main aspects: surface deposits and morphogenetic processes. The deposits are presented using an innovative legend scheme that combines age and type. The forms, or morphologies, are grouped by morphogenetic agent, including anthropomorphism. The maps are illustrated using geomorphological cross sections in order to describe and explain the morphostructural setting, climate, substratum (by rock type) and slopes. The base of the geomorphological map is a topographic sheet. In this example, the rock substratum is not shown on the main map, where the focus is only on recent superficial deposits.

The digitisation of these maps has been carried out within project MAGNA (Mapa Geologico Nacional) at a scale of 1:50,000 for updating and conservation of geological maps of Spain (http://www.igme.es/internet/default.asp). The digitisation programme began in 1971 with the aim of developing a homogeneous geological and geomorphological mapping product for the nation (Rodríguez Fernández, 2005).

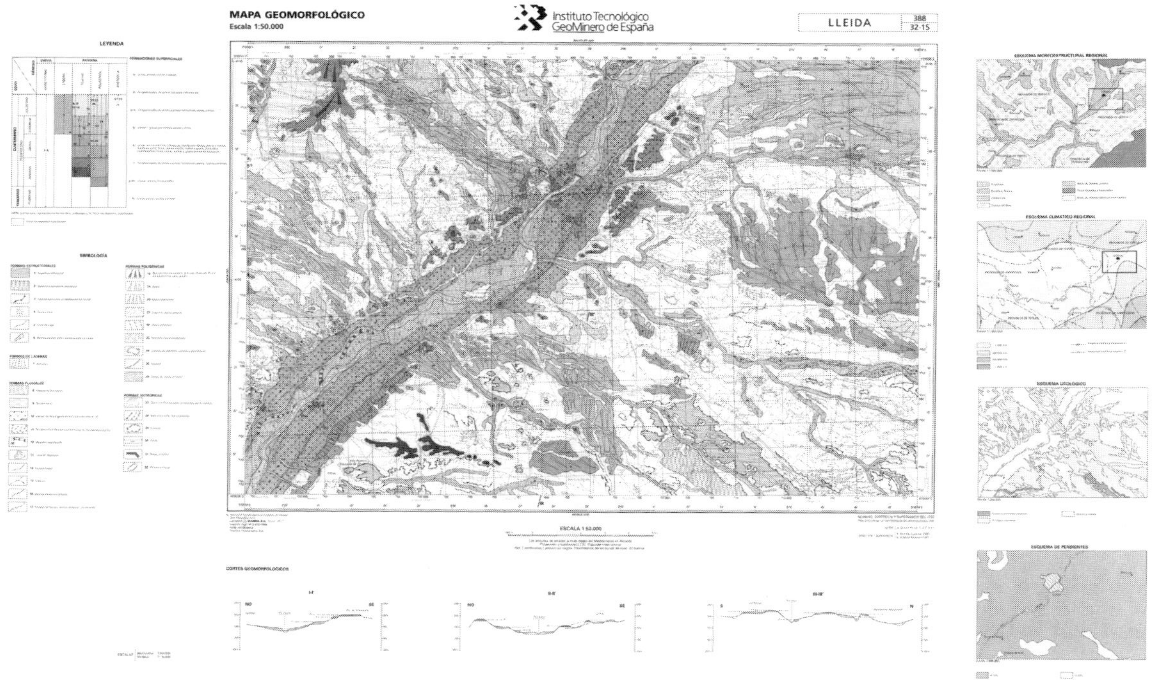

Figure 4.4 Example of a Spanish geomorphological map (Lleida) at a scale of 1:50,000. *From http://www.igme.es/internet/cartografia/cartografia/datos/Geomorfologico_50/jpg/d3_jpg/d388/Editado_Geomorfologico50_388.jpg, accessed on 13 August 2010.*

4.3 The Netherlands

In the Netherlands a nationally consistent programme of geomorphological mapping began in the 1960s. From the 1980s onwards, Dutch geomorphological and geological mapping progressed further with the development of a very detailed GIS incorporating large quantities of surface and sub-surface data leading to the TNO (the Netherlands Organisation for Applied Scientific Research) database (http://www.dinoloket.nl/en/DINOLoket.html) which incorporated all boreholes and other sources of geological and geomorphological data. This allowed the creation of detailed 3D models of the surface and sub-surface of the country which is primarily defined by marine, coastal and fluvial processes.

Given the low-lying nature of the terrain, the recent availability of a high spatial resolution digital elevation model (DEM) derived from a nationwide light detection and ranging (LiDAR) survey (AHN, Actueel Hoogtebestand van Nederland, http://www.ahn.nl/) was a major breakthrough in the definition and mapping of the many low-relief landforms. The national coverage is made up of 62 maps at the scale of 1:50,000 and stems from the work of Koomen and Maas (2005) available at the Alterra website (http://content.alterra.wur.nl/Webdocs/PDFFiles/Alterrarapporten/AlterraRapport1039.pdf).

In 2003 the Geomorphological Map of the Netherlands was available digitally; since that time, an increasing number of the 1:50,000 scale sheets have become available via the Internet, although only in Dutch (http://www.aardkunde.nl/).

The combination of very detailed topography and accurate geomorphological survey also allows a number of geo-visualisations such as draping geomorphological data over DEM data (see Figure 4.5).

4.4 Italy

In Italy there is a longstanding tradition of academic production of geomorphological maps. Efforts to produce a nationally consistent geomorphological map by the Geological Survey of Italy (Servizio Geologico d'Italia) started within the last national geological mapping project (CARG) in 1988, and aimed at producing 652 geological and thematic maps at a scale of 1:50,000, covering the entire national territory. After the survey and map creation, a dedicated digitisation process following detailed guidelines took place. Guidelines for the geomorphological mapping have also been produced (Servizio Geologico Nazionale, 1994).

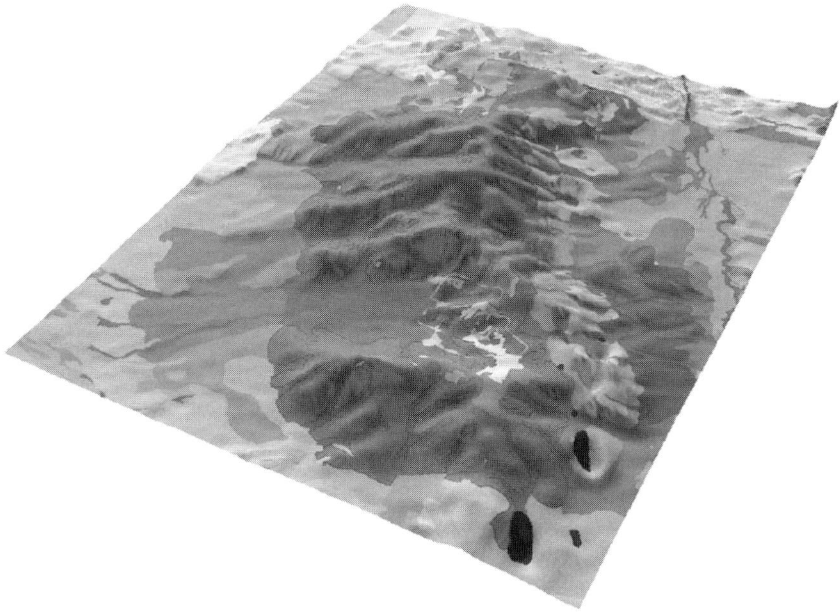

Figure 4.5 Draping of geomorphological information on the LiDAR-derived DEM. *From http://www.aardkunde.nl/.*

The Italian geomorphological maps contain information about the topography, hydrology, lithology of the substratum and of the superficial deposits, tectonics, morphogenesis, morphochronology of landforms and morphoevolution (active or dormant and fossil landforms). They are complex cartographic products with many colours, symbols and complex legends. A screenshot of a WebGIS for a portion of northern Italy is illustrated in Figure 4.6.

A dedicated WebGIS interface (GeoMapViewer, see Figure 4.7) allows the display of most of the elements contained in the geological and geomorphological maps. The raw data can also be downloaded for use in a GIS (http://sgi.isprambiente.it/geoportal/catalog/content/carg.page).

In Italy some very active regional mapping agencies have also carried out extensive geomorphological programmes. For example, the Geological, Seismic and Soil Survey of the Region Emilia-Romagna (http://www.regione.emilia-romagna.it/geologia_en/) has produced several geothematic maps derived from their geological and geomorphological mapping (http://www.regione.emilia-romagna.it/wcm/geologia/canali/cartografia.htm); these include maps on geo-environmental hazards, soil erosion, aquifer vulnerability, slope stability, landslides and coastal erosion.

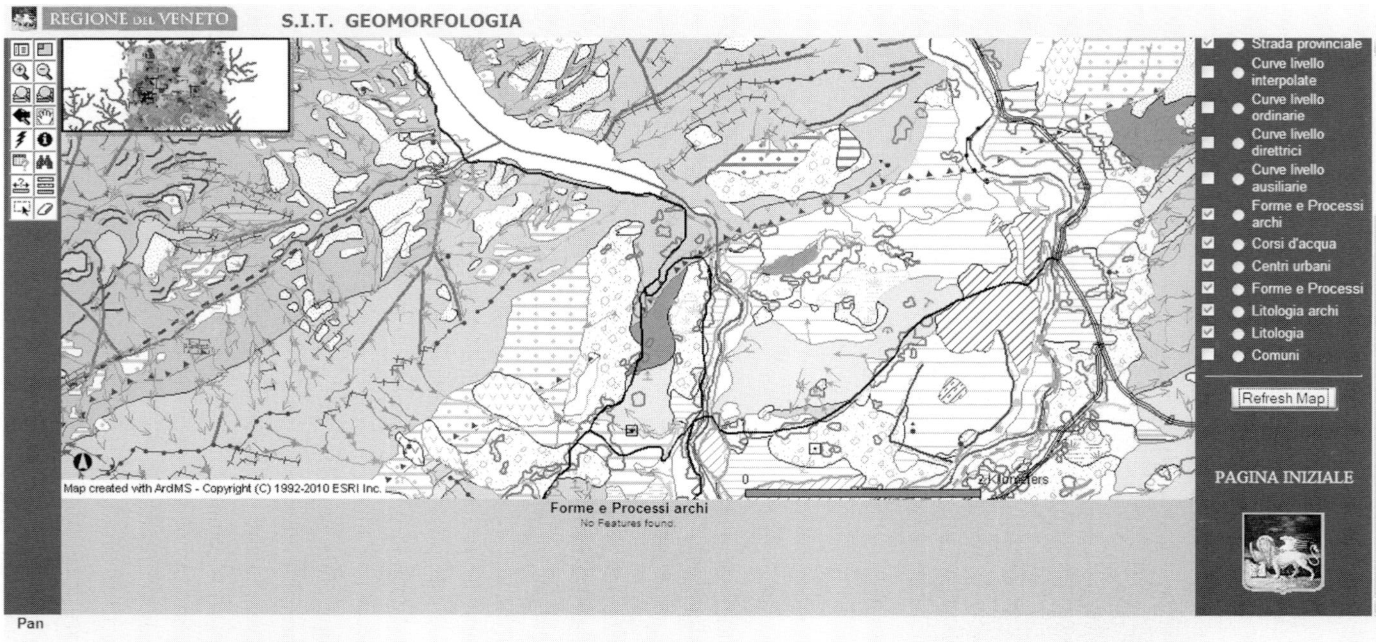

Figure 4.6 Excerpt from the geomorphological map of the Regione Veneto at an original scale of 1:50,000. For the legend, see the link to the handbook on geomorphological mapping. *From http://gisgeologia.regione.veneto.it/Website/sit_geomorf-1/viewer.htm.*

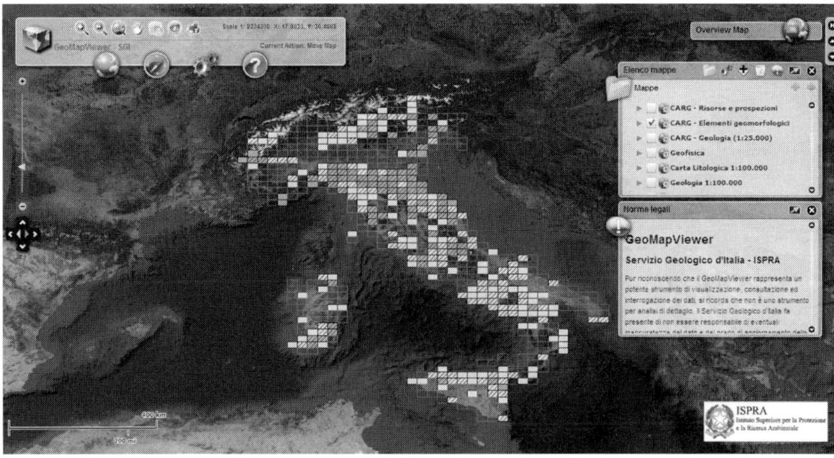

Figure 4.7 Screenshot of the Italian GeoMapViewer.
From http://sgi1.isprambiente.it/GeoMapViewer/index.html.

4.5 Romania

In Romania geomorphological mapping was undertaken during the period 1976–1990 by the Bucharest-based Institute of Geography. They produced a set of fifty 1:200,000 scale maps covering the entire country (Buza, 1997; see Figure 4.8) that have recently been digitised. The legend contains 167 elements (Badea and Sandu, 1992), although it has not been published in hardcopy. The Institute of Geography also produced larger scale maps of Romania, at a scale of 1:25,000–1:50,000, which is currently ongoing. Figure 4.9 provides an example of the 1:25,000 map (Buza, 1997).

These detailed maps present a legend showing geological substratum, tectonic elements, broad units of relief and nine different geomorphological 'reliefs' (denudational, fluvial, lacustrine and marine, glacial and periglacial, karst, aeolian, volcanic, structural and anthropic) (Buza, 1997).

4.6 Australia

Australia represents an exception to the general lack of national level mapping in Anglo-Saxon countries where it was a forerunner in morphological mapping particularly in remote territories. It was Australian scholars who developed the land-systems mapping method in the 1950s and 1960s for subdividing territory into homogeneous compound units for applied purposes (Beckett and Webster, 1965; Christian and Steward, 1968; Ollier, 1977). More recently a regolith-landform mapping programme has covered

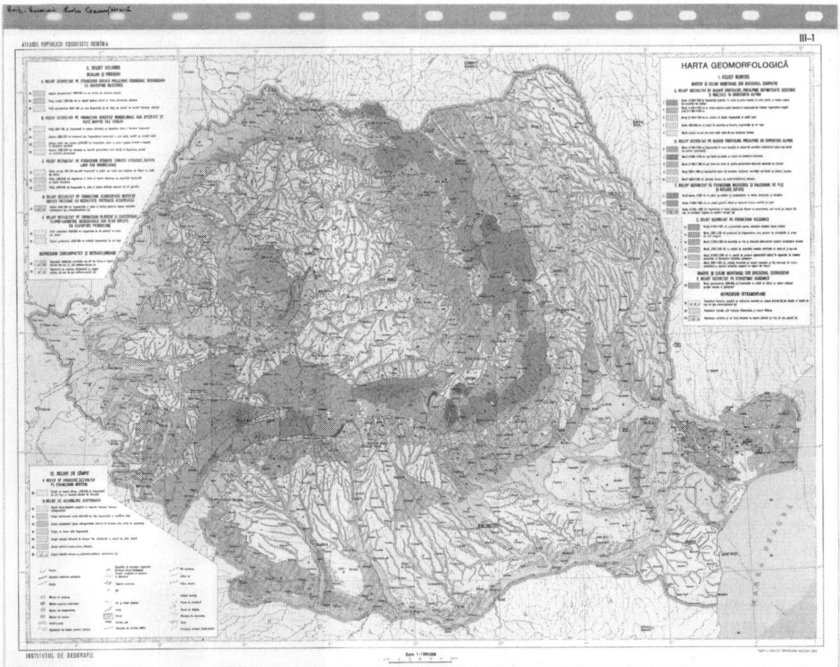

Figure 4.8 Geomorphological map of Romania, 1:1,000,000. *From http://geomorf.rosa.ro/index.htm.*

most of the red continent's superficial deposits and landforms (CRC LEME Programme, Anand and de Broekert, 2004) and from this baseline CRC LEME has developed further themes of research such as mineral exploration in areas of regolith; environmental applications in the field of geochemistry and contaminant diffusion through regolith; mapping, assessment and prediction of salinity stores and discharges in both regolith materials and groundwater (http://crcleme.org.au/Research/programs.html; Scott and Pain, 2008). The main aim of this programme was mapping the economically exploitable minerals contained in the regolith of Australia (Pain et al., 2007; Pain, 2008).

Regolith-terrain mapping was primarily based on the identification of unique dominant regolith-landform associations and led to the creation of 205 unique mapping units for 1:100,000 scale maps and derived from these were further maps at a scale of 1:250,000. The mapping scheme underlying the regolith-terrain mapping approach has been described in detail by Pain et al. (1991, 2001, 2008). Pain and Kilgour (2003) defined regolith mapping units as the 'real landscape units that can be conveniently mapped, and their

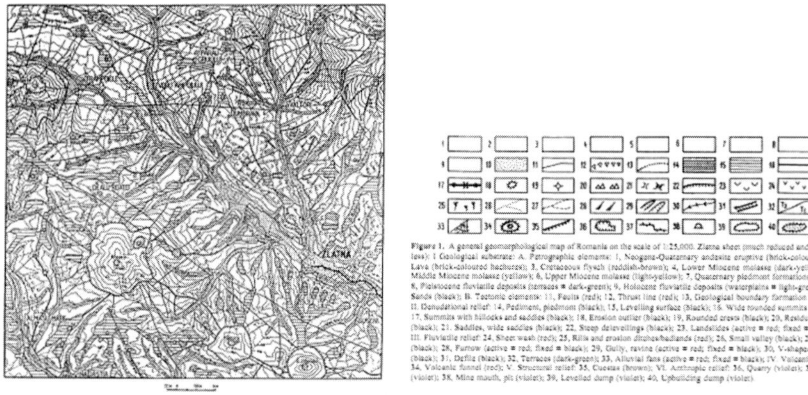

Figure 4.9 Extract from the 1:25,000 Zlatna map. *From Buza (1997).*

definition will therefore depend on the scale of the map. The more detailed the map scale, the more pure the mapping units will be', that is the closer to the regolith classification unit it will be. Therefore, the regolith mapping units are irrespective of the chronology or morphogenetic processes of regolith formation (Pain and Kilgour, 2003), but they accurately describe the geometry and location of the soil-regolith occurrences and also provide 3D information where possible. Geomorphic symbols indicate the location and type of geomorphic activity. Besides the use of remote sensing and field surveys, airborne gamma-ray spectrometry and electromagnetics were utilised for the understanding of spatial variations of surface deposits and their mapping (Scott and Pain, 2008; Smith and Pain, 2009).

4.7 China

The Geomorphological Atlas of the People's Republic of China is a major effort at documenting the vast landmass, using a combination of field work, remote sensing and GIS. The output is a homogeneous mapping product at a scale of 1:1,000,000, utilising a hierarchical approach, comprising the following seven layers of information: basic morphology, genesis, sub-genesis, morphology, sub-morphology, slope and aspect, material composition and lithology. About 1300 types of morphogenetic processes and 300 types of morphostructures are included in this huge cartographic effort. The entire Chinese territory is covered by 74 sheets (Figure 4.10) and also compiled in a GIS, with a general legend system in Chinese and English (Cheng et al., 2011). Figure 4.10 shows an example of a geomorphological sheet.

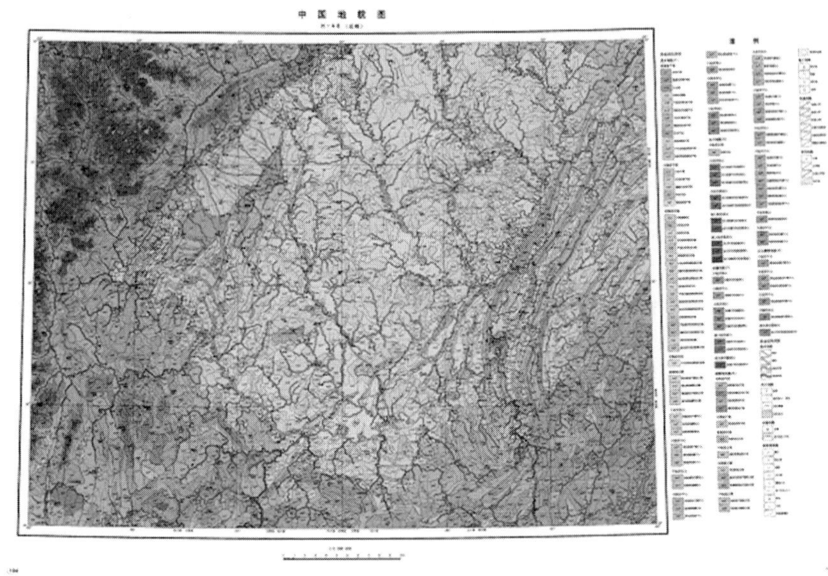

Figure 4.10 Example of 1:1,000,000 sheet from the Chinese Atlas.

4.8 Brazil

During the period 1971–1985, RADAM BRASIL, a joint NASA–CNEA programme, coordinated by IBGE (Brazilian Institute of Statistics and Geography), was developed to map the Amazon region (RADar AMazonia), but in fact carried out extensive systematic mapping of the whole country using side-looking airborne radar and field surveys. The project mapped geology, geomorphology, vegetation and land use of this immense territory (more than 8 million square kilometres) and acted also as an incubator for the first Brazilian school of geomorphological mapping (Ab'Saber, 1969; Barbosa, 1984) following the influence of Tricart and Cailleaux (1956). The first products were later updated using more recent remote sensing data sets and for the production of 49 geomorphological sheets at a scale of 1:1,000,000 (ftp://geoftp.ibge.gov.br/mapas/tematicos/sistematizacao/geomorfologia/).

The core of the Brazilian geomorphological methodology includes (IBGE, 2009) (a) geological substratum, (b) precise landform identification and delineation, (c) morphostructural and morphoclimatic domains, (d) morphogenetic processes and (e) recent superficial deposits. The systematic mapping of the whole national territory lead to the creation of a hierarchical legend made up of four hierarchical units (IBGE, 2009): morphostructural domain, geomorphological region, geomorphological unit and landforms.

An example of the final output is given in Figure 4.11, where the geomorphology of Cuiabá region in Mato Groso do Sul region is presented.

5. USERS

The primary supporters of applied geomorphological mapping are the hydrocarbon industry, civil engineering, and the environmental consultancy and planning community, particularly in Anglo-Saxon countries. This is paradoxical as many of these countries have no state programmes of geomorphological mapping.

Since the 1970s (Anonymous, 1972), the collaboration between geomorphologists, engineers and other professionals has resulted in a joint approach to the solution of complex civil, environmental and water problems, as well as leading to savings in time and money. Some of the applications are in oil pipeline alignment (Fookes et al., 2001), civil engineering construction (Birch, 1989; Birch and Griffiths, 1996), natural hazard assessment (Hearn, 1995), water resources (Jones et al., 2007), planning (Griffiths and Abraham, 2008), soil erosion (Charman and Griffiths, 1993), groundwater recharge (Barsch and Mausbacher, 1979), precision farming (Ciba Foundation, 1997; MacMillan et al., 2000), geoheritage (Catani et al., 2002) and so on.

The distance between academic makers and objective-driven users still remains wide, though there are increasing amounts of collaboration (Brunsden et al., 1975; Doornkamp et al. (1975); Cooke and Doornkamp, 1990; Fookes, 1997; Griffiths, 2001, 2002; Fookes et al., 2007; Smith, 2011). Academics are often interested in displaying the full morphogenetic evolution, including forms, processes and deposits leading to an understanding of the actual landform setting. Applied users may not need such complexity and demand simpler task-driven maps, which nevertheless must be prepared by expert geomorphologists incorporating the full geomorphological history of the area mapped. A similar approach has been also proposed by Fookes (1997) for the Total Geological Approach. Griffiths and Abraham (2008) present an excellent example of the dichotomy between the academic and client communities, showing the stages of generalisation from an 'academic' approach through to the client-driven needs for a simple visualisation of the processes acting on natural or anthropic landforms.

Across the spectrum of users, the primary message that is evident is the need for more effective dialogue between the various professional

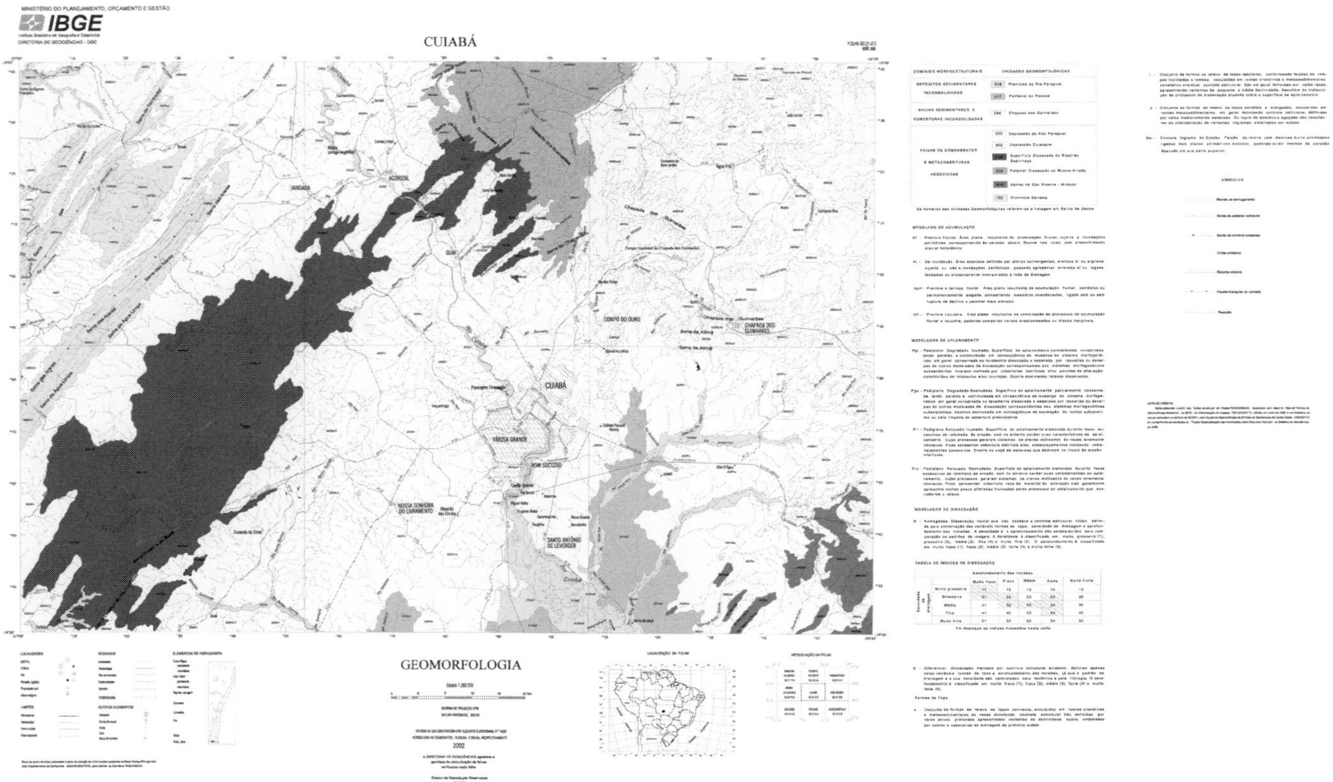

Figure 4.11 Brazilian geomorphological map for Cuiabá at a scale of 1:1,000,000.

communities in order to educate each other on the needs, lexicon and requirements of each field of application. In particular, land planners, engineers, architects and, probably most urgently, the personnel in the technical departments of various ministries (environment, land, water, planning, infrastructure and so on) need to be aware, and capable, of understanding maps aimed at showing, for instance, the natural hazards of a particular area of their country. The academic geological and geomorphological communities need to dedicate further resources to rigorously translate difficult concepts into terms that can be easily understood by the non-geoscientist.

The Dutch programme of 'Map4Planners' is an earth science example of what could potentially be achieved in geomorphology. Experts at TNO make a number of spatial data sets available to the users' community and, where necessary, run interactive consultation workshops where the group explores and analyses the spatial data in order to answer user-oriented queries (http://www.tno.nl/content.cfm?context=markten&content=product&laag1=188&laag2=392&item_id=1521).

With such data availability, it has been easy for the private sector and the ecological community to integrate geomorphological mapping into their work.

An example of this type of collaboration is provided by emergency mapping organisations where, for example, UNOSAT (http://unosat.web.cern.ch/unosat/) and the Respond Consortium (http://www.respond-int.org/) have a mandate to prepare maps to facilitate the aid intervention in post-conflict areas. They rely on regularly updated remotely sensed images, which are then used to prepare simple and clear maps for areas most affected that inform emergency and rescuing teams. These include, for example, volcanic eruptions (Figure 4.12) and flood extent (Figure 4.13) and provide clear examples of simple outputs derived from geomorphological and remote sensing expert knowledge.

Global scale work has seen the application of geomorphometric principles from a global DEM data set (SRTM 90 m) and lead to the creation of the world SOil and TERrain Digital database (SOTER; http://eusoils.jrc.ec.europa.eu/projects/soter/index.htm). SOTER is a UN–ISRIC joint programme, in which global SRTM digital elevation data are used to derive landform units and terrain information (Dobos et al., 2005). From this process, a new large-scale terrain-unit data set has been created utilising a unique combination of physiographic and soil characteristics that constitute a new baseline for digital soil mapping. Here geomorphometry (in this specific case a combination of slope gradient, terrain roughness,

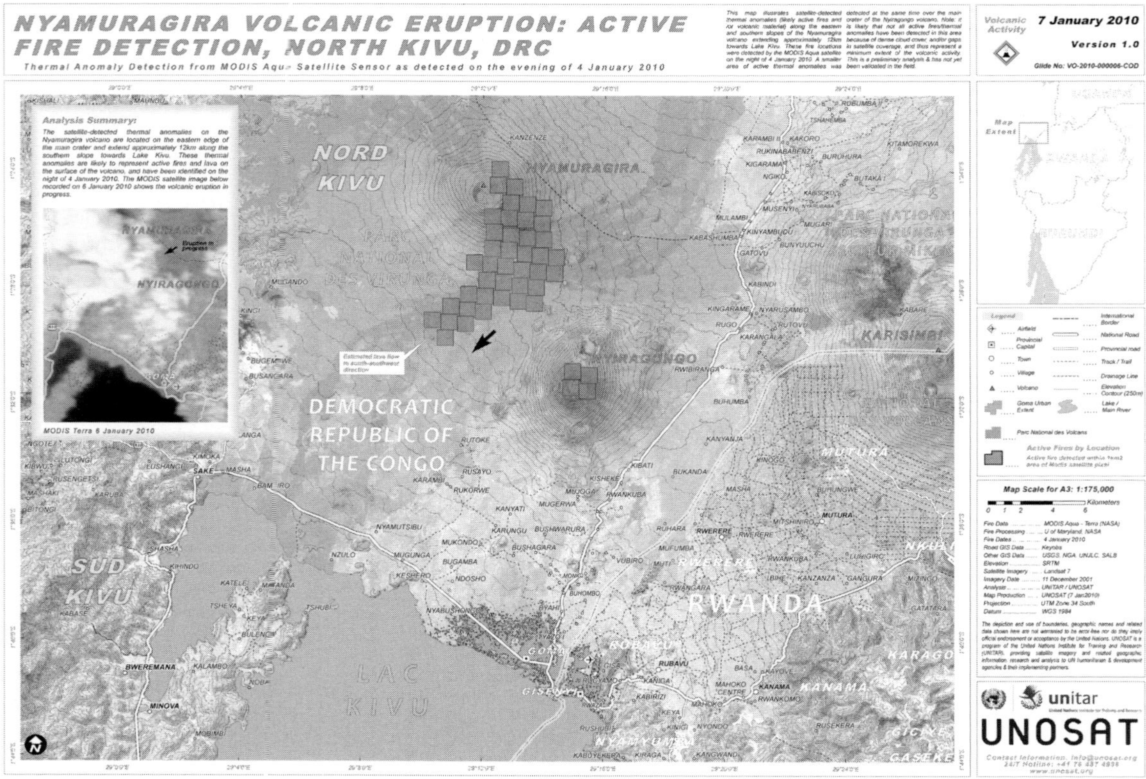

Figure 4.12 Volcanic fires affecting an area in Eastern Congo – North Kivu region in January 2010 (UNOSAT map).

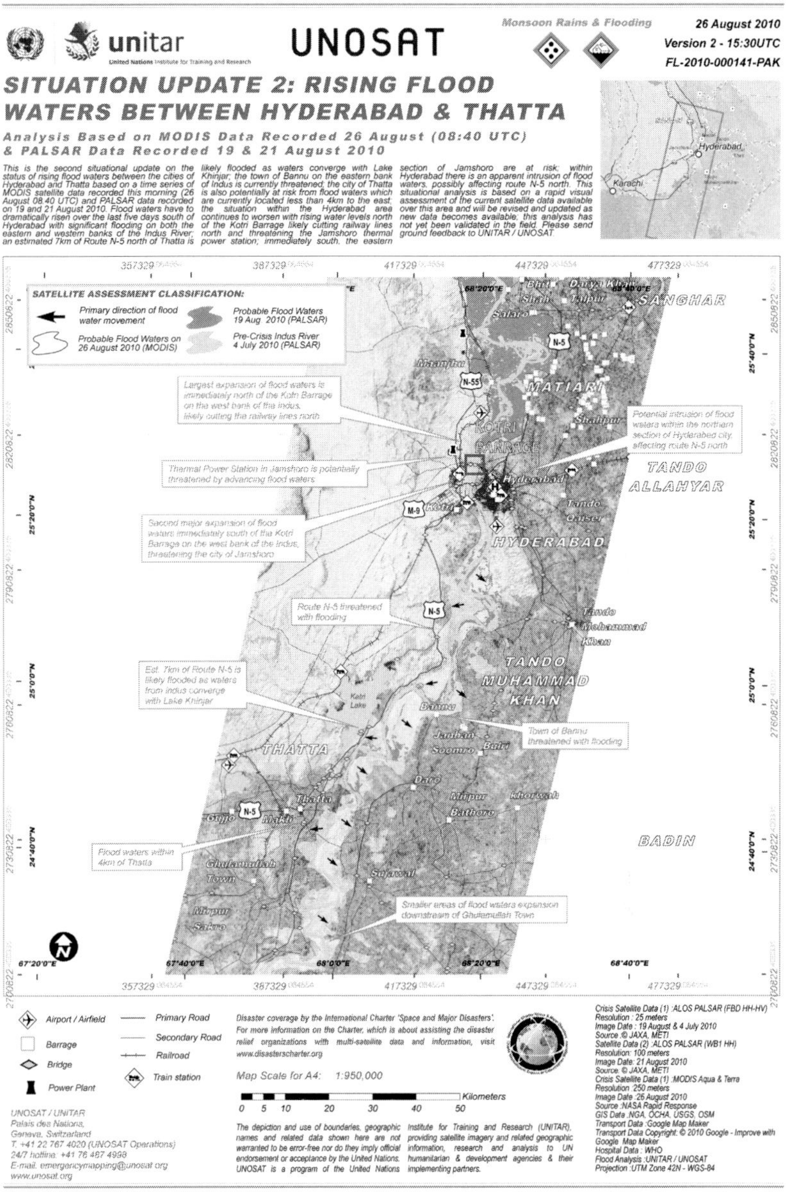

Figure 4.13 Flood-affected areas in Pakistan during the floods of August 2010 (UNOSAT map).

hypsometry and degree of dissection; Dobos and Montanarella (2004)) plays a major role, informing pedologists of the location of different soil types. As a global data set, the map scale is at 1:5,000,000, although scales of the final products will vary from 1:5M to 1:500,000. The SOTER data set constitutes the baseline for applications such as assessment of soil degradation and soil vulnerability to pollution and will contribute to larger global programmes such as the Land Degradation Assessment project (http://www.fao.org/nr/lada/).

5.1 The Special Role of the Reinsurance Companies

The reinsurance companies (Munich Re, Swiss Re, AoN and others) play an increasingly important role in collecting information about natural hazards and in zoning territories according to the degree of impact of a variety of natural hazards. For example, the Munich Re Disaster prevention programme (http://www.munichre-foundation.org/StiftungsWebsite/Topics/DisasterPrevention/) focuses on statistical analysis of the impact of natural hazards on a more populated world and on ways to prevent these impacts. They have set up and maintain one of the largest databases of global natural hazards that is instrumental in formulating risk zonation of different parts of the world to various hazards.

In addition to meteorological hazards, geomorphological and geological hazards are increasingly important in insurance company statistics (Munich Re, 2010; Figure 4.14). This can be seen as a positive driver for the mapping and zonation of countries in terms of natural hazards, where applied geomorphological mapping plays a fundamental role. The process combines geological and geomorphological mapping principles with modelling to enable the delineation of the most likely areas to be affected by, for example, cyclones, tsunamis, landslides, floods, avalanches and volcanic flows. The following section presents an example for a developing nation (Uganda) where a study on landslide risks was performed but not endorsed by the appropriate stakeholders. In this case, the maps defining the affected areas could have played a major role in informing the communities of the imminent danger. An after-event assessment of the map/modelling has proved that this mapping was fit-for-purpose.

Reinsurance companies could play a driving role in enabling individual nations to develop proper hazard cartography, including applied geomorphological mapping.

5.2 An Example of Landslide Mapping in Uganda

The footslopes of Mount Elgon in East Uganda are known to be vulnerable to rainfall-triggered landslides. Settlements increasingly encroach on the forested slopes as part of the expansion of small-scale agriculture. In 2006 a study was conducted documenting the characteristics and causal factors of historical landslides in the Manjiya study area (Knapen et al. 2005). In total, 98 recent landslides that displaced approximately 11 million cubic metres of slope material were mapped and investigated. By statistically comparing topographical characteristics from landslide sites with those from the whole study area, it was shown that landslides occur predominantly on steep concave slopes that are oriented towards the main rainfall direction (northeast) and at a relatively large distance from the water divide. Furthermore, the Manjiya area was divided into different zones based on landslide type (rotational or translational) and frequency of occurrence. Expanding upon these results, Claessens et al. (2007) investigated the suitability of the LAPSUS-LS landslide model (Claessens et al., 2005) to delineate zones which are prone to landsliding in general and to group the observed landslides into a specific landslide type and hazard category. Furthermore, an attempt was made to revisit the main causal factors for landsliding in Manjiya and to use the model to simulate possible future landslide scenarios with resulting sediment yields and geomorphic impacts for the region. By constructing a landslide hazard map (Figure 4.15) and simulating future landslide scenarios with the model, slopes in Manjiya County were identified as inherently unstable, and volumes of soil redistribution were predicted to yield four times higher than currently observed. More than half of this quantity could be deposited in the stream network, possibly damming rivers and causing major damage to infrastructure, as well as the siltation and pollution of streams. It was concluded that the combination of a high population density, land shortage and a high vulnerability to landslides would likely prolong the issue of population sustainability for the region. Unfortunately, the landslide hazard map, whilst published, was never disseminated to populations living in the area of potential landslide hazards. On 1 March 2010, heavy rainfall caused widespread landsliding in the region causing loss of more than 300 lives and property.

It is desirable that insurance and reinsurance companies leverage governments of the developing and developed world to define the most at-risk areas, making good use of their global and local knowledge of areas under risk and thus saving human life. Preventive geomorphological mapping, like the case of Mount Elgon, will play a major role in the territory zonation.

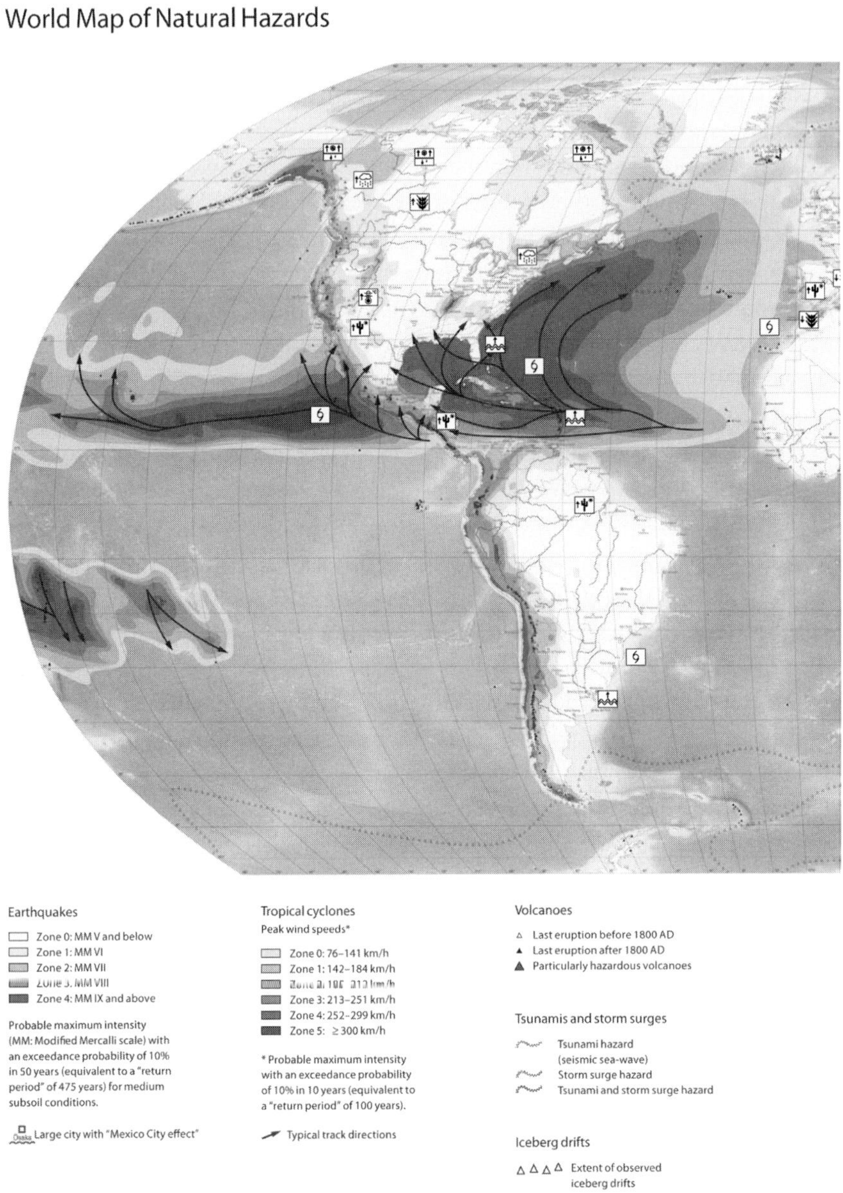

Figure 4.14 Synthetic global natural catastrophe map for 2009 (Munich Re, 2010).

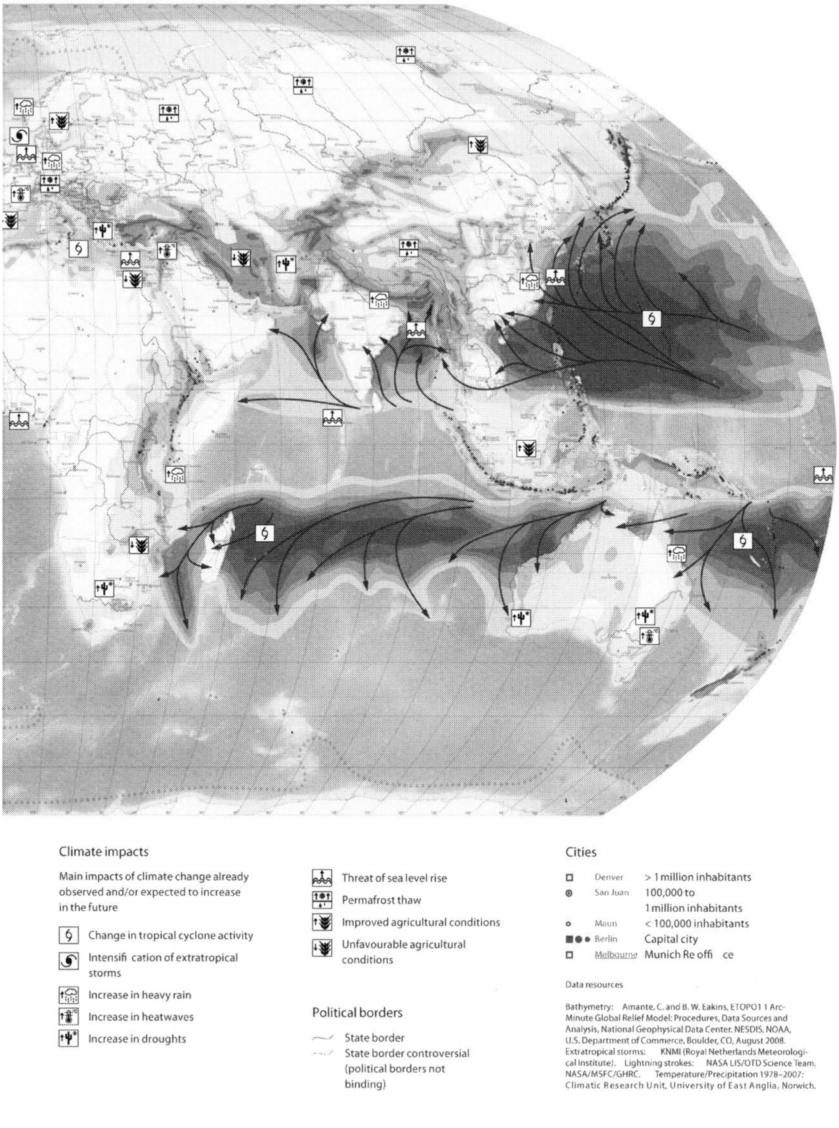

Figure 4.14 *(Continued)*

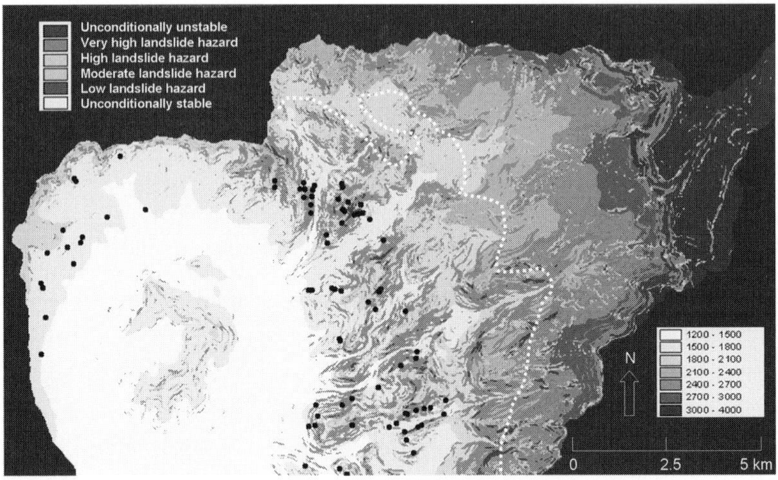

Figure 4.15 Landslide hazard map for the Manjiya study area on the footslopes of Mount Elgon, Uganda. The map was produced with the LAPSUS-LS landslide model. Landslide hazard classes (colours) are projected on the digital elevation model (grey shades). The black dots represent historical landslides mapped in the study by Knapen et al. (2005). The white dotted line is the border of Mount Elgon National Park. *From Claessens et al. (2007).*

6. CONCLUSIONS

In 1979 Barsch and Mausbacher stated that 'It is quite common (. . .) that a potential user has difficulties to appreciate the information furnished by a geomorphological map, because he is not able to find his way through the multiplicity of the different horizons of data presented. As a result he may be frustrated using a geomorphological map'. Griffiths and Abraham (2008) stated that 'A geomorphological map created by academic geomorphologists for applied purposes can be a complex document that requires interpretation and simplification if it is to meet the requirements of end-users'. It seems that in the last 30 years the perspectives of academics and end-users have not yet met.

Part of the problem relates to the collection, simplification and visualisation of complex data; this is, in part, being addressed by access to modern survey equipment and visualisation and analysis techniques available in a GIS. It is now possible to collect large amounts of detailed and quantitative data leading to a more meaningful and accurate understanding of the landscape and its evolution. At the same time, it is also easier to display only relevant data or to interpolate more dense information into simpler visualisations that can be used by the non-specialist.

This chapter has outlined the need and importance for an academic understanding of landscape evolution and that it is vital for governments to develop a national capacity in the preparation of applied geomorphological maps in order to inform and augment civil protection and planning projects. These derived and thematic maps could also serve the private sector and, more specifically, the reinsurance industry could accelerate the further derivation of simpler information. It seems that there is a need for stronger feedbacks between the makers and users of geomorphological maps and surveys.

The recent popularisation of geography, thanks to media such as Google Earth, Bing Maps, and NASA World Wind, could play a major role in informing and stimulating non-experts so that they can formulate more precise questions to the community of geomorphologists and Earth scientists. These knowledge communities should benefit more and more from these new dissemination tools to educate the wider lay community on the potential of geomorphological mapping for practical purposes (Tooth, 2009), particularly for hydro-meteorologically linked natural hazards that pose a threat to increasing and encroaching populations (Bates et al., 2008).

REFERENCES

Ab'Saber, A.N., 1969. Problemas do mapeamento geomorfológico no Brasil. Geomorfologia, Universidade de São Paulo, Instituto de Geografia, São Paulo, 6, 1–16.

AGI, 2004. In: Thomas, W.A. (Ed.), Meeting Challenges with Geological Maps. American Geological Institute, Alexandria, VA, pp. 1–69.

Anand, R.R., de Broekert, P. (Eds.), 2004. Regolith landscape evolution across Australia. A compilation of regolith-landscape case studies and landscape evolution models. <http://crcleme.org.au/Pubs/Monographs/RegLandEvol.html>. (accessed 17.12.10).

Anonymous, 1972. The preparation of maps and plans in terms of engineering geology. Q. J. Eng. Geol. 5, 297–367.

Badea, L., Sandu, M., 1992. The general geomorphological map of Romania on a medium scale (1:200,000). Revue Roumaine de Géographie 36.

Barbosa, G.V., 1984. Evolução da metodologia para mapeamento geomorfológico do projeto RADAM BRASIL. Projeto RADAM BRASIL, Salvador, 187 pp. (Boletim técnico do Projeto RADAM BRASIL. Série Geomorfologia).

Barsch, D., Mausbacher, R., 1979. Geomorphological and ecological mapping. GeoJournal 3.4, 361–370.

Bates, B.C., Kundzewicz, Z.W., Wu, S., Palutikof, J.P. (Eds.), 2008. Climate change and water. Technical Paper of the Intergovernmental Panel on Climate Change, IPCC Secretariat, Geneva, 210 pp.

Beckett, P.H.T., Webster, R., 1965. A classification system for terrain. Mil. Eng. Expt. Estab., Christchurch, Report no. 872, p. 29.

Birch, G.P., 1989. Applications of geomorphology to small hydro schemes. Q. J. Eng. Geol. 22, 231–239.

Birch, G.P., Griffiths, J.S., 1996. Engineering geomorphology. In: Harris, C.S., Hart, M.B., Varley, P.M., Warren, C.D. (Eds.), Engineering Geology of the Channel Tunnel. Thomas Telford, London, pp. 64–75. Chapter 4.

Brunsden, D., Doornkamp, J.C., Fookes, P.G., Jones, D.K.C., Kelly, J.M.H., 1975. Large-scale geomorphological mapping and highway engineering design. Q. J. Eng. Geol. 8, 227–253.

Brunsden, D., Doornkamp, J.C., Jones, D.K.C., 1979. The Bahrain Surface Materials Resource Survey and its application to regional planning. Geogr. J. 145 (1), 1–35.

Budel, J., 1980. Climatic and Climatomorphic geomorphology. Z. Geomorph. NF Suppl., 36: 1–8.

Buza, M., 1997. A general geomorphological map of Romania on the scale of 1:25,000, Zlatna sheet. Geogr. J. 41 (1), 85–91.

Catani, F., Fanti, R., Moretti, S., 2002. Geomorphologic risk assessment for cultural heritage conservation. In: Allison, R.J. (Ed.), Applied Geomorphology. John Wiley & Sons, Chichester, pp. 303–316.

Charman, J.H., Griffiths, J.S., 1993. Terrain evaluation methods for predicting relative hazard from soil erosion, mass movement, and flooding in the developing world. In: Merriman, P.A., Browitt, C.W.A. (Eds.), Natural Disasters: Protecting Vulnerable Communities. Thomas Telford, London, pp. 167–183.

Cheng, W., Zhou, C., Chai, H., Zhao, S., Liu, H., Zhou, Z., 2011. Research and compilation of the Geomorphologic Atlas of the People's Republic of China (1:1,000,000). J. Geogr. Sci. 21 (1), 89–100.

Chorley, R. J., Kennedy, B. A., 1971. *Physical Geography: A systems approach*. London: Prentice-Hall International.

Christian, C.S., Steward, G.A., 1968. Methodology of integrated surveys. Aerial survey and integrated studies. Proceedings of Toulouse Conference, UNESCO, Paris, pp. 233–280.

CIBA Foundation, 1997. Precision agriculture: spatial and temporal variability of environmental quality. CIBA Foundation Symposium, John Wiley & Sons, UK, p. 210.

Claessens, L., Heuvelink, G.B.M., Schoorl, J.M., Veldkamp, A., 2005. DEM resolution effects on shallow landslide hazard and soil redistribution modelling. Earth Surf. Process. Landforms 802 (30), 461–477.

Claessens, L., Knapen, A., Kitutu, M.G., Poesen, J., Deckers, J.A., 2007. Modelling landslide hazard, soil redistribution and sediment yield of landslides on the Ugandan footslopes of Mount Elgon. Geomorphology 90, 23–35.

Cooke, R.U., Doornkamp, J.C., 1990. Geomorphology in Environmental Management. A New Introduction. second ed. Oxford University Press, Oxford, 410 p.

Demek, J., 1972. Manual of Detailed Geomorphological Mapping. Academia, Prague.

Demek, J., Embleton, C. (Eds.), 1978. Guide to Medium Scale Geomorphological Mapping. IGU, Stuttgart.

Dobos, E., Montanarella, L., 2004. The development of a quantitative procedure for soilscape delineation using digital elevation data for Europe. Digital Soil Mapping Workshop. Montpellier, France, 14–17 September.

Dobos, E., Daroussin, J., Montanarella, L., 2005. An SRTM-based Procedure to Delineate SOTER Terrain Units on 1:1 and 1:5 Million Scales. Office for Official Publications of the European Communities, Luxembourg, EUR 21571 EN, 55 pp.

Doornkamp, J.C., Brunsden, D., Jones, D.K.C., Cooke, R.U., Bush, P.R., 1975. Rapid geomorphological assessments for engineering. Q. J. Eng. Geol. 12, 189–204.

Fookes, P.G., 1997. Geology for engineers: the geological model, prediction and performance. Q. J. Eng. Geol. 30, 293–424.

Fookes, P.G., Lee, E.M., Sweeney, M., 2001. Pipeline route selection and ground characterization, Algeria. In: Griffiths, J.S. (Ed.), Land Surface Evaluation for Engineering

Practice. Geological Society, London, pp. 115–122. Engineering Geology Special Publication, 18.

Fookes, P.G., Lee, E.M., Griffiths, J.S., 2007. Engineering Geomorphology, Theory and Practice. Whittles Publishing, Scotland, 279 pp.

Galon, R., 1962. Instruction to the detailed geomorphological map of the Polish Lowland. Polish Academy of Science, Geography Institute of Geomorphology and Hydrography of the Polish Lowland at Torun.

Gregory, K.J., 2010. The Earth's Land Surface. SAGE, London, 348 pp.

Griffiths, J.S., 2001. Engineering geological mapping. In: Griffiths, J.S. (Ed.), Land Surface Evaluation for Engineering Practice. Geological Society, London, pp. 39–42. Engineering Geology Special Publication, 18.

Griffiths, J.S., 2002. Mapping and engineering geology: an introduction. In: Griffiths, J.S. (Ed.), Mapping in Engineering Geology, 1. The Geological Society, Bath, pp. 1–5. Key Issues in Earth Sciences.

Griffiths, J.S., Abraham, J.K., 2008. Factors affecting the use of applied geomorphology maps to communicate with different end-users. J. Maps 2008, 201–210.

Griffiths, J.S., Brunsden, D., Lee, E.M., Jones, D.K.C., 1995. Geomorphological investigations for the Channel Tunnel Terminal and Portal. Geogr. J. 161, 275–284.

Gustavvson, M., Kolstrup, E., Seijmonsbergen, A.C., 2006. A new symbol-and-GIS based detailed geomorphological mapping system: renewal of a scientific discipline for understanding landscape development. Geomorphology 77, 90–111.

Hayden, R.S., 1986. Geomorphological mapping. In: Short, N.M., Blair, R.W.J. (Eds.), Geomorphology from Space. NASA, Greenbelt, MD.

Hearn, G.J., 1995. Engineering geomorphological mapping and open-cast mining in unstable mountains: a case study. Trans. Inst. Min. Metall. Sect. A Miner. Ind. 104, A1–A17.

IBGE, Instituto Brasileiro de Geografia e Estatistica, 2009. Manual Tecnico de Geomorfologia, second ed. Manuais Tecnicos de Geosciecas, numero 5, 175 pp.

Jones, A.F., Brewer, P.A., Johnstone, E., Macklin, M.G., 2007. High-resolution interpretative geomorphological mapping of river valley environments using airborne LiDAR data. Earth Surf. Process. Landforms 32, 1574–1592.

Klimaszewski, M., 1956. The principles of the geomorphological survey of Poland. Przeglad Geograficzny 28 (Suppl.), 32–40.

Klimaszewski, M., 1982. Detailed geomorphological maps. ITC J. 1982-3, 265–271.

Knapen, A., Kitutu, M.G., Poesen, J., Breugelmans, W., Deckers, J., Muwanga, A., 2005. Landslides in a densely populated county at the footslopes of Mount Elgon (Uganda): characteristics and causal factors. Geomorphology 73 (1–2), 149–165.

Koomen, A.J.M., Maas, G.J., 2005. Geomorfologische Kaart Nederland (GKN). Achtergronddocument bij het landsdekkende digitale bestand. Wageningen, Alterra, Alterra Report no. 1039, 38 pp.

Lee, 2001. Geomorphological mapping. In: Griffiths, J.S. (Ed.), Land Surface Evaluation for Engineering Practice. Geological Society, London, pp. 53–56. Engineering Geology Special Publication, 18.

MacMillan, R.A., Pettapiece, W.W., Nolan, S.C., Goddard, T.W., 2000. A generic procedure for automatically segmenting landforms into landform elements using DEMs, heuristic rules and fuzzy logic. Fuzzy Sets Syst. 113 (1), 81–109.

Munich Re (2010). TOPICS GEO natural catastrophes 2009. Analyses, assessments, positions. <http://www.munichre.com/touch/naturalhazards/en/natcatservice/default.aspx>. (accessed 17.12.10).

NAP, 2010. Landscape on the edge. New Horizons for Research on Earth's Surface. Edited by Committee on Challenges and Opportunities in Earth Surface Processes. The National Academies Press, Washington, DC.

Ollier, C.D., 1977. Terrain classification, principles and applications. In: Hails, J.R. (Ed.), Applied Geomorphology. Elsevier, Amsterdam, pp. 277–316.
Pain, C., Chan, R., Craig, M., Hazell, M., Kamprad, J., Wilford, J., 1991. RTMAP BMR regolith database field handbook. BMR Record 1991/29.
Pain, C.F., Craig, M.A., Gibson, D.L., Wilford, J.R., 2001. Regolith-landform mapping: an Australian approach. In: Bobrowsky, P.T. (Ed.), Geoenvironmental mapping, method, theory and practice A.A. Balkema, Swets and Zeitlinger. Publishers, The Netherlands, pp. 29–56.
Pain, C.F., 2008. Field Guide for Describing Regolith and Landforms. CRC LEME Publishing, Canberra, 108 p.
Pain, C., 2008. Field guide for describing Regolith and Landforms. CRC-LEME, Camberra, Australia, 96 pp. ISBN 978-0-9806030-0-2.
Pain, C.F., Kilgour, P., 2003. Regolith mapping – a discussion. In: Roach, I.C. (Ed.), Advances in Regolith. CRC LEME Publishing, Canberra, pp. 309–313.
Pain, C.F., Chan, R., Craig, M., Gibson, D., Kilgour, P., Wilford, J., 2007. RTMAP regolith database field book and users guide. CRC LEME Publishing, Canberra, Open File Report 231.
Rodríguez Fernández., L.R., 2005. El Plan MAGNA: evolución histórica y perspectivas futuras. Boletín Geológico y Minero, 116 (4): 281–289. ISSN: 0366-0176
Rodriguez Garcia, J.A., Perez Cerdan, F., 2006. Normas de organizacion de la informacion del Mapa Geomorfologico Nacional digital. Version 1.0. TSIG, Instituto Geologico ey Minero de Espana, Madrid, Spain, 42 pp.
Savigear, R.A.G., 1965. A technique of morphological mapping. Ann. Ass. Am. Geog. 53, 514–538.
Schmitz, G., 1980. A rural development project for erosion control in Lesotho. ITC J. 2, 349–363.
Scott, K.M., Pain, C.F., 2008. Regolith Science. CSIRO Publishing, Collingwood, 472 p.
Servizio Geologico, Nazionale, 1994. Carat Geomorfologica d'Italia – 1:50,000. Guida al rilevamento. Quaderno n.3 del SGN, a cura del Gruppo di Lavoro per la Cartografia Geomorfologica. Istituto Poligrafico e Zecca dello Stato 42.
Smith, M.J., 2011. Digital mapping: visualisation, interpretation and quantification of landforms. In: Smith, M.J., Paron, P., Griffiths, J. (Eds.), Geomorphological Mapping: A Handbook of Techniques and Applications. Elsevier, Amsterdam.
Smith, M.J., Clark, C.D., 2005. Methods for the visualisation of digital elevation models for landform mapping. Earth Surf. Process. Landforms 30, 885–900.
Smith, M.J., Pain, C.F., 2009. Applications of remote sensing in geomorphology. Progr. Phys. Geogr. 33 (4), 568–582.
St-Onge, D.A., 1968. Geomorphic maps. In: Fairbridge, R.W. (Ed.), Encyclopedia of Geomorphology. Reinhold, New York, pp. 388–403.
St-Onge, D.A., 1981. Theories, paradigms, mapping and geomorphology. The Can. Geogr. XXV (4), 307–315.
Thornes, J.B., Brunsden, D., 1977. Geomorphology and time. Methuen, London, 208 pp.
Tooth, S., 2009. Invisible geomorphology. Earth Surf. Process. Landforms 34, 752–754.
Tricart, J., 1965. Principes et Méthodes de la Geomorphologie. Masson, Paris, 496 p.
Tricart, J., Cailleaux, A., 1956. Introduction à la Géomorphologie Climatique. Sedes, Paris.
Verstappen, H.T., 1970. Introduction to the ITC system of geomorphological survey. Koninklijk Nederlands Aadrijkkunding Genootschap, Geografisch Niewe Reeks, 4.1, pp. 85–91.
Verstappen, H.T., 1983. Applied Geomorphology: Geomorphological Surveys for Environmental Development. Elsevier, Amsterdam.
Verstappen, H.Th, van Zuidam, R.A., 1968. ITC textbook of Photo-Interpretation, VII:2 –ITC system of geomorphological survey. ITC, Delft, The Netherlands.

CHAPTER FIVE

Geomorphological Contributions to Landslide Risk Assessment: Theory and Practice

Gareth J. Hearn and Andrew B. Hart
URS Scott Wilson Ltd, Scott House, Alencon Link, Basingstoke, UK, RG21 7PP

Contents

1. INTRODUCTION

Landslides, debris flows and floods pose an ever-increasing risk to communities and infrastructure in many parts of the world. This apparent increase in risk is fuelled primarily by the expansion of development and infrastructure into more hazard-prone areas. Changing land use and drainage patterns can lead to increased levels of hazard, while population expansion and the investment in higher value land uses result in potentially increased levels of risk (see discussions in Cascini et al., 2005 and Petley, 2010). For example, typhoons Ondoy and Pepeng hit the island of

Developments in Earth Surface Processes, Volume 15
ISSN: 0928-2025, DOI: 10.1016/B978-0-444-53446-0.00005-7

Figure 5.1 Typical damage to roads in the Central Cordillera of the Philippines following typhoons Ondoy and Pepeng in 2009 (Hearn 2011).

Luzon in the Philippines in September and October 2009, with the former depositing an unprecedented 450 mm of rain in 12 h in Manila, and the latter being responsible for a total of 850 mm of torrential rains in Baguio, located in the Central Cordillera further north (GFDRR, 2009). Several landslides occurred in the vicinity of Baguio giving rise to numerous fatalities and major damage to infrastructure (Figure 5.1). The road that links Baguio with the rice terraces of Banaue to the north (Hart et al., 2002) – a lifeline for a large number of rural communities and roadside industries – became blocked by landslides in 30 locations over a 20 km section. More significantly, the foundation of the road itself failed in eight locations over the same 20 km section. Typhoons that have swept through the area during the last 20 years have generally caused less damage, but engineers, planners and civil defence authorities can expect recurrent future damage or destruction to infrastructure, and injury or loss of life, during such events.

When planning and maintaining infrastructure, and attempting to protect communities and the public from geo-hazards of this nature, the principal questions that need to be answered are as follows:

What geo-hazards are present?

Where are the current high-hazard areas?

What are the controlling factors?
Where might they occur in the future?
What degree of damage or loss will occur?
When will they occur?
How frequently will they occur?
Are there any means of forewarning?
What are the effective means of mitigation?

In the Philippines, as with many countries, despite the frequency with which landslide, debris flow and flood hazards recur, there is generally inadequate information with which to answer these questions other than to say where they have occurred in the past and what have been the most likely causes. From this, suppositions can be drawn as to where they might occur in the future, from what causes and with what impacts. On the opposite side of the Philippine Sea, Hong Kong does have an accumulated knowledge and event database with which to make significant inroads into answering these questions as it is one of the most documented areas of the world in terms of topography, ground conditions and geo-hazard. However, even in Hong Kong, attempts to quantify and predict landslide hazard and risk are also hampered by information gaps, and Ho and Lau (2010) describe a number of instances where problems have occurred as a result of a failure to fully understand and design for the ground conditions encountered and the slope failures that actually take place.

Another factor that serves to limit the reliability with which outcomes can be predicted is the often extremely localised nature of rainfall and storm events. In mountain areas, and particularly those in the tropics and subtropics, it is this meteorological uncertainty that poses one of the greatest challenges for geo-hazard preparedness at any particular moment in time. Although substantial investments might be made in improving geological and geotechnical databases for slope stability assessments, it will be the unpredictable distribution of rainfall from one catchment or slope to the next that will generally be the ultimate factor in controlling where the greatest damage occurs, and particularly over short time frames.

Despite these constraints, it is imperative that the most reliable assessment of landslide hazard is undertaken and fully utilised, whether for planning purposes spanning several decades or engineering decision-making over much shorter time frames. This chapter outlines some of the methods of landslide hazard and risk assessment that have been developed, and discusses and illustrates their use on three projects at different application scales and programme stages. Importantly in all three illustrations:

- limited data availability was a constraining factor;
- mapping outputs were required for decision-making over short time frames;
- maximum use was made of geomorphological mapping techniques and simple interpretative methods to yield the required outputs.

2. LANDSLIDE SUSCEPTIBILITY, HAZARD AND RISK

Much has been written on the subject of landslide assessment for planning and engineering purposes (Turner and Schuster, 1996; Lee and Jones, 2004; Glade et al., 2005), and the concepts of landslide susceptibility, hazard and risk are most commonly referred to (Box 5.1), with quantified risk assessment being the ultimate goal for decision-making (see, for example, the reviews contained in Aleotti and Chowdhury, 1999, and Guzzetti et al., 1999, and the procedural guidelines set out in AGS, 2007,

Box 5.1

Landslide susceptibility is defined as the relative extent to which a particular slope might be more prone to failure than another.

Landslide hazard is defined as the potential posed by an existing or possible future landslide to cause damage or loss (economic and social). Hazard combines size, probability and intensity; parameters that are determined by magnitude–frequency relationships (Cascini et al., 2005), areal extent and depth of failure, and speed of movement. Commonly, hazard is considered as the product of magnitude (including intensity) and probability of movement in a given area over a given time, such that:

$$Hazard\ (H) = Magnitude\ (M) \times Probability\ (P)$$

Landslide risk is defined as the actual or potential damage or loss that may occur as a result of a landslide movement taking place. Risk combines hazard (H) with the value of the assets (engineering, environmental and social) at risk and their vulnerability (degree of loss) to the landslide movement should it take place. Risk is therefore commonly considered as the product of hazard and value and vulnerability, such that:

$$Risk\ (R) = Hazard\ (H) \times Value\ (Va) \times Vulnerability\ (Vu)$$

and Fell et al., 2008). The risk posed by landslides that have already been documented, and from ground movements that are active and observable, can usually be defined with reasonable confidence from historical records and the actual damage and injuries or fatalities that have resulted. The difficulty usually occurs where landslides are identified in areas where there are no reliable records of ground movement or impact and where there is limited or non-existent geotechnical information available with which to assess factors of safety and the likelihood of future reactivated movement. Moreover, the hazard and risk posed by future, first-time landslides, i.e. landslides that have not yet occurred, can also rarely be fully evaluated because there are:

- multiple and commonly indeterminate parameters that ultimately dictate hazard, including:
  - conditioning and triggering factors responsible for the initiation of slope movement,
  - areal extent and depth of movement when it does occur,
  - speed, frequency and timing of movement,
  - displacement distance, or run-out in the case of mudflows, debris flows and avalanches.
- multiple assets at risk, including engineering structures, traffic, land use, land resources, population and social infrastructure,
- multiple vulnerabilities to hazard, including damage, partial loss or complete loss, either repairable or irreparable (replacement required) in the case of infrastructure and land resources.

3. EXPERIENCE FROM INDUSTRY

The authors have been engaged on engineering and planning-related projects over the past 20 years where assessment has been required of the hazard posed by landslides to structures and land use. Many of these projects have been in remote locations where desk study data are limited, and information on existing and past landslide events varies from sketchy to non-existent. In addition, information on ground conditions is often extremely limited, and so geological models for slope analysis become reliant on what can be interpreted from an inspection of the ground surface. Furthermore, the time and resources available with which to carry out landslide susceptibility, hazard and risk assessment are often limited by programme and budgetary constraints. In nearly all cases, there have been no published landslide records or mapping of any kind to fall back on, and hence the assessments have had to be made from first principles using remote sensing and field mapping (Hearn, 2011). In these circumstances, and these are considered to prevail in many parts of the world, it is geomorphology that offers the greatest potential in yielding an interpretation for design and decision-making within the constraints of limited data, limited time and limited budget.

This chapter describes how geomorphology has been used to yield the information required to assess susceptibility, hazard and risk for engineering and planning projects using three case histories. It reviews the scope of the approaches adopted in the light of the procedural guidelines contained in Fell et al. (2008), and it comments on the value of the output as a function of the time and resources available to carry out the work. Other published studies are also reviewed in terms of their ability to yield mapping outputs for planning and engineering purposes when faced with varying levels of input data.

The first case study concerns the development of landslide susceptibility, hazard and risk maps for infrastructure and land use planning purposes in Nepal, relying principally on desk study and landslide inventory and broad geological and geomorphological indicators. The second and third involve assessments for linear infrastructure and regional planning purposes in the Russian Far East and Cyprus, respectively, and take greater account of engineering geological, geomorphological and geotechnical considerations.

4. LANDSLIDE HAZARD AND RISK MAPPING FOR RURAL INFRASTRUCTURE PLANNING IN NEPAL

Rural infrastructure development in Nepal and much of the Himalayan region below approximately 4000 m and above 250 m asl is affected significantly by landslides. Road access represents an increasingly important element in the sustainability of the rural economy and the provision of health care to remote mountain communities. Unfortunately, roads are commonly disrupted by landslides, debris flows and earthworks failures. In order to assist in the planning of rural infrastructure to help

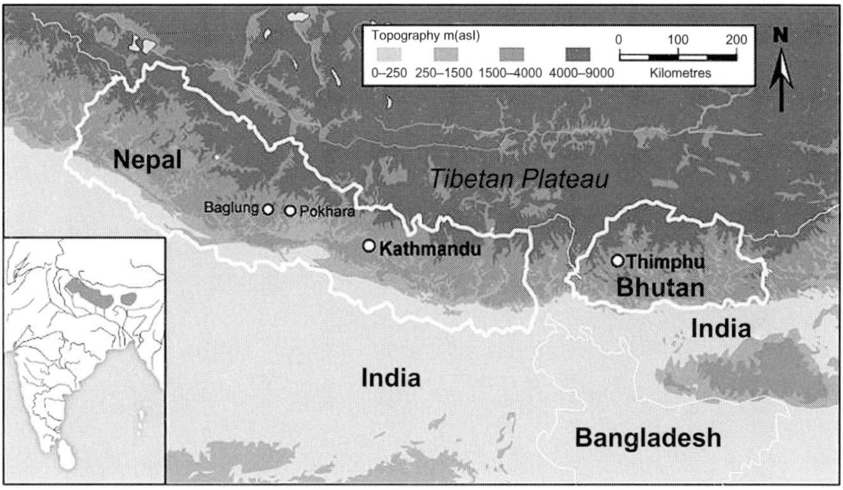

Figure 5.2 Location of the Baglung study area in Nepal.

minimise landslide impacts, a series of mapping studies was undertaken at scales of 1:25,000 and 1:50,000 on behalf of the Ministry of Local Development with funding from Department for International Development (DFID), United Kingdom. Three areas were selected in Nepal, one of which was in the Baglung District in the west of the country (Figure 5.2).

The Baglung study area covered almost 530 km^2 and was located in predominantly hilly to mountainous terrain at elevations commonly above 1500 m asl. The annual rainfall is almost 2 m, more than 80% of which falls between June and September. Landslides cause regular damage to roads and agricultural land and have destroyed villages, schools and other community assets (Figure 5.3). Using stereo aerial photographs, Landsat and SPOT satellite imagery and ground verification, an inventory of over 230 landslide scarps (source areas) and deposits (run-out areas) was developed (Figure 5.4).

Standard spatial and attribute query functions within geographical information system (GIS) software were used to compare the distribution of mapped landslides with a number of factors that were considered likely to influence slope stability (Table 5.1).

The distribution of landslides with respect to each of these factors was analysed using the Chi^2 (χ^2) statistics (Hammond and McCullagh, 1978).

Figure 5.3 Typical landslide affecting land use and road alignments in the Baglung District.

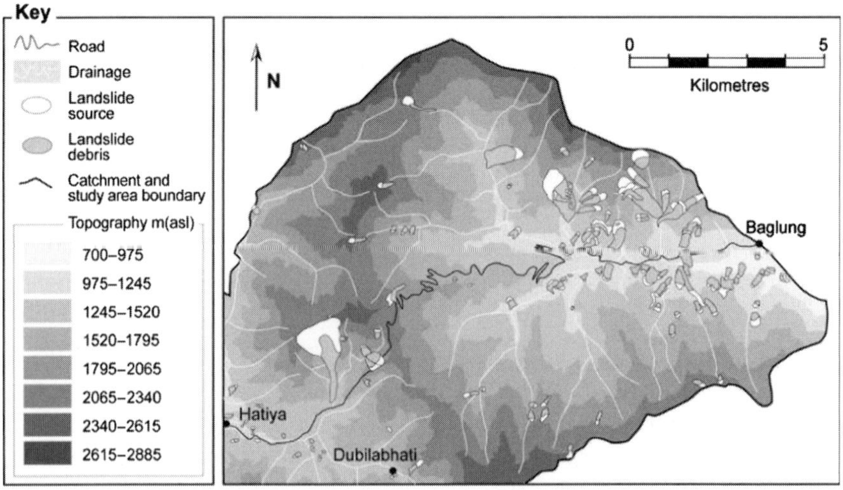

Figure 5.4 Part of the landslide map for the Baglung study area (original scale 1:50,000).

Table 5.1 Factors Examined in Relation to the Distribution of Mapped Landslides

Factor	Source of Data
Slope angle	Digitised contours from published map
Rock type	Digitised from published map
Wet areas (water table at surface)	Tonal appearance on air photos and spectral reflectance in Landsat imagery
Slope angle and aspect versus bedding/foliation dip and dip direction (kinematic feasibility)	Published geological map and field observations
Slope aspect	GIS polygons from digitised contours
Rainfall distribution	From daily records for three recording stations
Earthquake distribution	Downloaded from the website of the USGS Earthquake Hazards Program (http://earthquake.usgs.gov/earthquakes/)
Proximity to faults and folds	Measured on digitised geological map
Relative relief	Digitised contours from published map
Areas of erosion	Interpreted from air photos and spectral reflectance in Landsat imagery
Terrain classification	Derived from air photo interpretation
Land use	Digitised from published map

The criteria used to select which of the factors listed in Table 5.1 to include in the susceptibility mapping are listed below:

- More than 75% of total existing landslides lie within the highest two susceptibility classes in a fivefold classification,
- Less than 20% of the total existing landslides lie within the lowest two susceptibility classes,
- The χ^2 value is significant at the 0.001 confidence level,
- The landslide density increases with susceptibility class,
- The relationships so derived make sense geotechnically.

The comparison between the observed (O) number of landslides within a given factor class, such as rock type or slope angle interval, with that expected (E) had the distribution been random, provides a useful indicator of the susceptibility of each factor class to slope failure, with the higher O/E values indicating greater susceptibility. Rock type and slope angle were found to be most significantly correlated with the landslide distribution (Tables 5.2 and 5.3). Most of the other factors analysed were either:

- not significantly correlated, or
- correlated but with anomalous results, probably brought about by auto-correlation amongst factors.

Table 5.2 Observed/Expected Landslide Distribution According to Rock Type

Formation Name	Major Lithology	O/E Susceptibility
Dandagaon Phyllite	Phyllite with subordinate quartzite beds	2.04
Benighat Slates	Carbonaceous slates with calcareous beds	0.92
Nourple Formation	Variegated phyllite, quartzite and limestone	0.9
Malekhu Limestone	Dolomitic to argillaceous limestone	0.44
Dhading Dolomite	Thick to massive stromatolitic dolomite	0.41
Robang Phyllite	Phyllite with intercalation of quartzite	0.08
Raduwa Formation	Garnet schist with micaceous quartzite	0.00
Kuncha Formation	Phyllite and quartzite/phyllitic quartzite	0.00

Table 5.3 Observed/Expected Landslide Distribution According to Slope Angle

Slope Angle (°) Range	O/E Susceptibility
50+	1.14
46–50	2.41
41–45	1.30
36–40	0.90
31–35	1.18
26–30	0.72
21–25	0.87
16–20	0.51
11–15	0.47
0–10	0.22

Furthermore, the quality and completeness of the data sets across the study area varied according to contour accuracy on topographic maps, errors in the published geological maps, cloud cover on aerial photographs and difficult access where data sets were reliant on field observations for their derivation.

As the susceptibility of any given slope angle class varies with lithology (Figure 5.5), the distribution of landslides was then analysed according to slope angle classes within each lithology and grouped according to observed susceptibility. Figure 5.6 shows how landslide density varies according to the final landslide susceptibility classes. The combined rock type–slope angle susceptibility classes were then tested using landslides mapped from aerial photographs of the Arun Valley in east Nepal (Figure 5.2) where approximately the same rock types are exposed. The susceptibility classes were more than 90% successful in predicting the

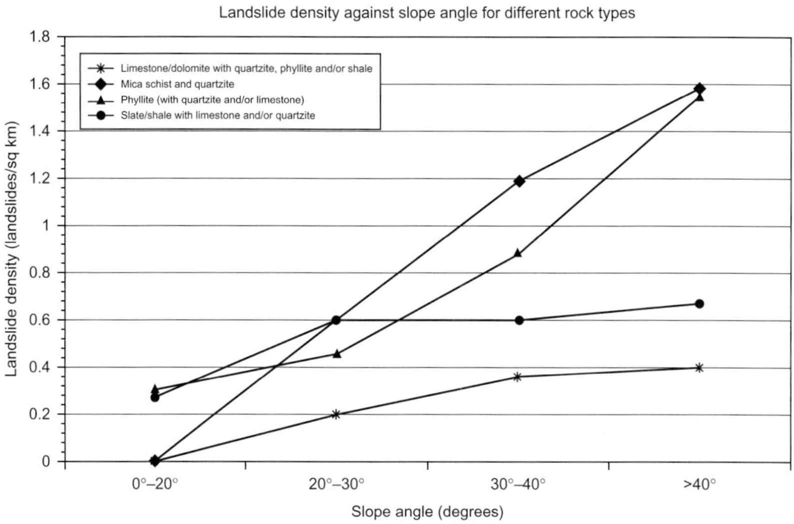

Figure 5.5 Landslide density against slope angle for different rock type groups in the Baglung study area.

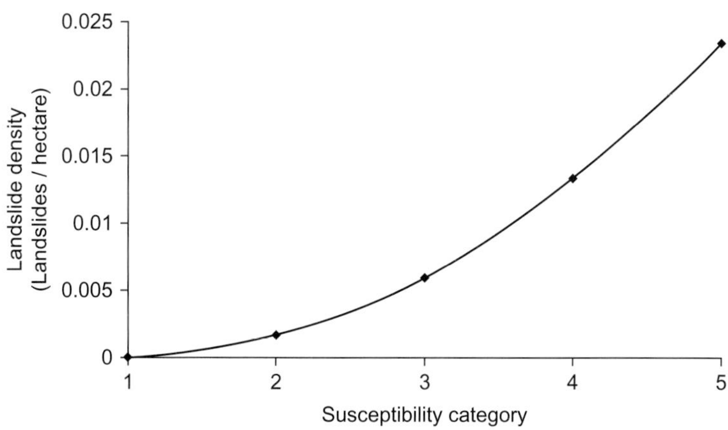

Figure 5.6 Landslide density versus landslide susceptibility class.

distribution of mapped landslides, thus demonstrating the validity of the model.

An extract of the landslide susceptibility map produced from this distribution of rock type–slope angle classes is shown in Figure 5.7, covering an area of approximately 16 km^2. This figure also shows the computed

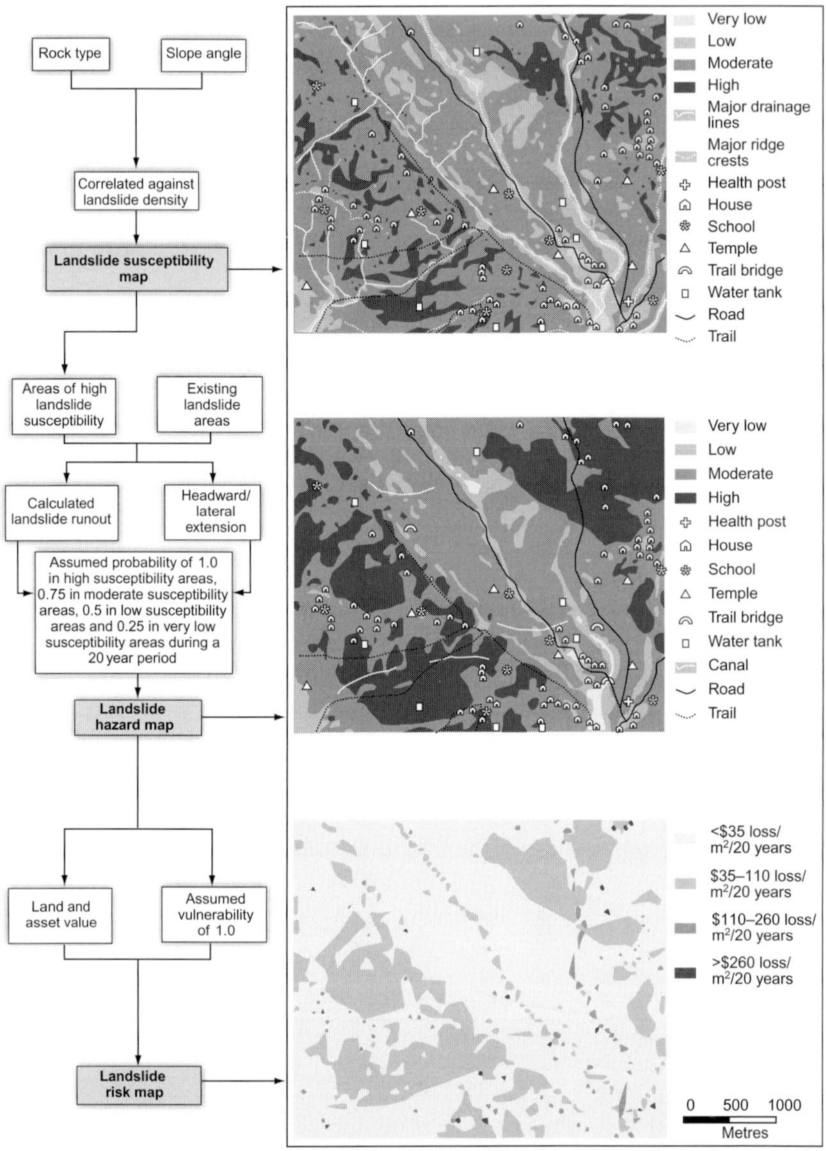

Figure 5.7 Extract of landslide susceptibility, hazard and risk map for the Baglung study area (from Hearn, 2011).

displacement and run-out of future landslides from the high-susceptibility areas. These run-out areas are derived from the empirical relationships between landslide volume and run-out distance for mapped landslides

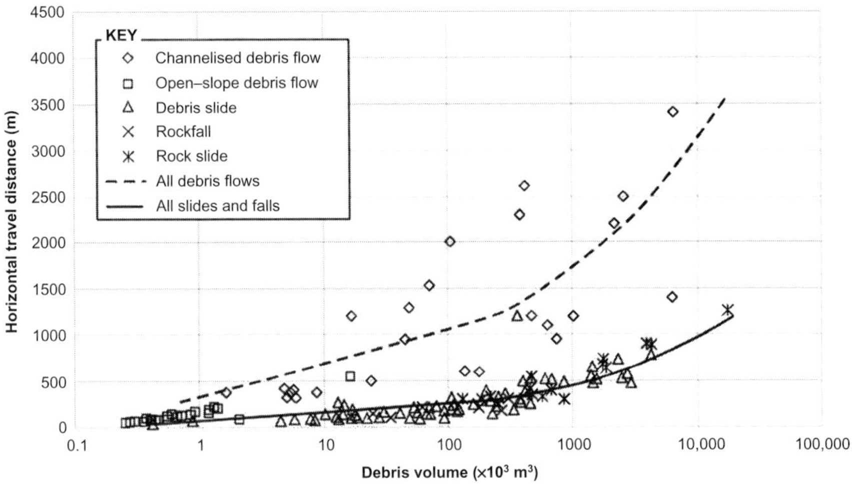

Figure 5.8 Displacement/run-out curves for mapped landslides (from Hearn, 2011).

(Figure 5.4) according to landslide type (Figure 5.8). Future landslide volumes were assessed by averaging the mapped volumes of previous landslide mechanisms and deciding upon the most prevalent mechanism for each rock type.

During fieldwork, local communities were consulted with regard to their knowledge of the timing of landslide events and their impact on lives, land use and infrastructure. A total of 40 landslides were dated in this way (the earliest being 1923) and comparisons were made with the record of seismicity. An assessment of the earthquake record and comparison with work undertaken by Keefer (1984) and Giardini (1999) led to the conclusion that the probability of an earthquake of sufficient magnitude to generate widespread slope instability in the Baglung area was low. None of the communities consulted had any recollection of earthquake-induced landslides, and none of the dated landslides occurred during years when earthquakes were known to have occurred. With regard to rainfall-induced landslides, Caine and Mool (1982) suggested a landslide threshold of 100 mm/24 h for Nepal. In the study area, this rainfall intensity occurs on average between 0.41 and 1.15 times each year, and therefore landslides might be expected to occur every 1–2 years in the high-susceptibility areas shown on Figure 5.7.

In the absence of any reliable means with which to assign probability, it was assumed that slopes located in the high-susceptibility areas would fail during a 20 year period. This period is considered appropriate to the life cycle of rural infrastructure planning and probabilities of 1.0, 0.75, 0.5 and 0.25 were applied to slope failure from high-, moderate-, low- and very low-susceptibility areas, respectively. In terms of vulnerability, it was assumed that a section of road, a building or an area of cultivated land, for example, that was located in the source or run-out areas of future landslides would be destroyed during the event, i.e. their vulnerability to the event would be 1.0. Clearly this is an oversimplification in many cases as slow ground movements beneath a road, for example, do not necessitate its complete reconstruction, whereas debris run-out onto a cultivated field might destroy a year's crop production, but the material can generally be ploughed and cultivation can be resumed during following years. The market value of agricultural land and the reconstruction costs associated with buildings and roads were calculated and the total economic loss likely to arise as a result of landslides occurring during a nominal 20 years was determined. A risk map showing the areal distribution of these losses is shown in Figure 5.7. This map shows asset losses only and does not include injury or fatality to the local population. Although calculations were made of value of life according to age group, these were not included in the calculation of loss because they provoked too much emotive controversy when discussed with local development authorities and community representatives.

5. SAKHALIN 2, PHASE II OIL AND GAS PIPELINE IN RUSSIA

Twin oil and gas pipelines run almost the entire north–south length of Sakhalin Island, located off the east coast of Russia, and convey hydrocarbons from the point where they are brought onshore at Nogliki in the north to an all-season port at Prigorodnoye, in the far south of the island (Figure 5.9). Approximately 120 km of this alignment is located within the Makarov Mountains. These mountains range up to 1600 m asl and are underlain by tightly folded Cretaceous mudstone and Tertiary sandstone, siltstone, coal measures and conglomerate with lavas and tuffs at certain horizons. Through the interpretation of stereo aerial photography, a total of 414 landslides (Figure 5.10) were mapped at a scale of 1:5000 and subsequently ground-truthed within a 500 m wide corridor. The Engineering,

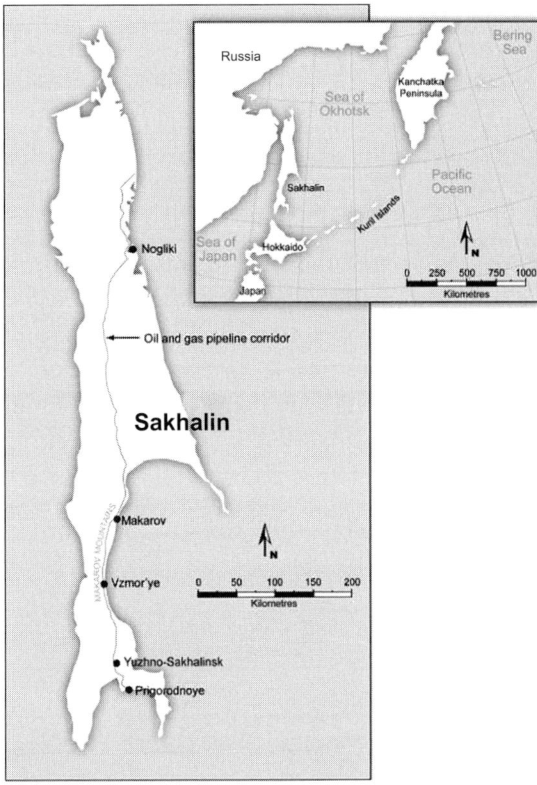

Figure 5.9 Sakhalin Island.

Figure 5.10 Typical landslide morphology (winter).

Procurement and Construction (EPC) contractor developed the detailed horizontal alignment to avoid these landslides wherever possible and adjusted the vertical alignment in order to construct the pipelines beneath the failure zone of those landslides that could not be avoided.

A hazard register was required to be produced for all known landslides and areas considered to be the potential locations for future first-time failures (Hearn et al., 2012). Although the EPC contractor undertook a ground investigation as part of the design, this had to be supplemented with detailed aerial photograph interpretation, geomorphological mapping (Figure 5.11) and field observations during right of way and pipeline trench excavations in order to enable an assessment to be made of the potential hazard posed by landslides to the pipeline corridor. The first stage in this assessment was to develop a proximity check in order to be able to select those landslides that were in close enough geographical proximity to be of potential relevance to the stability and security of the

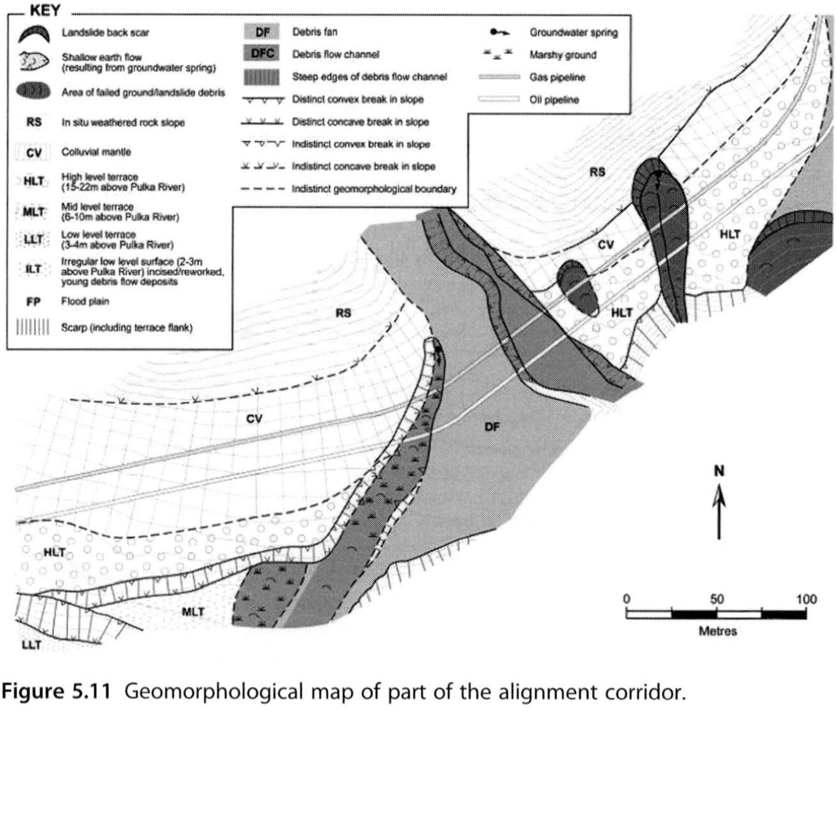

Figure 5.11 Geomorphological map of part of the alignment corridor.

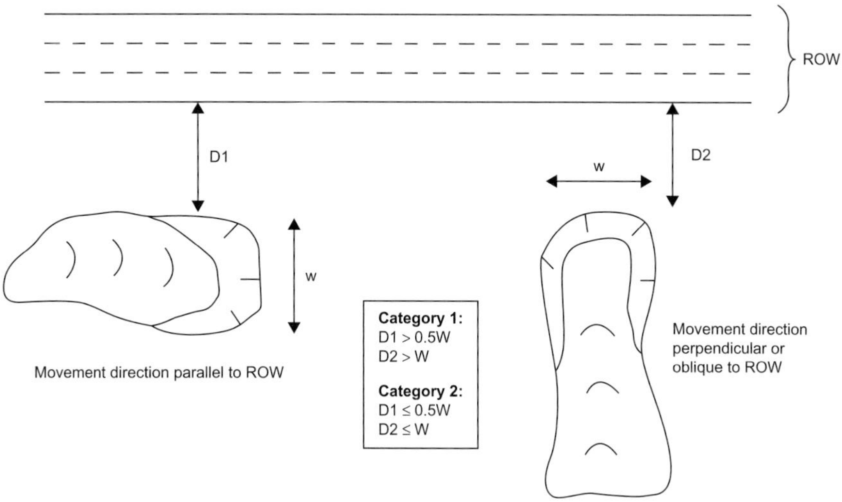

Figure 5.12 Proximity check for mapped landslides.

pipelines. The criteria used in this proximity check are defined below using the dimensions shown in Figure 5.12:

- D1 > 0.5W or D2 > W, the landslide was considered too small or remote and there was no perceived hazard – Category 1,
- D1 < 0.5W or D2 < W, the landslide was considered a potential hazard and a geometry check was required – Category 2 (Figure 5.13).

For the close proximity cases derived from Figure 5.12, a further classification was applied to define the design pipeline location in relation to landslide failure surfaces. An additional set of criteria was applied to each of the cases A, B and C (in Figure 5.13) in order to identify a condition of potential hazard that would require a slope stability analysis to be performed. This applied to all cases where:

- landslide failure surfaces were considered to be located beneath or within the pipeline trenches (A)
- there was a computed regression potential (B) or
- the thickness of landslide deposits on slopes above the pipeline trenches was considered sufficient to generate a surcharge load should further ground movement take place (C) in these upslope locations.

In each case, geological cross sections were derived using the EPC contractor's ground investigation data, geomorphological mapping and logging of all natural and construction exposures within the vicinity. Slope stability back analyses were undertaken and the parameters derived

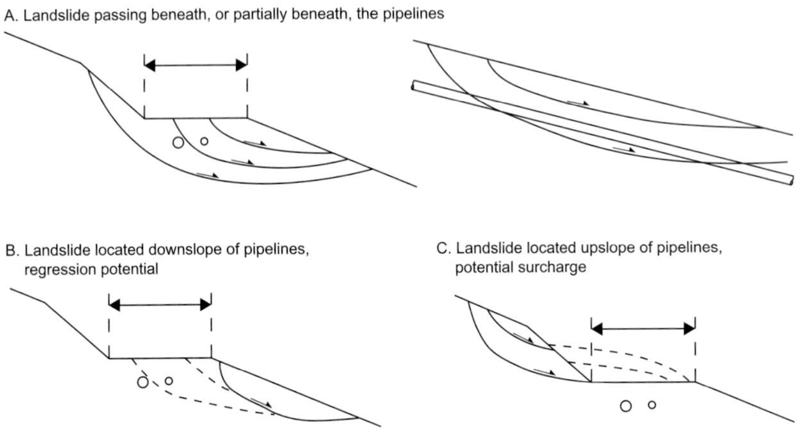

Figure 5.13 Geometry check for landslides within or in close proximity to the pipeline corridor.

were used in the forward analysis of the design, taking into account the EPC contractor's proposals for pipeline burial depth, ground lowering and removal of driving loads. Sensitivity analyses involved the inclusion of the 1 in 20 year seismic loading and the maximum snow melt condition with a groundwater table at the slope surface.

A hazard rating was derived for each entry in the landslide hazard register based on the following criteria:

- Landslides considered too remote or too small from the proximity check to constitute a hazard were automatically assigned a low hazard rating,
- Geometry checks that showed existing or projected regressive failure surfaces to be vertically above the as-built location of the pipelines were assigned a low hazard rating,
- Where existing or projected regressive failure surfaces posed a hazard to the integrity of the pipelines, the forward analysis factor of safety (FoS) determined the hazard rating:
  - FoS > 1.1 was given a moderate hazard rating,
  - FoS < 1.1 was given a high-hazard rating.

Figure 5.14 shows how the hazard classification was derived and applied using an extract from the hazard register for existing slope failures. The hazard posed by potential first-time failures and cut slope failures was assessed separately (Hearn et al., 2012) but also summarised in the register.

Hazard (impact potential on pipelines)

			Likelihood of occurrence		
Landslide category			Low	Mod	High
Cat 1/Cat 1A		Hazard	L	L	L
Cat 2	A1		L	L	L
	A2		L	M	H
	B1		L	L	L
	B2		L	L	M
	C1		L	L	L
	C2		M	H	H
	Cat 3				M

Pipelines notes:

1. Cat 1/Cat 1A are either too remote or small or abandoned without an oversteepened back scar to pose anything other than a low impact potential,
2. A1, B1, C1 by definition have no impact on the pipelines,
3. A2 and B2 have the potential to deform the pipeline, but the A2 class has the potential to rupture the pipeline, and therefore this class has been assigned a higher impact potential,
4. C2 represents the case where a failure surface already exists beneath or through the pipeline trench and therefore is assigned the highest impact potential to rupture the pipeline,
5. Cat3 represents the case where debris flows occur across the pipeline. Their likelihood of occuring over the design life is considered high, but their depth or scour potential is not known and so an impact potential of moderate is assigned.

Existing landslides (EL) review									
Landslide category	EL landslide class		Landslide length, width, depth (m)	Length of right of way (RoW) final profile affected (m)	Hazard (impact potential)				Priority action recommended
	Oil	Gas			RoW (Oil)	RoW (Gas)	Pipeline (Oil)	Pipeline (Gas)	Y/N
2	C1	C1	700,120,5	110	L	L	L	L	N
3	B3	B3	?,?,4	50	M	M	M	M	N
1	NA	NA	60,60,4	NA	L	L	L	L	N
1	NA	NA	100,70,4	NA	L	L	L	L	N
3	B3	B3	?,?,3	50	M	M	M	M	N
–	–	–	–	–	–	–	–	–	–
2	C2	C2	300,160,6	80	H	H	H	H	Y
3	B3	B3	?,?,4	70	M	M	M	M	N

Figure 5.14 Extract from the hazard register for existing landslides.

For landslides with a high-hazard rating, outline remedial measures were proposed (including earthworks, drainage and slope retention measures). A prioritised monitoring scheme was developed for slopes of moderate and high hazard that combined routine visual and photographic observation and proforma-based records with piezometric and inclinometric movement monitoring.

As far as potential future first-time failures were concerned, a study was undertaken to identify those hitherto unfailed slopes that might become unstable in the future due to toe erosion, elevated groundwater during snow melt or summer rain and seismicity (Hearn et al., 2012). A fourfold approach that combined back analysis and sensitivity analysis, landslide susceptibility analysis, quantitative factor analysis and geological and geomorphological judgement was applied. The slopes identified in this way were checked in the field and those that were considered to pose a potential hazard were included in the hazard register.

6. LANDSLIDE MAPPING FOR LAND USE PLANNING IN CYPRUS

The Paphos District contains the most landslide-prone terrain in Cyprus (Pantazis, 1969; Northmore et al., 1986, 1988). A combination of high relief, steep slopes, intense winter rainfalls and periodic earthquakes coincide with exceptionally weak rocks, including melange and bentonite clay overlain by a capping chalk aquifer. The resultant slope instability has had a significant impact on the local population of the area with a number of villages being relocated, roads damaged or destroyed and loss of productive agricultural land (Pantazis, 1969; Hadjigeorgiou et al., 2006). This situation is exacerbated by the rapid increase in tourist and retirement home development and associated infrastructure in recent years, which is increasingly encroaching onto unstable or potentially unstable slopes, thereby increasing the risk posed to life and infrastructure in the region.

The Geological Survey Department (GSD) of Cyprus identified that part of the reason for this situation was the lack of detailed information about which slopes:

- had been affected by landslides in the past,
- were affected by ongoing landslide activity (Figure 5.15),
- had the potential to become unstable in the future or be affected by landslide run-out.

Figure 5.15 Typical failed slopes in the Cyprus study area (landslide in middle distance).

Therefore, the GSD decided to create a GIS-based inventory cataloguing all of the landslides affecting the region at a scale of 1:25,000. Such information and mapping could then be used for planning purposes, with the aim of either avoiding landslide-prone areas altogether or, where necessary, applying appropriate landslide mitigation measures in areas of existing risk. The GSD selected three test areas within the Paphos District that covered a total of almost 550 km^2.

Over 1800 landslides were mapped using aerial photography and high-resolution satellite imagery held by the GSD (Hart et al., 2010). For each landslide, the location and geometric details were recorded within a GIS-based landslide inventory. Landslide identification and classification were recorded according to Cruden and Varnes (1996), while the structure of the landslide inventory was designed following the International Working Party on the World Landslide Inventory (WP/WLI, 1993, and references therein). The remote sensing interpretation and the landslide details contained in the landslide inventory were verified by regular field visits which enabled the following activities to be undertaken:

- Review of the mapped outline of back scarp and failed material and the interpreted failure mechanism,

- Review of how the mapped landslides may have changed since the imagery was taken,
- Mapping of any landslides that had occurred since the imagery was taken,
- Observation of the ground conditions within the landslide areas that could not be detected from the imagery, including rock discontinuity data and slope drainage conditions.

Once the remote sensing interpretation was complete, and as a final stage in the field verification process, detailed geomorphological mapping was undertaken at 20 landslide locations; a predetermined number fixed by the GSD on the basis of their programme and budgetary constraints. Twenty landslides were selected to reflect the maximum range of differing geological conditions and landslide failure mechanisms present within the study areas. While this represented only 1% of the total number of identified landslides, each of the 20 posed a potential hazard to elements of land use or infrastructure and were therefore of immediate interest to the GSD. The field mapping allowed the causes, mechanisms and activity of each landslide to be assessed further and provided the basis for the scheduling of an intrusive ground investigation and programme of laboratory testing. This investigation comprised *in situ* density testing, groundwater observations, geological logging of recovered core, index tests and ring shear residual strength tests on disturbed samples. A small number of piezometers and inclinometers were also installed so that water levels and slope movements could be monitored. A ground model was developed for each of these landslides and back analyses carried out to confirm strength parameters and most likely failure scenarios. Various options for mitigation were also reviewed using these failure scenarios.

A terrain classification was developed (Figure 5.16 and Table 5.4) which subdivided the study areas according to topography, geomorphology, geology and drainage conditions, and formed the basis for all other mapping outputs. Spatial and statistical analysis of the landslide inventory data allowed landslide activity across the study areas and within each of the identified terrain units to be classified and patterns in the spatial distribution of landslide activity to be identified (Hart et al., 2010) based on geological unit, slope angle class and failure mechanism. From this a map was derived that highlighted those slopes that have been affected, are currently affected, or could be affected, by landslide activity in the future (i.e. a map of landslide susceptibility). Statistical analysis of the landslide inventory data also allowed a number of landslide run-out curves (similar

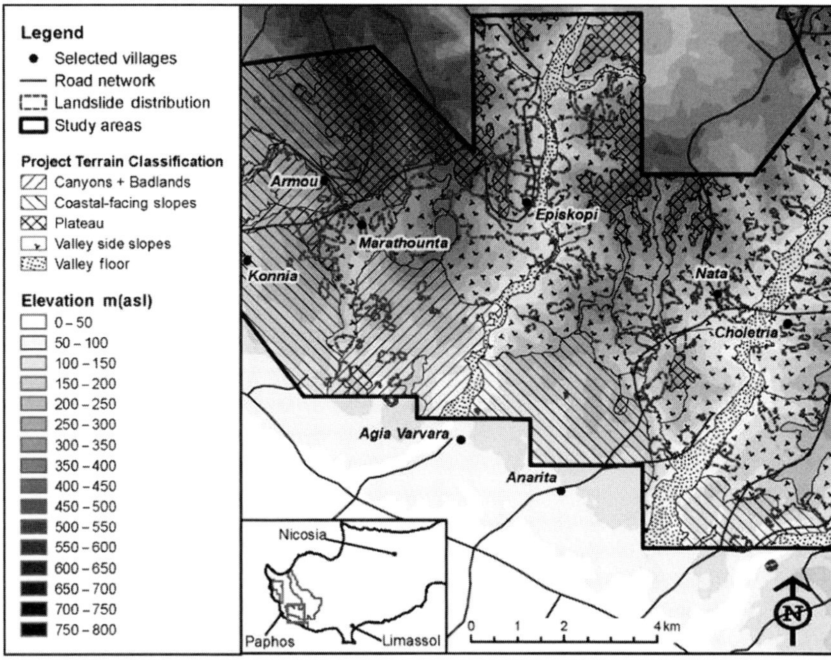

Figure 5.16 Terrain classification map for the three Paphos study areas.

to Figure 5.8) to be generated for different rock type and landslide mechanism combinations. Field observations of active landslides enabled landslide regression rates to be estimated. This information was added to the terrain unit descriptions and used to identify areas that could potentially be impacted by either landslide run-out or regression.

There are reports and publications that link a small number of landslides within the study areas to certain periods of (apparently) higher than average rainfall or specific earthquake events (Pantazis, 1969). Unfortunately, much of this linkage appears to be based on anecdotal or circumstantial evidence and therefore can only provide limited information for assessing the frequencies of triggering events. This lack of detailed historical and observational data highlights the problems and issues that need to be addressed when attempting to assess levels of landslide susceptibility, hazard and risk across large areas. It was agreed with the GSD that terrain classification, combining the existing landslide distribution with an assessment of future

Table 5.4 Terrain Unit Descriptions

Unit Name	Geology and Topography	Typical Landslide Activity	Landslide Density
Plateau	• Chalk and limestone forming broad plateau areas with distinctly rounded terrain that highlights the sub-horizontal bedding • Localised fault zones and structural lineaments influencing hydrogeological flow paths: some sinkholes and minor karst features evident • Localised 'Mushrooms' of reef limestone outliers, commonly at lower elevations	• Very limited landslide activity • Small-scale rockfalls along edges of bedding in incised drainage lines • Potential for larger failures to cut into plateau areas where terrain is incised by drainage lines, possibly related to (gently) folded and weaker beds (gypsum?)	0.6 landslides per km^2
Coastal-facing slopes	• Marine terraces overlying chalk or limestone • Slopes formed by wave action and marine erosion during periods of previous high sea levels • Rounded terrain	• Limited landslide activity • Predominantly, rockfalls along incised drainage lines	2.0 landslides per km^2
Valley-side slopes	• Chalk and/or limestone overlying melange or igneous material • Overlain in some places by alluvial terraces • Commonly incised by minor streams and drainage channels	• Landslide activity includes rotational slides, block slides, flows, rockfalls and topples	4.2 landslides per km^2

	• Landscape characterised by more competent 'cap rock' overlying significantly 'weaker'/'softer' material (e.g. chalk or limestone overlying melange material)		
Canyons and badlands	• Canyons characterised by steep/vertical cliffs of rock material – typically marine terraces, chalk, limestone or igneous material • Badlands formed by the erosion of melange material, bentonitic clays and highly weathered igneous material • Badlands characterised by subdued rounded terrain	• Canyons affected by small-scale rockfalls and topples • Badlands affected by slides and flows that erode very quickly • Where the clays are overlain by harder rock (forming escarpments), landslides include rotational slides, block slides, rockfalls and topples	4.4 landslides per km^2
Valley floor	• Alluvial and fluvial deposits relating to Quaternary development of drainage network • Relatively level ground related to the main river and drainage lines of the study area	• Occasionally, undercutting of valley-side slopes can lead to formation of small-scale rockfalls and topples. These are commonly related to meander bends cutting into rock slopes • Some relict meander bends can appear to resemble landslide blocks	0.5 landslides per km^2

potential source areas, regression areas and run-out zones, was the most useful for planning and monitoring purposes.

7. DISCUSSION

The three case studies described illustrate how pragmatic approaches, based on the maximum utilisation of geomorphological information and interpretation, have been used to yield the required outputs for planning and decision-making. However, data limitations have served to restrict the extent to which full hazard and risk assessment can be performed.

This discussion commences with a comparison of the approaches adopted by the three case studies (a) in relation to one another and (b) as far as the procedural guidelines contained in Fell et al. (2008) are concerned. The discussion ends with a review of some other published examples and how these have been able to yield assessments of landslide susceptibility, hazard and risk under conditions of varying input data.

7.1 Case Studies

The three case studies described illustrate the methods used to derive outputs that met end-user requirements for engineering and planning purposes. Table 5.5 summarises the approximate resources and programme time needed to derive the required outputs. The greatest investment of the three was made on the Sakhalin oil and gas pipeline study because detailed work was required to evaluate landslide hazard as part of the final design during the construction phase. The project was also financed by a private sector client, whereas the other two case studies were both public sector supported with limited budgets. The Sakhalin study was not able to quantify hazard, but it was able to derive quantitative and analytical surrogates that enabled the required design assessments to be made. The Cyprus land use planning study covered approximately the same area in square kilometres as the infrastructure planning study of Nepal, though the resource provision was approximately double. In the Cyprus case, the number of landslides mapped was significantly more and the end product would become used in its entirety as a means of hazard assessment for reviewing future planning and development proposals. In the Nepal case, the objective was to identify broadly stable corridors within which more detailed studies could later follow, and therefore the required investment

Table 5.5 Comparison of Resource Inputs and Outputs of the Three Case Studies

Case Study	Area Covered (km^2) and No. of Landslides	Outputs	Approximate Stafftime Required (months)	Approximate Programme Time (months)
Nepal rural infrastructure planning	530: susceptibility map; 16: hazard and risk map; No. of landslides, 230	Landslide database, landslide susceptibility map for entire area, hazard and risk (economic loss map for extract)	14	6
Sakhalin oil and gas pipeline slope stability hazard assessment	60; No. of landslides (existing natural, cut slope and potential first-time failures), 860	Landslide database, landslide hazard register, geomorphological field mapping and exposure logging, geological ground models, slope stability checks, as-built records, monitoring/maintenance manual	40	16
Cyprus land use planning	550; No. of landslides, 1800	Landslide database, terrain classification, landslide susceptibility mapping, engineering geological field mapping, ground investigation and lab testing, slope analysis	25	12

in data collection and analysis was less. In all three cases, the outputs were judged to be fit-for-purpose and were approved by the client concerned.

Table 5.6 assesses the levels of investigation undertaken in each of the three case studies against the levels of detail defined in the procedural guidelines contained in Fell et al. (2008). It should be pointed out that the column headings are simplified from the susceptibility, hazard and risk parameters contained in these procedural guidelines.

As far as the Nepal case study is concerned, all activities qualify as *basic* according to the Fell et al. (2008) criteria. This is considered to be fit-for-purpose, bearing in mind the intended outcome of the mapping. The inventory and susceptibility mapping outputs were based on the interpretation of small-scale aerial photographs and field observations, both of which were heavily dependent upon geomorphological interpretation. Run-out predictions were based on the analysis of the landslide inventory data and the mapping of source areas and run-out zones. The use of a test area, independent of the study area, enabled the susceptibility model to be verified. The inability to properly assess probability, and hence risk, without the need for major assumptions, is symptomatic of many applications. Had the study been taken to a more detailed level following the selection of an alignment corridor, hazard would have been evaluated using qualitative and deterministic methods, namely geomorphological and engineering geological mapping, ground investigation and slope analysis, i.e. areal-based statistical methods would probably not have been undertaken. Risk assessment would have been reliant on considerations of factor of safety combined with engineering judgement.

In the Cyprus study, all activities, again, qualify as *basic*. The purpose of the study was to develop an inventory of existing landslides and susceptibility maps to assist in the development of future planning initiatives. Again, lack of data prevented a full hazard and risk assessment from being undertaken. Instead, a qualitative assessment of hazard and risk based on terrain classification enabled the client to consider the basic principles for planning purposes. Risk, of course, cannot be fully assessed until the proposed infrastructure or change in land use is known and its vulnerability assessed, and this information was not available to the study.

In the Sakhalin case, the landslide inventory and susceptibility assessment qualified as *advanced* or *intermediate/advanced* due to the fact that the route corridor for the pipelines had already been selected by the time the work was undertaken. The production of a landslide hazard register for a narrow pipeline corridor was the principal aim of the study at the outset

Table 5.6 Comparison of Case Study Investigations with the Procedural Guidelines in Fell et al. (2008)

Case Study	Qualifying Level of Investigation Against Fell et al. (2008) Benchmarks						
	Scale of Application	**Landslide Inventory**	**Landslide Susceptibility**	**Travel Distance and Velocity**	**Frequency of Movement (Existing Landslides)**	**Probability (First-Time Failures)**	**Vulnerability and Risk**
Nepal rural infrastructure planning	Local/Regional	Basic	Basic with some intermediate	Basic with some intermediate	Basic	Basic with inconclusive outcome	Basic with some intermediate (with some major assumptions)
Sakhalin oil and gas pipeline slope stability	Site	Advanced	Intermediate with some advanced (Hearn et al., 2012)	None	Basic	None, no historical data	Basic with some intermediate based on factor of safety
Cyprus land use planning	Local/Regional	Basic	Basic with some intermediate	Basic with some intermediate	Basic	Basic with inconclusive outcome	Basic

and the activities undertaken were designed and streamlined to achieve this. The qualitative hazard ratings were based on a combination of feature geometry and its relation to pipeline trench and right of way proximity, both in a horizontal and vertical sense, and used factor of safety to assist in the hazard assessment. The geomorphological contributions to this exercise focused on the identification of landslides and their detailed mapping in relation to the pipeline corridor, and particularly with respect to the assessment of regression potential. The approach was a pragmatic one, given the number of features identified and the need to adopt screening that enabled residual hazards to be defined and assessed with focused geotechnical analysis. The resources required to derive these outputs (Table 5.5) reflect the need to devote sufficient time and effort to the modelling of slope stability and the assessment of landslide hazard for engineering purposes. The work extended into and contributed to the construction period, and a greater degree of detailed study and analysis was required over a much smaller area than the other two case studies. The hazard register contained entries for residual hazard from existing landslides and the hazard posed by potential future first-time slope failures. The presentation of these areas both in the form of a register and as a series of maps in conjunction with as-built details provided a very useful platform for the design of slope monitoring and slope inspections during pipeline operation.

This comparison between the three case studies and the procedural guidelines demonstrates the difficulty in achieving levels of analysis and output data that can be classed as anything other than *basic* or *intermediate.* In all three cases, and in the case of projects where even less resources and programme time are available, a pragmatic approach is relied upon to yield the required interpretation for decision-making. There is no formula for this pragmatic approach, but it must be based on maximum utilisation of available information, combined with geological, geomorphological and geotechnical interpretation and judgement to bridge the gap between inadequate data and decision-making. A review of some of the published literature allows these observations to be viewed in terms of what has been achieved in other studies.

7.2 Landslide Hazard and Risk Studies

Many significant advances have been made in the last decade or so, whereby data sets of recorded hazard and vulnerability, or loss, have been utilised in risk assessment. For example, Hardingham et al. (1998),

Malone (1998) and Reeves et al. (1998) described early illustrations of some Hong Kong practice in quantitative risk assessment (QRA), whereas Jaiswal et al. (2010) used landslide records spanning a period of 21 years on the Nilgiri Railway in southern India to develop QRA for hazard and risk management purposes. Nevertheless, although there is an extensive body of literature that describes the concept and procedure of landslide hazard mapping and risk zoning, few published examples actually depict these in quantitative terms. Among the most notable exceptions to this, for example, is the work undertaken by Bonachea et al. (2009) where risk maps showing potential economic losses resulting from damage to infrastructure, buildings and land use were prepared for parts of northern Spain for various landslide frequency scenarios, using vulnerability data collected over 50 years of record. Michael-Leiba et al. (2005) combined elements at risk in order to yield a risk map for planning purposes in the Cairns region of Australia. However, several limitations were noted in the data and methodology adopted, including the judgement-based assessment of vulnerability.

Cascini et al. (2005) described and illustrated a number of landslide and risk maps but noted that significant errors can be made when there is insufficient information available to properly evaluate hazard intensity and probability. Huabin et al. (2005) and Lee (2009) made similar observations with the latter concluding that landslide probability cannot be calculated, and the process of risk assessment is left to judgement, leading to a range of estimates by different specialists even when using the same data. Corominas and Moya (2008) pointed out some of the difficulties associated with the development of a magnitude–frequency record of landslides for probability assessment due to limited data on large magnitude–low-frequency events and the fact that small landslides may become removed from the landscape by erosion.

Glade and Crozier (2005, p. 71) noted that 'if none of the information sources is available, (landslide) impacts to elements at risk have to be estimated based on examples from other regions, or even other processes (e.g. earthquakes and floods)'. In the case of earthquakes, rainstorms and floods, long-term and continuous records enable the frequency, and hence probability, of events with different magnitudes to be ascertained. If it can be demonstrated that significant landsliding is triggered by an earthquake or a rainstorm of a certain intensity or threshold (Caine, 1980; Corominas et al., 2002; Ahrendt and Zuquette, 2003; Dai and Lee, 2003; Guzzetti et al., 2004, 2008; Dahal and Hasegawa, 2008; Jaiswal and van Westen, 2009; Wu and Chen, 2009), then the likelihood of a landslide occurring over a given time period can be approximated through associated probability. However, the required data are commonly unavailable to make these linkages with any degree of certainty. Dai et al. (2002, p. 82) provided a summary of international practice and concluded that 'there are few reliable techniques available for assessing landslide hazard ... (and it is) ... virtually impossible to forecast the location, magnitude and timing of specific future events.

7.3 Landslide Susceptibility Mapping Studies

If quantitative landslide hazard and risk mapping is constrained by a lack of data, then planners and civil engineers may find that it is landslide susceptibility mapping that is able to yield sufficient information to assist in important decision-making. The least robust of these techniques rely on judgement to apply scores to various conditioning or triggering factors in the derivation of a composite susceptibility rating. The most robust are based on the correlation between recorded landslides and mapped variability in controlling factors. Weights-of-evidence analysis is a technique that calculates the weight (level of importance) of each mapped or measured factor based on the presence or absence of landslides within each mapping unit (van Westen et al., 2003; Mathew et al., 2007; Dahal et al., 2008).

Dahl et al. (2010) used mapped landslides to create a susceptibility map based on threshold slope angles for failure in soils developed on basaltic rocks. An independent set of landslide locations was used to validate the model, and it was found to be 69% successful. The authors noted that the accuracy could have been improved had other factors, such as geology, soil depth, slope aspect and land use, been analysed. Wu and Chen (2009) used six factors, including slope angle, geology, vegetation, soil moisture, road development and historical record of landslides to develop a landslide susceptibility threshold rating. This rating was then combined with a rainfall factor that included 24 h total and 10 day antecedent rainfall. The resulting susceptibility map comprised three classes: lower than; slightly above and significantly above the threshold. Validation using an independent set of landslide locations demonstrated a significant relationship with these three classes. Other studies that have included a range of factors in the susceptibility analysis are described, for example, by Dai and Lee (2003), Guinau et al. (2005), Mathew et al. (2007), Dahal et al. (2008), Rossi et al. (2010) and Jimenez-Peralvarez et al. (2011).

The latter study combined terrain units, incorporating gradient, aspect and roughness, with lithology, bedding orientation and land use to develop susceptibility models using linear discriminant analysis, quadratic discriminant analysis, logistic regression and neural network analysis and a data set of landslides recorded between 1941 and 1996. The models were compared against landslide locations that occurred between 1997 and 2005, and the highest prediction rates obtained were in excess of 90%.

The work by Rossi et al. (2010) is particularly noteworthy as it has benefited from a long record of landslide events and has rigorously analysed their distribution using factors that have direct and meaningful relevance to landslide potential. However, some other studies have developed landslide susceptibility maps based on factors that do not have a direct physical relationship with slope failure and issues such as chance relationships and auto-correlation between factors can result. Furthermore, unless the boundaries of mapping units that make up the susceptibility map are based on real variability in controlling factors, significant generalisations and errors can result. Das et al. (2010) compared the outcome of landslide susceptibility mapping using logistic regression analysis with site-based assessments of rock slope stability using rock mass classification systems and kinematic considerations. Although there was significant spatial correlation between the two techniques, due to the generalisations in the statistical method, the susceptibility mapping was found to miss some slopes that were considered to be only marginally stable from a geotechnical perspective.

7.4 Landslide Run-out

This review so far has focused on source area susceptibility. Once a landslide is initiated, it can travel over significant distances thus exposing a much greater area to potential hazard. Methods devised to model landslide run-out are based on empirical, analytical and simulation approaches (Dai et al., 2002) and are described, for example, in Scheidegger (1973), Hsü (1975), Hutchinson (1986), Corominas et al. (1988), Sousa and Voight (1991), Corominas (1993), Evans and Hungr (1993), Hearn (1995a, 2002b, 2004), Hungr (1995), Corominas (1996), Lau and Woods (1997), Evans and King (1998), Hadley et al. (1998), Chen and Lee (2000), Dai et al. (2002), Hungr et al. (2005), Fell et al. (2007), Hürlimann et al. (2008), Fannin and Bowman (2010) and Dahl et al. (2010). The simplest and probably most conservative of the empirical approaches are based on the angle of reach. For example, Dahl et al. (2010)

found that the angle of reach was able to predict 92% of landslide run-out distances. Relationships have been found between run-out length and the volume and mechanism of slope failure (Corominas, 1996), though some studies indicate that only volume becomes significant above a certain threshold (Scheidegger, 1973; Dahl et al., 2010). Difficulties therefore remain in the modelling of run-out with the accuracy and confidence required for planning and engineering purposes, particularly when landslide source areas are also difficult to predict, both in terms of location, size and timing.

7.5 The Contribution of Geomorphology

Geomorphology has been used frequently to provide qualitative hazard and risk assessments for engineering and planning purposes where quantitative assessments are not possible (Baynes and Lee, 1998). Attempts to add numbers to the assessment, primarily through the use of inventories (see, for example, Hearn, 1995b, for mining and road projects, Ko Ko et al., 2004, for railway projects and AGS, 2007, for general practice) add objectivity to the assessment. Maximising the use of engineering geology and geomorphology in these numerical assessments is the only way of ensuring that they can be relied upon to provide robust indicators of hazard and risk.

The largest contribution that the discipline of geomorphology has made to planning and engineering is through mapping landforms and processes. Geomorphological maps have been prepared at national, regional, local and site level for a range of applications and are illustrated, for example, by Anonymous (1972), Brunsden et al. (1975), Demek and Embleton (1978), Verstappen (1983), Varnes (1984), Griffiths (2001), Hearn (1995a, 2001, 2002a,b), Fookes (1997) and Fookes et al. (2005). Detailed geomorphological maps record the morphological details of individual landslides and are commonly produced at scales greater than 1:10,000, i.e. at the local and site-specific zoning scales defined by Fell et al. (2008). These maps assist in the delineation of failed areas, the assessment of failure depth and landslide activity (Hearn and Massey, 2009). They can contribute directly to the development of an engineering geological ground model for stability analysis and the review and design of mitigation works.

Hearn (1995a) used geomorphological mapping to develop a data set of past landslide events and run-out distances. Failure volumes and angle of reach were determined for various failure mechanisms and the results compared with the relationships obtained by Scheidegger (1973), Hsü (1975), Davies (1982), Li (1983), Ikeya (1989) and Nicoletti and Sorriso-Valvo (1991). In the majority of cases, run-out was significantly overestimated by these relationships, and it was concluded that this was most likely due to the fact that the data set contained small-volume landslides; the majority were less than 50,000 m^3. Instead, linear and log regression relationships were developed combining slope geometry in the failure and run-out zones with run-out distance. The analyses were undertaken with the data set first differentiated according to landslide volume and then landslide mechanism. The models were then used to predict run-out distances for potential future first-time failures based on anticipated failure volume and mechanism derived from geomorphological mapping and landslide susceptibility analysis. Mitigation measures were proposed accordingly.

8. CONCLUSIONS

A wide range of published studies describe landslide susceptibility, hazard and risk assessment for research, planning and engineering purposes. Some of these studies benefit from high quality data sets that include landslide event and consequence records that extend over long periods of time, thus allowing landslide hazard and risk to be assessed with reasonable confidence. Unfortunately, in the authors' experience, and as illustrated in much of the published literature, there are usually insufficient data available with which to make these assessments. The work of AGS (2007) and Fell et al. (2008) is to be commended for its progress in the standardisation of procedures but it seems unlikely that, given this lack of necessary data, most attempts at hazard and risk mapping will achieve much beyond the *basic* level of output. Landslide susceptibility mapping may also be difficult to progress beyond the *basic* level of output in many cases.

This general observation needs to be considered in the context of scale. Where small-scale mapping applications are required, the ensuing generalisation may be reasonable and acceptable for the decision-making that has to be made at that scale. By contrast, when dealing with large-scale mapping (and something much more relevant to basic infrastructure, e.g. roads, railways, pipelines and housing), the predictions may be too vague or too inaccurate to be of any real use. In an attempt to resolve this, some studies have resorted to the inclusion of tenuous indicators of landslide potential in the development of susceptibility and hazard models. Others have benefited from detailed and long-period data sets and have been able to develop models that closely reflect landslide initiation and movement, and it is these that offer the proven strategies for future practice (Fell et al., 2008). Unfortunately, lack of suitable data is the norm rather than the exception in many parts of the world and therefore a pragmatic approach is usually required to achieve these outcomes that maximise the combined use of geology, geomorphology and geotechnical assessment.

The case studies presented here were carried out with varying degrees of data availability, data quality, time and budget, and, in each case, the approach adopted allowed the objectives to be realised. Geomorphology, as part of an holistic approach, played a key role in ensuring that the important controls on stability were included in the assessment. Opportunities were taken, wherever possible, to include multiple assessment methods, combining analytical, statistical and judgement-based approaches in order to yield the most representative of outputs. Each case study involved a discussion of the outputs with the end-users to ensure that outputs matched expectations and that decision-making was undertaken in the light of the limitations and uncertainties with the available data. Although a review of some of the published literature illustrates that significant advances have been made, this lack of data remains a key limitation that can only be resolved by the development of long-term data sets of event magnitude and frequency and risk and vulnerability outcomes. Unfortunately, some countries have a much longer way to go in achieving this than others, and the main observations to take forward in this regard are that:

- susceptibility mapping, based on available landslide data sets and geological, geomorphological and geotechnical control on the factor analysis, offers the greatest value to planning and engineering decision making,
- hazard and risk assessment, at the moment at least, is best left to specialist judgement utilising all available data and an holistic approach that maximises geomorphological interpretation,
- end-user liaison is imperative to ensure that outputs match expectations and that the limitations inherent in the assessment are fully recognised and allowed for in decision-making

These conclusions may be obvious to most readers, but they need to be seriously borne in mind when developing and applying landslide assessment for planning and engineering projects under the typical conditions of limited data, limited time and geotechnical uncertainty.

ACKNOWLEDGEMENTS

The authors wish to thank the Ministry of Local Development of Nepal, Sakhalin Energy Investment Company Ltd and the Geological Survey Department of Cyprus for the opportunity to carry out the case studies described and permission to publish the extracts contained herein. The work in Nepal was funded by the UK DFID. The authors would also like to thank D. Petley, I. Hodgson, P. Ward, W. Crick, A. Bhandari, B. Upadhaya and S. Tiwari in relation to the Nepal case study, P. Fookes, C. Morgan, C. Massey, D. Wise, J. Conway, P. Quinlan, M. Ruse, A. Kaldy, G. Pettifer, J. Mitchell, F. Baynes and N. O'Donnell in relation to the Sakhalin case study and M. Ruse, P. Quinlan, J. Mitchell, P. Hobbs (British Geological Survey) and M. Efthymiou and K. Hadjicharalambous (Cyprus GSD) for their contributions to the Cyprus case study. The authors would also like to thank F. Guzzetti, J. Corominas and T. Hunt for their useful comments on the draft paper and K. Jones for assistance with the drawings.

REFERENCES

AGS, 2007. A national landslide risk management framework for Australia. J. News Aust. Geomech. Soc. 42 (1), 182.

Ahrendt, A., Zuquette, L.V., 2003. Triggering factors of landslides in Campos do Jordao city, Brazil. Bull. Eng. Geol. Environ. 62, 231–244.

Aleotti, P., Chowdhury, R., 1999. Landslide hazard assessment: summary review and new perspectives. Bull. Eng. Geol. Environ. 58 (1), 21–44.

Anonymous, 1972. The preparation of maps and plans in terms of engineering geology. Q. J. Eng. Geol. 5, 297–367.

Baynes, F.J., Lee, E.M., 1998. Geomorphology in landslide risk analysis, an interim report. In: Moore, D., Hungr, O. (Eds.), Proceedings of the Eighth International IAEGE Conference, Vancouver, Canada, 21–25 September. A.A. Balkema, Rotterdam, pp. 1129–1136.

Bonachea, J., Remondo, J., Gonzalez-Diaz, A., Diaz de Taran, J.R., Cendrero, A., 2009. Landslide risk modelling: an experience from northern Spain. In: Malet, J.P., Remaitre, A., Bogaard, T. (Eds.), Landslide Processes from Geomorphological Mapping to Dynamic Modelling. Strasbourg European Centre on Geomorphological Hazards (CERG), Strasbourg, pp. 259–264.

Brunsden, D., Doornkamp, J.C., Fookes, P.G., Jones, D.K.C., Kelly, J.M.H., 1975. Large scale geomorphological mapping and highway engineering design. Q. J. Eng. Geol. 8, 227–253.

Caine, N., 1980. The rainfall intensity-duration control of shallow landslides and debris-flows. Geogr. Ann. 62A, 23–27 .

Caine, N., Mool, P.K., 1982. Landslide in the Kolpu Khola Drainage, Middle Mountains, Nepal. Mt. Res. Dev. 2 (2), 157–173.

Cascini, L., Bonnard, C., Corominas, J., Jibson, R., Montero-Olarte, J., 2005. Landslide hazard and risk zoning for urban planning and development. In: Hungr, O., Fell, R., Couture, R., Eberhardt, E. (Eds.), Landslide Risk Management. Proceedings of the International Conference on Landslide Risk Management, Vancouver, Canada. A.A. Balkema, Taylor Francis Group, London, 199–235.

Chen, H., Lee, C.F., 2000. Numerical simulation of debris flows. Canadian Geotechnical Journal 37, 146–160.

Corominas, J., 1993. Spatial prediction of landslides. In: Nemec, J., Nigg, J.M., Siccardi, F. (Eds.), Prediction and Perception of Natural Hazards. Kluwer Academic Publishers, Dordrecht, 125–132.

Corominas, J., 1996. The angle of reach as a mobility index for small and large landslides. Can. Geotech. J. 33 (2), 260–271.

Corominas, J., Moya, J., 2008. A review of assessing landslide frequency for hazard zoning purposes. Eng. Geol. 102, 193–213.

Corominas, J., Penaranda, R., Baeza, C., 1988. Identificacion de factores que condicionan la formacion de movimentos superficiales en los valles altos del Llobregat y Cardener. II Symp Taludes y Laderas Inestables, Andorra La Vella, pp. 195–207.

Corominas, J., Moya, J., Hürlimann, M., 2002. Landslide rainfall triggers in the Spanish eastern Pyrenees. In: Proceedings of the Fourth Plinius Conference on Mediterranean Storms, European Geophysical Society, Mallorca, Spain.

Cruden, D.M., Varnes, D.J., 1996. Landslide types and processes. In: Turner, A.K., Schuster, R.L. (Eds.), Landslides Investigation and Mitigation. Special Report 247 of the Transport Research Board, National Research Council. National Academy Press, Washington, DC, 36–75.

Dahal, R.K., Hasegawa, S., 2008. Representative rainfall thresholds for landslides in the Nepal Himalaya. Geomorphology 100 (3–4), 429–443.

Dahal, R.K., Hasegawa, S., Nonomura, A., Yamanaka, M., Dhakal, S., Paudyal, P., 2008. Predictive modelling of rainfall-induced landslide hazard in the Lesser Himalaya of Nepal based on weights-of-evidence. Geomorphology 102, 496–510 .

Dahl, M.-P.J., Mortensen, L.E., Veihe, A., Jensen, N.H., 2010. A simple qualitative approach for mapping regional landslide susceptibility in the Faroe Islands. Nat. Hazards Earth Syst. Sci. 10, 159–170.

Dai, F.C., Lee, C.F., 2003. A spatiotemporal probabilistic modelling of storm-induced shallow landsliding using aerial photographs and logistic regression. Earth Surf. Processes Landforms 28, 527–545.

Dai, F.C., Lee, C.F., Ngai, Y.Y., 2002. Landslide risk assessment and management. An overview. Eng. Geol. 64, 65–87.

Das, I., Sahoo, S., van Westen, C., Stein, A., Hack, A., 2010. Landslide susceptibility assessment using logistic regression and its comparison with a rock mass classification system, along a road section in the northern Himalayas (India). Geomorphology 114, 627–637.

Davies, T.R.H., 1982. Spreading of rock avalanche debris by mechanical fluidisation. Rock Mech. 15, 9–24.

Demek, J., Embleton, C., 1978. Guide to Medium-Scale Geomorphological Mapping. IGU, Stuttgart.

Evans, N.C., King, J.P., 1998. The Natural Terrain Landslide Study. Debris Avalanche Susceptibility. Technical Note No. TN 1/98. Geotechnical Engineering Office, Hong Kong, 96 pp.

Evans, S.G., Hungr, O., 1993. The assessment of rock fall hazards at the base of talus slopes. Can. Geotech. J. 30, 620–636.

Fannin, R.J., Bowman, E.T., 2010. Debris flows – entrainment, deposition and travel distance. In: Williams, A.L., Pinches, G.M., Chin, C.Y., McMorran, T.J., Massey, C.I. (Eds.), Geologically Active. Proceedings of the Eleventh IAEGE Congress, Auckland, New Zealand, 5–10 September. Taylor and Francis Group, London, pp. 1111–1118.

Fell, R., Glastonbury, J., Hunter, G., 2007. Rapid landslides: the importance of understanding mechanisms and rupture surface mechanics. The Eighth Glossop Lecture. Q. J. Eng. Geol. Hydrogeol. 40, 9–27.

Fell, R., Corominas, J., Bonnard, C., Cascini, L., Leroi, E., Savage, W.Z., 2008. Guidelines for landslide susceptibility, hazard and risk zoning for land use planning. JTC-1 Joint Technical Committee on Landslides and Engineered Slopes. Eng. Geol. 102, 85–111.

Fookes, P.G., 1997. Geology for engineers: the geological model, prediction and performance. The First Glossop Lecture. Q. J. Eng. Geol. 30, 293–424.

Fookes, P.G., Lee, E.M., Milligan, G., 2005. Geomorphology for Engineers. Whittles Publishing, Dunbeath, Caithness, 851 pp.

GFDRR, 2009. Philippines: Typhoons Ondoy and Pepeng. Post-Disaster Needs Assessment. Global Facility for Disaster Reduction and Recovery, Philippines, 26 November 2009, 62 pp.

Giardini, D., 1999. The global seismic hazard assessment program (GSHAP), 1992/1999. Ann. Geofis. 42, 857–876.

Glade, T., Crozier, M.J., 2005. The nature of landslide hazard impact. In: Glade, T., Anderson, M., Crozier, M.J. (Eds.), Landslide Hazard and Risk. John Wiley & Sons, Chichester, 43–74.

Glade, T., Anderson, M., Crozier, M.J. (Eds.), 2005. Landslide Hazard and Risk. John Wiley & Sons, Chichester, , 802 pp.

Griffiths, J.S. (Ed.), 2001. Land Surface Evaluation for Engineering Practice. Geological Society, London, Engineering Geology Special Publications, 18, The Geological Society of London, London.

Guinau, M., Pallas, R., Vilaplana, J.M., 2005. A feasible methodology for landslide susceptibility assessment in developing countries: a case study of NW Nicaragua after Hurricane Mitch. Eng. Geol. 80, 316–327.

Guzzetti, F., Cardinali, M., Reichenbach, P., Cipolla, F., Sebastian, C., Galli, M., Salvati, P., 2004. Landslides triggered by the 23 November 2000 rainfall event in the Imperia Province, Western Liguria, Italy. Eng. Geol. 73, 229–245.

Guzzetti, F., Carrara, A., Cardinali, M., Reichenbach, P., 1999. Landslide hazard evaluation: a review of current techniques and their application in a multi-scale study, Central Italy. Geomorphology 31, 181–216.

Guzzetti, F., Peruccacci, S., Rossi, M., Stark, C.P., 2008. The rainfall intensity-duration control of shallow landslides and debris flows: an update. Landslides 5, 3–17.

Hadjigeorgiou, J., Kyriakou, E., Papanastasiou, P., 2006. A road embankment failure near Pentalia in Southwestern Cyprus. In: International Symposium on Stability of Rock Slopes in Open Pit Mining and Civil Engineering. The South African Institute of Mining and Metallurgy, Cape Town, 343–352.

Hadley, D., Hearn, G.J., Taylor, G.R., 1998. Debris flow assessments for the Foothills Bypass, Hong Kong. In: Li, K.S., Kay, J.N., Ho, K.K.S. (Eds.), Slope Engineering in Hong Kong. Proceedings of the Annual Seminar on Slope Engineering in Hong Kong, Hong Kong Institution of Engineers. A.A. Balkema, Rotterdam, 153–162.

Hammond, R., McCullagh, P.S., 1978. Quantitative Techniques in Geography: An Introduction. Clarendon Press, Oxford, UK.

Hardingham, A.D., Ho, K.K.S., Smallwood, A.R.H., Ditchfield, C.S., 1998. Quantitative risk assessment of landslides: a case history from Hong Kong. In: Li, K.S., Kay, J.N., Ho, K.K.S. (Eds.), Slope Engineering in Hong Kong. Proceedings of the Annual Seminar on Slope Engineering in Hong Kong, Hong Kong Institution of Engineers. A.A. Balkema, Rotterdam, 145–151.

Hart, A.B., Ruse, M.E., Hobbs, P.R.N., Efthymiou, M., Hadjicharalambous, K., 2010. Development of a landslide inventory to assess landslide hazard in Paphos District, Cyprus. In: Williams, A.L., Pinches, G.M., Chin, C.Y., McMorran, T.J., Massey, C.I. (Eds.), Geologically Active. Proceedings of the Eleventh IAEGE Congress, Auckland, New Zealand, September. Taylor and Francis Group, London, 229–239.

Hart, J., Hearn, G., Chant, C., 2002. Engineering on the precipice: mountain road rehabilitation in the Philippines. Q. J. Eng. Geol. Hydrogeol. 35, 223–231.

Hearn, G.J., 1995a. Engineering geomorphological mapping and opencast mining in unstable mountains: a case study. Trans. Inst. Min. Metall. A104, A1–A18.

Hearn, G.J., 1995b. Landslide and erosion hazard mapping at Ok Tedi copper mine, Papua New Guinea. Q. J. Eng. Geol. 28, 47–60.

Hearn, G.J., 2001. Low-cost road construction and rehabilitation in unstable mountain areas. In: Griffiths, J.S. (Ed.), Land Surface Evaluation in Engineering Practice, vol. 18. Geological Society of London Special Publication, London, 135–141.

Hearn, G.J., 2002a. Engineering geomorphology for road design in unstable mountainous areas: lessons learnt after 25 years in Nepal. Q. J. Eng. Geol. Hydrogeol. 35, 143–154.

Hearn, G.J., 2002b. Natural terrain hazard assessment: the art of applied science. In: Proceedings of the Conference on Natural Terrain a Constraint on Development? Institution of Mining and Metallurgy, Hong Kong Branch, 39–60.

Hearn, G.J., 2004. The role of geology in landslide risk assessment for civil engineering purposes. In: Jardine, R.J., Potts, D.M., Higgins, K.G. (Eds.), Advances in Geotechnical Engineering. Institution of Civil Engineers, Thomas Telford, London. The Skempton Conference, vol. 2. 1316–1329.

Hearn, G.J., 2011. Slope Engineering for Mountain Roads. Geological Society of London Special Publication, No 24, London, in press.

Hearn, G.J., Massey, C.I., 2009. Engineering geology in the management of roadside slope failures: contributions to best practice from Bhutan and Ethiopia. Q. J. Eng. Geol. Hydrogeol. 42 (4), 511–528.

Hearn, G.J., Hart, A.B., Morgan, C., Wise, D., O'Donnell, N., 2012. Assessing the potential for future first-time slope failures to impact the oil and gas pipeline corridor through the Makarov Mountains, Sakhalin Island, Russia. Q. J. Eng. Geol. Hydrogeol. in press.

Ho, K.K.S., Lau, J.W.C., 2010. Learning from slope failures. Q. J. Eng. Geol. Hydrogeol. 43 (1), 33–68.

Hsü, K.J., 1975. Catastrophic debris streams (sturzstroms) generated by rockfalls. Geol. Soc. Am. Bull. 86, 129–140.

Huabin, W., Gangjun, L., Weiya, X., Gonghui, W., 2005. GIS-based landslide hazard assessment: an overview. Prog. Phys. Geogr. 29 (4), 548–567.

Hungr, O., 1995. A model for the runout analysis of rapid flow slides, debris flows and avalanches. Can. Geotech. J. 32, 610–623.

Hungr, O., Corominas, J., Eberhardt, E., 2005. Estimating landslide motion mechanism, travel distance and velocity. In: Hungr, O., Fell, R., Couture, R., Eherhardt, E. (Eds.), Landslide Risk Management. Taylor and Francis Group, London, 99–128.

Hürlimann, M., Rickenmann, D., Medina, V., Bateman, A., 2008. Evaluation of approaches to calculate debris-flow parameters for hazard assessment. Eng. Geol. 102, 153–163.

Hutchinson, J.N., 1986. A sliding-consolidation model for flowslides. Canadian Geotechnical Journal 23, 115–126.

Ikeya, H., 1989. Debris flow and its countermeasures in Japan. Bull. Int. Assoc. Eng. Geol. 40, 15–33.

Jaiswal, P., van Westen, C.J., 2009. Estimating temporal probability for landslide initiation along transportation routes based on rainfall thresholds. Geomorphology: an international journal of pure and applied geomorphology, 12, 1-2, 96–105.

Jaiswal, P., van Westen, C.J., Jetten, V., 2010. Quantitative assessment of landslide risk along transportation lines in southern India. In: Williams, A.L., Pinches, G.M., Chin, C.Y., McMorran, T.J., Massey, C.I. (Eds.), Geologically Active. Proceedings of the Eleventh IAEGE Congress, Auckland, New Zealand, 5–10 September. Taylor and Francis Group, London, 1185–1193.

Jimenez-Peralvarez, J.D., Irigaray, C., Hamdouni, R.El., Chacon, J., 2011. Landslide susceptibility mapping in a semi-arid mountain environment: an example from the southern slopes of the Sierra Nevada (Granada, Spain). Bulletin of Engineering Geology and the Environment 70 (22), 265–278.

Keefer, D.K., 1984. Landslides caused by earthquakes. Geol. Soc. Am. Bull. 95, 406–421.

Ko Ko, C., Flentje, P., Chowdhury, R., 2004. Landslide qualitative hazard and risk assessment method and its reliability. Bull. Eng. Geol. Environ. 63, 149–165.

Lau, K.C., Woods, N.W., 1997. Review of methods for predicting the travel distance of debris from landslides on natural terrain. Technical Note, TH 7/97. Geotechnical Engineering Office, Hong Kong.

Lee, E.M., 2009. Landslide risk assessment: the challenge of estimating the probability of landsliding. Q. J. Eng. Geol. Hydrogeol. 42 (4), 445–458.

Lee, E.M., Jones, D.K.C., 2004. Landslide Risk Assessment. Thomas Telford, London.

Li, T., 1983. A mathematical model for predicting the extent of a major rock fall. Z. Geomorphol. NF 27, 473–482.

Malone, A.W., 1998. Risk management for slope safety in Hong Kong. In: Li, K.S., Kay, J.N., Ho, K.K.S. (Eds.), Slope Engineering in Hong Kong. Proceedings of the Annual Seminar on Slope Engineering in Hong Kong, Hong Kong Institution of Engineers. A.A. Balkema, Rotterdam, 3–17.

Mathew, J., Jha, V.K., Rawat, G.S., 2007. Weights-of-evidence modelling for landslide hazard zonation mapping in part of Bhagirathi valley, Uttarakhand. Curr. Sci. 92 (5), 628–638.

Michael-Leiba, M., Baynes, F., Scott, G., Granger, K., 2005. Quantitative landslide risk assessment of Cairns, Australia. In: Glade, T., Anderson, M., Crozier, M.J. (Eds.), Landslide Hazard and Risk. John Wiley & Sons, Chichester, 621–642.

Nicoletti, P.G., Sorriso-Valvo, M., 1991. Geomorphic controls of the shape and mobility of rock avalanches. Geol. Soc. Am. Bull. 103, 1365–1373.

Northmore, K.J., Charalambous, M., Hobbs, P.R.N., Petrides, G., 1986. Engineering geology of the Kannaviou, Melange and Mamonia Complex formations – Phiti/Statos area, SW Cyprus: engineering geology of cohesive soils associated with ophiolites, with particular reference to Cyprus. Report of the EGARP Research Group British Geological Survey, No. EGARP-KW/86/4; Report of the Geological Survey Department of Cyprus, No. G/EG/15.

Northmore, K.J., Hobbs, P.R.N., Charalambous, M., Petrides, G., 1988. Complex landslides in the Kannaviou Melange and Mamonia formation of South-West Cyprus. In: Bonnard, C. (Ed.), Landslides Glissements de Terrain, Proceedings of the Fifth International Symposium on Landslides, vol. 1. A.A. Balkema Publishers, Brookfield, WI, 263–268.

Pantazis, T.M., 1969. Landslide in Cyprus. Geological Survey Department, Ministry of Commerce and Industry, Bulletin No. 4. 1–20.

Petley, D.N., 2010. On the impact of climate change and population growth on the occurrence of fatal landslides in South, East and SE Asia. Q. J. Eng. Geol. Hydrogeol. 43 (4), 487–496.

Reeves, A., Chan, H.C., Lam, K.C., 1998. Preliminary quantitative risk assessment of boulder falls in Hong Kong. In: Li, K.S., Kay, J.N., Ho, K.K.S. (Eds.), Slope Engineering in Hong Kong. Proceedings of the Annual Seminar on Slope Engineering in Hong Kong, Hong Kong Institution of Engineers. A.A. Balkema, Rotterdam, 185–191.

Rossi, M., Guzzetti, F., Reichenbach, P., Mondini, A.C., Peruccacci, S., 2010. Optimal landslide susceptibility zonation based on multiple forecasts. Geomorphology 114, 129–142.

Scheidegger, A.E., 1973. On the prediction of the reach and velocity of catastrophic landslides. Rock Mech. 5, 231–236.

Sousa, J., Voight, B., 1991. Continuum simulation of flow failures. Geotechnique 41, 515–535.

Turner, A.K., Schuster, R.L. (Eds.), 1996. Landslides Investigation and Mitigation. Special Report 247. Transport Research Board, National Research Council, National Academy Press, Washington, DC.

van Westen, C.J., Rengers, N., Soeters, R., 2003. Use of geomorphological information in indirect landslide susceptibility assessment. Natural hazards: journal of the international society for the prevention and mitigation of natural hazards, 30° 3, 399–419.

Varnes, D.J., 1984. Landslide Hazard Zonation: a Review of Principles and Practice. International Association of Engineering Geology, Commission on Landslides and Other Mass Movements on Slopes, UNESCO, Paris.

Verstappen, H.T., 1983. Applied Geomorphology: Geomorphological Surveys for Environmental Development. Elsevier, Amsterdam.

WP/WLI, 1993. A suggested method for describing the activity of a landslide. UNESCO Working Party on World Landslide Inventory. Bull. Int. Assoc. Eng. Geol. 47, 53–57.

Wu, C.-H., Chen, S.-C., 2009. Determining landslide susceptibility in Central Taiwan from rainfall and six site factors using the analytical hierarchy approach. Geomorphology 112, 190–204.

SECTION 2

Techniques in Applied Geomorphological Mapping

CHAPTER SIX

Geomorphological Field Mapping

Jasper Knight[a], Wishart A. Mitchell[b] and James Rose[c,d]
[a]School of Geography, Archaeology and Environmental Studies, University of the Witwatersrand, Johannesburg, South Africa
[b]Department of Geography, Durham University, Durham, UK
[c]Department of Geography, Royal Holloway University of London, Egham, Surrey, UK
[d]British Geological Survey, Keyworth, Nottingham, UK

Contents

1. INTRODUCTION

This chapter deals with the techniques and methodology of geomorphological field mapping. Such mapping serves both as a means of collecting field observations and in generating and organising a spatial database, such as through a geographic information system (GIS) that reflects the landform distribution in a specific area from which morphometric and other properties can be derived (Demek and Embleton, 1978; St Onge, 1981; Gardiner and Dackombe, 1983). This means that geomorphological field mapping has to be purposeful and with a clearly defined and articulated set of aims and objectives. This includes the scale of investigation, techniques to be used and the types of landforms that are the focus of the project. It is also important at the outset to distinguish between morphological and geomorphological mapping as the two methods have

Developments in Earth Surface Processes, Volume 15
ISSN: 0928-2025, DOI: 10.1016/B978-0-444-53446-0.00006-9

fundamentally different purposes, although they both involve representation of surface form.

Morphological mapping sets out to describe changes of slope character across a land surface by identifying inflections and sharp breaks of slope, describing convex or concave slope profiles and quantifying these by measuring slope attributes such as angle and direction of maximum surface slope (Waters, 1958; Savigear, 1965; Crofts, 1981). Such breaks of slope are the fundamental tools by which a landscape can be described in terms of the morphological attributes that make it up. The first stage of describing these attributes is to identify, map and interpret slope elements. Morphological mapping is a tool that provides a graphical representation of the form of a terrain without any genetic implications.

Geomorphological mapping, in contrast, seeks to identify, interpret and represent the landforms according their form (morphology) and formational processes (Hubbard and Glasser, 2005). Geomorphological mapping records not only the nature of the individual landforms observed in a landscape but also the materials of which they are composed and an indication of the process—response geomorphic systems associated with their formation (Cooke and Doornkamp, 1974; St Onge, 1981; Klimaszewski, 1990; Rhoads and Thorn, 1996). Although large-scale (1:10,000 or 1:25,000) geomorphological field mapping has long been a fundamental means of data collection in the geological sciences (Barsch and Liedtke, 1980), production of specific geomorphological maps has had less emphasis than the mapping of superficial deposits. Thus many studies are only concerned with producing low-resolution geomorphological maps with little attempt to accurately record landforms in terms of detailed morphology and spatial distribution. In some cases, this has reflected the lack of detailed topographic base maps available, which are a prerequisite for high-quality field mapping, although experienced mappers can produce detailed geomorphological maps using basic surveying procedures.

Undertaken correctly, geomorphological field mapping can distinguish between description and interpretation. Morphological maps are a visual representation of spatial patterns of breaks of slope, and as such they build together to delimit the land's form (relief) rather than individual landforms. Geomorphological maps, by contrast, result from the correct genetic interpretation of specific landforms and so have wider application to environment management and planning. However, morphological mapping alone does not provide the necessary conditions required for accurate interpretation because of the naturally variable morphological range of all landform types and due to equifinality (Rhoads and Thorn,

1996). Accurate interpretation also depends on supporting data, including topography, geology and sedimentology.

Production of comprehensive geomorphological maps has been developed in a number of European countries as part of national surveys (Demek and Embleton, 1978; Barsch and Liedtke, 1980; Ten Kate, 1983; Klimaszewski, 1990). However, geomorphological maps can also be drawn for specific research projects concentrating on thematic landforms. In the United Kingdom, for example, geomorphological mapping for the purpose of identifying the distribution of late Pleistocene glacial landforms became an important research tool from the 1960s onwards with the greater availability of aerial photographs and topographic maps. This can be exemplified in the work of Sissons (1972, 1974, 1977a,b, 1979, 1980) who mapped glacial and periglacial landforms to allow reconstructions of former glaciers and ice retreat patterns in the Scottish Highlands. The use of air photos in geomorphological mapping developed mainly in the 1950s and 1960s, where landforms were mapped for spatial patterns and morphometric properties and based on a spatial scale of around 1:10,000 (Svensson, 1964; Welch and Howarth, 1968). Satellite-derived digital data, from the 1970s onwards, enabled more remote areas to be mapped, higher resolution data to be obtained on different spatial scales and repeat-pass imagery used to examine landscape change over time (Welby, 1976). More recently, the advent of higher resolution elevational data of increasingly smaller grid size to produce high-resolution digital elevation models (DEMs), such as NextMap and LiDAR/InSAR (see Oguchi et al., 2011 for discussion), means that resultant DEMs can be used to produce high-quality geomorphological maps more quickly than by field survey. Such remote mapping alone, however, cannot substitute for the experience and outcomes of field mapping, and remote observations should be ground-truthed in test areas. Many recent papers also use GIS software presentation and analysis to produce geomorphological maps in different physical environments (Greenwood and Clark, 2010; Clark et al., 2011). However, it has been shown that a combination of NextMap DEM, GIS and field survey produces the best maps with the highest ground truth (*cf.* Smith et al., 2006; Gustavsson et al., 2006, 2008). Thus field mapping must remain an important technique for geomorphology allowing practitioners to gain real-world experience.

An important paradigm within geomorphological field mapping is the concept of landsystems, whose origin lies in large-scale geomorphological and soils mapping programmes undertaken across various terrains particularly in the Australian CSIRO reports (Crofts, 1981; Eyles, 1983; Evans,

2003). Landsystems refer to integrated landform-sediment assemblages that occur within discrete geomorphic systems and can be used to explain morphogenetic relationships between distinctive constructional landforms and sediments and thus allowing interpretation of depositional environments (Eyles, 1983; Evans, 2003). Use of landsystems' concepts provides an interpretive context for geomorphological field mapping and facilitates mapped landforms to be linked together more effectively. This can also be achieved by consideration of supporting data including bedrock geology and structure, sediments, lithostratigraphy, radiometric dating, soils, ecology and archaeology. A landsystems approach tends to lead to a field mapping procedure that maps all landscape elements in the region of interest. This is because all these landscape elements can potentially be linked to one another through the landsystems approach (Evans et al., 2009). However, most geomorphological mapping campaigns are selective or targeted, in-as-much as they are focused on mapping only particular features, not everything, within the region of interest. For example, this includes mapping of landslides (Jarman, 2007) and moraines (Lukas and Lukas, 2006) and is discussed in detail in this chapter.

This chapter considers the major procedures and protocols of geomorphological field mapping. We first describe the context and methodology of geomorphological field mapping and their application to different depositional environments. We focus specifically on environments affected by glacial, fluvial and mass movement processes in upland terrain. We focus on these environments because relationships are linked in upland terrain between climate change and geomorphic evolution, and mapping and monitoring changes in upland terrain are important for evaluating geohazards and quantifying environmental resources. This can also be facilitated by the use of DEMs that are based on high-resolution topographic data from satellites and displayed within a GIS. Although described with respect to upland terrain, these methods are valid for other landscapes including lowlands and coasts.

2. PROCEDURES AND PROTOCOLS OF GEOMORPHOLOGICAL FIELD MAPPING

Geomorphological mapping, from either field or remotely sensed observations or a combination of both, involves two sequential stages.

First, morphological features are delimited using standard morphological mapping symbols. Second, the features thus delimited are interpreted with respect to their origin, environmental significance and spatial relationships to one another. This refers specifically to geomorphological mapping. These two stages are described below in more detail.

Morphological mapping is based upon recording the outline shape of morphological features, usually delimited by a basal enclosing convex and concave breaks of slope that enables one morphological form to be distinguished from an adjacent one, using a set of standardised and commonly accepted morphological symbols that are unambiguous, clear and reproducible (Figure 6.1). The symbols most commonly used in morphological field mapping are based upon those of Savigear (1965) and focus on identifying the types of slope break which, when integrated together, delimit the outer margins of individual landforms (Waters, 1958). The use of these symbols is somewhat subjective: there are no definitive rules as to what an angular or a smooth break of slope looks like, to what extent small (<1 m high) undulations in the landscape can or should be mapped and where a break of slope ceases to be present. The use of these symbols is also dependent on spatial scale, such that features that are significant or mappable locally may not be significant or mappable regionally. Contour information is generally a useful guide, although cannot substitute, for morphological mapping (Elvhage, 1980); as topographic maps become more detailed and accurate, geomorphological mapping can be generally enhanced (Rose and Letzer, 1975; Elvhage, 1980). Use of remote sensing data and GIS in computer-based mapping commonly means that morphological mapping symbols are not used, but basal outlines of landforms and their crestlines are generally marked with continuous lines (Gustavsson et al., 2006).

Although morphological mapping generally employs standard mapping symbols that are underpinned by an identification of breaks of slope, geomorphological mapping seeks to identify particular landforms. As a result, geomorphological mapping symbols are more extensive and diverse, which reflects their different purpose. Geomorphological mapping is concerned with delimiting different landforms within a formational classification associated with process—response systems (Verstappen, 2011; Otto et al., 2011). Generally, landform margins coincide with significant breaks of slope, but this is not always the case in practice. As a result, geomorphological mapping symbols are interpretative rather than solely descriptive. Examples of typical symbols that can be used in mapping of glaciated

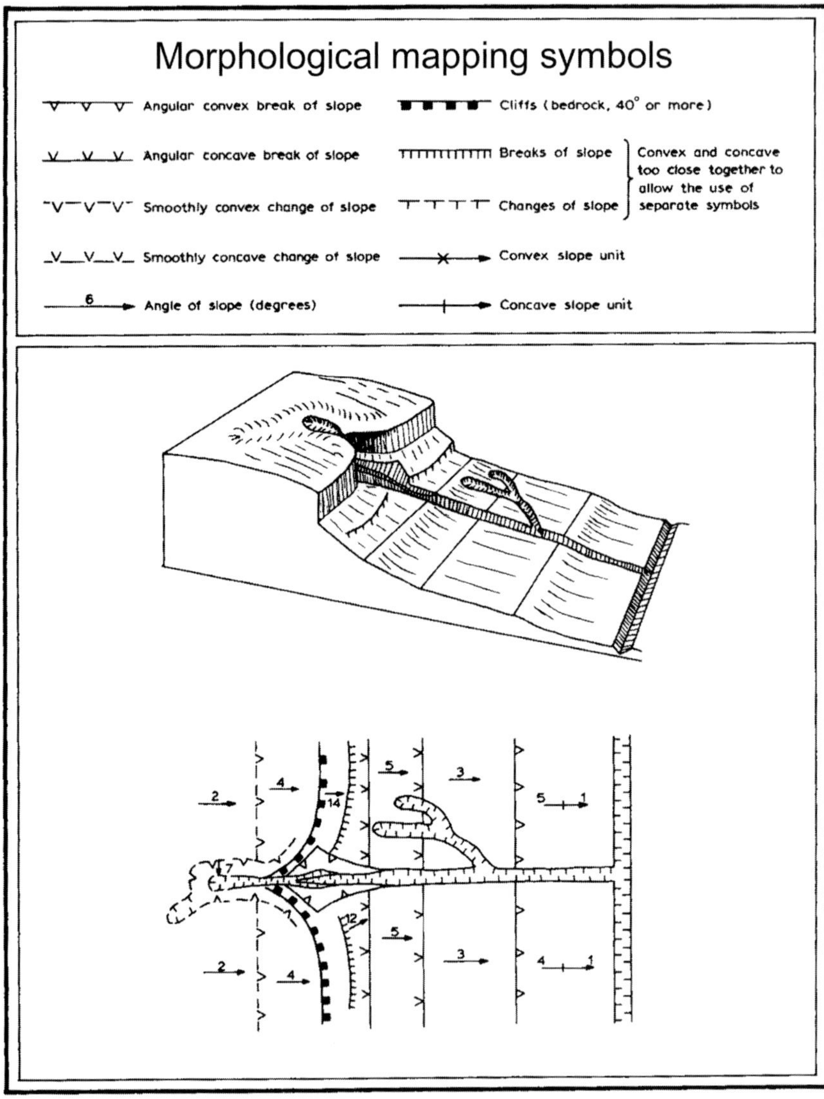

Figure 6.1 Basic morphological mapping symbols. *From Cooke and Doornkamp (1974).*

terrain are shown in Figure 6.2, and an example of these symbols applied to an area of upper Swaledale, northwest England (Rose, 1980) is shown in Figure 6.3. In addition, it is important to recognise that different researchers may use different symbol sets as well as different shadings and

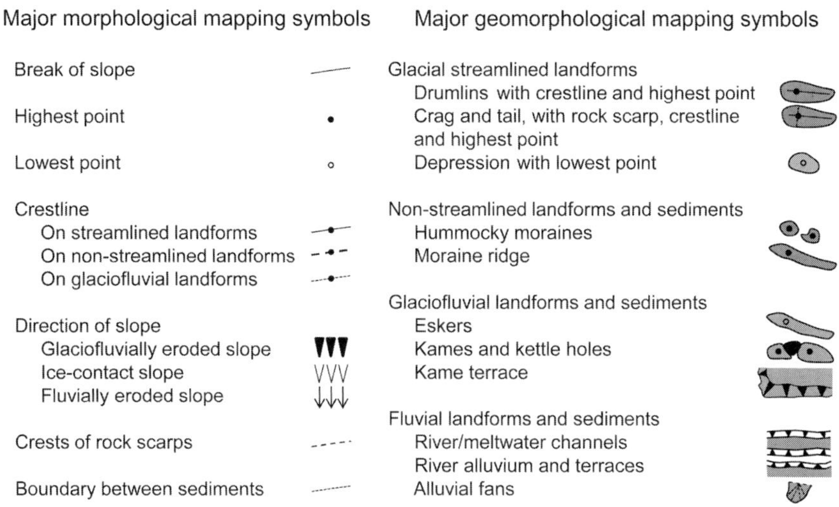

Figure 6.2 Typical morphological mapping symbols (left) and examples of geomorphological mapping symbols used in upland terrain (right).

colours. It is also the case that maps and mapping symbols employed in the field may be different to those used on a final, published map. In all cases, the technique of field mapping is most effective in regions where the landforms have a discrete recognisable shape and have not been significantly modified by subsequent processes. In the United Kingdom, it has been most effectively applied to areas within the margins of the last ice sheet, such as the drumlin areas of northern England (Rose and Letzer, 1977; Mitchell, 1994; Mitchell and Riley, 2006) or the glacial landforms associated with Loch Lomond Stadial glaciers and ice caps within the British mountains (Sissons, 1974, 1979, 1980).

The process of field mapping at scales of 1:3000 to 1:25,000 (most commonly at 1:10,000 scale) is laborious and time consuming (Crofts, 1981; Mitchell, 1991a,b), requiring detailed examination of the landscape under investigation by walking over all the ground and viewing landforms from several directions. A typical workflow model for field mapping is shown in Table 6.1, which identifies the key tasks to be undertaken before, during and after the field mapping period. During mapping, breaks of slope (and landform margins) are mapped by standing on them and traversing them to minimise inconsistencies that can arise due to a perspective gained from only observing the landform from one point, thereby leading to marking inaccurate boundaries on the field slip leading

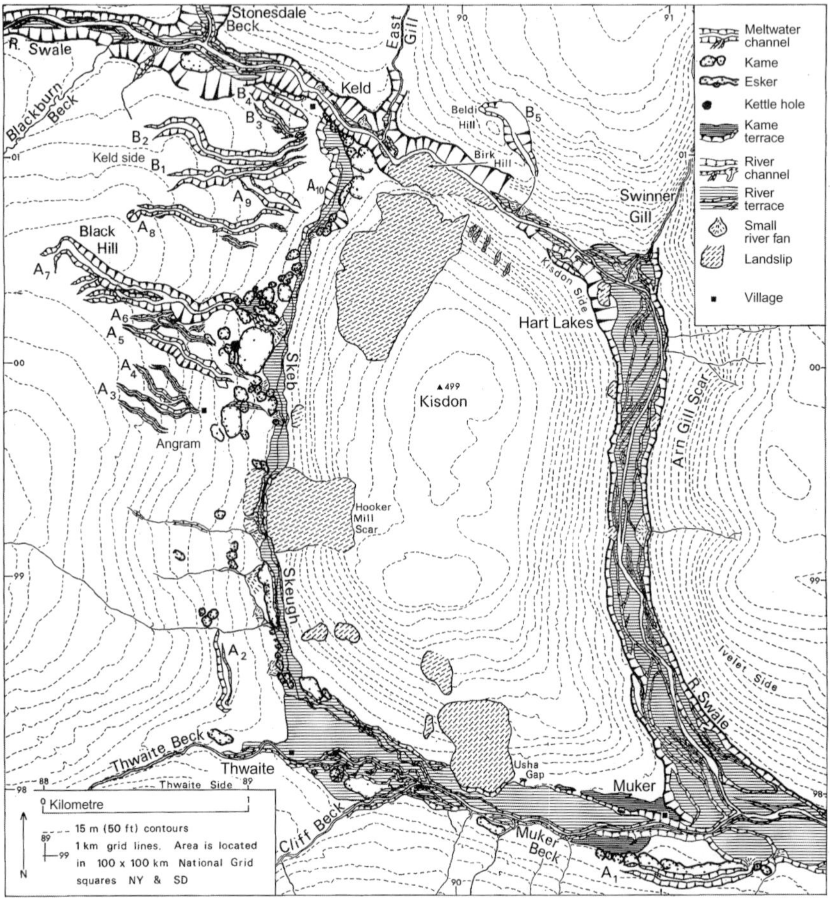

Figure 6.3 An example of geomorphological mapping in part of a glaciated upland region, Kisdon, upper Swaledale, Lake District, northwest England. *From Rose (1980).*

to distortion and poorly defined landforms (Mitchell, 1991b; Smith et al., 2006; Rose and Smith, 2008). Field mapping and plotting of breaks of slope can also be facilitated by the use of a global positioning system (GPS) which can provide digital data suitable for input into a GIS (Dykes, 2008). In order to achieve this, GPS waypoints need to be taken as the field mapper is actually standing on the break of slope or landform margin, which may not be possible in all cases. Some technological problems also arise in using a GPS in areas of woodland or high relief where interference reduces the accuracy of the signal.

Table 6.1 Table Showing a Workflow Model for Undertaking Geomorphological Field Mapping

Time Period	Activity
Pre-mapping	• Identify the geographical region of interest • Identify and articulate the purpose or goal of mapping • Identify and obtain remote sensing data, including topographic survey data, stereo air photos, satellite imagery, topographic maps and DEMs • Design and create a GIS database using digital and digitised remote sensing data • Identify and articulate the field mapping protocol to be used, including the purpose of field mapping • Map major morphological forms using remotely sensed data as indicative tools • Create paper field maps at a suitable scale for field mapping (1:10,000 or 1:5000 scale) • Obtain permission for access to the mapping region, where necessary • Conduct a risk assessment for the planned mapping activities • Ensure whether appropriate information on weather, tide times and so on is available
During mapping	• Conduct field mapping following the agreed protocol, including walking the area effectively, using morphological mapping symbols, confirming any breaks of slope and landforms identified using the remote sensing • Use of hand-held GPS to mark tracks or waypoints • Write notes and take photos, which should be positioned using GPS • Adhere to health and safety issues and/or update the risk assessment
Post-mapping	• Download and integrate GPS data with the existing GIS database • Compare field and remote sensing mapping data in order to validate remotely sensed observations • Write up notes, integrate written notes and field photos to locations within the GIS • Produce a final geomorphological map • Draw final geomorphological map, using analogue or digital cartographic symbols • Write/present (digital) explanatory notes accompanying the map • Apply geomorphological map output to issues in identifying and interpreting landscape patterns, identifying geohazards and considering the sensitivity of relict landscapes to external forcing

Not all possible activities are shown and not all steps shown here are appropriate in all situations.

In field mapping, a potential problem for the mapping process, and the subsequent interpretation of such maps, is the small number of identifiable geometric forms that occur within a landscape. Ridges, mounds and hollows together with distinctive linear breaks of slope that define different slope facets within a landscape can occur in a number of distinctive environmental settings and do not necessarily reflect the operation of specific process–response systems. Hence, interpretation of the morphology can only be concluded where distinctive forms occur in juxtaposition allowing an interpretation to be proposed. In some situations this is far from simple, with polygenic constructional ridges of similar morphology occurring in similar upland environments being wrongly interpreted, which can have far-reaching consequences. A good example of this is ongoing scientific discussion between ridges that have been interpreted as moraines, protalus ramparts and landslides (Mitchell, 1991b,c, 1996; Shakesby and Matthews, 1996; Shakesby, 1997; Wilson, 2004, 2009). This means that accurate mapping and interpretation of complex landscapes are facilitated by expert knowledge, experience of the field mapper and where exposures of subsurface sediments are present.

2.1 Geomorphological Mapping in Upland Terrain

There is a long history of geomorphological mapping in upland terrain affected by glacial, fluvial and mass movement processes. For example, most early studies in glacial landscapes simply stippled or shaded areas where moraines or glacial lake deposits are located (Reade, 1893; Charlesworth, 1928, 1929). Drumlins and eskers were mapped in particular, largely because their upstanding nature and clear margins are easily defined (Sollas, 1896; Dryer, 1901; Wright, 1912; Fairchild, 1929). More recently, geomorphological field mapping has focused on more complex glaciated landscapes, including mountains (Mitchell et al., 2007; Sahlin and Glasser, 2008), piedmonts/mountain valleys (Mitchell, 1994; Mitchell and Riley, 2006; Rose and Smith, 2008) and glacier forefields (Kjaer et al., 2008; Evans et al., 2009), where landforms are morphologically diverse, may be superimposed and record several climatic phases.

The process and outcomes of geomorphological field mapping are best demonstrated in upland terrain for two main reasons. First, many constructional landforms in these areas were dominantly shaped during or following the late Pleistocene glaciation, and as such the landforms are geomorphically fresh and of generally high relief. This tends to make morphological mapping easier and allows accurate location of breaks of slope and landform interpretation. Second, the dominant role of glacial, periglacial and other upland geomorphological processes in these landscapes means that these landforms are well developed and have not been significantly modified by other processes. Landscapes whose landforms are dominated by a single formational environment or set of processes tend to be more easily interpreted than landscapes that have been formed over long time periods, that are palimpsest, or that have been affected by multiple climate cycles and concomitant changes in formational environment. However, many upland landscapes generally exhibit a strong geologic, structural and topographic control, and so accurate geomorphological mapping can help distinguish between these different factors in their influence on the evolution of their component landforms.

3. EXAMPLES OF GEOMORPHOLOGICAL FIELD MAPPING IN UPLAND TERRAIN

We here describe examples of geomorphological field mapping in upland terrain, focusing on glacial, fluvial and mass movement processes. These examples illustrate the range of geomorphological processes that are present in upland terrain. (Other processes such as periglacial processes are also present, but are not considered in detail here.) The interplay between different processes in upland terrain results in the formation of different landforms and the juxtaposition of landforms of different origins.

3.1 Landforms that Result from Glacial Processes

Many upland terrains at the present time and in the recent past have been glaciated. Furthermore, glacierised catchments are significantly different to adjacent non-glaciated catchments in terms of their relief, sediment dynamics and hydrological regime (Bartsch et al., 2009). In particular, the glacial processes that contribute to substrate erosion and deposition, and patterns and processes of ice retreat, are of greatest significance in upland terrain. This is because small valley glaciers and ice caps are sensitive to climate change, evidence for which can be seen in mapped patterns of their component landforms. Drumlins and moraines are described here because they are common features of glaciated landscapes.

Drumlins have been extensively mapped in the field (Evans et al., 2005). The process of mapping in drumlin landscapes is relatively straightforward since drumlin outer margins are generally clearly delimited by a concave break of slope that commonly coincides with field boundaries. Accurate mapping of drumlin margins is important because it enables length, width, area and shape properties to be calculated (Chorley, 1959; Reed et al., 1962; Smalley and Unwin, 1968; Rose and Letzer, 1975). These are very useful measures that reflect the dynamics of the overlying glacier (Hill, 1973; Mitchell, 1994). For example, drumlins characteristically become smaller and more closely spaced nearer the ice margin as a result of a decrease in driving stress (Trenhaile, 1975; Karczewski, 1976; Aario, 1977; Smalley and Warburton, 1994). Examples from early studies in drumlin field mapping during the early twentieth century were described by Charlesworth (1957). More complex drumlin forms have been identified in areas of upland relief where drumlin formation on slopes has altered the simple planform (Figure 6.4; Mitchell, 1991a,b, 1994). Where drumlins are located on a hill flank (Figure 6.4a), their crests lie parallel to the slope and are located on the upslope side of the drumlin. Where drumlins are located on flatter terrain (Figure 6.4b), drumlin forms are generally better developed but more complex superimposed and cross-cutting morphologies may be present especially where drift thickness is greater (Rose and Letzer, 1977; Mitchell, 1994; Knight, 1997, 2010). Drumlins can be superimposed, aligned *en echelon*, or be composed of several geomorphological elements that are fused together (Figure 6.5). This means that accurate and meaningful measurements of geometric properties cannot be so easily made. Superimposed, cross-cutting and overprinted bedform patterns are useful, however, because they characteristically reflect successive ice flow stages with different directional components, which means that such geomorphic patterns commonly have high interpretive power (Rose and Letzer, 1977; Boots and Burns, 1984; Hättestrand et al., 1999; Mitchell and Riley, 2006). Mapping where fluvial erosion and undercutting have influenced drumlin shape also helps identify areas of potential land surface instability and geohazards. Geomorphological field mapping has also been a significant tool in the reinterpretation of subglacial landform patterns (Rose and Smith, 2008). For example, in north-central Ireland nested patterns of elongate ridges were previously interpreted as ice-marginal moraines, reflecting stages of ice retreat (Charlesworth, 1924). More detailed geomorphological mapping, including identifying areas of surface streamlining, shows

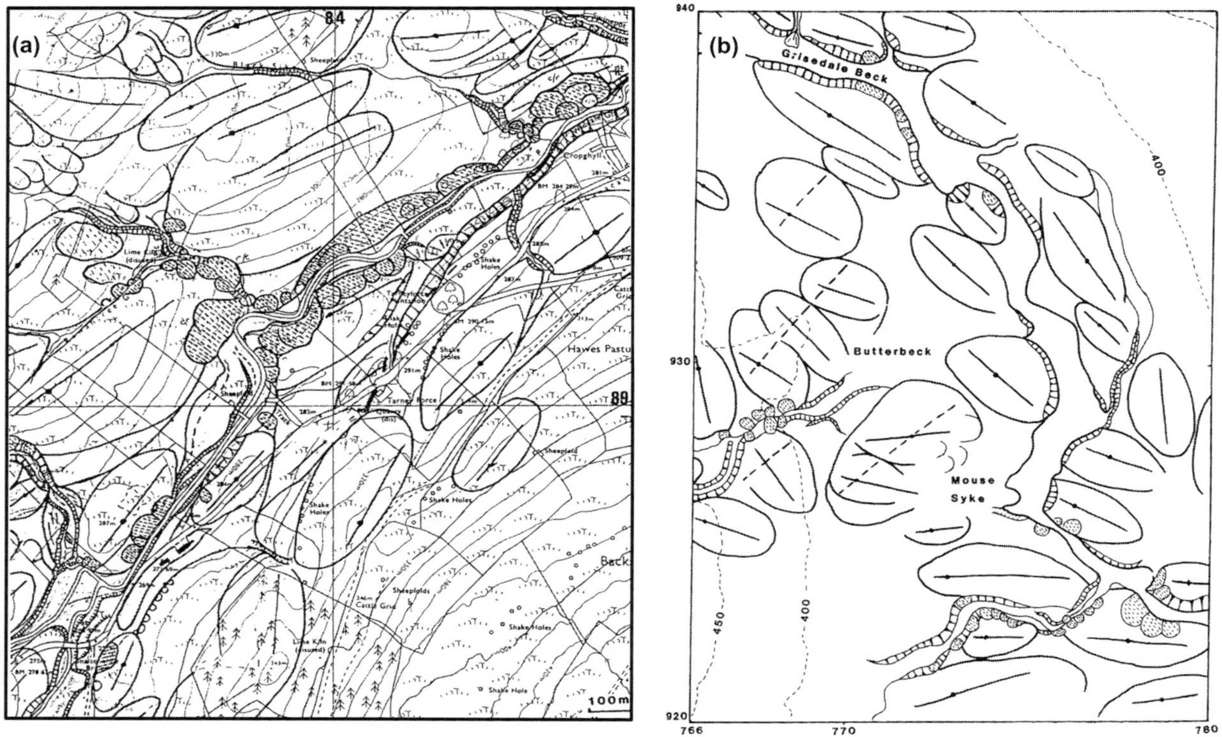

Figure 6.4 Examples of drumlin mapping in different landscape settings, Lake District, northwest England (mapping by W.A. Mitchell). (a) Copy of a field slip showing geomorphological mapping in mid-Widdale. Drumlins are located along hill flanks, and drumlins around river margins show fluvial erosion and slope failure. (b) Geomorphological mapping in flatter terrain in Grisedale, showing superimposed drumlin forms.

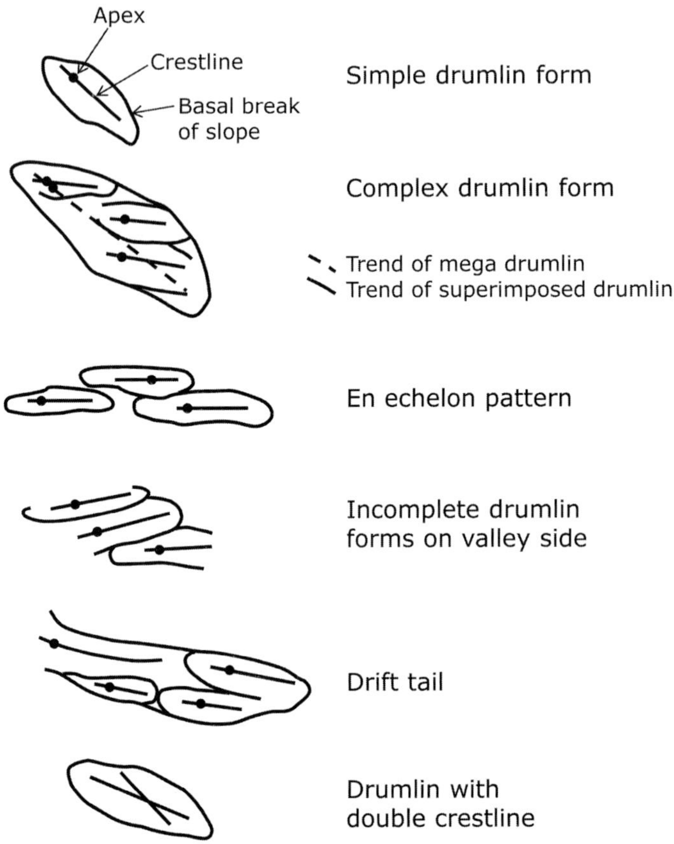

Figure 6.5 Examples of the typical outline morphology of common drumlin types, showing crestline position and drumlin apex (see Figure 6.4 for identification of these types in the field).

that these landforms are better interpreted as Rogen or ribbed moraines formed subglacially and behind the ice margin (Knight and McCabe, 1997). This illustrates the power of accurate and detailed geomorphological field mapping to allow for better interpretation of depositional landforms and therefore former glacial processes in upland terrain.

Ice-marginal moraines occurring in upland glaciated terrain are significant glacigenic landforms because, when interpreted correctly, they can be used in the reconstruction of glacier extent and patterns and processes of ice retreat. Accurate and detailed geomorphological field mapping of distinct ridge forms is therefore fundamental for their correct interpretation as terminal or lateral moraines. Such ice-marginal moraines that have a ridge form with a crestline parallel to the former ice margin can be most clearly mapped in the field, and their relationship to the ice margin position can then be inferred. However, some ice-marginal moraines do not display a simple ridge form but comprise a number of partly connected mounds that have complex basal outlines, and undulating long- and cross profiles that may be superimposed or overlapping. Such 'hummocky moraines' have been identified in particular in western Scotland, Canada and Svalbard and linked to a range of formative processes including areal stagnation (Sissons, 1967), englacial and proglacial thrusting and stacking of debris bands (Hambrey et al., 1997; Dyke and Savelle, 2000; Lukas, 2005), soft-sediment deformation (Eyles et al., 1999) and subglacial meltwater erosion (Munro and Shaw, 1997). This clearly shows that the geomorphological field mapping of such complex landforms can have significant implications for their resultant interpretation. Bennett (1994) described how views of the formation and interpretation of hummocky moraines in northwest Scotland have changed over time, with active and stagnant ice models proposed at different times. What was originally thought to be a chaotic pattern (Sissons, 1967) is today better mapped as distinctive moraine ridges that demonstrate active and ordered glacier recession rather than disordered areal stagnation (Lukas and Benn, 2006). This shows that accurate geomorphological mapping of moraines is very significant with respect to correct interpretation of ice retreat patterns and subglacial conditions. Hummocky moraines occur most commonly in association with former corrie and valley glaciers. These moraines are of generally high relief and comprise a mixture of drift and upstanding bedrock that is partly intact and partly thrust up as rafts into the moraine. Moraine elements with a high rock and/or debris content tend to have a high relief, but meltout of buried ice can cause re-sedimentation and inversion of relief. Hummocky moraine may therefore reflect a complex history of variable synformational glacitectonic and post-formational gravity and mass movement processes. As a result, the geomorphic patterns of hummocky moraines are a palimpsest of geomorphological processes and variable geologic and glaciological controls.

Deglaciation from the Younger Dryas readvanced ice limit in northwest Scotland resulted in extensive areas of hummocky moraine formation within bedrock valleys and on valley sides (Bennett and Boulton, 1993; Lukas and Lukas, 2006) (Figure 6.6). In detail, these geomorphic patterns comprise three separate components: (1) small drumlins and

Figure 6.6 Photo of typical hummocky moraines at Glen Grudie, northwest Scotland, illustrating their morphological diversity.

flutes that may be present beneath a variable morainal cover, (2) recessional cross-valley moraines that correspond to successive ice margin positions and (3) non-aligned hummocky moraines that reflect periods of stagnation and sediment reworking. Although the mapping of individual hummocks is problematic – due to the relationship between adjacent hummocks being uncertain because of the absence of a clear moraine crestline – they can be linked together as landform assemblages that correspond to a zone of ice-marginal deposition (Hambrey et al., 1997). As such, hummocky moraines that are mapped on a regional scale can reveal patterns of ice retreat that are not apparent on the local scale. Patterns of hummocky moraines, where associated with other geomorphological features formed at or near an ice margin, can be used to construct recessional patterns of ice retreat (Figure 6.7).

3.2 Landforms that Result from Fluvial Processes

Fluvial catchments in upland terrain are typically steep, gravel dominated, contain braided river systems, terraces and are characterised by seasonally variable discharge that results in geomorphic change, both within the uplands and in adjacent downstream river valleys. Fluvial catchments in upland terrain are strongly driven by the interplay between tectonic

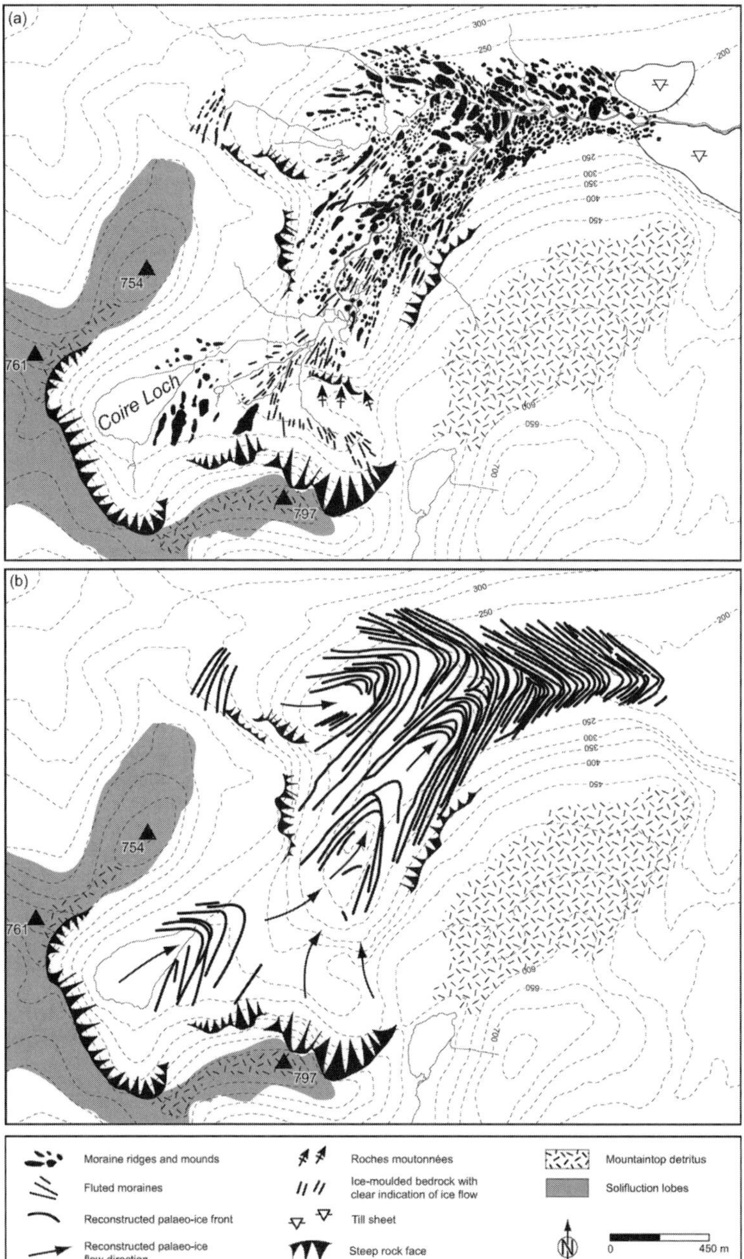

Figure 6.7 (a) Geomorphological map of landforms in Coire na Phris, northwest Scotland, showing the crestlines of hummocky moraines; (b) interpreted patterns of ice front positions and ice flow direction during ice retreat, identified by joining the crestlines of moraines. *From Lukas and Benn (2006).*

uplift, climate and fluvial downcutting (Hovius et al., 2004; Korup and Clague, 2009; Bridgland et al., 2010). Decreased temperatures caused by tectonic uplift can also contribute to increased sediment supply through more active physical weathering (Kirkby, 1995). Changes in base level, leading to river downcutting and formation of strath and aggradational terraces, are enhanced in mountains and upland terrain where uplift by tectonics and isostasy affects mountain headwaters. In the Himalaya this results in steep river valley sides incised into bedrock, with large thicknesses (several kilometres) of fluvial gravels contained within fault-bounded basins. Such steep and unstable bedrock slopes can also increase the sediment flux into these catchments by high-magnitude landslides and other catastrophic mass failures (Mitchell et al., 2007; Jarman, 2009). Landforms that are most commonly mapped within upland terrain are river terraces and braided channel bars. These are now examined in turn.

River terraces are formed as a result of sediment deposition on floodplains or adjacent to channels by overbank sedimentation during flood events. Fluvial incision into these unconsolidated sediments, which can be driven by changes in base level or high river discharge, causes the development of a steep scarp face that delimits the proximal edge of the flat terrace surface (Bridgland and Westaway, 2007; Bridgland et al., 2010). As such, river terraces are relatively easy to map in plan view, but distinguishing between different generations of terraces requires accurate measurement of terrace surface elevation. These different terraces are formed by successive episodes of floodplain aggradation followed by incision. Terrace fragments are commonly paired and nested within valleys such that the oldest terrace has the highest elevation and younger terraces occur at successively lower elevations closer to the river. Although terrace margins can be mapped relatively easily and with high precision because their scarped edges are generally well defined, small terrace fragments may be too small to map at the chosen scale. Slope attributes of very gently sloping terrace surfaces are more difficult to map because significant breaks of slope are uncommon. As steep terrace faces are commonly inaccessible in the field, terraces are most commonly mapped from a high landscape position or from air photos. More recently, LiDAR and GPS have been used (Jones et al., 2007; Meikle et al., 2010). Other analytical methods that enhance these field observations include sedimentology in section, coring, radiometric dating of organic and inorganic materials (using radiocarbon, cosmogenic and luminescence techniques), pollen and faunal analysis and ground-penetrating radar (Cotton et al., 1999).

Much work has been done on the late Pleistocene–Holocene development of river systems in northern England (Passmore et al., 1993; Howard et al., 2000; Jones et al., 2007; Bridgland et al., 2010) and New Zealand (Berryman et al., 2000; Kasai et al., 2001; Litchfield and Berryman, 2005), where river dynamics have been strongly affected by the impacts of glaciation (*cf.* Church and Ryder, 1972). In northern England, river terrace development shows a complex relationship with glacigenic sediment supply and glacio-isostatic unloading history (Bridgland et al., 2010; Mitchell et al., 2010). Glacial and periglacial processes are contributory factors to the provision of high sediment supply in river headwaters. For example, in the basin of the River Till, a tributary of the River Tweed in northeast England, postglacial river patterns were established as the drainage was downcut into glaciolacustrine delta deposits, forming multiple paired terrace sequences that reflect downstream sediment reworking (Figure 6.8) (Passmore and Waddington, 2009). Similar paired terrace sequences occur in eastern North Island, New Zealand, where they reflect multiple episodes of sediment aggradation. Regional climate or base-level controls on terrace formation are most likely because similar dated periods of aggradation occur in different catchments (Litchfield and Berryman, 2005). However, in detail, base-level controls are more common in downstream locations, and climate and deforestation controls are more common in upstream locations. An important point is that terrace morphology is the same irrespective of forcing factor, thus that morphology alone cannot be used to interpret the evolution of these forms. Figure 6.9 shows the complex geomorphic relationships between partially preserved terraces and bedrock valley sides around the eastern Grand Canyon, Arizona, United States, during the late Pleistocene (Pederson et al., 2006). Here, fluvial incision into bedrock provides the sediment source for river terraces, but the patterns of incision and deposition are not spatially or temporally uniform. In addition, landforms that are morphologically similar to terraces can also occur in these piedmont environments, including solifluction sheets, alluvial fans and glaciolacustrine deltas.

Geomorphological mapping of braided river environments in upland terrain has focused on the dynamics of within-channel bars, including their spatial and temporal evolution. Bridge (1993) showed how channel planform patterns evolve as a result of variations in river discharge and sediment supply that cause changes in bar morphology. Accurate morphological mapping can aid the calculation of braid channel ratio, channel wavelength and sinuosity, which vary with water depth and therefore

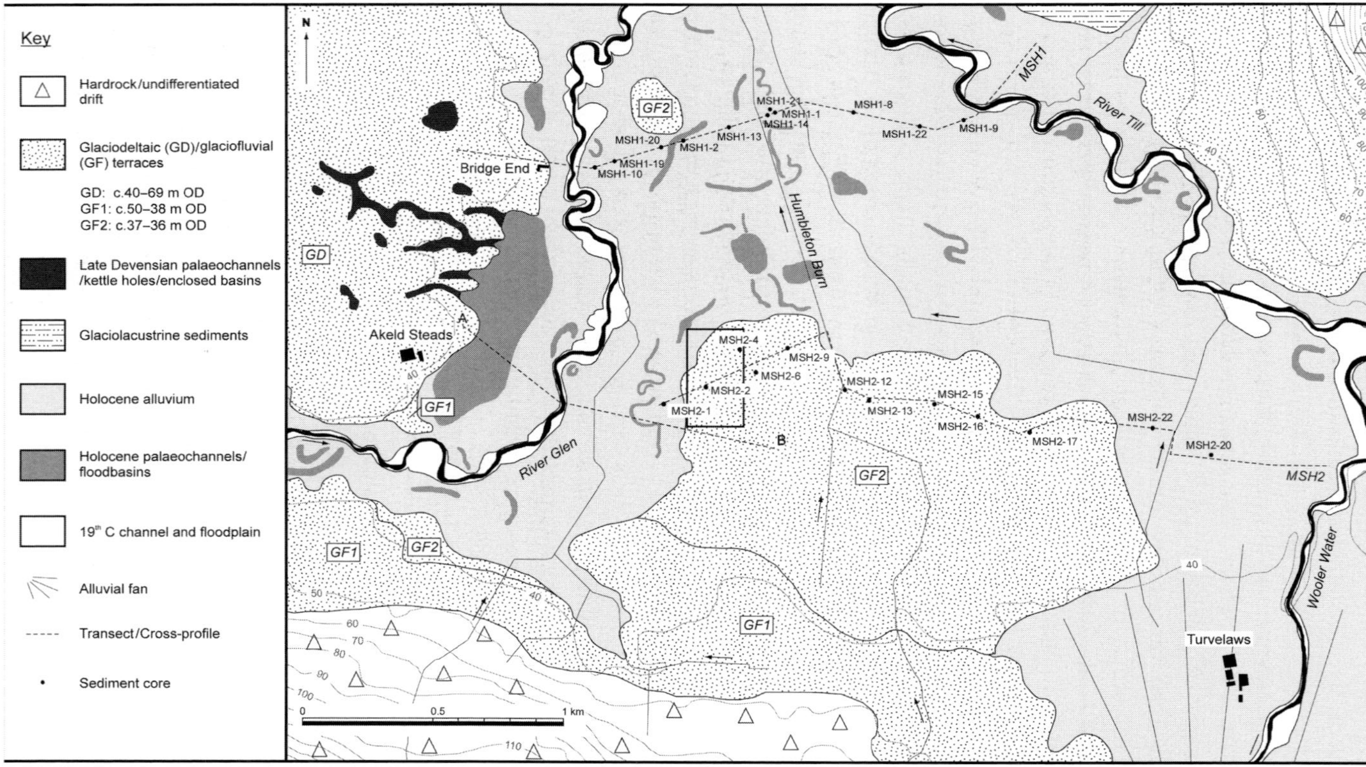

Figure 6.8 Simplified geomorphological map of part of the River Till floodplain, northeast England, showing fluvial terraces of different ages. *From Passmore and Waddington, (2009).*

Figure 6.9 (a, b) Views of terrace deposits along the Colorado River, Arizona, United States, showing the positions of dated sediments. (c) Composite cross section showing the terrace stratigraphy and radiometric ages. *From Pederson et al. (2006).*

flood state. Bertoldi et al. (2009) showed that there is a relationship between braiding morphology, water discharge and stream power. In addition, higher river discharge leads to increased network complexity (i.e. greater bar dynamism by erosion and deposition). Passmore et al. (1993) used historic maps and field mapping from the Llandinam experimental catchment in the Upper Severn River, central Wales, in order to examine changes in braided river morphology in the period 1890–1983 (Figure 6.10). Geomorphological mapping can identify changes in bar position within the channel, and changes in channel margins over time, from which patterns of erosion and deposition can be determined. Brewer and Passmore (2002) described how variations in bar morphology can be used to calculate variations in sediment budget at different points through the river system. They first use their geomorphological maps from different time periods to identify different channel elements, including the bar platform and chute channel. The volumetric change of each geomorphological element over the time period is first calculated. The volumetric change for all elements is then summed in order to calculate changes in the total sediment budget. In this environment, if bars aggregate together then sediment export is reduced along with the river's wetted perimeter. Channel form and position then become more fixed, and channel bar dynamic behaviour decreases.

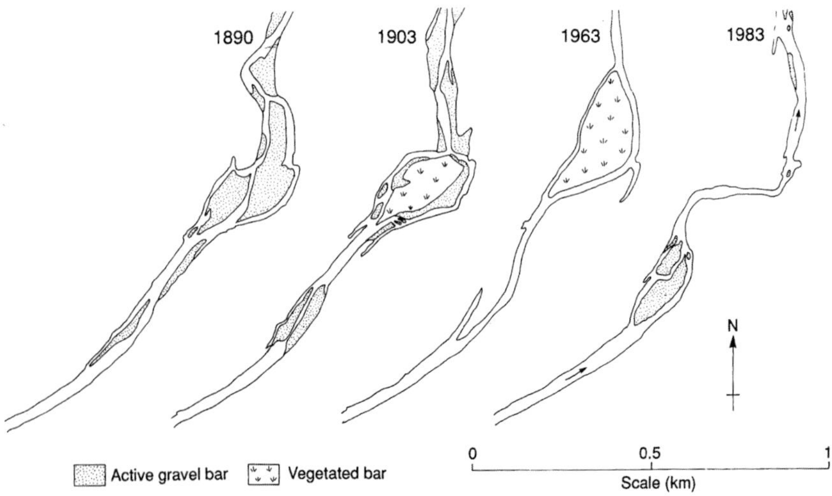

Figure 6.10 Maps of channel and bar morphology at different time periods at Llandinam, Upper Severn River, central Wales (from Passmore et al., 1993). See text for discussion of how geomorphological and sediment budget changes are calculated.

3.3 Landforms that Result from Mass Movement Processes

The limited number of morphological units (ridge, mound, depression) that occur within a landscape can be interpreted in different ways depending on the possible explanation of geomorphological environment. This means that there is a possibility of incorrect interpretation of such forms and that many landforms may have a composite nature (Wilson, 2009). This has proved to be particularly true in mountain and upland areas where the juxtaposition of different cold climate process—response systems has seen an over reliance on the role of glaciation to interpret constructional ridges and mounds found in such environments as moraines; there is now increased appreciation of the interaction of different non-glacial processes associated with the transition of a landscape from glacial to non-glacial conditions, in particular mass movement processes (Hewitt, 2006; Wilson, 2009).

Such landscape disturbance will be reflected in the operation of periglacial processes, particularly associated with permafrost as the spatial extent of ice is reduced with time. Also the increasing exposure of rock slopes previously buried beneath a glacier will lead to stress changes that may initiate rock slope failures (RSFs) and other types of landslides (Hewitt, 2006). Much of this has been explained by overextending the original definition of 'paraglacial' related to sediment flux within glacially influenced catchments (Church and Ryder, 1972) to encompassing geomorphic processes associated with landscape adjustment to the termination of a period of glaciation at a variety of temporal and spatial scales (Ballantyne, 2002). However, this is an oversimplification, generally at the expense of an appreciation of the role of periglacial processes during this time of important landscape change (André, 2009). Rather, attention should focus on mountain and upland periglacial and nival processes, particularly the presence of permafrost and its role in controlling mass movements.

Mapping of RSF and landslides is a more difficult challenge than many glacial landsystems, given that mass movements characteristically construct smaller more complex forms that cannot be picked up on presently available DEMs in upland areas and which are commonly areas for which LiDAR is unavailable; thus this area of geomorphology may still require field mapping to allow for detailed maps to be constructed. To adequately present such landforms requires detailed high-resolution maps, generally at scales larger than 1:10,000 and now using GPS to determine precise location. However, this can become complicated where tree cover (woodland/forest) disrupts the GPS signal.

Mapping mass movements begins with the identification of two distinct subsystems associated with the failure scars and resultant landslide debris leading to consideration of different failure mechanisms (*cf.* Cooper, 2007; Jones and Lee, 1994). Although it is generally straightforward to quickly ascertain the slope area affected by slope failure, many landslides are complex forms and have a subsequent superimposition of later material and constructional forms on earlier event features. From a mapping perspective, many studies have commenced with morphological mapping to identify key breaks of slope allowing the identification of morphological elements of scarps, benches, steep slopes, mounds and lobes (Jarman, 2006). This then allows an interpretation of the style of landslide, as the different failure mechanisms (fall, topple, rotation, slide and flow) (Dikau et al., 1996; Cooper, 2007). This can result in distinctive morphological units within different parts of the landslide system, with the upper slopes dominated by scarps and rotated rock blocks associated with mass loss, mid-slopes associated with sediment transportation resulting in increased disaggregation of rock and the commencement of flow and the lower slopes reflecting deposition in the forms of thick debris lobes, commonly with evidence of compression. Mapping will incorporate bedrock exposures along exposed scarps and, where exposed, displaced transported debris to determine the presence/absence of rotation. It may also be important to distinguish areas of active slope failure on the resultant geomorphological map in comparison to areas on a slope that have become stabilised with time.

Most published geomorphological maps of landslide systems do not attempt to map the fine detail of the different morphological elements in detail; rather, symbols are employed to indicate key breaks of slope that are characterised by specific landforms related to specific slope processes (Guzzetti et al., 2000; Cardinali et al., 2002; Hervás et al., 2003; Jarman, 2007). Some of the earlier maps relate to investigations that were carried out on the Dorset coast, southern England (Brunsden and Jones, 1972), showing the major breaks of convex and concave slopes as well as zones of present-day slope movement (Figure 6.11); more detail may in fact obscure overall morphological patterns as well as being extremely hazardous on liquefied moving sediment. In fact, many schematic maps of landslides only identify the main scarp and the overall spatial extent of the resultant landslide debris (*cf.* Cooper, 2007).

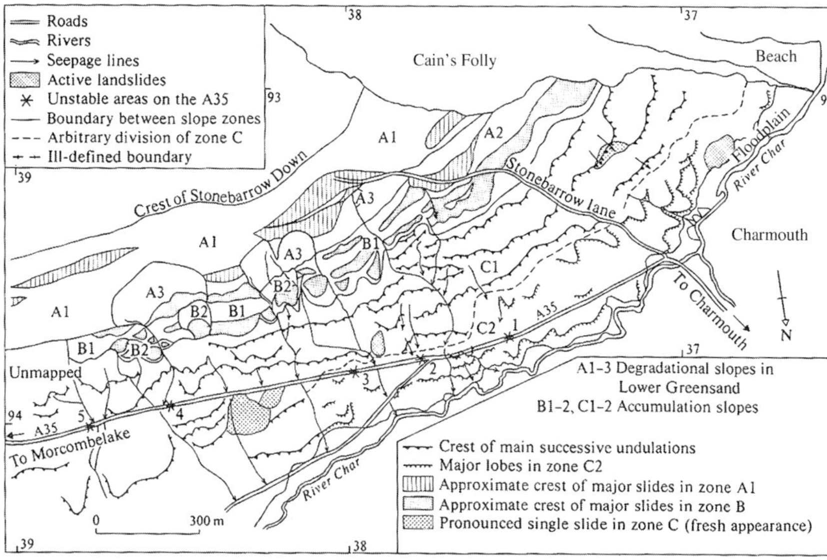

Figure 6.11 Geomorphological map of the Stonebarrow Hill area, Dorset, southern England. *From Goudie (1981).*

Use of aerial photographs has been successfully employed to produce geomorphological maps of a number of large-scale (>0.25 km^2) RSFs within many parts of the Scottish Highlands (*cf.* Jarman, 2006, 2007). Here, detail is limited to allow visual appreciation of the key morphological elements that have been generalised to characterise the different parts of the RSF area (Figure 6.12). This example from Sgurr na Ciste Duibhe in the Scottish Highlands illustrates the primary features of a RSF geometry with the main scarps that characterise the upper slopes allowing quantification of the result cavity. Attention is focused on the important antiscarps that characterise the upper zone southeast of the mountain summit, with less attention focused on the detail of the lower slopes (Jarman, 2003).

Use of aerial orthophoto enlargements, to a nominal scale of 1:3000, as base maps allows the construction of a detailed geomorphological map of different landslides in conjunction with the latest digital imagery and using a GPS. Such large-scale mapping is time consuming but does produce a map illustrating the wealth of surface forms that reflect the complexity of mass movement processes related to displacement and movement of a slope in response to extension and compression. An example of this mapping methodology can be demonstrated from one of the complex landslides that

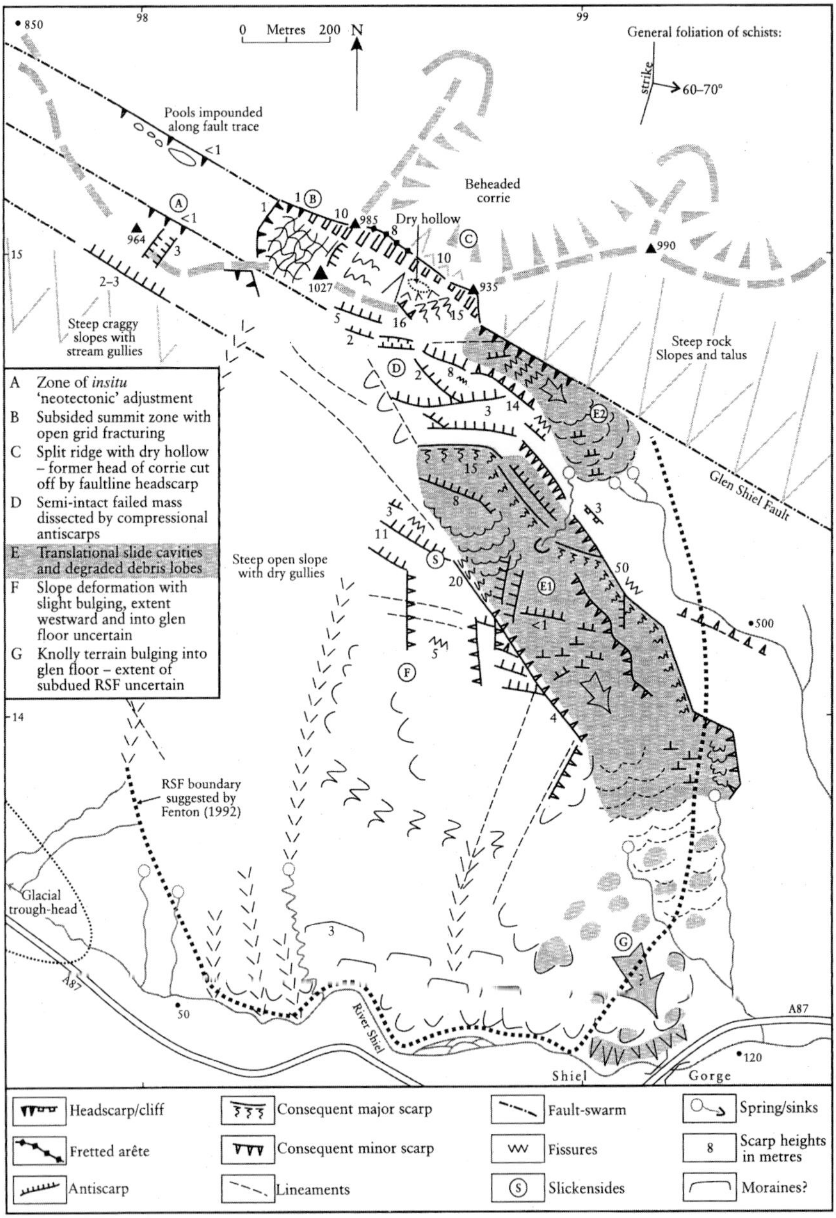

Figure 6.12 Annotated geomorphological map of the Sgurr na Ciste Duibhe rock slope failure, Scotland. *From Jarman (2007).*

characterise the through valley of Mallerstang in the western Pennines of northern England. Here, the west-facing slopes of Hangingstone Scar show a complex sequence of small scarps and ridges that can be mapped for about 1 km of the slopes towards the valley bottom (Mitchell, 1991b). Much of this area is covered in large angular boulders of the local Carboniferous gritstone, which have been removed from the map except where they form discrete and relevant lobes, particularly on the upper slopes and in the lower compressional toe zone. From this detailed mapping, a number of discrete events can be determined as mudslides stack on top of each other associated with ongoing failure of sections of the head scarp.

In many parts of the world, lack of reliable topographic maps and aerial photographs may require the production of a geomorphological map using simple field mapping techniques of compass orientation/triangulation and a GPS. Such mapping can be used to supplement mapping using satellite imagery and topographic data. For example, within a large (12 km^2) catastrophic rock avalanche at Keylong Serai in the Indian Himalaya, only the largest ridges within the carapace of large angular blocks could be mapped to delimit a series of large lobes (Figure 6.13, Mitchell et al., 2007). Such large-scale rock avalanches form important geomorphological elements within many high mountain areas and their correct interpretation and distinction from glacial deposits forms ongoing research into sediment flux within high mountain environments (*cf.* Hewitt, 2006). The production of detailed geomorphological maps of individual rock avalanches will aid our appreciation of their formation, their significance within mountain landscape development and role in hazard assessment. The increased identification of mass movements at a variety of spatial scales as a consequence of high resolution imagery has been important in advancing models of landscape transition and development in such upland and mountain terrains and their role in geomorphic change (Dikau et al., 1996).

4. DISCUSSION

Geomorphological mapping is important as an investigative process to better understand landform patterns and genesis. It is the first and probably the most important stage in quantifying landscape attributes and

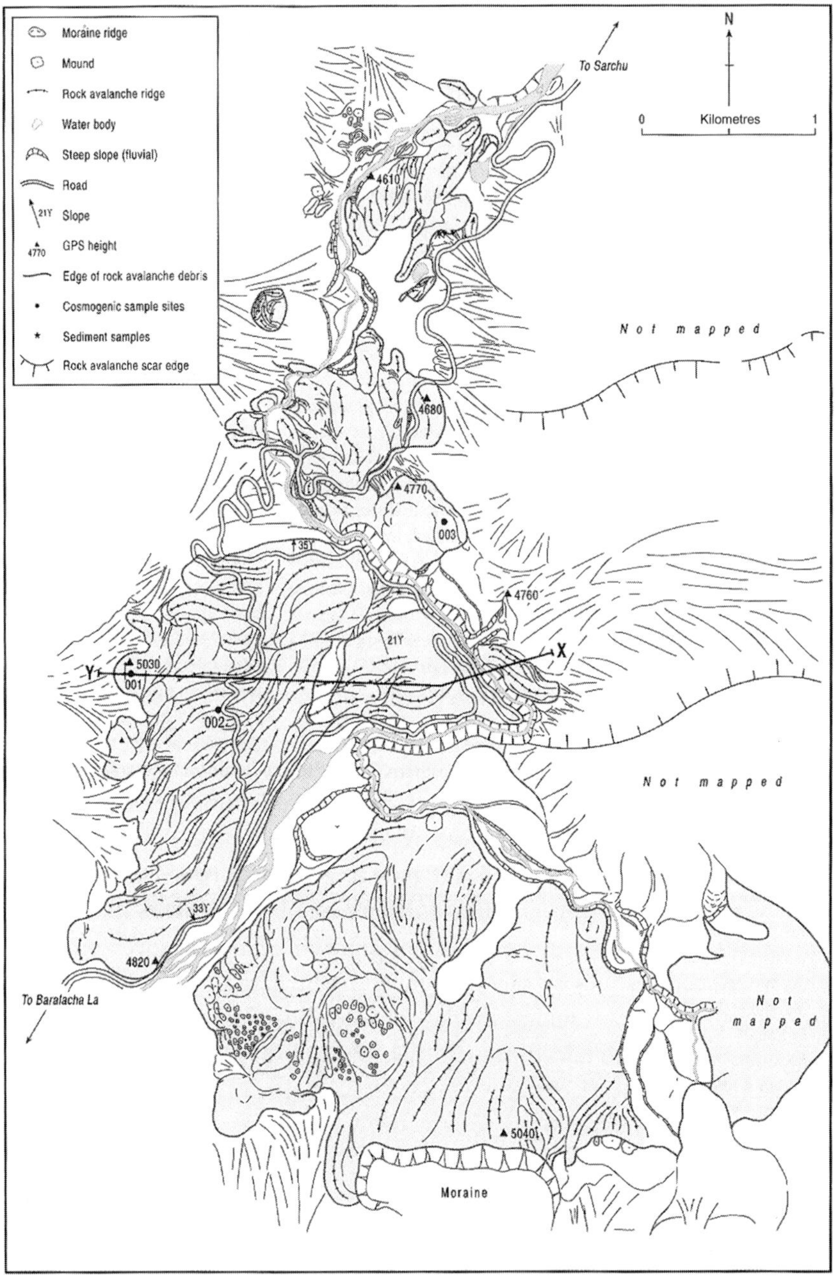

Figure 6.13 Geomorphological map of the Keylong Serai rock avalanche, northwest Indian Himalaya. *From Mitchell et al. (2007).*

resources, and so is closely related to the practice of environmental management. It is strongly driven by technological change in remote sensing and GIS, which has enabled more powerful analyses of landform morphometric properties and spatial patterns to take place. However, computer-based interpretations are only as good as the geomorphological knowledge and field experience of the operator; thus ground truth and field checking become increasingly important as emphasis changes to GIS. The application of geomorphological mapping in upland terrain includes

1. *identifying and interpreting landscape patterns*, Geomorphological mapping can help identify and interpret landform properties, patterns and resources, including their physical and cultural attributes. The physical properties of landscapes include the morphometrics and spatial patterns of their component landforms. Accurate geomorphological mapping is therefore an important prerequisite for identifying the commonness/rarity of certain landform types, their spatial distribution and their relationships to adjacent landforms. An important outcome of this mapping is that landscape planning, resource and environmental management can be better informed by an understanding of landscape properties (Knight et al., 1999). More practically, geomorphological mapping underpins assessment of regional geodiversity, which is the basis for geoconservation (Gordon, 2010) and for engineering and planning purposes (Brunsden, 2002).

   Landforms are the building blocks of landscapes, so measuring the morphometric properties of landforms can help evaluate how and why the physical attributes of landscapes vary over time and space, and landform morphometry, based on geomorphological mapping, can therefore help distinguish between landscapes of different regions and their relative values. For example, Leopold (1969) used 46 morphometric, biological, landuse and human properties in order to calculate the 'uniqueness factor' of upland fluvial landscapes. Ergin et al. (2006) proposed a similar scheme to distinguish between different coastal landscapes. The 'uniqueness factor' can be linked directly to landscape spatial scale and qualitative measures of landscape scenic value such as its beauty and spectacle and relationships to cultural aspects of the landscape such as archaeological patterns (Knight, 2001).
2. *identifying geomorphological processes and events that result in geohazards*, Geomorphological mapping can be used to identify evidence for past

geohazard events and to identify those locations that may be susceptible to future geohazards. Geomorphological evidence for past events includes oversteepened or ice-contact slopes, landforms such as RSFs and landslides and evidence for reworking of previously deposited sediment by debris or mass flows (Kienholz, 1977; Rupke et al., 1988). When coupled with data on the spatial and temporal patterns of past events, risk analysis can take place, which considers the likelihood of occurrence of an event of a certain magnitude and the range of possible impacts that such an event may have (Fuchs, 2009). Accurate geomorphological mapping can yield a better understanding of past geohazards and their impacts. Where data sources (maps, air photos, satellite imagery) are available for different time periods, spatial and temporal patterns of geohazard occurrence can be established. As such, geomorphological mapping can be used as an effective management tool to identify past, monitor present and predict future environmental change.

3. *evaluating the sensitivity of relict landscapes to climate or external forcing*, The high-resolution, digital remote sensing data now available for many upland terrains mean that geomorphic features can be mapped quickly and effectively. Protocols for automated ('objective') mapping using remotely sensed data have been established for different physical environments (Smith and Pain, 2009). Where repeat-pass data are available, spatial and temporal changes in landscape geomorphology can also be mapped. Geomorphological mapping now can be used as a tool to monitor landscape responses to external forcing by climate or human activity. This is an exciting trajectory for geomorphological mapping because, when matched with contemporary climate records, it can reveal the sensitivity of the landscape to external forcing and the magnitude and/or time lags of response to hazardous events. Remote sensing enables generally remote areas affected by rapid or catastrophic landscape change to be imaged quickly and accurately and can be used as input into geomorphological or geophysical models of landscape response to jökulhlaups, glacial lake outburst floods, rock mass failures or landslides.

5. CONCLUSIONS AND OUTLOOK

Geomorphological mapping has evolved from a purely field-based exercise aimed at accurate depiction of landforms on a map with limited interpretation to highly interpretive maps at different scales that are mainly based on digital remotely sensed data with limited ground-truthing in the field in some places. Modern geomorphological mapping is strongly set within the technical capabilities of multimedia, digital data and imagery at different scales, including presentation of geomorphological mapping using a GIS, DEMs, and in colour (Gustavsson et al., 2006). This facilitates the interpretation of landscape mapping with respect to the relationships between geomorphology and other landscape variables such as underlying geology. However, given the complexity of present-day landscapes and their wealth of geological evidence for former geomorphological systems, correct interpretation based on field observations is still a prerequisite in landform interpretation. These relationships highlight the application of geomorphological mapping for landscape, environmental and resource evaluation and management.

This increased use of technology has also meant that the use of standard morphological mapping techniques and symbols has decreased over time. This has also resulted from geomorphological mapping moving away from national surveys towards more focused and localised studies centred on environmental management and monitoring. A future imperative is to ensure that the integrity and objectivity of morphological and geomorphological mapping are maintained for the user whilst integrating supporting data effectively within real landscapes.

ACKNOWLEDGEMENT

We thank Sven Lukas, Harry Seijmonsbergen and Mike Smith for their comments on a previous version of this chapter.

REFERENCES

Aario, R., 1977. Associations of flutings, drumlins, hummocks and transverse ridges. GeoJournal 6, 65–72.

André, M.-F., 2009. From climatic to global change geomorphology: contemporary shifts in periglacial geomorphology. In: Knight, J., Harrison, S. (Eds.), Periglacial and Paraglacial Processes and Environments. Geological Society, London, Geological Society Special Publication No. 320, pp. 5–28.

Ballantyne, C.K., 2002. A general model of paraglacial landscape response. Holocene 12, 371–376.

Barsch, D., Liedtke, H., 1980. Principles, scientific values and practical applicability of the geomorphological map of the Federal Republic of Germany at a scale of 1:25000 (GMK 25) and 1:100000 (GMK 100). Z. Geomorphol. 46, 296–313.

Bartsch, A., Gude, M., Gurney, S.D., 2009. Quantifying sediment transport processes in periglacial mountain environments at a catchment scale using geomorphic process units. Geogr. Ann. 91A, 1–9.

Bennett, M.R., 1994. Morphological evidence as a guide to deglaciation following the Loch Lomond Readvance: a review of research approaches and models. Scott. Geogr. Mag. 110, 24–32.
Bennett, M.R., Boulton, G.S., 1993. A reinterpretation of Scottish 'hummocky moraine' and its significance for the deglaciation of the Scottish Highlands during the Younger Dryas or Loch Lomond Stadial. Geol. Mag. 130, 301–318.
Berryman, K, Marden, M., Eden, D., Mazengarb, C., Ota, Y., Moriya, I., 2000. Tectonic and paleoclimatic significance of Quaternary river terraces of the Waipaoa River, east coast, North Island, New Zealand. N. Z. J. Geol. Geophys. 43, 229–245.
Bertoldi, W., Zanoni, L., Tubino, M., 2009. Planform dynamics of braided streams. Earth Surf. Process. Landforms 34, 547–557.
Boots, B.N., Burns, R.K., 1984. Analyzing the spatial distribution of drumlins: a two-phase mosaic approach. J. Glaciol. 30, 302–307.
Brewer, P.A., Passmore, D.G., 2002. Sediment budgeting techniques in gravel-bed rivers. In: Jones, S.J., Frostick, L.E. (Eds.), Sediment Flux to Basins: Causes, Controls and Consequences. Geological Society, London, Geological Society Special Publication No. 191, pp. 97–113.
Bridge, J.S., 1993. The interaction between channel geometry, water flow, sediment transport and deposition in braided rivers. In: Best, J.L., Bristow, C.S. (Eds.), Braided Rivers. Geological Society, London, Geological Society Special Publication No. 75, pp. 13–71.
Bridgland, D.R., Westaway, R., 2007. Climatically controlled river terrace stairways: a worldwide Quaternary phenomenon. Geomorphology 98, 285–315.
Bridgland, D.R., Westaway, R., Howard, A.J., Innes, J.B., Long, A.J., Mitchell, W.A., et al., 2010. The role of glacio-isostasy in the formation of post-glacial river terraces in relation to the MIS2 ice limit: evidence from northern England. Proc. Geol. Assoc. 121, 113–127.
Brunsden, D., 2002. Geomorphological roulette for engineers and planners: some insights into an old game. Q. J. Eng. Geol. Hydrogeol. 35, 101–142.
Brunsden, D, Jones, D.K.C., 1972. The morphology of degraded landslide slopes in south-west Dorset. Q. J. Eng. Geol. 5, 205–222.
Cardinali, M., Reichenbach, P., Guzzetti, F., Ardizzone, F., Antonini, G., Galli, M., et al., 2002. A geomorphological approach to the estimation of landslide hazards and risks in Umbria, Central Italy. Nat. Hazards Earth Syst. Sci. 2, 57–72.
Charlesworth, J.K., 1924. The glacial geology of the north-west of Ireland. Proc. R. Ir. Acad. 36 (B), 174–314.
Charlesworth, J.K., 1928. The glacial retreat from central and southern Ireland. Q. J. Geol. Soc. 84, 293–344.
Charlesworth, J.K., 1929. The South Wales end-moraine. Q. J. Geol. Soc. 85, 335–358.
Charlesworth, J.K., 1957. The Quaternary Era, 2 vols. Edward Arnold, London.
Chorley, R.J., 1959. The shape of drumlins. J. Glaciol. 3, 339–344.
Church, M., Ryder, J.M., 1972. Paraglacial sedimentation: a consideration of fluvial processes conditioned by glaciation. Geol. Soc. Am. Bull. 83, 3059–3071.
Clark, C.D., Hughes, A.L.C., Greenwood, S.L., Jordan, C., Sejrup, H.-P., 2011. Pattern and timing of retreat of the last British-Irish Ice Sheet. Quat. Sci. Rev.
Cooke, R.U., Doornkamp, J.C., 1974. Geomorphology in Environmental Management. Clarendon, Oxford.
Cooper, R.G., 2007. Mass Movements in Great Britain. Geological Conservation Review Series, 33. Joint Nature Conservation Committee, Peterborough.
Cotton, J.A., Heritage, G.L., Large, A.R.G., Passmore, D.G., 1999. Biotic response to late Holocene floodplain evolution in the River Irthing catchment, Cumbria. In: Marriott, S.B., Alexander, J. (Eds.), Floodplains: Interdisciplinary Approaches.

Geological Society, London, Geological Society Special Publication No. 163, pp. 163–178.
Crofts, R.S., 1981. Mapping techniques in geomorphology. In: Goudie, A.S. (Ed.), Geomorphological Techniques. Allen and Unwin, London, pp. 66–75.
Dikau, R., Brunsden, D., Schrott, L., Ibsen, M-L. (Eds.), 1996. Landslide Recognition. Wiley, Chichester.
Demek, J., Embleton, C., 1978. Guide to Medium-Scale Geomorphological Mapping. IGU Commission on Geomorphological Survey and Mapping, Brno.
Dryer, C.R., 1901. Certain peculiar eskers and esker lakes of northeastern Indiana. J. Geol. 9, 123–129.
Dyke, A.S., Savelle, J.M., 2000. Major end moraines of Younger Dryas age on Wollaston Peninsula, Victoria Island, Canadian Arctic: implications for paleoclimate and for formation of hummocky moraine. Can. J. Earth Sci. 37, 601–619.
Dykes, A.P., 2008. Geomorphological maps of Irish peat landslides created using hand-held GPS. J. Maps 2008, 258–276.
Elvhage, C., 1980. An experimental series of topo-geomorphological maps – with an example from a deglaciated mountain area in Jämtland, Sweden. Geogr. Ann. 62A, 105–111.
Ergin, A., Williams, A.T., Micallef, A., 2006. Coastal scenery: appreciation and evaluation. J. Coast. Res. 22, 958–964.
Evans, D.J.A. (Ed.), 2003. Glacial Landsystems. Arnold, London.
Evans, D.J.A., Clark, C.D., Mitchell, W.A., 2005. The last British Ice Sheet: a review of evidence utilised in the compilation of the Glacial Map of Britain. Earth Sci. Rev. 70, 253–312.
Evans, D.J.A., Twigg, D.R., Rea, B.R., Orton, C., 2009. Surging glacier landsystems of Tungnaárjökull, Iceland. J. Maps 2009, 134–151.
Eyles, N., 1983. Glacial geology: a landsystems approach. In: Eyles, N. (Ed.), Glacial Geology. An Introduction for Engineers and Earth Scientists. Permagon, Oxford, pp. 1–18.
Eyles, N., Boyce, J.I., Barendregt, R.W., 1999. Hummocky moraine: sedimentary record of stagnant Laurentide Ice Sheet lobes resting on soft beds. Sediment. Geol. 123, 163–174.
Fairchild, H.L., 1929. New York drumlins. Proc. Rochester Acad. Sci. 7, 1–37.
Fuchs, S., 2009. Susceptibility versus resilience to mountain hazards in Austria – paradigms of vulnerability revisited. Nat. Hazards Earth Syst. Sci. 9, 337–352.
Gardiner, V., Dackombe, R.V., 1983. Geomorphological Field Manual. Allen and Unwin, London.
Gordon, J.E., 2010. Scottish Landform Example–41: the geological foundations and landscape evolution of Scotland. Scott. Geogr. J. 126, 41–62.
Goudie, A.S., 1981. Geomorphological Techniques. George Allen and Unwin.
Greenwood, S.L., Clark, C.D., 2010. The sensitivity of subglacial bedform size and distribution to substrate lithological control. Sedimentary Geology 232, 130–144.
Gustavsson, M., Kolstrup, E., Seijmonsbergen, A.C., 2006. A new symbol-and-GIS based detailed geomorphological mapping system: renewal of a scientific discipline for understanding landscape development. Geomorphology 77, 90–111.
Gustavsson, M., Seijmonsbergen, A.C., Kolstrup, E., 2008. Structure and contents of a new geomorphological GIS database linked to a geomorphological map – with an example from Liden, central Sweden. Geomorphology 95, 335–349.
Guzzetti, F., Cardinali, M., Reichenbach, P., Carrara, A., 2000. Comparing landslide maps: a case study in the Upper Tiber River basin, central Italy. Environ. Manage. 25, 247–263.

Hambrey, M.J., Huddart, D., Bennett, M.R., Glasser, N.F., 1997. Genesis of 'hummocky moraines' by thrusting in glacier ice: evidence from Svalbard and Britain. J. Geol. Soc. 154, 623–632.

Hättestrand, C., Goodwillie, D., Kleman, J., 1999. Size distribution of two cross-cutting drumlin systems in northern Sweden: a measure of selective erosion and formation time length. Ann. Glaciol. 28, 146–152.

Hervás, J., Barredo, J.I., Rosin, P.L., Pasuto, A., Mantovani, F., Silvano, S., 2003. Monitoring landslides from optical remotely sensed imagery: the case history of Tessina landslide, Italy. Geomorphology 54, 63–75.

Hewitt, K., 2006. Disturbance regime landscapes: mountain drainage systems interrupted by large rockslides. Prog. Phys. Geogr. 30, 365–393.

Hill, A.R., 1973. The distribution of drumlins in County Down, Ireland. Ann. Assoc. Am. Geogr. 63, 226–240.

Hovius, N., Lague, D., Dadson, S., 2004. Processes, rates and patterns of mountain-belt erosion. In: Owens, P.N., Slaymaker, O. (Eds.), Mountain Geomorphology. Arnold, London, pp. 109–131.

Howard, A.J., Macklin, M.G., Black, S., Hudson-Edwards, K.A., 2000. Holocene river development and environmental change in Upper Wharfedale, Yorkshire Dales, England. J. Quat. Sci. 15, 239–252.

Hubbard, B., Glasser, N., 2005. Field Techniques in Glaciology and Glacial Geomorphology. John Wiley & Sons, London.

Jarman, D., 2003. The Glen Shiel rock slope failure cluster. In: Tipping, R. (Ed.), The Quaternary of Glen Affric and Kintail; Field Guide. Quaternary Research Association, London, pp. 165–183.

Jarman, D., 2006. Large rock slope failures in the Highlands of Scotland: characterisation, causes and spatial distribution. Eng. Geol. 83, 161–182.

Jarman, D., 2007. Sgurr na Ciste Duibhe. In: Cooper, R.G. (Ed.), Mass Movements in Great Britain. Joint Nation Conservation Committee, Peterborough, Geological Conservation Review Series, 33, pp. 62–69.

Jarman, D., 2009. Paraglacial rock slope failure as an agent of glacial trough widening. In: Knight, J., Harrison, S. (Eds.), Periglacial and Paraglacial Processes and Environments. Geological Society, London, Geological Society Special Publication No. 320, pp. 103–131.

Jones, D.K.C., Lee, E.M., 1994. Landsliding in Great Britain. HMSO, London.

Jones, A.F., Brewer, P.A., Johnstone, E., Macklin, M.G., 2007. High-resolution interpretative geomorphological mapping of river valley environments using airborne LiDAR data. Earth Surf. Process. Landforms 32, 1574–1592.

Karczewski, A., 1976. Morphometric features of drumlins in western Pomerania. Quaestiones Geogr. 3, 35–42.

Kasai, M., Marutani, T., Reid, L.M., Trustrum, N.A., 2001. Estimation of temporally averaged sediment delivery ratio using aggradational terraces in headwater catchments of the Waipaoa River, North Island, New Zealand. Earth Surf. Process. Landforms 26, 1–16.

Kienholz, H., 1977. Kombinierte Geomorphologische Gefahrenkarte 1:10000 von Grindelwald. Catena 3, 265–294.

Kirkby, M.J., 1995. A model for variations in gelifluction rates with temperature and topography: implications for global change. Geogr. Ann. 77A, 269–278.

Kjaer, K.H., Korsgaard, N.J., Schomacker, A., 2008. Impact of multiple glacier surges – a geomorphological map from Brúarjökull, East Iceland. J. Maps 2008, 5–20.

Klimaszewski, M., 1990. Thirty years of detailed geomorphological mapping. Geogr. Pol. 56, 11–18.

Knight, J., 1997. Morphological and morphometric analyses of drumlin bedforms in the Omagh Basin, north central Ireland. Geogr. Ann. 79A, 255–266.

Knight, J., 2001. A geocultural classification of landscapes in Northern Ireland: implications for landscape management and conservation. Tearmann 1, 113–124.

Knight, J., 2010. Basin-scale patterns of subglacial sediment mobility: implications for glaciological inversion modelling. Sediment. Geol. 232, 145–160.

Knight, J., McCabe, A.M., 1997. Identification and significance of ice flow-transverse subglacial ridges (Rogen moraines) in northern central Ireland. J. Quat. Sci. 12, 519–524.

Knight, J., McCarron, S.G., McCabe, A.M., Sutton, B., 1999. Sand and gravel aggregate resource management and conservation in Northern Ireland. J. Environ. Manage. 56, 195–207.

Korup, O., Clague, J.J., 2009. Natural hazards, extreme events, and mountain topography. Quat. Sci. Rev. 28, 977–990.

Leopold, L.B., 1969. Landscape esthetics. Nat. Hist. 78, 36–45.

Litchfield, N.J., Berryman, K.R., 2005. Correlation of fluvial terraces within the Hikurangi Margin, New Zealand: implications for climate and baselevel controls. Geomorphology 68, 291–313.

Lukas, S., 2005. A test of the englacial thrusting hypothesis of 'hummocky' moraine formation: case studies from the northwest Highlands, Scotland. Boreas 34, 287–307.

Lukas, S., Benn, D.I., 2006. Retreat dynamics of Younger Dryas glaciers in the far NW Scottish Highlands reconstructed from moraine sequences. Scott. Geogr. J. 122, 308–325.

Lukas, S., Lukas, T., 2006. A glacial geological and geomorphological map of the far NW Highlands, Scotland. J. Maps 2006, 43–55.

Meikle, C., Stokes, M., Maddy, D., 2010. Field mapping and GIS visualisation of Quaternary river terrace landforms: an example from the Río Almanzora, SE Spain. J. Maps 2010, 531–542.

Mitchell, W.A., 1991a. Glaciation of Upper Wensleydale and Adjoining Watershed Regions. Unpublished Ph.D. Thesis. University of London, London.

Mitchell, W.A., 1991b. Geomorphological mapping. In: Mitchell, W.A. (Ed.), Western Pennines: Field Guide. Quaternary Research Association, London, pp. 19–23.

Mitchell, W.A., 1991c. Loven Scars and Hangingstone Scar. In: Mitchell, W.A. (Ed.), Western Pennines: Field Guide. Quaternary Research Association, London, pp. 82–90.

Mitchell, W.A., 1994. Drumlins in ice sheet reconstruction, with reference to the western Pennines, northern England. Sediment. Geol. 91, 313–331.

Mitchell, W.A., 1996. Significance of snowblow in the generation of Loch Lomond Stadial (Younger Dryas) glaciers in the western Pennines, northern England. J. Quat. Sci. 11, 233–248.

Mitchell, W., Riley, J.M., 2006. Drumlin map of the western Pennines and southern Vale of Eden, northern England, UK. J. Maps 2006, 10–16.

Mitchell, W.A., McSaveney, M.J., Zondervan, A., Kim, K., Dunning, S.A., Taylor, P.J., 2007. The Keylong Serai rock avalanche, NW Indian Himalaya: geomorphology and palaeoseismic implications. Landslides 4, 245–254.

Mitchell, W.A., Bridgland, D.R., Innes, J.B., 2010. Late Quaternary evolution of the Tees–Swale interfluve east of the Pennines: the role of glaciation in the development of river systems in northern England. Proc. Geol. Assoc., 121, 410–422.

Munro, M.J., Shaw, J., 1997. Erosional origin of hummocky terrain in south-central Alberta, Canada. Geology 25, 1027–1030.

Oguchi, T., Hayakawa, Y., Wasklewicz, T., in press. Data sources. In: Smith, M.J., Paron, P., Griffiths, J. (Eds.), Geomorphological Mapping: A Handbook of Techniques and Applications. Elsevier, London, pp. 189–224.

Otto, J.-C., Gustavsson, M., Geilhausen, M., 2011. Cartography: design, symbolisation and visualisation of geomorphological maps. In: Smith, M.J., Paron, P., Griffiths, J. (Eds.), Geomorphological Mapping: A Handbook of Techniques and Applications. Elsevier, Amsterdam.

Passmore, D.G., Macklin, M.G., Brewer, P.A., Lewin, J., Rumsby, B.T., Newson, M.D., 1993. Variability of late Holocene braiding in Britain. In: Best, J.L., Bristow, C.S. (Eds.), Braided Rivers. Geological Society, London, Geological Society Special Publication No. 75, pp. 205–229.

Passmore, D.G., Waddington, C., 2009. Paraglacial adjustment of the fluvial system to Late Pleistocene deglaciation: the Millfield Basin, northern England. In: Knight, J., Harrison, S. (Eds.), Periglacial and Paraglacial Processes and Environments. Geological Society of London, Special Publication, 320, 145–164.

Pederson, J.L., Anders, M.D., Rittenhour, T.M., Sharp, W.D., Gosse, J.C., Karlstrom, K.E., 2006. Using fill terraces to understand incision rates and evolution of the Colorado River in eastern Grand Canyon, Arizona. J. Geophys. Res. 111, F02003, doi:10.1029/2004JF000201.

Reade, T.M., 1893. The Drift Beds of the Moel Tryfaen area of the North Wales coast. Proc. Liverpool Geol. Soc. 7, 36–79.

Reed, B., Galvin Jr., C.J., Miller, J.P., 1962. Some aspects of drumlin geometry. Am. J. Sci. 260, 200–210.

Rhoads, B.L., Thorn, C.E., 1996. Towards a philosophy of geomorphology. In: Rhoads, B.L., Thorn, C.E. (Eds.), The Scientific Nature of Geomorphology. John Wiley & Sons, Chichester, pp. 115–143.

Rose, J., 1980. Landform development around Kisdon, upper Swaledale, Yorkshire. Proc. Yorkshire Geol. Soc. 43, 201–219.

Rose, J., Letzer, J.M., 1975. Drumlin measurements: a test of the reliability of data derived from 1:25,000 scale topographic maps. Geol. Mag. 112, 361–371.

Rose, J., Letzer, J., 1977. Superimposed drumlins. J. Glaciol. 18, 471–480.

Rose, J., Smith, M.J., 2008. Glacial geomorphological maps of the Glasgow region, western central Scotland. J. Maps 2008, 399–416.

Rupke, J., Cammeraat, E., Seijmonsbergen, A.C., Van Westen, C.J., 1988. Engineering geomorphology of the Widentobel catchment, Appenzell and Sankt Gallen, Switzerland. A geomorphological inventory system applied to geotechnical appraisal of slope stability. Eng. Geol. 26, 33–68.

Sahlin, E.A.U., Glasser, N.F., 2008. A geomorphological map of Cadair Idris, Wales. J. Maps 2008, 299–314.

Savigear, R.A.G., 1965. A technique of morphological mapping. Ann. Assoc. Am. Geogr. 55, 514–538.

Shakesby, R.A., 1997. Pronival (protalus) ramparts: a review of forms, processes, diagnostic criteria and palaeoenvironmental implications. Prog. Phys. Geogr. 21, 394–418.

Shakesby, R.A., Matthews, J.A., 1996. Glacial activity and paraglacial landsliding activity in the Devensian Lateglacial: evidence from Craig Cerrig-gleisiad and Fan Dringarth, Forest Fawr (Brecon Beacons), South Wales. Geol. J. 31, 143–158.

Sissons, J.B., 1967. The Evolution of Scotland's Scenery. Oliver and Boyd, Edinburgh.

Sissons, J.B., 1972. The last glaciers in part of the south-east Grampians. Scott. Geogr. Mag. 88, 168–181.

Sissons, J.B., 1974. A Late-glacial ice cap in the central Grampians, Scotland. Trans. Inst. Br. Geogr. 62, 95–114.

Sissons, J.B., 1977a. The Loch Lomond Readvance in the northern mainland of Scotland. In: Gray, J.M., Lowe, J.J. (Eds.), Studies in the Scottish Lateglacial Environment. Pergamon Press, Oxford, pp. 45–59.

Sissons, J.B., 1977b. The Loch Lomond Advance in southern Skye and some palaeoclimatic implications. Scott. J. Geol. 13, 23–36.
Sissons, J.B., 1979. The Loch Lomond Advance in the Cairngorm Mountains. Scott. Geogr. Mag. 95, 66–82.
Sissons, J.B., 1980. The Loch Lomond Advance in the Lake District, northern England. Trans. R. Soc. Edinb. Earth Sci. 71, 13–27.
Smalley, I.J., Unwin, D.J., 1968. The formation and shape of drumlins and their distribution and orientation in drumlin fields. J. Glaciol. 7, 377–390.
Smalley, I., Warburton, J., 1994. The shape of drumlins, their distribution in drumlin fields, and the nature of the sub-ice shaping forces. Sediment. Geol. 91, 241–252.
Smith, M.J., Pain, C.F., 2009. Applications of remote sensing in geomorphology. Prog. Phys. Geogr. 33, 568–582.
Smith, M.J., Rose, J., Booth, S., 2006. Geomorphological mapping of glacial landforms from remotely sensed data: an evaluation of the principal data sources and an assessment of their quality. Geomorphology 76, 148–165.
Sollas, W.J., 1896. A map to show the distribution of eskers in Ireland. Sci. Proc. R. Dublin Soc. 5, 785–822.
St Onge, D.A., 1981. Theories, paradigms, mapping and geomorphology. Can. Geogr. 4, 307–315.
Svensson, H., 1964. Aerial photographs for tracing and investigating fossil tundra ground in Scandinavia. Biul. Perygl. 14, 321–325.
Ten Cate, J.A.M., 1983. Detailed systematic geomorphological mapping in the Netherlands and its applications. Geol. Mijnb. 62, 611–620.
Trenhaile, A.S., 1975. The morphology of a drumlin field. Ann. Assoc. Am. Geogr. 65, 297–312.
Verstappen, H., 2011. Old and new trends in geomorphological and landform mapping. In: Smith, M.J., Paron, P., Griffiths, J. (Eds.), Geomorphological Mapping: A Handbook of Techniques and Applications. Elsevier, Amsterdam.
Waters, R.S., 1958. Morphological mapping. Geography 43, 10–18.
Welby, C.W., 1976. LANDSAT-1 imagery for geologic evaluation. Photogramm. Eng. Remote Sens. 42, 1411–1419.
Welch, R., Howarth, P.J., 1968. Photogrammetric measurements of glacial landforms. Photogramm. Rec. 6, 75–96.
Wilson, P., 2004. Relic rock glaciers, slope failure deposits, or polygenic features? A reassessment of some Donegal debris landforms. Ir. Geogr. 37, 77–87.
Wilson, P., 2009. Rockfall talus slopes and associated talus-foot features in the glaciated uplands of Great Britain and Ireland: periglacial, paraglacial or composite landscapes? In: Knight, J., Harrison, S. (Eds.), Periglacial and Paraglacial Processes and Environments. Geological Society, London, Geological Society Special Publication No. 320, pp. 133–144.
Wright, W.B., 1912. The drumlin topography of south Donegal. Geol. Mag. 9, 153–159.

CHAPTER SEVEN

Data Sources

Takashi Oguchi[a], Yuichi, S. Hayakawa[a] and Thad Wasklewicz[b]
[a]Center for Spatial Information Science, University of Tokyo, Kashiwa, Japan
[b]Department of Geography, East Carolina University, Greenville, NC, USA

Contents

1. INTRODUCTION

Spatial data are fundamental for any mapping activities and they can be classified into two types: raw and derived. For geomorphological mapping, raw data include information about the distribution of height such

Developments in Earth Surface Processes, Volume 15
ISSN: 0928-2025, DOI: 10.1016/B978-0-444-53446-0.00007-0

as contour lines and spot heights on a topographic map and a raster digital elevation model (DEM). In a sense, the acquisition of such elevation data can be called geomorphological mapping. In addition, thematic maps showing the spatial distribution of landform units are typical geomorphological map products for which both raw and derived data are used. Derived data include DEM derivatives such as slope angle, curvature and aspect. Results of the visual interpretation of topographic maps and aerial/satellite images are also derived data useful for applied geomorphological mapping.

Another common binary classification of spatial data is analogue versus digital. Classic spatial data are in an analogue format such as printed maps and handwritten illustrations in field notes. Although analogue data have contributed to the development of geomorphology, the qualitative and subjective nature of these data make them difficult to analyse directly with computers. Since the 1980s, digital maps have largely superseded traditional analogue data sources, and topographic maps have been replaced by DEMs, with their analysis facilitated by related technologies such as fast personal computers and geographical information systems (GIS).

This chapter deals with the various types of spatial data used for geomorphological mapping in both analogue and digital formats. Like the other fields of science and technology, the shift from analogue to digital data is historically important in geomorphology. Therefore, we first describe analogue data and then digital data. The basic characteristics, historical background and some examples of geomorphological mapping using the data are described.

2. ANALOGUE DATA

Analogue data for mapping have been important tools within society. Ancient maps such as those carved on clay tablets in ancient Mesopotamia and those drawn on papyrus in ancient Egypt were based on analogue data representing people's knowledge about geographical locations. By comparison, digital data are recent phenomena – developed in the mid-twentieth century when information storage and management using electronic data processing became possible. Despite this recent development, there are many cases where data for mapping are not always available in a digital format. Therefore, it is important to review the nature of analogue data usable for geomorphological mapping.

2.1 Text Descriptions

Information about the characteristics of landforms is generally written in text. A typical example is the description of field sites, recorded as notes during a field survey. The location of each site must be known in order to use such information for geomorphological mapping. One common way for recording locations in the field is to carry a printed topographic map and document site locations on it directly. Another common way in recent years is to use the global positioning system (GPS) to record geographic coordinates whenever a text description is made.

Text descriptions are not only used to describe particular sites but they also play an important role in recording observed relationships between different landforms. Any reader of geomorphological research articles will be familiar with statements such as *geomorphological mapping was conducted based on field surveys*, although such mapping is usually supported by other methods including map reading and aerial-photograph interpretation (Bocco et al., 2005; Gutiérrez-Santolalla et al., 2005; Van der Schriek et al., 2008). In these cases, text descriptions in field notes played a certain role in the mapping, although they tend to be subjective and qualitative. Text descriptions included in published material may also be used for geomorphological mapping if locational information for each description is available.

2.2 Hand-Drawn Illustrations

Geomorphological descriptions in a field note may also include hand-drawn illustrations or field sketches. These usually take the form of scaled, schematic diagrams and are used to illustrate the characteristics and distributions of landforms; sedimentological sections and exposures; and simple maps depicting the approximate distribution of landforms and sites.

Hand-drawn sketches of scenery were a major source of geomorphological descriptions when the use of photography was limited. Such black line sketches were also suitable for publications when printing technology was limited, and many early articles and books in geomorphology included such sketches (Miller, 1883; Barbour, 1933). To this day, the ability to draw accurate, representative field sketches remains an important skill for geomorphologists. A merit of such drawings is to provide details of target landforms and neglect information about non-essential elements. However, this approach is inevitably subjective because the individual has intentionally focused on a landform at the detriment of the landscape and the image reflects the individual's interpretation of the landform. The subjectivity of the drawings brings into question the comparability of this data source from location to location.

Some old maps also contain handwritten illustrations of landforms to represent topographic relief. For example, one of the oldest maps of the United States (L'America Settentrionale, published in 1677) shows the locations of mountain chains with cartoon-like symbols (hill profiles). Similarly, manual map-drawing techniques such as hachures and hill shading were developed to illustrate topographic relief, although their production required tedious work by a skilled person until methods were automated by computers. Such information on old maps may be useful even today if topographic conditions in the past need to be considered for a particular geomorphological mapping. However, their credibility is relatively low and their quantitative interpretation is difficult.

2.3 Analogue Photographs and Videos for Visual Interpretation

Geomorphological articles and monographs published in the early twentieth century generally included analogue photographs taken in the field (Huntington, 1907; Jones, 1924). Although photography was invented in 1826 by Nicephore Niépce and commercial cameras became available in Europe in the mid-ninteenth century, the use of cameras for geomorphological studies was limited until their availability and portability increased in the twentieth century. Ground photographs have commonly been used for realistic geomorphological descriptions in publications. If the locations of taking photographs are recorded accurately, their information such as small-relief features becomes more useful. However, the role of ground photographs as the basic material of geomorphological mapping is relatively minor because their viewpoints close to the ground led to geometrically distorted images, and their field of vision is usually narrow.

Aerial photographs have been more commonly used for geomorphological mapping than ground photographs. In 1858 Gaspard-Félix Tournachon took a picture of a broad area of Paris from a balloon at a height of *ca.* 80 m, which is considered the first aerial photograph. In the late ninteenth and early twentieth centuries, photographs from a kite or a bird were also tried. After the invention of the airplane by the Wright Brothers in 1903, aerial photography developed rapidly particularly during World War I because it was found to be effective for reconnaissance work. The quality of cameras, films and camera-mounting systems for aerial photography was also improved, and aerial photographs taken by governmental agencies became available. For example, about two-thirds of the conterminous United States was photographed from air by the beginning of the 1940s (Smith, 1941).

The potential of aerial photographs for geoscientific mapping was recognised in the early twentieth century (Tieje, 1929). Smith (1941) evaluated the various applications of aerial photographs in geomorphology and indicated that effective mapping and studies are possible if contour maps and information obtained by field surveys are used along with aerial photographs. Stereo viewing of a pair of photographs to visually understand relief features also contributed to geomorphological mapping. Increasing usage of colour and near-infrared aerial photographs in the late twentieth century has enabled the detection of minor features such as natural levees and abandoned channels on floodplains.

Visual interpretation of aerial photographs has been a major method for manual description, classification and mapping of landforms (Speight, 1977; Easterbrook and Kovanen, 1999). Advantages of aerial photographs are their high resolution, high availability in many places and relatively low cost. Even today, aerial photographs play a significant role in mapping landforms, despite other types of remote data having become available. Recent examples include morpho-tectonic research (Modenesi-Gauttieri et al., 2002), creation of landslide inventories (Van Westen et al., 2008) and analysis of past glacial flow (Jansson et al., 2002).

Videos taken on the ground and from the sky have applications similar to that of ground and aerial photographs. Although the resolution of videos tends to be lower than that of photographs, videos may be suitable for preliminary mapping based on a reconnaissance field survey using a vehicle or a train. In addition, videos and stop-motion photographs can be used for recording and mapping earth-surface movement such as landslides.

The term 'remote sensing' was coined by an American geographer, Evelyn L. Pruitt, around 1960 (Estes and Senger, 1974). The Corona satellite programme (1960–1972) was conducted as the first systematic trial of remote sensing from the space (Ruffner, 2005). A significant difference between current space satellites and the Corona satellites is the type of sensors. The Corona satellites carried analogue cameras for panchromatic films, not digital sensors. Although the Corona programme was mostly successful and high-resolution (1.8–7.5 m) photographs were acquired, all of them were classified for military purposes. Therefore, geomorphological studies based on the satellite images did not develop during the 1960s. An exceptional case study by Verstappen and Van Zuidam (1970) utilised photographs taken by Gemini and Apollo astronauts, as well as low-resolution telemetered images from meteorological satellites including Nimbus and Environmental Science Services Administration (ESSA), to create a generalised geomorphological map of the Sahara. In 1995 the photographs acquired by the Corona satellites were declassified and have been sold at reasonable prices. They have been used for mapping and understanding surface conditions before the Landsat era (Grosse et al., 2005; Takagi et al., 2007).

2.4 Data from Classical Ground Surveying

A plane table has long been used as a measurement instrument since the sixteenth century, and after the establishment of triangulation by Willebrord Snellius in the Netherlands in the early seventeenth century, it became a popular tool for surveying. Although the classical plane-table method has been mostly replaced with the modern methods as described later, it is still often taught in schools and universities as a basic method of surveying.

Based on the principle of triangulation, the plane-table survey uses a portable table, a tape to measure the lengths of a few base lines and a theodolite (also referred to as transit) to measure the horizontal angle between two target points. An alidade, often equipped as a part of a theodolite, can measure vertical angles to obtain height differences. The method has been applied to geomorphological and archaeological field surveys for small areas (Low, 1952; He and Oguchi, 2008).

Levelling survey, using a small hand level or a larger one on a tripod, is another classic surveying method to obtain height differences between points. It can be employed to produce topographic sections for the rapid identification and classification of geomorphic surfaces (Oguchi et al., 2008).

2.5 Topographic Data from Plane-Table and Analogue Photogrammetry

In the mid-ninteenth century, Dominique Francois Jean Arago and Aime Laussedat in France proposed the concept and a prototype of photogrammetry for topographic surveys (Mellor, 1999). The method is called plane-table photogrammetry, which is an extension of the conventional plane-table surveying (Konecny, 1985). Exposed photographs were oriented on a plane table and directions to different objects were transferred onto the map sheet. Cameras useful for plane-table photogrammetry and methods to acquire higher accuracy data were developed in the late nineteenth century, particularly by Albrecht Meydenbauer in Germany (Meyer, 1987). However, these trials were experimental, and practical usage of photogrammetry was very limited at that time.

As noted, aerial photography developed significantly in the early twentieth century. At the same time, mechanical instruments for stereoscopic plotting using overlapping photographs without particular association with a plane table were invented. The Zeiss company started selling such instruments, leading to the propagation of aerial photogrammetry. This method is called analogue photogrammetry and was found particularly useful for making contour maps of mountainous areas with limited access. For example, Finsterwalder (1931) and Petrie and Price (1966) conducted geomorphological mapping of glaciated areas using analogue photogrammetry. The method, however, required tedious work of a skilled operator, and only dealt with nearly vertical photographs with minimal distortion, captured by a metric camera. Consequently, in the field of geomorphology, data obtained directly from analogue photogrammetry were only infrequently used before more advanced photogrammetric methods became widely available in the 1980s. A few exceptions include Lewin and Weir (1977) who mapped landforms on a floodplain in Scotland. However, general topographic maps produced by governmental agencies with an aid of analogue photogrammetry became common in the mid-twentieth century (Bagley, 1941), and they served as basic data sources for numerous geomorphological studies.

2.6 Topographic Maps and Thematic Maps

Although very old maps indicate relative terrain height using symbols and illustrations such as hill profiles, they do not describe landforms quantitatively. Maps with topographic contours were first produced in the mid-eighteenth century to represent river depths in the Netherlands (Van den Brink, 2000; Figure 7.1) and the technique was soon applied to terrestrial areas in France (Friendly and Denis, 2005). In the late eighteenth century, the first multi-sheet topographic map series of an entire country was completed in France (Carte géométrique de la France). Since then many governmental institutes have been charged with making official

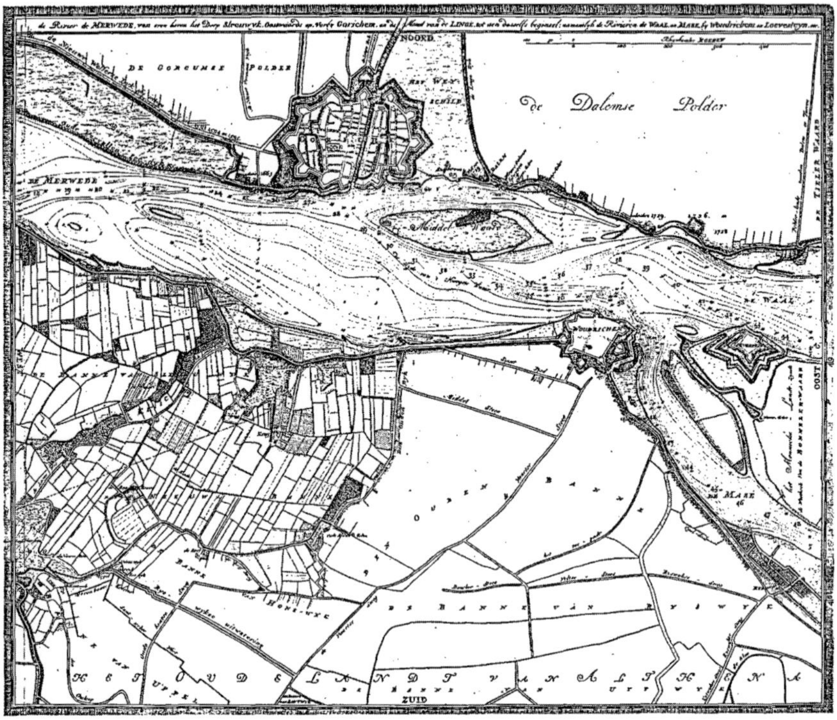

Figure 7.1 An eighteenth-century map showing contour lines of the riverbed in the Netherlands (Van den Brink, 2000).

topographic maps, such as the Ordnance Survey in the United Kingdom. This reflects the strong military demand for accurate topographic information. Such institutes produced numerous medium- to large-scale topographic maps (typically 1:100,000 to 1:20,000).

Governmental topographic maps have contributed significantly to geomorphology. For example, classic studies of quantitative landform analyses in the mid-twentieth century such as Horton (1945) and Strahler (1952) depended on such maps, and morphometric data were retrieved manually from contours and spot heights on the maps. Topographic maps have also supported many regional case studies of geomorphological mapping, particular before DEMs and GIS became widely available (De Graaff et al., 1987).

Thematic maps other than topographic maps may be also used for geomorphological mapping. For example, if landforms are strongly controlled by lithology, geological maps provide useful information on their distribution (Garrote et al., 2006). Land use/cover maps may also be useful for geomorphological mapping; for example, in lowlands, different land use/cover types may correspond to subtle height differences such as those between a natural leveé and an adjacent floodplain. However, such maps are much less frequently used for landform mapping compared to topographic maps because the correlation between proxy data and landforms is not always high.

3. DIGITAL DATA

Since the development of digital information storage in the mid-twentieth century, various types of digital topographic data have become available. Geomorphological research now relies heavily on digital topographic data collected from a variety of sources (Smith et al., 2006). The data vary in scale depending on the data source, and different scales of digital data are selected by geomorphologists dependent upon the scale of the features under investigation. Digital topography has also entered mainstream usage with the advent of digital 3D globes such as Google Earth (Google), World Wind (NASA) and Bing Maps (Microsoft) (Tooth, 2006).

3.1 Digital Ground/Aerial Photographs and Videos for Visual/Optical Interpretation

Since the mid-1990s, digital cameras and videos have been replacing analogue ones. Therefore, most photographs and videos recorded during field surveys are now in digital formats, permitting efficient data handling and storage using a computer. Digital cameras with GPS, or GPS track recorders whose locational records can be transferred to photographs taken by a digital camera based on time stamps, facilitate geomorphological mapping. In most cases, digital photographs are used for visual and qualitative interpretation such as analogue photographs. Graphic software enables the modification of digital photographs for better visual interpretation through the application of filters such as sharpening and edge enhancement as well as colour and histogram adjustment (Lillesand et al., 2008). Digital photographs can also be used for quantitative photogrammetric measurements. Analysis of digital video data using specific software facilitates reconnaissance surveys for geomorphological mapping (Sas et al., 2008).

Aerial photographs can also be taken using digital cameras, including both metric and non-metric types (Dugdale et al., 2010). For obtaining very high-resolution photographs, radio-controlled unmanned aerial vehicles and blimps with lightweight gas can also be used (Lejot et al., 2007; Vericat et al., 2009).

3.2 Digital Satellite Imagery for Visual/Optical Interpretation

The end of the Corona satellite project in 1972 coincided with the beginning of the new US satellite programme, Landsat (originally called ERTS, Earth Resources Technology Satellites). The Landsat satellites were equipped with digital electromagnetic sensors: Multi-Spectral Scanner (MSS) since the beginning of the programme (*ca.* 80 m resolution for Landsat-1 to -5) and later Thematic Mapper (TM), enhanced thematic mapper (ETM) and ETM+ (mostly 30 m resolution, for Landsat-4 to -7). The acquired images were transmitted to Earth using radio waves. Landsat images became widely available to researchers because the programme was designed for resource monitoring and scientific studies. The images were soon found to be useful for geomorphological research (e.g. geomorphology-related papers in Freden et al., 1973) and various applications were made during the 1970s particularly in relation to neotectonics (Welby, 1976; Kayan and Klemas, 1978), mass movement hazards (Ives et al., 1976; Cotecchia, 1978) and general geomorphological mapping (Verstappen, 1977; Johansson and Strömquist, 1978).

A marked advantage of Landsat and more recent satellite imagery is the availability of analysis of multi-frequency (band) data from the optical spectrum. The MSS sensor could take four-band images (green, red and two near-infrared bands) and the TM sensor seven-band images (blue, green, red, and one near-, two middle- and one thermal-infrared bands). Combining three of the bands, a variety of colour composite images can be derived with false and natural colours, enabling easier identification of landform components based on subtle differences in land cover (Figure 7.2) and ground moisture. Another merit of these satellite images is that data are in a digital format from the beginning, permitting direct processing by computers.

A wide range of earth resources and meteorological satellites have since been launched, resulting in a dramatic increase in the availability of satellite images. Meteorological satellites with advanced very high-resolution radiometer (AVHRR) sensors have been in operation since 1978. However, only a few geomorphological studies (Cuq, 1993) used AVHRR images

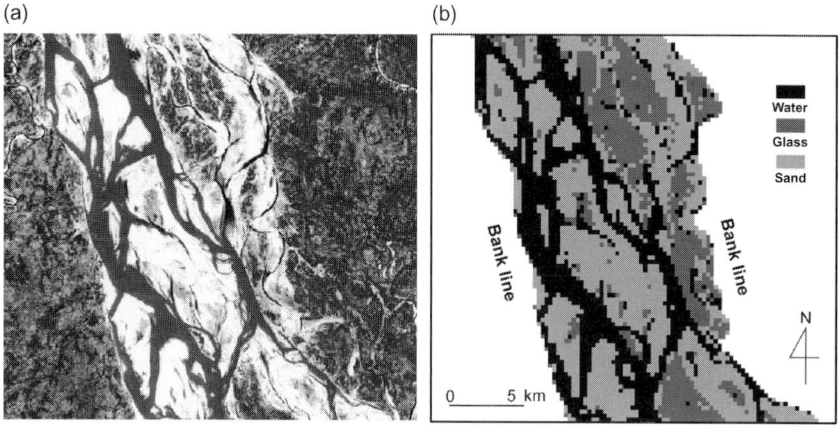

Figure 7.2 (a) Landsat image and (b) derived raster land cover for a part of the Brahmaputra River, Bangladesh (Takagi et al., 2007).

because their resolution (*ca.* 1 km) is low for non-meteorological applications. French SPOT (Satellite Pour l'Observation de la Terre) satellites have been collecting images since 1986 that are more suitable for geomorphological studies. The resolution of SPOT images (2.5–20 m) is better than that of Landsat images, giving an advantage for detailed geomorphological mapping (Callot et al., 1994). However, studies using SPOT images (Smith G.R. et al., 2000) have been limited probably because they are more expensive than Landsat images (Grasso, 1993). Since 2000, low-cost Terra/ASTER (Advanced Spaceborne Thermal Emission and Reflection Radiometer) images with 15 m resolution became available, which have been more frequently used for geomorphological applications including mapping (Fourniadis et al., 2007; Harrison et al., 2008).

A major advancement was the emergence of very high-resolution images from Ikonos in 1999 and Quickbird in 2000. Their *ca.* 1 m (panchromatic) or 3–4 m (multi-band) resolution is equivalent to aerial photographs and has changed the concept of satellite remote sensing (Figure 7.3). These digital data sources have been used for detailed geomorphological mapping and interpretations (Bacon et al., 2008; Siart et al., 2009) particularly for describing small features (Gupta and Liew, 2007). The WorldView and GeoEye satellites launched in the late 2000s have been providing higher resolution images (*ca.* 0.5 m for panchromatic, 2 m for multi-band), which are all the more suitable for detailed

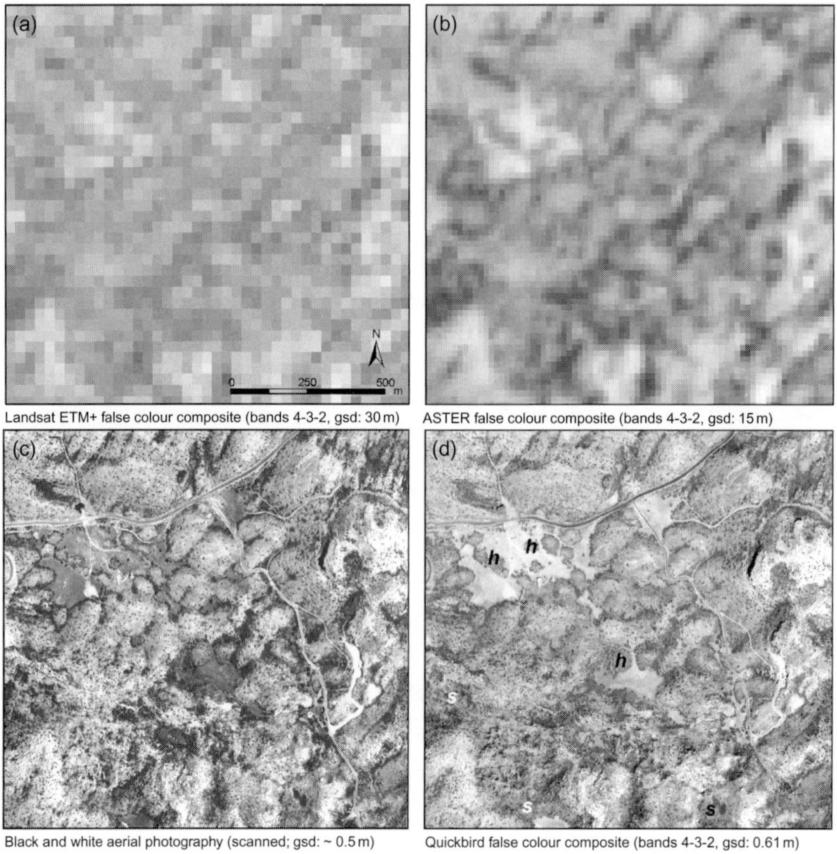

Figure 7.3 Comparison of different remote sensing data with regard to spatial resolution (Siart et al., 2009).

mapping. However, these images are expensive and might not be suitable for research for a wide area because of their large data size.

Another development of recent satellite remote sensing is the use of hyperspectral sensors which can capture data across hundreds of bands. For example, the EO-1 satellite launched in 2000 has a hyperspectral sensor called Hyperion, with 220 unique channels covering visible, near infrared and short-wave infrared. Hyperspectral data are suitable for identifying minerals in surface deposits or regolith (Papp and Cudahy, 2002). Although the use of hyperspectral satellite data in geomorphology is still limited, they allow the production of complex composite images useful for landform mapping based upon differences in surface materials (Wang et al., 2010).

The satellites and sensors described above are all for passive remote sensing, i.e. the detection of reflected or emitted electromagnetic radiation from natural sources. Active remote sensing using synthetic aperture radar (SAR) is another method of obtaining satellite imagery (Palmann et al., 2008). It became available first in 1978 (Seasat, which was only in operation for 3 months), then in the 1980s through the space-shuttle imaging radar (SIR). For example, Seasat data facilitated mapping drumlin fields (Ford, 1981), and SIR data contributed to the discovery and mapping of subsurface valleys (McCauley et al., 1982). Since the 1990s, several satellites with SAR have been launched: ERS-1 and -2, JERS-1, RADARSAT-1, Envisat, ALOS and TerraSAR-X; and visual/optical interpretation of their images has contributed to geomorphological mapping (Li et al., 1998; Glenn and Car, 2004). The major advantages of SAR images include (1) obtaining data through clouds, (2) sensing at both day and night, (3) sensing subsurface conditions and (4) high sensitivity to ground moisture conditions. The first point is particularly important for geomorphological mapping in tropical areas with frequent cloud cover (Haruyama and Shida, 2008).

Images from different satellites and sensors have both advantages and disadvantages. Therefore, many geomorphological studies deal with images from more than one satellite (Coulibaly and Gwyn, 2005; Glasser et al., 2008), and use each, or combinations, of them depending on the characteristics of the target landforms such as dimensions and spatial extent. Gilvear and Bryant (2003), Gupta (2003) and Smith and Pain (2009) reviewed and summarised various applications of remote sensing in geomorphology including geomorphological mapping.

3.3 Digital Aerial Imagery for Visual/Optical Interpretation

Digital sensors similar to those used in satellite remote sensing can be used for airborne remote sensing from fixed-wing aircraft and helicopters. Lower flying heights, compared to satellite remote sensing, provide higher resolution images. Like satellite remote sensing, the most common airborne measurement has been visible and near-infrared multi-spectral imagery. Indeed, airborne sensors similar to the Landsat MSS and TM (AMSS and ATM: A=Airborne) have been used for geoscientific applications since the 1980s (Belanger and Rencz, 1983; Saraf and Cracknell,

1989). Other airborne multi-spectral or hyperspectral sensors such as CASI (Compact Airborne Spectral Imager; Lillesand et al., 2008) have also been developed.

Applications of airborne digital images related to geomorphological mapping until the mid-1990s included landform identification in a remote area (Dean and Morrissey, 1988) and paleochannel mapping based on soil moisture content (Davidson and Watson, 1995). Recent studies deal with more quantitative aspects such as grain-size distribution of gravel-bed rivers (Carbonneau et al., 2004; Dugdale et al., 2010) and bathymetric characteristics based on optical relationships between water depth and reflectance levels (Winterbottom and Gilvear, 1997; Bryant and Gilvear, 1999). Compared to satellite remote sensing images, however, airborne images have been used less for geomorphological mapping, simply because the latter are not collected regularly.

In the United States, governmental 1 m resolution aerial images called digital orthophoto quarter quads (DOQQs) are available for almost the entire country and are free in most states. Although they are usually regarded as aerial photographs, they have either blue-green-red or green-red-near-infrared coverage. Similar multi-band data are provided in some countries instead of conventional aerial photographs, as a useful data source for geomorphological mapping.

3.4 Topographic Data from Modern Ground Surveying

Digital topographic data are often acquired using modern surveying techniques that became readily available in the late twentieth century, developed in and after the 1990s. Four major techniques are introduced below, which are well developed and commonly used for geomorphological mapping.

3.4.1 Global Navigation Satellite Systems

Global navigation satellite systems (GNSS) are satellite-based positioning systems which provide three-dimensional geodetic coordinates of a measurement point (Hofmann-Wellenhof et al., 2008). The GPS developed by the US Department of Defence is the most popular GNSS service. Parallel GNSS are also available or under development by a number of other countries and consortia including the GLONASS by Russia, Galileo by the European Union/European Space Agency and COMPASS (BeiDou) by China. The following description of these systems relates primarily to the GPS.

A GPS device receives signals via radio waves from a constellation of Earth-orbiting satellites. This signal provides information on the distance from the receiver to the satellite, through the travel time, and also additional information on the precise orbital location of the space vehicle. Measurements from at least four satellites are necessary to determine the *XYZ* coordinates of a ground measurement point. Positioning using a single GPS receiver without 'augmentation' usually results in error of >10 m.

A variety of methods have been developed to enhance the positioning accuracy. The most popular of these augmentation systems is differential GPS (DGPS) positioning, which uses two GPS receivers with one (the base) located at a fixed, known location. Collation of data from the base and mobile (rover) stations can be undertaken either in real time or through post-processing and can enhance 3D position accuracy down to centimetric precision. The satellite-based augmentation systems enables real-time differential correction, which is now widely available in most of the commercial GPS receivers giving position accuracy at several metres down to 1 m.

A further augmentation system is carrier-phase GPS (or tracking), which is based on the correspondence of radio wave pulses from different satellites (integral bias) giving better positional accuracy than DGPS. A static carrier-phase measurement, which requires tens of minutes, gives the best accuracy of about 5 mm. Kinematic measurement (a variant of the carrier-phase techniques) requires less measurement time (usually less than a minute) and is suitable for measuring many points with an accuracy in the range of centimetres. The real-time method (real-time kinematic GPS, RTK-GPS) using a wireless connection is a more recent advance that is designed to provide a quick and stable means for differentially correcting data collected with the carrier-phase approaches.

GPS with augmentation has been applied to geomorphological studies (Cornelius et al., 1994; Dykes, 2009). Higgitt and Warburton (1999) mapped landforms in an upland fluvial system using DGPS (Figure 7.4) and found a trade-off of measurement accuracy and efficiency between DGPS and conventional methods. DGPS can be combined with other instruments such as a laser range finder (LRF) (Hayakawa and Tsumura, 2009) or data such as satellite imagery (Vassilopoulou et al., 2002). If carrier-phase GPS is used, repeated mapping of moving landforms including active landslides and glaciers can be performed at millimetre scale (Malet et al., 2002; Hubbard and Glasser, 2005). Tectonic and volcanic geomorphologists have also taken advantage of the carrier-phase GPS to

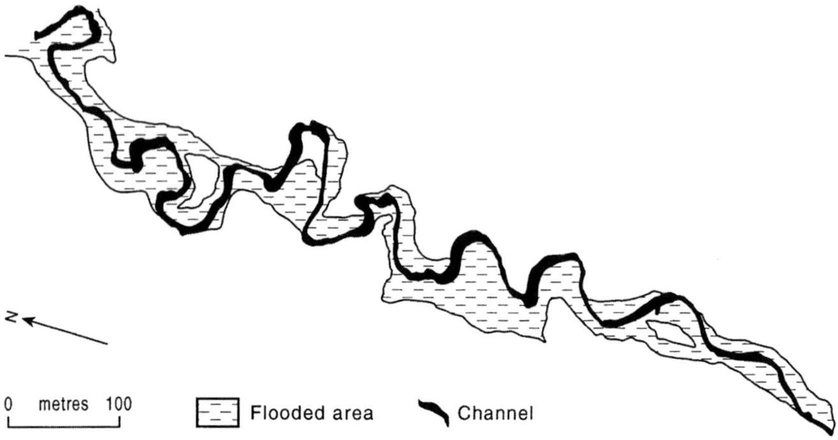

Figure 7.4 DGPS mapping of the extent of a flood of January 1997 at Swinhope Burn, United Kingdom. Flow is from right to left (Higgitt and Warburton, 1999).

investigate fault scarps, volcanoes and uplifted terrace mapping (Sonnette et al., 2010).

3.4.2 Total Station

A total station (TS) is often used for modern land surveying (McCormac, 2003). It is an advance over tape measures, plane tables and theodolites and is currently the standard instrument employed for combined measurement of distance and angles efficiently. A TS emits a laser or radio pulse towards a target, and the travel time of the reflected pulse is converted to the distance from the device to the target. The distance and the horizontal and vertical angles of the emission give *XYZ* coordinates of a target point. A prism reflector is set as a target, and a robotic TS automatically tracks the target for rapid surveying (Kvamme et al., 2006). A typical accuracy of the distance measurement using a TS is 2–3 mm over 1 km and that of angle measurement is 3–5 arc seconds. A TS collects a 'point cloud' of locational data, permitting the creation of a grid DEM for relatively small areas (Mottershead et al., 2008; Yakar, 2009) and is useful for high-resolution geomorphological mapping.

3.4.3 Laser Range Finder

An LRF enables rapid distance measurement between objects of interest at a study site. Some LRFs incorporate an inclinometer, digital compass and an angle encoder and can thus measure vertical and horizontal angles.

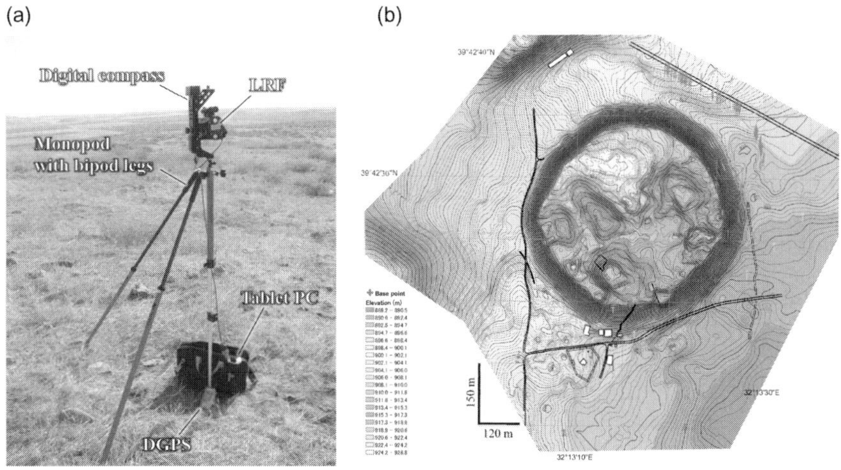

Figure 7.5 (a) LRF instrument combined with DGPS, and (b) LRF-derived topographic map with contour lines at 50 cm interval over 1 m resolution DEM around Hacıtuğrul Tepe, Turkey (Hayakawa and Tsumura, 2009).

An LRF is more portable and lighter than a TS, and a target reflector is usually unnecessary for the measurement of relatively short distances. Although the accuracy of measurement by an LRF (decimetres) is inferior to that of a TS especially in terms of angular measurement, the former is more suitable for rapid and mobile surveying. In addition, an LRF is usually much cheaper than a TS.

High-resolution grid DEMs and triangular irregular networks (TINs) for small areas can be produced from a point cloud collected with an LRF (Hayakawa et al., 2007). If DGPS is combined with an LRF, an accuracy of the final DEM can be better than 1 m (Hayakawa and Tsumura, 2009; Figure 7.5). Topographic profiles along slopes and river channels have also been accurately captured with LRFs (Kogure et al., 2006). LRFs may be effective for rapid geomorphological mapping for a small area if other larger and higher quality devices are unavailable.

3.4.4 Terrestrial Laser Scanning

Terrestrial laser scanning (TLS), also referred to as terrestrial LiDAR (light detection and ranging) or topographic LiDAR, acquires *XYZ* coordinates of numerous points on land by emitting laser pulses toward these points and measuring the distance from the device to the target (Vosselman and Maas, 2010). The number of measurable points within a certain period is much larger than those of TS and LRF devices: a modern TLS device can measure 10^4–10^6 points per second with an accuracy of 10^{-1}–10^0 cm. Bespoke software packages are generally required for managing and analysing the data because of the large amount of data stored in a TLS point cloud. A point cloud may be converted into a grid DEM to facilitate topographic mapping and spatial analyses.

TLS instruments are commonly broken into three categories based on the distance the laser light can travel to record a point in a field-of-view: short-, medium- and long-range scanners. TLS devices optimised for a long range (several hundreds of metres to kilometres) have been applied to measuring spatially larger areas (Hunter et al., 2003; Abellán et al., 2006), whereas shorter range scanners measure spatially smaller areas (up to several hundred metres) in greater detail and accuracy (Heritage and Large, 2009), reflecting a trade-off between the pulse rate and energy of laser light. For short-range scanners, the interval between adjacent measurement points can be up to 1 mm, although such densities are not practical for all but the smallest areas. A potential limitation to TLS approaches in geomorphology is the weight of the instrument (>20 kg including the battery), but as with many technologies lighter devices are being developed.

TLS use in geomorphology has been driven by the need to produce rapid topographic data that are accurate and precise (Heritage and Large, 2009). The precision and accuracy of TLS techniques permit scientists to conduct repeat surveys that are vital to unravelling complex space–time variations in landforms and landscapes. This, in conjunction with data describing process-mechanics, provides strong linkages between processes and forms that are needed to detect environmental change. This has been employed in a number of scenarios. Research in hillslope–channel coupling has combined hydrological and topographical changes in alpine drainages to provide an unprecedented view of channel changes (McCoy et al., 2010). This work has built upon TLS techniques that capture digital micro-topographic data used to analyse channel response to debris flow events (Wasklewicz and Hattanji, 2009; Figure 7.6). Similar approaches have been applied to other geomorphic features including gravel-bed rivers (Hodge et al., 2009) and fault surfaces (Candela et al., 2009; Sagy et al., 2009). Compared to airborne laser scanning, described later, the application of TLS to geomorphology is a relatively recent advancement that has concentrated on smaller spatial extents of the landscape (Heritage and Hetherington, 2007; Schaefer and Inkpen, 2010).

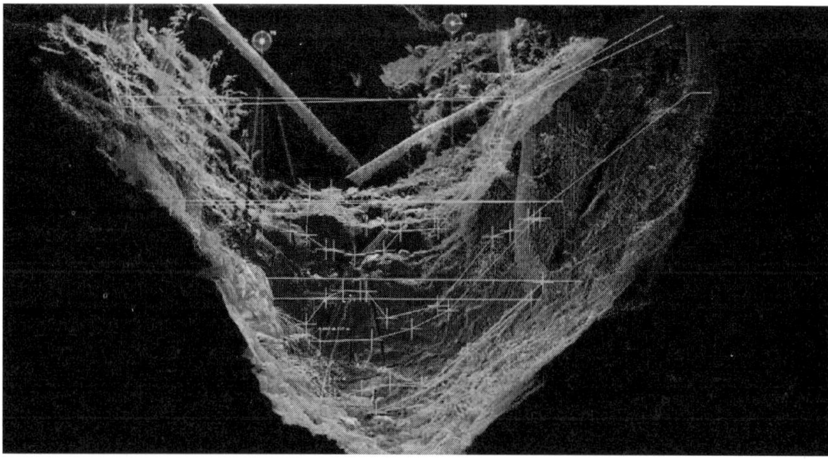

Figure 7.6 A point-cloud image of a headwater channel prior to debris flow event in Ashio, Japan (Wasklewicz and Hattanji, 2009).

3.5 Analytical and Digital Photogrammetry

In the mid-twentieth century, analogue photogrammetry was enhanced by the methods of computational geometry such as bundle adjustment for faster and more accurate mapping using projective equations. This technique is called analytical photogrammetry and it became a common method for deriving topographic maps during the 1970s and 1980s with the development of analytical plotters and computers. Analytical photogrammetry can deal readily with photographs with large distortions including oblique ones and those taken by non-metric cameras. The output is a digital data set, which can be used directly for quantitative analyses using GIS. Data from analytical photogrammetry have been used frequently for geomorphological mapping. In particular, data acquisition for more than one period enables detailed mapping of topographic changes due to mass movement (Fraser, 1983; Chandler and Brunsden, 1995) and fluvial processes (Welch and Jordan, 1983; Lane, 1998). Both aerial and ground photographs as well as satellite images have been used as source data.

Plane-table, analogue and analytical photogrammetry all use printed photographs or images as input data. Photogrammetry using scanned images of photographs and based only on mathematical procedures in a computer is called digital photogrammetry. Its concept was first proposed in the 1950s (Rosenberg, 1955), and specific apparatuses for it (digital photogrammetric workstation) were designed in the early 1980s (Sarajakoski, 1981; Case, 1982). However, real digital photogrammetry without analogue procedures only became available in the mid-1990s (Walker, 1995). Digital photogrammetry has made a significant contribution to geomorphological mapping, especially in the measurement of topographic change over small areas. Target landforms include gravel bars (Heritage et al., 1998), rock glaciers (Berthling et al., 1998), dunes (Brown and Arbogast, 1999) and mountain slopes (De Rose et al., 1998). Digital photogrammetry has also facilitated the use of oblique ground photographs (Chandler et al., 2002).

Although analytical/digital photogrammetry allows systematic mapping using various types of photography (Mikhail et al., 2001), landform dimensions and topographic changes can be approximately measured with ground photographs and relatively simple measurement tools (Graf, 1985; Butler and DeChano, 2001; Maas et al., 2006). Development of this kind of methodology is necessary because typical analytical/digital photogrammetry requires a large amount of indoor work.

3.6 Height Data from Airborne LiDAR and Airborne/Satellite InSAR

The principle and technology of range measurement using airborne LiDAR (Baltsavias, 1999) are the same as those of terrestrial LiDAR (Vosselman and Maas, 2010). Measurement from the sky instead of the ground is suitable for obtaining height distribution in a relatively broad area. To enable stable measurement from a moving airplane or a helicopter, a laser scanner is integrated with RTK-GPS and an inertial measurement unit. Data are typically acquired at a low relative height (200–2000 m from the ground), producing swathes of survey observations of similar width to flying height. The collected point cloud of height includes the height of objects on the ground such as trees and buildings. In the case of trees, it may be possible to obtain both tree and ground surfaces by measuring both first and last returns from a single pulse; the first return should represent the canopy top and the last return should penetrate the tree cover and may represent the ground surface. Conversion of the point-cloud data, including the subtraction of object height, yields a grid DEM with a typical resolution of 0.5–5 m.

Data collected with airborne LiDAR have significantly facilitated geomorphological studies since the turn of the twenty-first century (Lohani and Mason, 2001; Woolard and Colby, 2002). The capability to collect

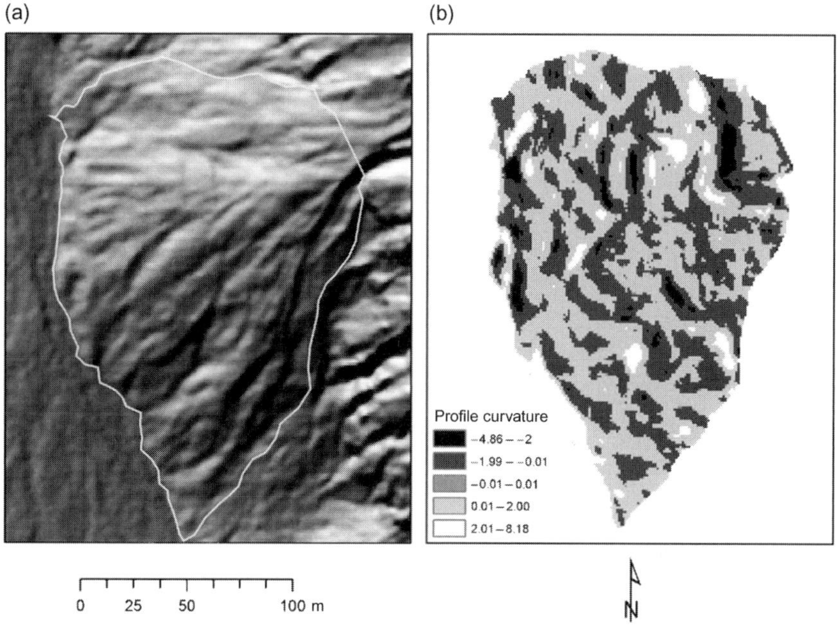

Figure 7.7 (a) Shaded relief and (b) profile curvature maps of an airborne LiDAR-derived DEM for an alluvial fan in Death Valley, United States (Staley et al., 2006).

very high-resolution data has enabled a step change in the science of geomorphometry. The targets of detailed geomorphological mapping using LiDAR DEMs include floodplain features (Jones et al., 2007; Chiverrell et al., 2008; Notebaert et al., 2009), alluvial fans (Staley et al., 2006; Volker et al., 2007; Wasklewicz et al., 2008; Figure 7.7), tectonically deformed landforms (Chan et al., 2007; Hilley and Arrowsmith, 2008), glacial landforms (Smith et al., 2006; Salcher et al., 2010) and landslides (Ardizzone et al., 2007; Booth et al., 2009). Even riverbed bathymetry below water (Hilldale and Raff, 2008) and shallow sea floor (Finkl et al., 2008) can be mapped using multi-band LiDAR sensors, typically green and near infrared. As in the case of aerial photogrammetry, repeated measurements using airborne LiDAR allow the mapping of topographic change (Zhang et al., 2005; Rumsby et al., 2008).

A technique of interferometric SAR (InSAR) enables the construction of topographic data, including DEMs, based on phase differences of multiple SAR images from a satellite or an aircraft acquired at slightly different positions (Graham, 1974; Zebker and Goldstein, 1986; Smith, 2002).

Images for InSAR can be obtained through single-pass or repeat-pass measurements depending on the type of mission (Bürgmann et al., 2000). Compared to airborne LiDAR, satellite InSAR provides a DEM with a coarser resolution but covering a much broader area. A representative example is the global Shuttle Radar Topography Mission (SRTM) DEM, compiled from the InSAR data obtained during the single-pass measurement of SRTM (in February 2000; Rabus, 2003; Kobrick, 2006). Its original resolution is 1 arc second (*ca.* 30 m); however, outside North America, the product is downgraded to a resolution of 3 arc seconds. The SRTM DEM has been used for geomorphological mapping for broad areas (Iwahashi and Pike, 2007; Ehsani and Quiel, 2008). A new global DEM with *ca.* 12 m resolution will be produced using SAR images from the TerraSAR-X and TanDEM-X satellites flying in constellation. InSAR using airborne radar images provides higher resolution DEMs; for example, 5 m NextMap DEMs for all of Europe and the conterminous United States.

Another important product from InSAR data is a map of elevation change based on the technique called differential InSAR (DInSAR; Gabriel et al., 1989). Such maps showing detailed topographic change, or displacement, have contributed to various fields in geomorphology including mass movement (Cascini et al., 2010), tectonic deformation (Fialko et al., 2005), volcanic deformation (Massonnet et al., 1995), ground subsidence (Castañeda et al., 2009), aeolian processes (Liu et al., 2001) and fluvial erosion/deposition (Smith L.C. et al., 2000).

3.7 Compiled Height Information

As noted, modern surveying, photogrammetry and active remote sensing provide digital elevation data, but the raw data from these methods (point clouds) are usually not directly used for geomorphological research. Typically, the data are rearranged by interpolation and compiled into a DEM with a fixed grid size. Topographic information on analogue maps can also be converted into a DEM by interpolating digitised contours and spot heights (Cole et al., 1990). The obtained grid DEMs permit quantitative analysis using GIS software and terrain representation in the form of elevation tints and shaded relief maps (Thelin and Pike, 1991).

Many countries have national DEMs covering the whole territory. For example, the National Elevation Dataset (NED) compiled by the USGS has facilitated the study of landforms in the United States. Some regional/global DEMs have also been released; in 1996 the GTOPO30 DEM with a resolution of 30 arc seconds (*ca.* 1 km) became available, and during the last decade, the 1 or 3 arc seconds SRTM data, the 1 arc second ASTER GDEM and the 5 m NextMap DEM were also released. Numerous researchers have analysed these DEMs for geomorphological mapping and analyses (see Chapter 8 by Smith and Chapter 10 by Seijmonsbergen et al.). They are particularly useful for studies in developing countries which do not have access to high-quality national DEMs.

3.8 Digital Topographic Maps and Thematic Maps

Topographic maps published by governmental agencies and private companies are generally available in both analogue and digital formats. However, most 'digital' topographic maps are only images, with each layer of map components (such as contours) unavailable. The usefulness of such images for geomorphological mapping does not differ significantly from that of analogue topographic maps. Even where contours and spot-height data in vector format are available, they are usually converted into a grid DEM before analysis (Takahashi et al., 2003). Although some methods of geomorphological analyses using vector contours have been proposed (Mizukoshi and Aniya, 2002), their applications have been limited.

Thematic maps such as geology and land cover/use maps are commonly available in the digital vector or raster format. When geology or land cover/use corresponds to specific landform types, such digital data facilitate geomorphological mapping using GIS. Although not essential, digital data showing basic components in a geographical space such as administrative boundaries, major roads and city locations can be added to geomorphological maps to increase map readability.

4. RECENT TRENDS, PROBLEMS AND FUTURE PERSPECTIVES

Although manual cartographic methods using analogue data are still in some places employed for geomorphological mapping, digital data and GIS are currently the most common for mapping activities. The shift from analogue to digital has various advantages such as much reduced time for map production, flexible modification of pre-production maps, easy application of effective visualisation techniques such as hill shading and direct reflection of the result of quantitative morphometric analysis. The drastic decrease in the cost for digital mapping has also facilitated this shift. Free or low-cost digital data are now widely available and the price of GIS software and fast computers have decreased markedly.

Moreover, there is also a fundamental shift in the scale and resolution of digital data now available. For example, although detailed DEMs were commonly produced using aerial photographs and analytical/digital photogrammetry where fine-scale topographic data for a relatively small area were needed, DEMs from airborne LiDAR can now not only fulfil this niche but also provide comparable quality data over much larger areas. The main reason for this shift is that the photogrammetric workflow required to produce a point-cloud or elevation contours from photographs is unnecessary if LiDAR data are used. However, both photogrammetric and LiDAR data require a common conversion procedure: from a digital surface model to a DEM. The conversion generally requires tedious manual work, although significant effort is currently being invested in the development of reliable automated methods (Sohn and Dowman, 2008). This example indicates that the production of base geomorphological mapping still tends to be difficult.

As noted, LiDAR and InSAR/DInSAR allow the detailed detection and analysis of temporal change of topography. However, only very recent topographic changes can be dealt with because these methods only became available in the late twentieth century. The ability to detect longer term topographic changes using additional data for a longer period is important and requires a broader discussion by geomorphologists to determine how to attain this needed information in a manner that is comparable to currently used, and future, digital data sets. To obtain such data with a resolution comparable to that of LiDAR and SAR data, photogrammetry applied to high-quality aerial photographs is effective. For example, Dewitt et al. (2008) examined the movement of landslides since the mid-twentieth century using both photogrammetric and LiDAR DEMs. Such approaches have still been limited, partly because it is challenging to compile and handle data from different sources. It will be some time before LiDAR and SAR data can be used to reconstruct long-term topographic changes, and more research combining mixed data sources will be required to bridge the gap in the time required to compile enough information. Mitasova et al. (2009) have provided a glimpse of the potential advantages of multi-temporal airborne LiDAR data sets.

The increased availability of digital data generally provides opportunities to select the most suitable data from various sources. Although the most precise, accurate and updated data are appropriate in many cases, this is not always true for geomorphological mapping and analyses. For example, mapping small-scale landforms in alluvial lowlands such as natural levees and paleochannels requires high-resolution data, whereas mapping general lowland topography for large-scale geomorphological and hydrological modelling may be made better with lower resolution data where there is no need to reflect minor topographic configurations. The same is true for hillslope mapping – whether to map details such as rock boulders on a slope or not determines the type of data to be used. Quantitative knowledge about the relationship between data scale or resolution and the dimension of mappable landform units (Van Asselen and Seijmonsbergen, 2006) is important to select proper spatial data for geomorphological mapping.

The quality of data (including errors) should also be taken into account when making inferences from map data. Locational error is a common problem with all spatial data, and relevant error metrics such as the standard deviation of observations or the possible maximum observations should be reported by the data provider. In the case of DEMs, two specific types of errors may significantly reduce the quality of geomorphological mapping. One is speckle noise or local spikes/pits that commonly occur in DEMs from remote sensing. The other is 'terracing' in grid DEMs automatically derived from contour data, particularly in lowlands. Although methods to diminish these effects have been proposed (Carrara et al., 1997; Gousie and Franklin, 2005; Stevenson et al., 2010), complete removal without distorting the inherent data quality remains a challenge. Therefore, it is important to understand the characteristics of errors in DEMs and select a DEM based upon the idea of minimising errors for geomorphological mapping under consideration. For example, remotely sensed DEMs are more suitable for slope mapping in the lowland than DEMs derived from vector contour data with terracing artefacts (Hashimoto et al., 2008).

Digital data may need to be interpolated before geomorphological mapping. A common case is to change the coordinate system or projection to enable overlay with other data. Furthermore, the apparent increase in data resolution using interpolation may improve data visualisation such as DEM-based hill shading (Oguchi et al., 2003). It should be noted that interpolated data are secondary products and tend to include additional errors resulting from the interpolation procedure (Hu et al., 2009). The choice of an appropriate interpolation method for each data set is important to minimise such errors. However, researchers are warned against applying GIS software default settings without paying attention to the characteristics of the methods and data.

Although DEMs are important source data for geomorphological mapping, many studies also use other data such as remote sensing images and field descriptions. It may be also necessary to validate the result of DEM-based mapping through a comparison with field-based mapping (Figure 7.8). This reflects the fact that automated landform classification and mapping using only DEMs is still under development, despite recent case studies highlighting their importance (Iwahashi and Pike, 2007; Stepinski and Bagaria, 2009). Indeed, commonly used landform units such as alluvial fans, river terraces and glacial moraines are generally detected based on a heuristic approach including visual interpretation of maps and images; the detection of such units based solely on an automated method is still difficult. Geomorphological mapping has a long history since the era of analogue data and manual cartography, and heuristic approaches have played an important role because landforms are complex objects affected by various factors, and it is impossible to quantify all their characteristics.

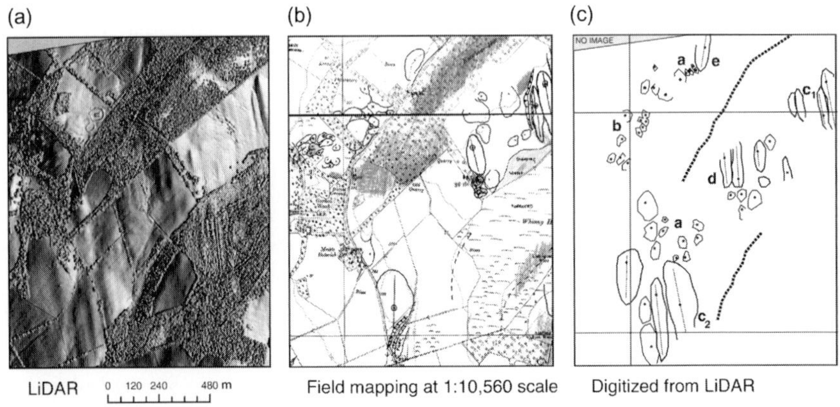

Figure 7.8 Comparison of LiDAR DEM imagery and field mapping (Smith et al., 2006).

Combining quantitative data like DEMs, qualitative data like aerial photographs and expert knowledge for geomorphological mapping is a challenging task. Although quantitative and objective approaches are recommended to provide geomorphological maps as reliable information sources (Guzzetti et al., 1999), a certain level of subjective reasoning is still required in many cases. Numerous examples of geomorphological mapping are based on both quantitative and qualitative data and/or both automated and heuristic approaches. However, few studies have discussed what kind of combination is the most effective for geomorphological mapping. There is no simple answer to this – it is commonly dependent upon the type of landforms and the mapping purpose. Concerning landform types, geomorphological mapping has many targets as shown by the examples illustrated earlier in this chapter. In relation to the purpose of mapping, geomorphological maps are created not only for scientific reasoning but also practical applications such as land use planning (Bocco et al., 2001). The combination of data sources and mapping system used (Gustavsson et al., 2006) should reflect the purpose of mapping. Summarising these complex aspects is difficult, but relevant knowledge needs to be accumulated through future studies for effective geomorphological mapping based on appropriate data sources.

ACKNOWLEDGEMENT

We thank M.J. Smith and J. Brasington for their helpful review comments that improved the manuscript.

REFERENCES

Abellán, A., Vilaplana, J.M., Martínez, J., 2006. Application of a long-range terrestrial laser scanner to a detailed rockfall study at Vall de Núria (Eastern Pyrenees, Spain). Eng. Geol. 88, 136–148.

Ardizzone, F., Cardinali, M., Galli, M., Guzzetti, F., Reichenbach, P., 2007. Identification and mapping of recent rainfall-induced landslides using elevation data collected by airborne Lidar. Nat. Hazards Earth Syst. Sci. 7, 637–650.

Bacon, S.N., McDonald, E.V., Baker, S.E., Caldwell, T.G., Stullenbarger, G., 2008. Desert terrain characterization of landforms and surface materials within vehicle test courses at U.S. Army Yuma Proving Ground, USA. J. Terramech. 45 (5), 167–183.

Bagley, J.W., 1941. Aerophotography and Aerosurveying. McGraw-Hill, New York, 324.

Baltsavias, E.P., 1999. Airborne laser scanning: basic relations and formulas. ISPRS J. Photogramm. Remote Sens. 54, 199–214.

Barbour, G.B., 1933. Pleistocene history of the Huangho. Geol. Soc. Am. Bull. 44, 1143–1160.

Belanger, J.R., Rencz, A.N., 1983. Prospecting in glaciated terrain-integrating airborne and Landsat MSS. Adv. Space Res. 3, 187–191.

Berthling, I., Etzelmiller, B., Eiken, T., Sollid, J.L., 1998. Rock glaciers on Prins Karls Forland, Svalbard. I: internal structure, flow velocity and morphology. Permafrost Periglacial Process. 9, 135–145.

Bocco, G., Mendoza, M., Velazquez, A., 2001. Remote sensing and GIS-based regional geomorphological mapping – a tool for land use planning in developing countries. Geomorphology 39, 211–219.

Bocco, G., Velazquez, A., Siebe, C., 2005. Using geomorphologic mapping to strengthen natural resource management in developing countries. The case of rural indigenous communities in Michoacan, Mexico. Catena 60, 239–253.

Booth, A.M., Roering, J.J., Perron, J.T., 2009. Automated landslide mapping using spectral analysis and high-resolution topographic data: Puget Sound lowlands, Washington, and Portland Hills, Oregon. Geomorphology 109, 132–147.

Brown, D.G., Arbogast, A.F., 1999. Digital photogrammetric change analysis as applied to active coastal dunes in Michigan. Photogramm. Eng. Remote Sens. 65, 467–474.

Bryant, R.G., Gilvear, D.J., 1999. Quantifying geomorphic and riparian land cover changes either side of a large flood event using airborne remote sensing; River Tay, Scotland. Geomorphology 23, 1–17.

Bürgmann, R., Rosen, P.A., Fielding, E.J., 2000. Synthetic aperture radar interferometry to measure Earth's surface topography and its deformation. Ann. Rev. Earth Planet. Sci. 28, 169–209.

Butler, D.R., DeChano, L.M., 2001. Environmental change in Glacier National Park, Montana: an assessment through repeat photography from fire lookouts. Phys. Geogr. 22, 291–304.

Callot, Y., Mering, C., Simonin, A., 1994. Analysis of sand hills massifs on high resolution images: an application to the Great. Western Erg (Algeria). Int. J. Remote Sens. 15, 3799–3822.

Candela, T., Renard, F., Bouchon, M., Brouste, A., Marsan, D., Schmittbuhl, J., et al., 2009. Characterization of fault roughness at various scales: implications of three-dimensional high resolution topography measurements. Pure Appl. Geophys. 166, 1817–1852.

Carbonneau, P.E., Lane, S.N., Bergeron, N.E., 2004. Catchment-scale mapping of surface grain size in gravel-bed rivers using airborne digital imagery. Water Resour. Res. 40, W07202, doi:10.1029/2003WR002759.

Carrara, A., Bitelli, G., Carla, R., 1997. Comparison of techniques for generating digital terrain models from contour lines. Int. J. Geogr. Inf. Sci. 11, 451–473.

Cascini, L., Fornaro, G., Peduto, D., 2010. Advanced low- and full-resolution DInSAR map generation for slow-moving landslide analysis at different scales. Eng. Geol. 112, 29–42.

Case, J., 1982. The digital stereo comparator/compiler (DSCC). Int. Arch. Photogramm. Remote Sens. 24, 23–29.

Castañeda, C., Gutiérrez, F., Manunta, M., Galve, J.P., 2009. DInSAR measurements of ground deformation by sinkholes, mining subsidence, and landslides, Ebro River, Spain. Earth Surf. Processes Landforms 34, 1562–1574.

Chan, Y.-C., Chen, Y.-G., Shih, T.-Y., Huang, C., 2007. Characterizing the Hsincheng active fault in northern Taiwan using airborne LiDAR data: detailed geomorphic features and their structural implications. J. Asian Earth Sci. 31, 303–316.

Chandler, J., Ashmore, P., Paola, C., Gooch, M., Varkaris, F., 2002. Monitoring river-channel change using terrestrial oblique digital imagery and automated digital photogrammetry. Ann. Assoc. Am. Geogr. 92, 631–644.

Chandler, J.H., Brunsden, D., 1995. Steady state behavior of the Black Ven mudslide: the application of archival analytical photogrammetry to studies of landform change. Earth Surf. Processes Landforms 20, 255–275.

Chiverrell, R.C., Thomas, G.S.P., Foster, G.C., 2008. Sediment-landform assemblages and digital elevation data: testing an improved methodology for the assessment of sand and gravel aggregate resources in north-western Britain. Eng. Geol. 99, 40–50.

Cole, G., MacInnes, S., Miller, J., 1990. Conversion of contoured topography to digital-terrain data. Comput. Geosci. 16, 101–109.

Cornelius, S.C., Sear, D.A., Carver, S.J., Heywood, D.I., 1994. GPS, GIS and geomorphological fieldwork. Earth Surf. Processes Landforms 19, 777–787.

Cotecchia, V., 1978. Systematic reconnaissance mapping and registration of slope movement. Bull. In. Assoc. Eng. Geol. 17, 5–37.

Coulibaly, L., Gwyn, Q.H.J., 2005. Integration of optical and radar satellite image data and of topographic data for geomorphologic mapping. Canadian J. Remote Sens. 31, 439–449.

Cuq, F., 1993. Remote sensing of sea and surface features in the area of Golfe d'Arguin, Mauritania. Hydrobiologica 258, 33–40.

Davidson, D.A., Watson, A.I., 1995. Spatial variability in soil moisture as predicted from airborne thematic mapper (ATM) data. Earth Surf. Processes Landforms 20, 219–230.

Dean, K.G., Morrissey, L.A., 1988. Detection and identification of Arctic landforms: an assessment of remotely sensed data. Photogramm. Eng. Remote Sens. 54, 363–371.

De Graaff, L.W.S., De Jong, M.G.G., Rupke, J., Verhofstad, J., 1987. A geomorphological mapping system at scale 1:10,000 for mountainous areas (Austria). Z. Geomorphol. 31, 229–242.

De Rose, R.C., Gomez, B., Marden, M., Trustrum, N.A., 1998. Gully erosion in Mangatu Forest, New Zealand, estimated from digital elevation models. Earth Surf. Processes Landforms 23, 1045–1053.

Dewitte, O., Jasselette, J.-C., Cornet, Y., Van Den Eeckhaut, M., Collignon, A., Poesen, J., et al., 2008. Tracking landslide displacements by multi-temporal DTMs: a combined aerial stereophotogrammetric and LIDAR approach in western Belgium. Eng. Geol. 99, 11–22.

Dugdale, S.J., Carbonneau, P.E., Campbell, D., 2010. Aerial photosieving of exposed gravel bars for the rapid calibration of airborne grain size maps. Earth Surf. Processes Landforms 35, 627–639.

Dykes, A.P., 2009. Geomorphological maps of Irish peat landslides created using hand-held GPS – second edition. J. Maps v2009, 179–185.

Easterbrook, D.J., Kovanen, D.J., 1999. Interpretation of Landforms from Topographic Maps and Air Photographs: A Laboratory Manual. Prentice Hall, Upper Saddle River, NJ.

Ehsani, A.H., Quiel, F., 2008. Geomorphometric feature analysis using morphometric parameterization and artificial neural networks. Geomorphology 99, 1–12.

Estes, J.E., Senger, L.W. (Eds.), 1974. Remote Sensing: Techniques for Environmental Analysis. Hamilton Publishing, Santa Barbara, CA.

Fialko, Y., Sandwell, D., Simons, M., Rosen, P., 2005. Three-dimensional deformation caused by the Bam, Iran, earthquake and the origin of shallow slip deficit. Nature 435 (7040), 295–299.

Finkl, C.W., Becerra, J.E., Achatz, V., Andrews, J.L., 2008. Geomorphological mapping along the upper southeast Florida Atlantic continental platform; I: Mapping units, symbolization and geographic information system presentation of interpreted seafloor topography. J. Coast. Res. 24, 1388–1417.

Finsterwalder, R., 1931. Geschwindigkeitsmessungen an Gletchern mittels Photogrammetrie. Z. Gletscherk. 19, 251–262.

Ford, J.P., 1981. Drumlin fields and glaciated mountains – A contrast in geomorphic perception from Seasat radar images. International Geoscience and Remote Sensing Symposium, Washington, DC, 8–10 June 1981, pp. 864–869.

Fourniadis, I.G., Liu, J.G.., Mason, P.J., 2007. Regional assessment of landslide impact in the Three Gorges area, China, using ASTER data: Wushan-Zigui. Landslides 4, 267–278.

Fraser, C.S., 1983. Photogrammetric monitoring of Turtle Mountains: a feasibility study. Photogramm. Eng. Remote Sens. 49, 1551–1559.

Freden, S.C., Mercanti, E.P., Becker, M.A. (Eds.), 1973. Symposium on Significant Results obtained from the Earth Resources Technology Satellite-1 Symposium. National Aeronautics and Space Administration, Washington, DC.

Friendly, M., Denis, D., 2005. The early origins and development of the scatterplot. J. Hist. Behav. Sci. 41, 103–130.

Gabriel, A.K., Goldstein, R.M., Zebker, H.A., 1989. Mapping small elevation changes over large areas – differential radar interferometry. J. Geophys. Res. 94, 9183–9191.

Garrote, J., Cox, R.T., Swann, C., Ellis, M., 2006. Tectonic geomorphology of the southeastern Mississippi Embayment in northern Mississippi, USA. Geol. Soc. Am. Bull. 118, 1160–1170.

Gilvear, D.J., Bryant, R., 2003. Aerial photography and other remotely sensed data. In: Kondolf, M., Piegey, H. (Eds.), Tools in Fluvial Geomorphology. Wiley, Chichester, pp. 211–247

Glasser, N.F., Jansson, K., Harrison, S., Kleman, J., 2008. The glacial geomorphology and Pleistocene history of South America between 38°S and 56°S. Quat. Sci. Rev. 27, 365–390.

Glenn, N.F., Carr, J.R., 2004. The effects of soil moisture on synthetic aperture radar delineation of geomorphic surfaces in the Great Basin, Nevada, USA. J. Arid Environ. 56, 643–657.

Gousie, M.B., Franklin, W.R., 2005. Augmenting grid-based contours to improve thin-plate DEM generation. Photogramm. Eng. Remote Sens. 71, 67–79.

Graf, W.L., 1985. Geomorphologic measurements from ground-based photographs. In: Pitty, A.F. (Ed.), Themes in Geomorphology. Croom Helms Publishers, London, pp. 211–225

Graham, L.C., 1974. Synthetic interferometer radar for topographic mapping. Proc. IEEE 62, 763–768.

Grasso, D.N., 1993. Applications of the IHS color transformation for 1:24,000-scale geologic mapping: a low cost SPOT alternative. Photogramm. Eng. Remote Sens. 59, 73–80.

Grosse, G., Schirrmeister, L., Kunitsky, V.V., Hubberten, H.-W., 2005. The use of CORONA images in remote sensing of periglacial geomorphology: an illustration from the NE Siberian coast. Permafrost Periglacial Process. 16, 163–172.

Gupta, A., Liew, S.C., 2007. The Mekong from satellite imagery: a quick look at a large river. Geomorphology 85, 259–274.

Gupta, R.P., 2003. Remote Sensing Geology. second ed. Springer-Verlag, Heidelberg.

Gustavsson, M., Kolstrup, E., Seijmonsbergen, A.C., 2006. A new symbol-and-GIS based detailed geomorphological mapping system: renewal of a scientific discipline for understanding landscape development. Geomorphology 77, 90–111.

Gutiérrez-Santolalla, F., Gutiérrez-Elorza, M., Marín, C., Desir, G., Maldonado, C., 2005. Spatial distribution, morphometry and activity of La Puebla de Alfindén sinkhole field in the Ebro River valley (NE Spain) applied aspects for hazard zonation. Environ. Geol. 48, 360–369.

Guzzetti, F., Carrara, A., Cardinali, M., Reichenbach, P., 1999. Landslide hazard evaluation: a review of current techniques and their application in a multi-scale study, Central Italy. Geomorphology 31, 181–216.

Harrison, S., Glasser, N.F., Winchester, V., Haresign, E., Warren, C., Duller, G.A.T., et al., 2008. Glaciar León, Chilean Patagonia: late-Holocene chronology and geomorphology. Holocene 18, 643–652.

Haruyama, S., Shida, K., 2008. Geomorphologic land classification map of the Mekong Delta utilizing JERS-1 SAR images. Hydrol. Process. 22, 1373–1381.

Hashimoto, A., Oguchi, T., Hayakawa, Y., Lin, Z., Saito, K., Wasklewicz, T.A., 2008. GIS analysis of depositional slope change at alluvial-fan toes in Japan and the American Southwest. Geomorphology 100, 120–130.

Hayakawa, Y.S., Tsumura, H., 2009. Utilization of laser range finder and differential GPS for high-resolution topographic measurement at Hacıtuğrul Tepe, Turkey. Geoarchaeology 24, 176–190.

Hayakawa, Y.S., Oguchi, T., Komatsubara, J., Ito, K., Hori, K., Nishiaki, Y., 2007. Rapid on-site topographic mapping with a handheld laser range finder for a geoarchaeological survey in Syria. Geogr. Res. 45, 95–104.

He, H., Oguchi, T., 2008. Late Quaternary activities of the Zemuhe and Xiaojiang faults in southwest China from geomorphological mapping. Geomorphology 96, 62–85.

Heritage, G., Hetherington, D., 2007. Towards a protocol for laser scanning in fluvial geomorphology. Earth Surf. Processes Landforms 32, 66–74.

Heritage, G., Large, A. (Eds.), 2009. Laser Scanning for the Environmental Sciences. Wiley-Blackwell, Chichester, , 278 pp..

Heritage, G.L., Fuller, I.C., Charlton, M.E., Brewer, P.A., Passmore, D.P., 1998. CDW photogrammetry of low relief fluvial features: accuracy and implications for reach-scale sediment budgeting. Earth Surf. Processes Landforms 23, 1219–1233.

Higgitt, D.L., Warburton, J., 1999. Applications of differential GPS in upland fluvial geomorphology. Geomorphology 29, 121–134.

Hilldale, R.C., Raff, D., 2008. Assessing the ability of airborne LiDAR to map river bathymetry. Earth Surf. Processes Landforms 33, 773–783.

Hilley, G.E., Arrowsmith, J.R., 2008. Geomorphic response to uplift along the Dragon's Back pressure ridge, Carrizo Plain, California. Geology 36, 367–370.

Hodge, R., Brasington, J., Richards, K.S., 2009. In situ characterization of grain-scale fluvial morphology using terrestrial laser scanning. Earth Surf. Processes Landforms 34, 954–968.

Hofmann-Wellenhof, B., Lichtenegger, H., Wasle, E., 2008. GNSS – Global Navigation Satellite Systems: GPS, GLONASS, Galileo & More. Springer, New York.

Horton, R.E., 1945. Erosional development of streams and their drainage basins: hydrophysical approach to quantitative morphology. Geol. Soc. Am. Bull. 56, 275–370.

Hu, P., Liu, X., Hu, H., 2009. Accuracy assessment of digital elevation models based on approximation theory. Photogramm. Eng. Remote Sens. 75, 49–56.

Hubbard, B., Glasser, N.F., 2005. Field Techniques in Glaciology and Glacial Geomorphology. Wiley, Chichester, 400 pp.

Hunter, G., Pinkerton, H., Airey, R., Calvari, S., 2003. The application of a long-range laser scanner for monitoring volcanic activity on Mount Etna. J. Volcanol. Geotherm. Res. 123, 203–210.

Huntington, E., 1907. Some characteristics of the glacial period in non-glaciated regions. Geol. Soc. Am. Bull. 18, 351–388.

Ives, J.D., Mears, A.I., Carrara, P.E., Bovis, M.J., 1976. Natural hazards in Mountain Colorado. Ann. Assoc. Am. Geogr. 66, 129–144.

Iwahashi, J., Pike, R.J., 2007. Automated classifications of topography from DEMs by an unsupervised nested-means algorithm and a three-part geometric signature. Geomorphology 86, 409–440.

Jansson, K.N., Kleman, J., Marchant, D.R., 2002. The succession of ice-flow patterns in North-central Québec-Labrador, Canada. Quat. Sci. Rev. 21, 503–523.

Johansson, D., Strömquist, L., 1978. Interpretation of geomorphology and vegetation of LANDSAT satellite images from semi-arid central Tanzania. Nor. Geol. Tidsskr. 32 (2), 49–54.

Jones, A.F., Brewer, P.A., Johnstone, E., Macklin, M.G., 2007. High-resolution interpretative geomorphological mapping of river valley environments using airborne LiDAR data. Earth Surf. Processes Landforms 32, 1574–1592.

Jones, O.T., 1924. The upper Towy drainage system. Q. J. Geol. Soc. London 80, 568–609.

Kayan, I., Klemas, V., 1978. Application of LANDSAT imagery to studies of structural geology and geomorphology of the Mentese Region of southwestern Turkey. Remote Sens. Environ. 7, 51–60.

Kobrick, M., 2006. On the toes of giants: how SRTM was born. Photogramm. Eng. Remote Sens. 72, 206–210.

Kogure, T., Aoki, H., Maekado, A., Hirose, T., Matsukura, Y., 2006. Effect of the development of notches and tension cracks on instability of limestone coastal cliffs in the Ryukyus, Japan. Geomorphology 80, 236–244.

Konecny, G., 1985. The International Society for Photogrammetry and Remote Sensing – 75 years old, or 75 years young: keynote address. Photogramm. Eng. Remote Sens. 51, 919–933.

Lane, S.N., 1998. The use of digital terrain modelling in the understanding of dynamic river channel systems. In: Lane, S.N., Richards, K.S., Chandler, J.H. (Eds.), Landform Monitoring. Modelling and Analysis. Wiley, Chichester, pp. 311–342.

Kvamme, K.L., Ernenwein, E.G., Markussen, C.J., 2006. Robotic total station for microtopographic mapping: an example from the northern Great Plains. Archaeol. Prospect. 13, 91–102.

Lejot, J., Delacourt, C., Piégay, H., Fournier, T., Trémélo, M.-L., Allemand, P., 2007. Very high spatial resolution imagery for channel bathymetry and topography from an unmanned mapping controlled platform. Earth Surf. Processes Landforms 32, 1705–1725.

Lewin, J., Weir, J.C., 1977. Morphology and recent history of the Lower Spey. Scott. Geog. Mag. 93, 45–51.

Li, W., Bénié, G.B., He, D.-C., Wang, S., Ziou, D., Gwyn, Q.H.J., 1998. Classification of SAR images using morphological texture features. Int. J. Remote Sens. 19, 3399–3410.

Lillesand, T.M., Keifer, R.W., Chipman, J., 2008. Remote Sensing and Image Interpretation. Wiley, New York.

Liu, J.G., Black, A., Lee, H., Hanaizumis, H., Moore, J.M., 2001. Land surface change detection in a desert area in Algeria using multi-temporal ERS SAR coherence images. Int. J. Remote Sens. 22, 2463–2477.

Lohani, B., Mason, D.C., 2001. Application of airborne scanning laser altimetry to the study of tidal channel geomorphology. ISPRS J. Photogramm. Remote Sens. 56, 100–120.

Low, J.W., 1952. Plane Table Mapping. Harper, New York, 365 pp.

Maas, H.-G., Dietrich, D., Schwalbe, E., Bäßler, M., Westfeld, P., 2006. Analysis of the motion behaviour of Jakobshavn Isbræ glacier in Greenland by monocular image sequence analysis. Int. Arch. Photogramm. Remote Sens. Spatial Inf. Sci. XXXVI (Part 5), 179–183.

Malet, J.P., Maquaire, O., Calais, E., 2002. The use of global positioning system techniques for the continuous monitoring of landslides: application to the Super-Sauze earthflow (Alpes-de-Haute-Provence, France). Geomorphology 43, 33–54.

Massonnet, D., Briole, P., Arnaud, A., 1995. Deflation of Mount Etna monitored by spaceborne radar interferometry. Nature 375 (6532), 567–570.

McCauley, J.F., Schaber, G.G., Breed, C.S., Grolier, M.J., Haynes, C.V., Issawi, et al., 1982. Subsurface valleys and geoarcheology of the eastern Sahara revealed by shuttle radar. Science 218, 1004–1020.

McCormac, J.C., 2003. Surveying. fifth ed. Wiley, Chichester.

McCoy, S.W., Kean, J.W., Coe, J.A., Staley, D.M., Wasklewicz, T.A., Tucker, G.E., 2010. Evolution of a natural debris flow: in situ measurements of flow dynamics, video imagery and terrestrial laser scanning. Geology 38, 735–738.

Mellor, J.P., 1999. Automatically recovering geometry and texture from large sets of calibrated images. Massachusetts Institute of Technology Artificial Intelligence Laboratory, A.I. Technical Report No. 1674, 133 pp.

Meyer, R., 1987. 100 years of architectural photogrammetry. Kompendium Photogrametrie 19, 183–200.

Mikhail, E.M., Bethel, J.S., McGlone, J.C., 2001. Introduction to Modern Photogrammetry. Wiley, New York.

Miller, H., 1883. River-terracing: its methods and their results. Proc. R. Phys. Soc., Edinburgh 7, 263–306.

Mitasova, H., Overton, M.F., Recalde, J.J., Bernstein, D.J., Freeman, C.W., 2009. Raster-based analysis of coastal terrain dynamics from multitemporal LiDAR data. J. Coast. Res. 25, 507–514.

Mizukoshi, H., Aniya, M., 2002. Use of contour-based DEMs for deriving and mapping topographic attributes. Photogramm. Eng. Remote Sens. 68, 83–93.

Mondenesi-Gauttieri, M.C., Hiruma, S.T., Riccomini, C., 2002. Morphotectonics of a high plateau on the northwestern flank of the continental rift of southeastern Brazil. Geomorphology 43, 257–271.

Mottershead, D.N., Duane, W.J., Inkpen, R.J., Wright, J.S., 2008. An investigation of the geometric controls on the morphological evolution of small-scale salt terrains, Cardona, Spain. Environ. Geol. 53, 1091–1098.

Notebaert, B., Verstraeten, G., Govers, G., Poesen, J., 2009. Qualitative and quantitative applications of LiDAR imagery in fluvial geomorphology. Earth Surf. Processes Landforms 34, 217–231.

Oguchi, T., Aoki, T., Matsuta, N., 2003. Identification of an active fault in the Japanese Alps from DEM-based hill shading. Comp. Geosci. 29, 885–891.

Oguchi, T., Hori, K, Oguchi, C.T., 2008. Paleohydrological implications of late Quaternary fluvial deposits in and around archaeological sites in Syria. Geomorphology 101, 33–43.

Palmann, C., Mavromatis, M., Sequeira, J., Brisco, B., 2008. Earth observation using radar data: an overview of applications and challenges. Int. J. Digital Earth 1, 171–195.

Papp, E., Cudahy, T., 2002. Hyperspectral remote sensing. In: Papp, E. (Ed.), Geophysical and Remote Sensing Methods for Regolith Exploration, CRCLEME Open File Report 144, pp. 13–21.

Petrie, G., Price, R.J., 1966. Photogrammetric measurements of the ice wastage and morphological changes near the Casement Glacier, Alaska. Can. J. Earth Sci. 3, 827–840.

Rabus, B., Eineder, M., Roth, A., Bamler, R., 2003. The shuttle radar topography mission – a new class of digital elevation models acquired by spaceborne radar. J. Photogramm. Remote Sens. 57, 241–262.

Rosenberg, P., 1955. Information theory and electronic photogrammetry. Photogramm. Eng. Remote Sens. 21, 543–555.

Ruffner, K.C. (Ed.), 2005. Corona: America's First Satellite Program. CIA History Staff, Center for the Study of Intelligence, Central Intelligence Agency, Washington, DC.

Rumsby, B.T., Brasington, J., Langham, J.A., McLelland, S.J., Middleton, R., Rollinson, G., 2008. Monitoring and modelling particle and reach-scale morphological change in gravel-bed rivers: applications and challenges. Geomorphology 93, 40–54.

Sagy, A., Brodsky, E.E., Axen, G.J., 2009. Evolution of fault-surface roughness with slip. Geology 35, 283–286.

Salcher, B.C., Hinsch, R., Wagreich, M., 2010. High-resolution mapping of glacial landforms in the North Alpine Foreland, Austria. Geomorphology 122, 283–293.

Saraf, A.K., Cracknell, A.P., McManus, J., 1989. Geobotanical application of airborne thematic mapper data in Sutherland, north-west Scotland. Int. J. Remote Sens. 10, 545–555.

Sarajakoski, T., 1981. Concept of a completely digital stereo plotter. Photogramm. J. Finland 8, 95–100.

Sas Jr., R.J., Sklar, L.S., Eaton, L.S., Davis, J., 2008. A method for developing regional road-fill failure hazard assessments using GIS and virtual fieldwork. Environ. Eng. Geosci. 14, 221–229.

Schaefer, M., Inkpen, R., 2010. Towards a protocol for laser scanning of rock surfaces. Earth Surf. Processes Landforms 35, 417–423.

Siart, C., Bubenzer, O., Eitel, B., 2009. Combining digital elevation data (ASTER/SRTM), high resolution satellite imagery (Quickbird) and GIS for geomorphological mapping: a multi-component case study on Mediterranean karst in Central Crete. Geomorphology 112, 106–121.

Smith, G.R., Woodward, J.C., Heywood, D.I., Gibbard, P.L., 2000. Interpreting Pleistocene glacial features from SPOT HRV data using fuzzy techniques. Comp. Geosci. 26, 479–490.

Smith, H.T.U., 1941. Aerial photographs in geomorphic studies. J. Geomorphol. 4, 171–205.

Smith, L.C., 2002. Emerging applications of interferometric synthetic aperture radar (InSAR) in geomorphology and hydrology. Ann. Assoc. Am. Geogr. 92, 385–398.

Smith, L.C., Alsdorf, D.E., Magilligan, F.J., Gomez, B., Mertes, L.A.K., Smith, N.D., et al., 2000. Estimation of erosion, deposition, and net volumetric change caused by the 1996 Skeidararsandur jokulhlaup, Iceland, from synthetic aperture radar interferometry. Water Resour. Res. 36, 1583–1594.

Smith, M.J., Pain, C.F., 2009. Applications of remote sensing in geomorphology. Prog. Phys. Geogr. 33, 568–582.

Smith, M.J., Rose, J., Booth, S., 2006. Geomorphological mapping of glacial landforms from remotely sensed data: an evaluation of the principal data sources and an assessment of their quality. Geomorphology 76, 148–165.

Sohn, G., Dowman, I.J., 2008. A model-based approach for reconstructing a terrain surface from airborne LIDAR data. Photogramm. Rec. 23, 170–193.

Sonnette, L., Angelier, J., Villemin, T., Bergerat, F., 2010. Faulting and fissuring in active oceanic rift: surface expression, distribution and tectonic–volcanic interaction in the Thingvellir Fissure Swarm, Iceland. J. Struct. Geol. 32, 407–422.

Speight, J.G., 1977. Landform pattern description from aerial photographs. Photogrammetria 32, 161–182.

Staley, D.M., Wasklewicz, T.A., Blaszczynski, J.S., 2006. Surficial patterns of debris flow deposition on alluvial fans in Death Valley, CA using airborne laser swath mapping data. Geomorphology 74, 152–163.

Stepinski, T.F., Bagaria, C., 2009. Segmentation-based unsupervised terrain classification for generation of physiographic maps. IEEE Geosci. Remote Sens. Lett. 6, 733–737.

Stevenson, J.A., Sun, X., Mitchell, N.C., 2010. Despeckling SRTM and other topographic data with a denoising algorithm. Geomorphology 114, 238–252.

Strahler, A.N., 1952. Hypsometric (area–altitude) analysis of erosional topography. Geol. Soc. Am. Bull. 63, 1117–1142.

Takagi, T., Oguchi, T., Matsumoto, J., Grossman, M.J., Sarker, M.H., Matin, M.A., 2007. Channel braiding and stability of the Brahmaputra River, Bangladesh, since 1967: GIS and remote sensing analyses. Geomorphology 85, 294–305.

Takahashi, A., Oguchi, T., Sugimori, H., 2003. Effects of digital elevation model resolution on topographic representation: a case study in the Tama area, western Tokyo. Geogr. Rev. Japan 76, 800–818 (in Japanese with English abstract).

Thelin, G.P., Pike, R.J., 1991. Landforms of the conterminous United States – a digital shaded-relief portrayal. USGS Map I–2206.

Tieje, A.J., 1929. The study of geology by aeroplane. Science 69 (1785), 301–302.

Tooth, S., 2006. Virtual globes: a catalyst for the re-enchantment of geomorphology? Earth Surf. Processes Landforms 31, 1192–1194.

van Asselen, S., Seijmonsbergen, A.C., 2006. Expert-driven semi-automated geomorphological mapping for a mountainous area using a laser DTM. Geomorphology 78, 309–320.

Van Den Brink, P., 2000. River landscapes: the origin and development of the printed river map in the Netherlands, 1725–1795. Imago Mundi 52, 66–78.

Van der Schriek, T., Passmore, D., Franco Mugica, F., Stevenson, A.C., Boomer, I., Rolão,, J., 2008. Holocene palaeoecology and floodplain evolution of the Muge tributary, Lower Tagus Basin, Portugal. Quat. Int. 189, 135–151.

Van Westen, C.J., Castellanos, E., Kuriakose, S.L., 2008. Spatial data for landslide susceptibility, hazard, and vulnerability assessment: an overview. Eng. Geol. 102, 112–131.

Vassilopoulou, S., Hurni, L., Dietrich, V., Baltsavias, E., Pateraki, M., Lagios, E., et al., 2002. Orthophoto generation using IKONOS imagery and high-resolution DEM: a case study on volcanic hazard monitoring of Nisyros Island (Greece). ISPRS J. Photogramm. Remote Sens. 57, 24–38.

Vericat, D., Brasington, J., Wheaton, J., Cowie, M., 2009. Accuracy assessment of aerial photographs acquired using lighter-than-air blimps: low-cost tools for mapping river corridors. River Res. Appl. 25, 985–1000.

Verstappen, H.Th., 1977. Remote Sensing in Geomorphology. Elsevier, Amsterdam, 214 pp..

Verstappen, H.Th., Van Zuidam, R.A., 1970. Orbital photography and the geosciences: a geomorphological example from the central Sahara. Geoforum 2, 33–47.

Volker, H.X., Wasklewicz, T.A., Ellis, M.A., 2007. A topographic fingerprint to distinguish alluvial fan formative processes. Geomorphology 88, 34–45.

Vosselman, G., Maas, H.-G., 2010. Airborne and Terrestrial. Laser Scanning. Whittles Publishing, Dunbeath.

Walker, A.S., 1995. Analogue, analytical and digital photogrammetric workstations: practical investigations of performance. Photogramm. Rec. 15 (85), 17–25.

Wang, J., Chen, Y., He, T., LV, C., Liu, A., 2010. Application of geographic image cognition approach in land type classification using Hyperion image: a case study in China. Int. J. Appl. Earth Obs. Geoinf. 12, S212–S222.

Wasklewicz, T.A., Hattanji, T., 2009. High-resolution analysis of debris flow-induced channel changes in a headwater stream, Ashio Mountains, Japan. Prof. Geogr. 61, 231–249.

Wasklewicz, T.A., Mihir, M., Whitworth, J., 2008. Surface variability of alluvial fans generated by disparate processes, eastern Death Valley, CA. Prof. Geogr. 60, 207–223.

Welby, C.W., 1976. Landsat-1 imagery for geologic evaluation. Photogramm. Eng. Remote Sens. 42, 1411–1419.

Welch, R., Jordan, T.R., 1983. Analytical non-metric close-range photogrammetry for monitoring stream channel erosion. Photogramm. Eng. Remote Sens. 49, 367–374.

Winterbottom, S.J., Gilvear, D.J., 1997. Quantification of channel bed topography within gravel bed-rivers using aerial photography and multi-spectral imagery. Earth Surface Processes and Landforms 13, 1–11.

Woolard, J.W., Colby, J.D., 2002. Spatial characterization, resolution, and volumetric change of coastal dunes using airborne LiDAR: Capte Gatteras, North Carolina. Geomorphology 48, 269–287.

Yakar, M., 2009. Digital elevation model generation by robotic total station instrument. Exp. Tech. 33 (2), 52–59.

Zebker, H., Goldstein, R., 1986. Topographic mapping from interferometric synthetic aperture radar observations. J. Geophys. Res. 91 (B5), 4992–4999.

Zhang, K., Whitman, D., Leatherman, S., Robertson, W., 2005. Quantification of beach changes caused by Hurricane Floyd along Florida's Atlantic coast using airborne laser surveys. J. Coast. Res. 21, 123–134.

CHAPTER EIGHT

Digital Mapping: Visualisation, Interpretation and Quantification of Landforms

Mike J. Smith
School of Geography, Geology and the Environment, Kingston University, Kingston upon Thames, Surrey, UK

Contents

1. INTRODUCTION

Geomorphology is that part of physical geography that deals with the form of the Earth's land surface and the processes that act upon and shape it. Geomorphological mapping specifically deals with recording the location and distribution of landforms of interest for the production of geomorphological maps in their own right or as observational inputs into modelling and, within the context of scientific investigation, dates from the mid-1800s (Close, 1867). Since the 1950s, remotely sensed data sources have become more widely available (Smith and Pain, 2009), initially through the rapid increase in the collection of aerial photography. Photography covers relatively large areas, vertically and in stereo, allowing

Developments in Earth Surface Processes, Volume 15
ISSN: 0928-2025, DOI: 10.1016/B978-0-444-53446-0.00008-2

the mapping and interpretation of morphology that would previously have been difficult from the ground, as well as access to remote or inhospitable regions. In addition, the use of remotely sensed data can increase the *rate* of geomorphological mapping by at least an order of magnitude. Rates of field mapping are ~2 km^2 per day (Knight et al., 2010) in comparison to ~10–100 km^2 per day for remote mapping. Combined, these factors drove the establishment of geomorphological mapping from remotely sensed data.

1.1 Development of Data and Approaches

Although aerial photography was generally acquired at medium resolutions (~1:5000–1:50,000), it was the advent of satellite-based imaging that allowed the study of regional geomorphology as exemplified by Short and Blair (1986). The emergence and use of satellite imagery within geomorphology occurred during a general period of transition to digital data and digital processing. Not surprisingly, there was considerable investment in digital cartography through, for example, the Experimental Cartography Unit (ECU; Rhind, 1988) in the United Kingdom. ECU research looked at the digital representation of terrain, with Evans (1972) reviewing the geomorphometric framework for the quantitative analysis of the land surface. Digital elevation models (DEMs) formed the primary inputs to geomorphometric studies and were principally digitised contours (Pike et al., 2008). It was not until the 1990s that a general increase in the amount of DEM data occurred through the proliferation of collection techniques and providers. Oguchi et al. (2011) document the use of digital photogrammetry, interferometric synthetic aperture radar (InSAR) and light detection and ranging (LiDAR) for the provision of data for geomorphological mapping, and any project should be aware of the different data sources and their potential limitations. The wide availability of satellite imagery has generally decreased the unit cost (per km^2) of acquisition, and whilst not strictly a geomorphological consideration, many nationally collected data sets, for example those funded by the federal government in the United States, are made available free from charge to the end-user.

1.2 Terminology and Data Format Used

There is a varied and inconsistent use of terminology throughout the geomorphometric literature and this impacts upon digital mapping. For consistency, this chapter uses the definitions of Pike et al. (2008) shown in Table 8.1. In addition to varied terminology, several data models exist

Table 8.1 Definitions of Geomorphometric Terms

Term	Definition
Landform	A feature related to a process (or process-complex), usually composed of several 'elementary forms'
Landscape	Generally refers to a broad area which can have both human and physical attributes
Terrain	Applied imprecisely (with both qualitative and quantitative applications) and is therefore used in the same manner as *landscape*, except that it specifically refers to the land surface
Topography	Generally refers to the physical shape of the landscape, but has broader connotations (e.g. topographic map)
Altitude	Refers to the vertical distance above the sea surface and is generally synonymous with *elevation* (although elevation is also used to denote *rates* of uplift)
Relief	Best used to mean a range in altitude, although again it is used interchangeably with topography

Source: After Pike et al. (2008).

for the storage of altitudinal data. The gridded raster, triangulated irregular network (TIN) and point cloud (Pike et al., 2008) are common data models. However, it is the raster grid that has become the *de facto* standard for geomorphometric analysis and is therefore used for the examples presented in this chapter.

1.3 Manual Mapping: Overview and Limitations

The availability of digital remotely sensed data allows two approaches to geomorphological mapping. First, *manual* mapping is based upon the expertise and experience of the interpreter to identify and outline landforms of interest in the same manner that has been used for interpreting analogue aerial photographs (Colwell, 1983). This is a subjective process using complex visual heuristics to develop relationships between features in the displayed image leading to feature identification and has been continuously extended with the introduction of new data sources such as thermal and radar data and digital imagery (Philipson, 1997). Interpretive techniques include the assessment of shape, size, tone, texture, shadow, pattern, location and association. A 'convergence of evidence' allows the successful identification of an object. Estes et al. (1983) ordered the above techniques, thereby providing a hierarchical framework to image interpretation methodology.

The second approach uses automated or semi-automated techniques to identify features of interest (Bue and Stepinski, 2006). This benefits from a consistent, repeatable method, potentially at the expense of human 'expertise'. Seijmonsbergen et al. (2011) discuss this in more detail, and it is therefore not considered further here.

Manual mapping requires the detection of individual landforms from remotely sensed imagery, recording their morphology on a base map. The *detectability* of landforms (Smith and Wise, 2007) may vary according to (i) the experience of the *individual interpreter* (which is considered further in Section 6) and (ii) the *data source* employed. The representation of a landform on a remotely sensed image (both satellite image and DEM) is dependent upon the characteristics of the process used to create the data set, be that a satellite image, LiDAR point cloud or contour lines. More specifically, it is the characteristic of the individual sensor used to capture the data, the characteristics of the individual landform and the visualisation methods applied that determine landform *detectability*. For satellite imagery, this may also involve the physical conditions at the moment of capture, particularly solar illumination and meteorological conditions. The interaction of these variables produces the following controls on landform detectability:

1. *Relative size* (Figure 8.1): the minimum resolvable landform, a function of landform size relative to the spatial resolution of the data set. The higher the spatial resolution of the data, the greater the ability to resolve smaller landforms,
2. *Azimuth biasing* (Figure 8.2): where there is a fixed solar azimuth (either on a satellite image or relief-shaded DEM), landform shape will be visually altered depending upon the relative difference between the illumination angle and orientation of an individual landform. This is particularly pronounced for linear and compound landforms (e.g. lineaments) where suites of landforms can seem to appear and disappear (Smith and Clark, 2005),
3. *Landform signal strength* (Figure 8.3): the amount of tonal and textural information that is available to visually distinguish individual landforms. For satellite imagery, these variations are primarily caused by differences between the surface cover of landforms and their surroundings. This can be dramatically modified through the acquisition of imagery at low solar elevations where the effect of relief causes high reflectance on fore slopes and shadowing on lee slopes (Slaney, 1981). Where there is a uniform surface cover (e.g. snow cover, lunar

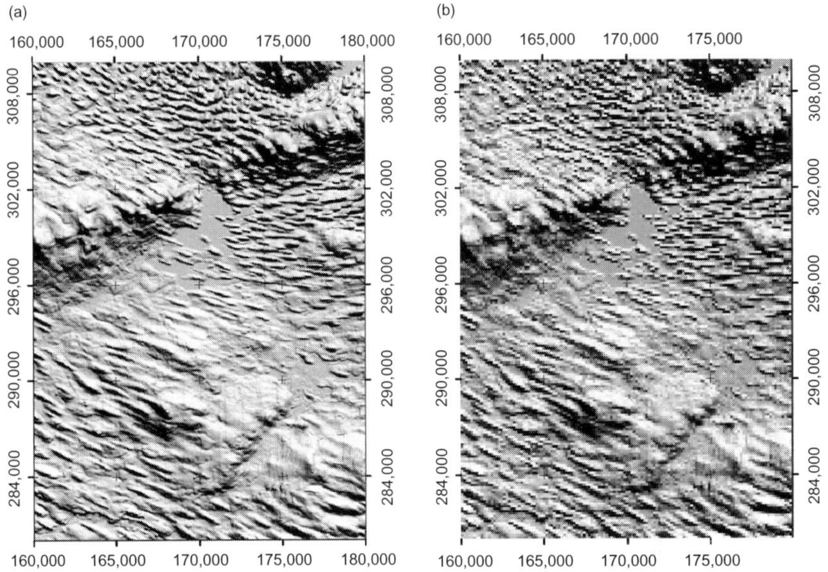

Figure 8.1 Illustration of the effects of relative size on the detectability of drumlins. Spatial resolution of the DEM is fixed at (a) 50 m and (b) 150 m. *Reproduced from Ordnance Survey Ireland, Copyright Permit MP001904.*

surface), the reflectance observed in the image is entirely a function of relief; this effect can actually be used to calculate a DEM, a process termed photoclinometry (Kirk et al., 2004). For relief-shaded DEMs, solar elevation has a limited effect as visual 'response' is modelled entirely from the relief, although as solar elevation approaches nadir (vertical) the image essentially represents changes in gradient (i.e. slope). It is also worth noting that SAR imagery has successfully been used to map landforms (Ford, 1984) as the oblique viewing geometry of the sensor highlights topography. SAR is not considered further here, but details on image processing for geomorphological mapping are presented by Vencatasawmy et al. (1998).

For satellite imagery, the greater the difference in reflectance properties of the surface cover of the landform when compared to surrounding terrain and the greater the *relief effect*, the greater the tonal differentiation of the landform on the image (Smith and Clark, 2005). Depending upon the landform of interest and the context in which it is being mapped, spectral differentiation may be applicable (Punkari, 1982).

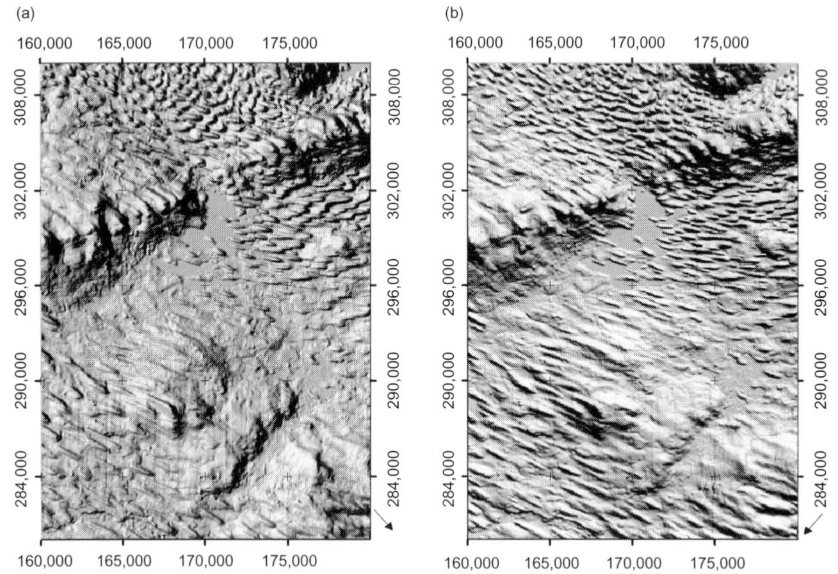

Figure 8.2 Illustration of the effects of azimuth angle on the detectability of drumlins from a relief-shaded DEM. (a) Azimuth angle parallel to the dominant drumlin orientation and (b) orthogonal to the principal drumlin orientation. Arrows indicate azimuth angle (see http://www.appgema.net/). *Reproduced from Ordnance Survey Ireland, Copyright Permit MP001904.*

In general, there is a minimum resolvable landform size and a range of landform orientations that an individual data source will be able to represent. In addition, the *definition* of these landforms is dependent upon the surface cover and the strength of the relief effect. With these general limitations in mind, this chapter is concerned with outlining an expert interpretive, or operator-based, approach to mapping landforms, with a specific focus upon methods for computer-based (or digital) mapping, data visualisation, landform quantification and sources of potential error.

2. MAPPING METHODS

Digital mapping, by definition, is performed through some kind of digital interface, typically a computer system with a graphical user interface (GUI). Whilst GUIs have been available for some considerable time, it is worth stressing that image interpretation requires *graphical* display and

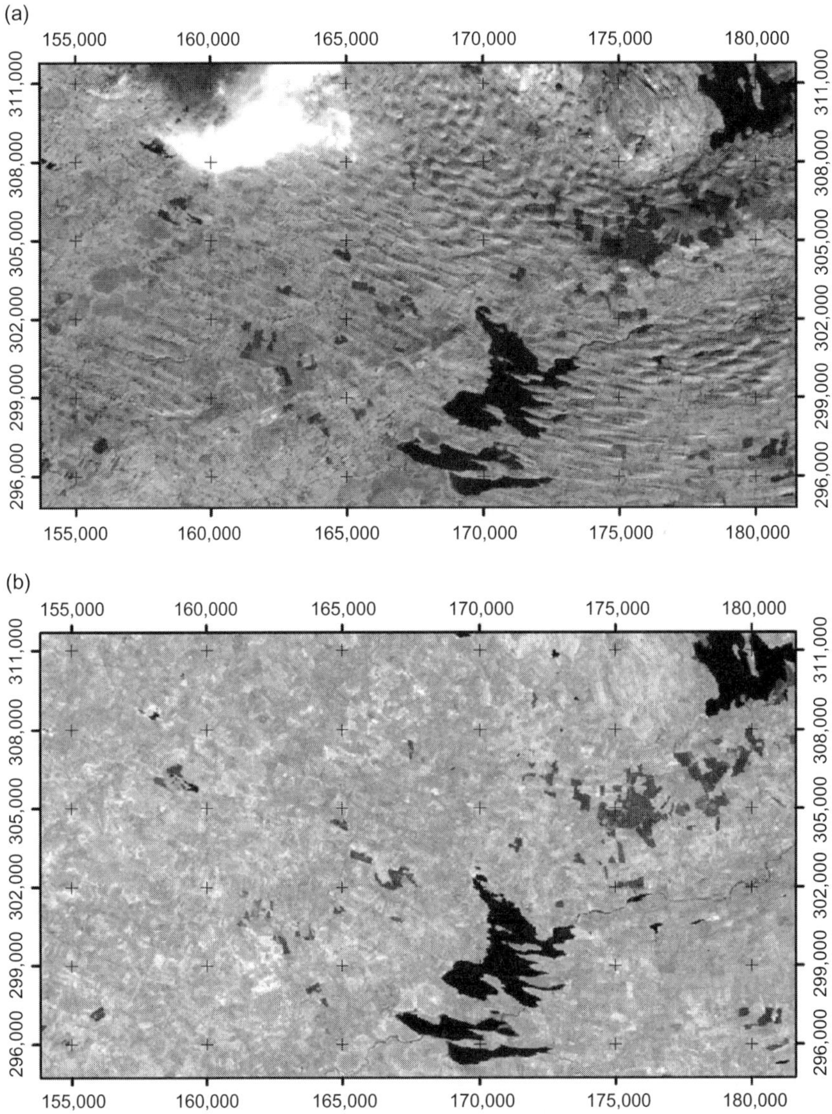

Figure 8.3 Illustration of the effects of landform signal strength through the use of Landsat TM imagery of the same location acquired on contrasting dates with (a) low solar elevation (11°) and (b) high solar elevation (48°).

the greater the size and number of pertinent displays, the easier interpretation potentially becomes. It is also essential for all work to be performed within a geographical information system (GIS) in order to ensure that input imagery and interpreted data sets maintain the same geographical coordinate system. This allows data export into other geographic products and facilitates accurate map production and quantitative analyses. Interpreters need to be familiar with the operation and use of a GIS, and familiarity with the principles of remote sensing is beneficial. Introductory texts describing GIS, remote sensing and image processing include Longley et al. (2006), Lillesand et al. (2008) and Mather (2004).

Primary *input data sets* used for digital geomorphological mapping include satellite imagery, DEMs and aerial photographs. These are typically raster data and, just like ordinary digital photos, are comprised of individual *pixels* (Figure 8.4) which are the minimum resolvable unit of information defined by a real-world area on the ground (termed *spatial*

(a)

(b)

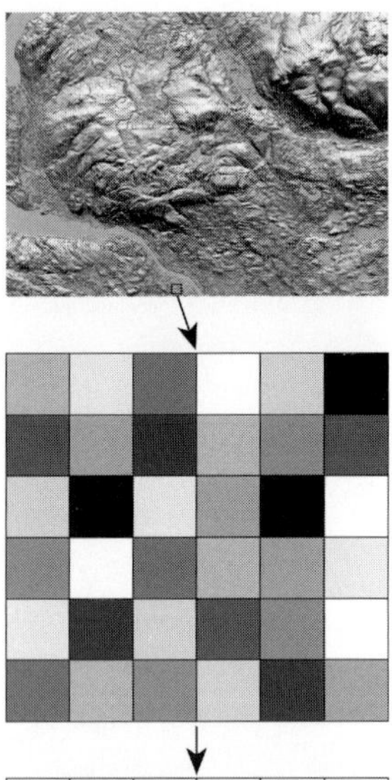

(c)

170	238	85	255	221	0
68	136	17	170	119	68
221	0	238	136	0	255
119	255	85	170	136	238
238	17	221	68	119	255
85	170	119	221	17	136

Figure 8.4 Satellite images and DEMs are raster data products. For example, (a) a relief-shaded DEM is a collection of (b) picture elements (pixels) shaded from black to white. (c) These reflect the underlying pixel value.

resolution). All recently collected data sets from these sources will be digital and usually supplied in a projected coordinate system. Only in the case of historical or legacy data will there be a requirement to convert from a paper-based analogue format in to a digital format. This may involve the scanning of, for example, aerial photography or the digitisation of contours from topographic maps (Oguchi et al., 2011).

Output data sets will include the interpreted geomorphological features identified on the input imagery. Within a GIS, this is conceptually the same as overlaying tracing paper on an aerial photograph and tracing the outlines of features of interest. Landform interpretation takes the complexity of the real world, as shown in the input image, and abstracts it to a meaningful representation for the end-user. *Morphological mapping* (or *morphographic*; Waters, 1958) concerns the determination of elementary forms in the landscape through the identification of breaks-of-slope; *geomorphological mapping* (Rose and Smith, 2008) uniquely identifies individual landforms, assigning an interpretation as to their genesis. All interpretations are digitally stored as vector data, comprised of points, lines or areas (more commonly termed *polygons*; Figure 8.5); these are generically termed feature types and, within a GIS, can be visually 'stacked' as layers.

Prior to beginning a mapping project, it is important to predetermine the features that are of interest (Clark et al., 2004; Sahlin and Glasser, 2008; Latocha, 2009); these should be grouped thematically (e.g. fluvial, glacial, peri-glacial and mass movement). For example, Clark et al. (2004)

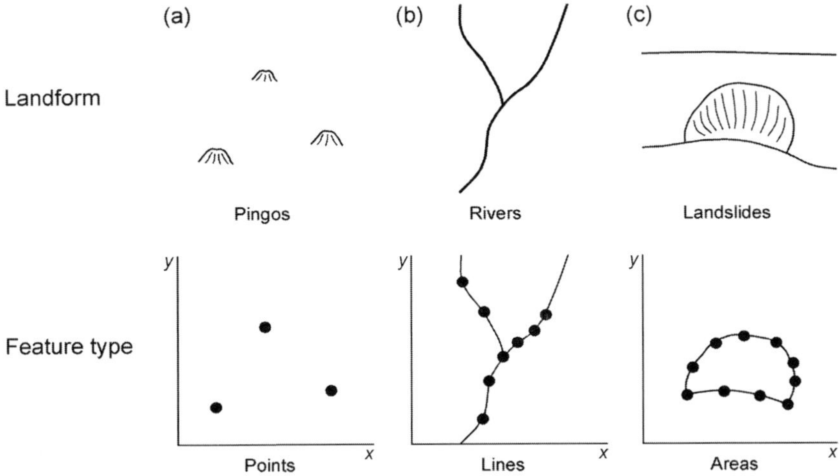

Figure 8.5 Vector data can be composed of three main feature types: points, lines and polygons.

used 20 thematic layers to represent a range of terrestrial and offshore glacial landforms. Within the file storage system, it is not usually possible to mix different vector feature types (e.g. points and polygons) and they therefore need to be created as separate layers. The choice of feature types will depend upon the landform being digitised and the scale of mapping, and it is important to remember that feature types are simplified categorisations of reality. Polygons approximate outlines of 2D features, or *area features*, such as drumlins or landslides. However, at bigger scales, lines and points can also be used to represent some area features. For example, regional scale mapping may represent drumlins as lines and landslides as points. Feature types are also selected based upon the nature of the source data and requirements of the project. As Smith et al. (2006) noted, mapping protocols 'need to be set-out as fit-for-purpose in terms of the objectives of the mapping exercise, and the resolution of the methods employed'. This approach to the categorisation of features to be mapped is best executed using a structured organisation of project data files.

Once the primary remotely sensed data sets have been acquired and processed, the features to be mapped decided upon and the data layers for mapping created, the actual process of digital mapping can begin. During manual interpretation, it is normal to use breaks-of-slope within the landscape, in conjunction with contextual information, to identify and outline individual landforms (Figure 8.6). Digitisation is the electronic filing of geographic coordinates, usually through a mouse click at the position identified on screen. Digitising a single vertex creates a new feature in a point layer, whilst 'strings' of vertices combine together to form lines. Area features again take 'strings' of vertices, joining them together, but also 'closing' the line feature by linking the start and end vertices. The GIS will simply store the points, lines and polygons using default symbols to represent the features on screen. Most GIS have a layout mode where spatial data and graphical symbols can be combined in the creation of maps; Otto et al. (2011) discuss this further.

Image interpretation and digitisation is often an iterative process, involving repeated mapping 'sessions', using a variety of data sources and visualisation techniques. During digitising, it is common to review and remap regions as the visual interpretative system is generally drawn to primary features. This should be performed at a variety of scales in order to identify landforms of different sizes, with the addition of alternative data sets and visualisation techniques in order to view the landscape using different terrain parameters and within different contexts.

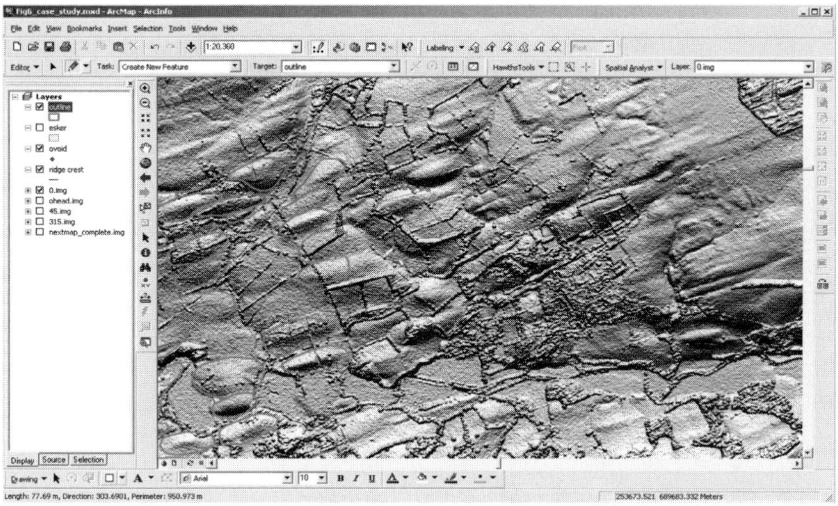

Figure 8.6 Screenshot illustrating the setup of thematic layers within ESRIs ArcGIS. Note that a polygon feature is currently being digitised, using the underlying raster DEM data as a backdrop.

Since repeated digitising can be both time consuming and tedious, various techniques have been developed in an attempt to increase the efficiency of the digitisation process and help reduce digitisation errors. For instance, an alternative to single mouse clicks, a GIS can be set to ‘stream’ points at pre-defined distance or time intervals to record movements of the pointer on screen. For certain types of work this can be beneficial, with trial and error suggested to ascertain optimum settings. Alternative input technologies include the use of a graphics tablet where a digital pen can be used to enter vertices or the use of a computer tablet where digitisation occurs directly on screen. Finally, many GIS can interpolate additional vertices and contain both smoothing and line generalisation algorithms to improve the visual appearance of the digitised features and reduce their complexity.

3. FILE FORMATS

Geomorphological mapping can be performed in a variety of free/open-source and commercial GIS, so different file formats for storing mapped interpretations of landscapes exist. Ideally, after the completion of a project, data files are archived so that they are available to potential future users. It is therefore pertinent to discuss the file formats used for data storage as this is central to the ability to share and archive any resultant mapping.

The ESRI Shapefile is the *de facto* standard for vector files that do not require support for topological relationships. Although the format is proprietary, the specification is published and widely supported. This makes it an ideal candidate for geomorphological mapping as it will remain accessible for future users, as well as easy to disseminate. Text files remain the most robust form of storage, and the Open Geospatial Consortium (OGC; www.opengeospatial.org) have defined an internationally agreed standard (ISO 19136:2007) for expressing geographical features called the Geography Markup Language (GML). This is recommended for long-term storage and archival.

For raster files, the Tagged Image File Format (TIFF) has become the most popular format due to its open specification and versatility. As a result, it has been adapted for geospatial needs through the addition of geocoding information (GeoTIFF) and now includes support for image files greater than 4 Gb (BigTIFF). Although this chapter is not specifically concerned with cartographic output, any vector features will have a default symbolisation applied within a GIS. Currently no standard exists for storing symbology, meaning that this information remains proprietary to the system being used. The OGC Styled Layer Descriptor standard may possibly be able to fulfil this role in the future; however, it is primarily designed for web mapping systems and support remains limited.

4. VISUALISATION

Primary inputs to the mapping process are remotely sensed digital data and these will be either satellite imagery or DEMs, usually delivered as raster grids. Prior to digitising landforms, it is necessary to process the imagery in order to optimise the visual information presented to the interpreter for the task in hand. This section will discuss methods of image manipulation and visualisation for satellite imagery and DEMs, respectively.

For mapping, satellite imagery should be acquired with a solar illumination (azimuth and elevation) that maximises the response from terrain. Smith and Wise (2007) recommended, where possible, solar elevation angles $<20°$. As polar orbiting satellites have orbits with fixed overpass times, it is not possible to specify acquisition at specific solar elevations; rather, acquisition needs to take advantage of seasonal variations. There is greater variation in solar elevation by season at higher latitudes, although requirements for snow- and cloud-free scenes may often limit the availability of suitable imagery. For persistent landscape features, this is less of a problem as there is a greater likelihood that an image archive may contain a suitable product. Acquiring imagery of ephemeral features is more difficult as the interpreter is restricted by the temporal resolution of the satellite and persistence of the feature in relation to the extent of image archive. For polar orbiting satellites that continuously capture imagery (e.g. Terra ASTER or Landsat ETM+), a significant archive is already available. This is not necessarily the case for commercial high-resolution satellites (e.g. GeoEye-1 or WorldView-1) which are generally 'tasked' to acquire imagery over specific areas and the conditions are unlikely to be optimal for maximising the terrain response. In addition to a fixed solar elevation, all VNIR (visual and near infrared) satellite imagery will have a solar azimuth fixed by conditions at acquisition. As image selection is primarily dictated by low solar elevation conditions, solar azimuth cannot be controlled for. Interpreters should therefore be aware of this potential bias and mitigate it through the use of additional or alternative data sources.

Once suitable imagery is obtained, the task of best presenting the image arises. General pre-processing routines exist (Mather, 2004; Lillesand et al., 2008), but specific aspects are particularly pertinent to geomorphological mapping. Clark (1997) notes that brightness variations (and so image structure) are more efficiently detected by the human eye from a greyscale image; he therefore recommends the use of monochrome images, experimenting with the different image bands available. Standard contrast enhancement techniques such as a linear stretch, histogram stretch or standard deviation stretch should be applied in order to maximise contrast within the image. Convolution (or kernel) filtering, in particular a high-pass filter, can provide useful enhancement (Lillesand et al., 2008).

Mapping in the visual spectrum can be augmented by other wavelengths of light recorded in satellite images. Punkari (1982) successfully took advantage of the tendency of moisture to collect in inter-drumlin areas. This variability in surface moisture affects surface cover (i.e. inter-drumlin regions become boggy) and allows differentiation between drumlin and inter-drumlin areas on the basis of their spectral signal or *spectral differentiation*. Jansson and Glasser (2005) found the use of false colour composites beneficial, incorporating both near infrared and thermal infrared bands. The greatest benefit was found when these were used in combination with relief-shaded DEMs.

For DEMs, the pixel value represents elevation above a fixed datum and is therefore directly related to relief. In comparison to satellite imagery it is equivalent to a single band, and in most GIS the default method used to visualise the data is a standard greyscale symbol palette (Figure 8.7a). This normally produces an image with poor contrast as the extremes in elevation saturate the image reducing contrast. Simple greyscale images are therefore not recommended. The most common form of visualisation is relief shading (Kraak and Ormeling, 2003); this uses an idealised light source to illuminate the landscape with the user specifying the azimuth and elevation (Figure 8.7b). The output produces a realistic depiction of the landscape and is subsequently easier to interpret.

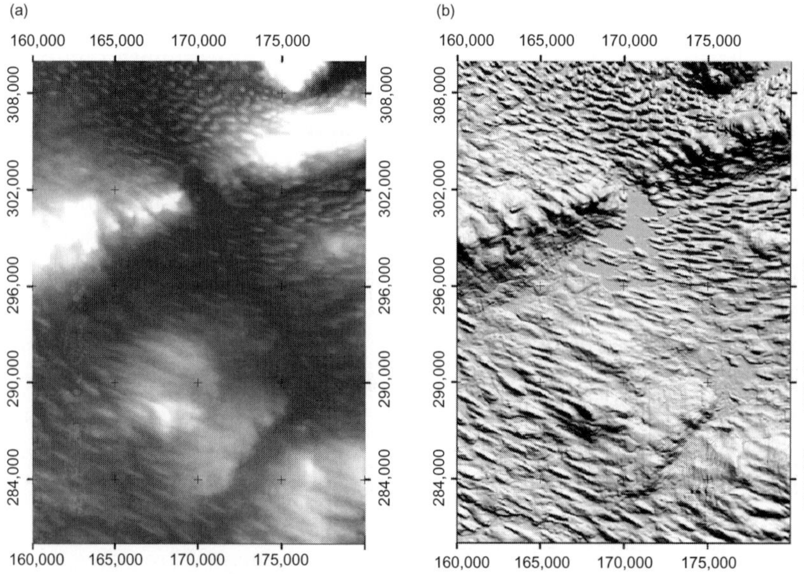

Figure 8.7 DEM visualisation using (a) greyscaling and (b) relief shading (illumination angle 20°). *Reproduced from Ordnance Survey Ireland, Copyright Permit MP001904.*

However, because an illumination source has been introduced, azimuth biasing will impact upon the detectability of landforms, with greater effects for more elongate landforms. Figure 8.2 illustrates this effect, contrasting drumlins that are illuminated parallel and orthogonal to their long axes; azimuth biasing is also effectively illustrated in DVD 8.1 (Multimedia 8.1), an animation that illuminates the same DEM at 5° azimuthal increments. This illustrates how suites of landforms appear and disappear although, of course, always existing in the DEM being visualised. It is the position of the breaks-of-slope in the direction of illumination that *change*, which has the effect of altering the apparent *shape* of area-based features. This reiterates the fact that what you see does not necessarily reflect the 'true' shape of the feature physically present in the landscape defined by some quantitative, objective and morphology-based criteria. This is an important point: although a DEM provides a consistent data source that represents the terrain surface, the method by which this is visualised can introduce inconsistencies.

As a result of the problems identified above, a variety of other manipulations of DEMs have been explored, many of which are reviewed by Smith and Clark (2005). These have been developed to process elevation data in order to highlight landforms, taking advantage of characteristic features or traits. Slopes make up the 'building blocks' of terrain as they control the gravitational force available for geomorphic processes (Evans, 1972) and, at a single point, are defined as a plane tangential to the terrain surface. This is characterised by the *steepness* (or *gradient*) and *orientation* (or *aspect*) of the tangent. Gradient is one of the most common manipulations as many landforms have relatively steep sides which should allow their identification (Figure 8.8a). Many interpreters find gradient to be less intuitive than relief-shaded imagery, so it is beneficial to view both data sets when mapping. One way to improve the visualisation quality of gradient is to use an inverted greyscale with increasing lightness for flatter areas. The resulting image has a similar appearance to relief-shaded terrain, without the illumination bias.

Although gradient measures the rate of change of elevation, curvature measures the rate of change of slope and is comprised of three elements (Schmidt et al., 2003): profile, planform and tangential curvature. Of particular interest is *profile curvature* as this measures downslope curvature and helps identify breaks-of-slope (Figure 8.8b). As breaks-of-slope are commonly used in the identification and mapping of landforms, this forms a

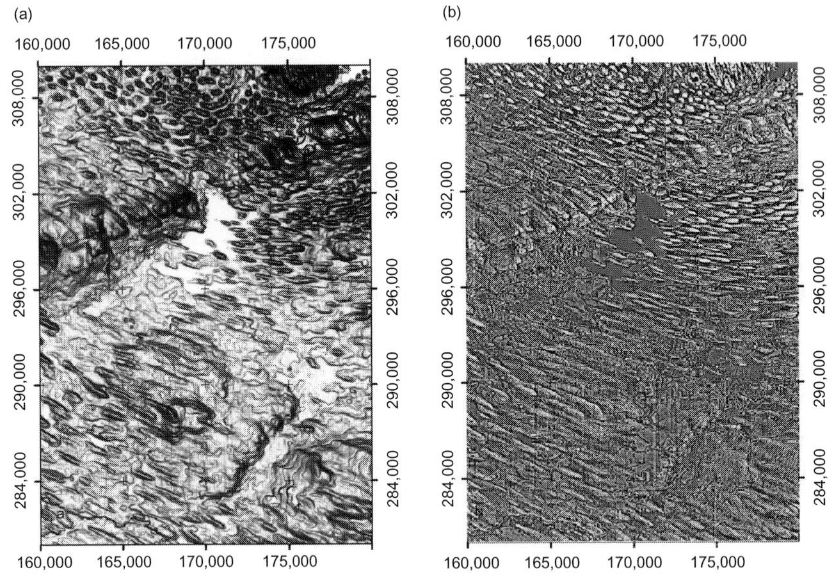

Figure 8.8 DEM visualisation using (a) gradient and (b) curvature. *Reproduced from Ordnance Survey Ireland, Copyright Permit MP001904.*

useful output. However, like gradient, it can also be difficult to interpret and should be used in conjunction with relief-shaded imagery.

Other visualisation techniques that have been used successfully for geomorphological mapping include local contrast stretch (LCS), residual relief separation (RRS) and spatial wavelet transform (SWT). LCS (Smith and Clark, 2005) uses the concept that landforms are distinct from the surrounding terrain as a result of a difference in elevation. A general linear contrast stretch is applied using the localised region to sample the elevation range around each pixel (Figure 8.9a), thereby significantly increasing the contrast. This approach is scale dependent and requires the user to specify a kernel size for the calculation. The value is dependent upon the spatial resolution of the raster grid and the general dimensions of the features of interest. For example in Figure 8.9a, a 50 m DEM formed the source data, with drumlins ranging in size from ~300 to 2500 m. Through trial and error, a 3 × 3 kernel was found to be most appropriate.

RRS (Hillier and Smith, 2008) takes a different approach using the notion that landscapes are formed of components each containing a class of feature (e.g. drumlins). The DEM may be separated into height (H) representing each component (e.g. $H\_DEM = H\_hills + H\_drumlins$).

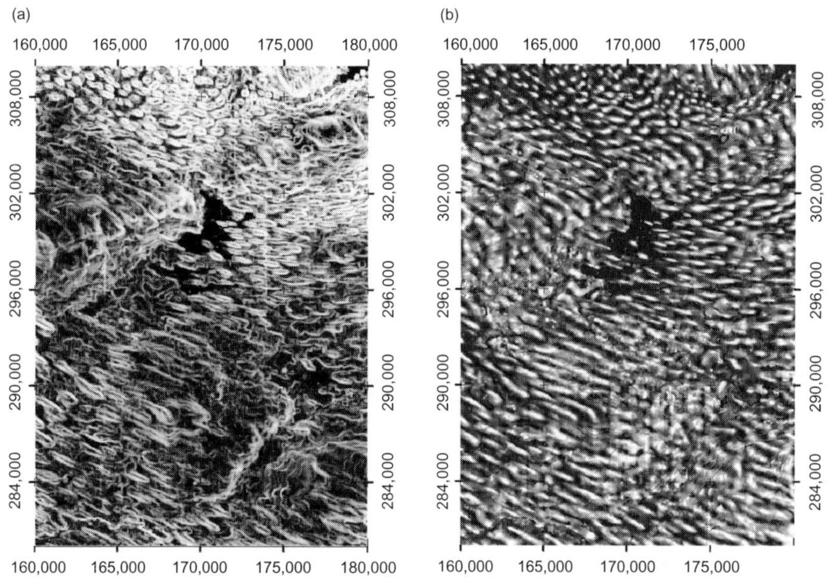

Figure 8.9 DEM visualisation using (a) LCS and (b) RRS. *Reproduced from Ordnance Survey Ireland, Copyright Permit MP001904.*

Hillier and Smith (2008) use the observation that classes of features occur at different width-scales to approximate the large-scale 'regional' relief and then extract the small-scale 'remainder' (or residual). The residual, now ideally containing only features of interest, can be analysed or visualised, with Hillier and Smith (2008) normalising it using a LCS (Figure 8.9b). The technique defines 'width-scales' to perform the separation and these must be specified in the form of kernels prior to the operation. As a result, the output will incorporate *all features* that are present at that width-scale, regardless of their origin. This could therefore incorporate anthropogenic feature (e.g. buildings).

Hillier (2008) used wavelets to isolate seamounts from the seabed, a process termed SWT (see also Hillier, 2011). Again, a component of the landscape containing features is isolated from the regional relief. A wavelet transform is computed, using appropriate coefficients, along a profile or a mesh of profiles across a raster DEM. Then, with the location and scale of each seamount known, the seamount boundaries are determined. The greatest advantage in this approach is that it is scale invariant and can identify multiple landforms at different sizes; however, it is yet to be applied to geomorphological mapping.

This section has introduced the main visualisation techniques for satellite imagery and DEMs currently used for geomorphological mapping, and its conclusions can be summarised as follows. Simple greyscale viewing is not appropriate and should be avoided during mapping and for the production of any maps. Currently, the most common technique is relief shading. It is widely supported by software packages, is very fast to compute and is easy to interpret. It also highlights subtle topographic features in the landscape. Unfortunately, the use of an illumination introduces azimuth biasing in output, altering the position of breaks-of-slope such that features may change shape, appear or disappear. To a certain extent this can be mitigated through the use of at least two relief-shaded images (shaded orthogonal to one another); however, this will not be correct for all errors. As a result, greater use has been made of techniques that produce imagery with no azimuth biasing. Along with elevation, slope and curvature form the basic attributes of a surface (Evans, 1972). Within the context of geomorphological mapping, it is gradient and profile curvature that are recommended for use in identifying landforms and whilst interpretation of the imagery may require some experience, there is no azimuth bias as the images are not illuminated. The remaining three techniques (LCS, RRS and SWT) successfully utilise a variety of methods to extract or isolate landforms from the underlying regional relief and again offer the advantage of no azimuth bias. Smith and Clark (2005) and Hillier and Smith (2008) review these methods and conclude that there is no single visualisation technique that is ideally suited to geomorphological mapping. Gradient, profile curvature, LCS, RRS and SWT are methodologically preferred as they do not introduce azimuth biasing. However, relief-shaded images are not only easy to interpret but are also able to highlight subtle topographic features. Best practise therefore involves initial mapping using a bias-free visualisation technique and, once complete, supplementing the work with mapping from relief-shaded imagery.

5. QUANTIFICATION

Although the primary concern of this chapter has been to outline the rationale for digital mapping, detailing the structural framework for digitisation, a specific focus has been upon satellite imagery and DEMs and how they are visualised. Primary outputs from mapping involve the cartographic presentation of maps and this is discussed in more detail by Otto et al. (2011); however, researchers may also want to quantitatively analyse the spatial attributes of landforms. A detailed review of such analyses for all areas of geomorphological mapping is beyond the scope of this chapter; however, the reader is directed to the case studies section and Hengl and Reuter (2008) where specific examples are presented. This section presents a brief overview, highlighting the basic characteristics from which further analyses can be performed.

The basic spatial attributes of features such as landforms (in addition to a count of the number of features) will depend upon the feature type used to map them; that is, whether a point, line or polygon is used to store landform information (Figure 8.10). Lines are 1D and it is possible to calculate line length and orientation. Polygons are 2D and the perimeter length and area can be calculated. For polygons that are elliptical, the major and minor axes can also be calculated (giving length and width), as well as both the elongation ratio and a preferred orientation.

If DEMs are being used for mapping, then it is also possible to make use of altitude to derive 3D attributes. This may be an elevation (point), a profile (line) or a volume (polygon), which can then be used in derivative calculations (e.g. length:height ratio). Smith et al. (2009) outlined a methodology, termed 'cookie cutter', for calculating material volumes of landforms; the process is illustrated in Figure 8.11, with an ArcGIS Python script available in DVD 8.2 (see http://www.appgema.net/). This

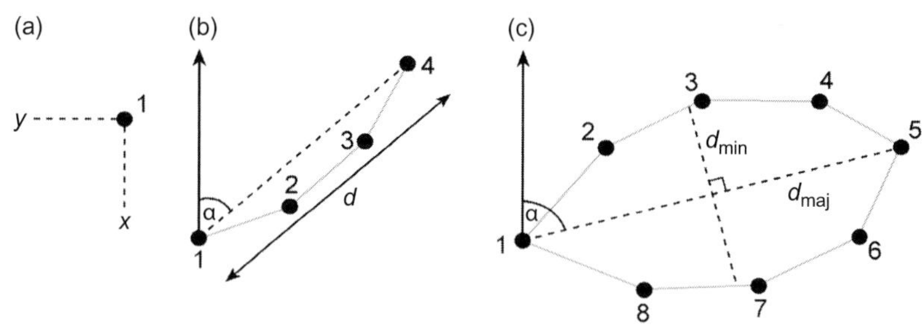

Figure 8.10 Basic spatial attributes of vector digitised landforms. (a) Location is known for points (0D); (b) vertices are known for lines (1D), with line length, *d*, and orientation, α, calculable and (c) locations of vertices are known for polygons (2D), with perimeter length and area calculable. For polygons that are elliptical, the major (d_{maj}) and minor (d_{min}) axes can be calculated giving length and width as well as both the elongation ratio and a preferred orientation.

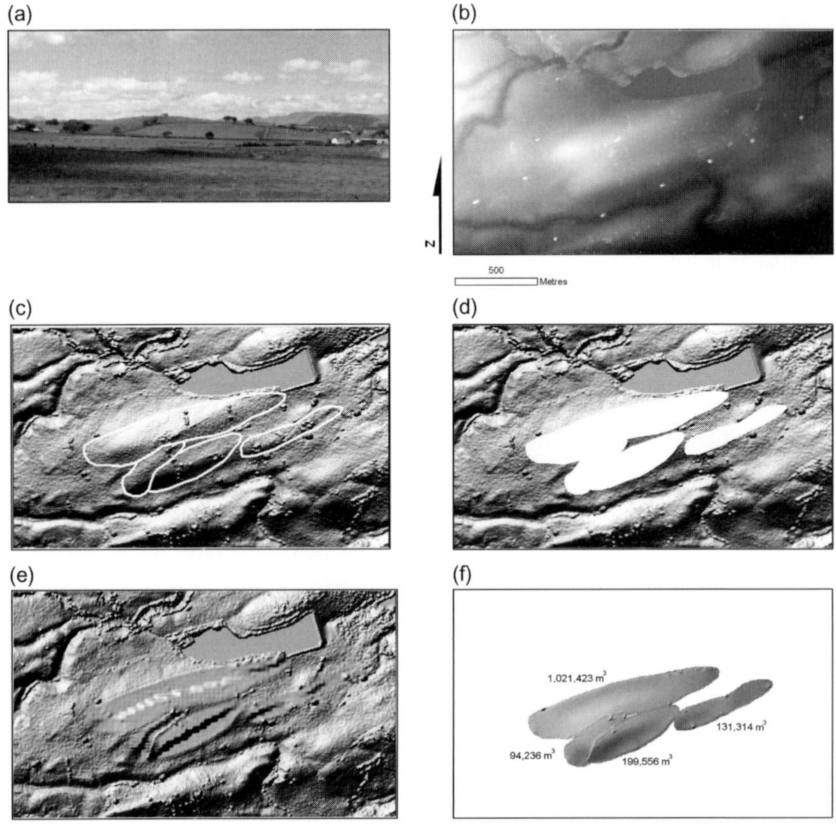

Figure 8.11 Workflow for the calculation of landform volume. The example is of a drumlin located at Bowridge (56.003085, −3.956384). (a) Example of a drumlin, (b) raw DEM data, (c) relief-shaded visualisation of terrain, with mapped drumlin outlines, (d) DEM voids, (e) interpolation of drumlin basal surfaces and (f) relief-shaded visualisation of drumlin volumes ($1.51\ m^3 \times 106\ m^3$). Note the 'stepping' in (e), a result of artefacts at the edges being interpolated across the basal surface.

utilises the polygons digitised during the landform mapping process to 'cookie cut' (or remove) the area of the landform from the underlying DEM leaving a void. Using the boundary elevation values, a 'new' base is interpolated across the void. When subtracted from the original data, *relative elevation* is computed allowing the subsequent calculation of volume. This process is applicable across a range of landform types both for positive (e.g. eskers) and negative (e.g. landslide scars) relief features. Material volumes can also be calculated using RRS. Once the regional relief has been calculated (Wessell, 1998; Hillier, 2007), the 'remainder' represents relative elevation. Landform volume is then computed using relative elevation *inside* the digitised outlines.

In addition to elevation, DEMs can also be used to calculate parameters for landforms, including gradient, aspect, profile/plan curvature and roughness. Hengl and Reuter (2008) provide a detailed outline of geomorphometry and its applications, including a review of parameter calculations within a variety of software products.

6. ERRORS

This final section briefly reviews potential errors that can occur during geomorphological mapping. The introductory section outlined the concept of *landform detectability* which is dependent upon both the data source(s) used and expertise of the individual interpreter. Three main errors result from this:

1. Completeness: the correct inclusion of all landforms that actually exist. There are two types of errors:
   - false negatives: failure to portray landforms where they actually exist,
   - false positives: portrayal of landforms where they do not actually exist.
2. Classification,
3. Locational accuracy.

The effect of completeness error is mitigated through appropriate selection and visualisation of source data. Care should also be taken to ensure consistency during mapping, whereas manual interpretation can also lead to landform classification errors. Locational accuracy is more complex as this can occur as a result of any, all or some of misdigitisation, the incorrect geolocation of the source data and the visualisation technique employed. These issues are considered further below.

As mapping is a subjective process, interpreter contributed error relates to expertise in terms of the manual identification of features of interest and an appraisal of their significance. The experience and skill of individual interpreters may vary considerably, and Siegal (1977) reports upon an experiment using five different interpreters where only 5% of geological lineaments were mapped as coincident between all interpreters and 50% were not coincident at all. In order to achieve consistent mapping that is comparable between different areas and different interpreters, it is necessary to employ a standardised mapping workflow. For smaller projects, consistency can be improved through the use of a single interpreter (Smith and Clark, 2005) and using a control area for re-digitisation and cross-checking. Validation requires the use of higher accuracy data. So, assuming that a project uses the best data available, validation, where performed, would involve checking landforms directly in the field.

Misdigitisation by an interpreter, although not related to landform identification, can lead to the introduction of error. This may involve the accidental digitisation of features such as vertices, lines or sliver polygons (Figure 8.12). 'Accidental' lines and vertices may be difficult to locate. Sliver polygons are more easily identified through querying the feature area; very small polygons are likely slivers and can be checked and deleted. Locational errors, particularly after any densification and smoothing, may have little impact upon the cartographic quality of the final output; however, it could significantly alter any variables that are subsequently computed. The greatest effect would be upon the calculation of volume where small planform variations can potentially lead to the inclusion or exclusion of large volumes of terrain. If volume is an important product, then greater care must be taken in digitising polygons.

In terms of source data, the visualisation technique used will have direct impact upon both the *completeness* of the mapping and the locational accuracy of any digitised landforms. Care should therefore be taken when deciding upon a mapping workflow in order to maximise the terrain response, and minimise the azimuth bias, in the techniques employed. For example, relief shading may cause landforms to appear or disappear and will cause displacement of breaks-of-slope (Mather, 2008), thereby affecting locational accuracy.

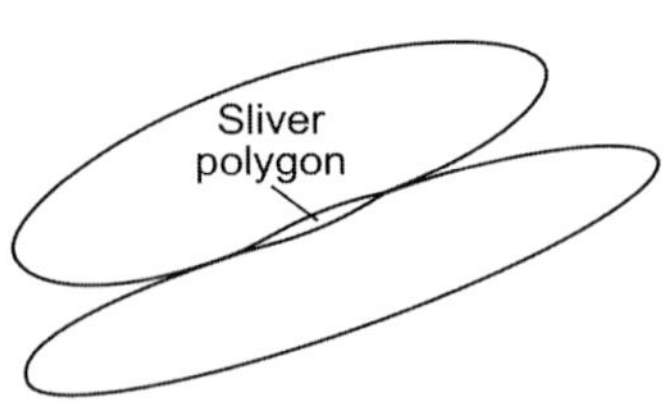

Figure 8.12 Creation of erroneous 'sliver' polygons through misdigitisation.

The introduction also noted how the source data will impact upon relative size and landform signal strength, thereby affecting detectability. Further impact will result from surface features in the landscape not related to terrain. Most notably, these are anthropogenic features (e.g. buildings) and vegetation. Where present, these obscure the terrain surface making the retrieval of landform shape more difficult. Where volumetric outputs are required, they will erroneously introduce extra volume in to the calculations. Certain DEM products may be able to mitigate against this; for example, LiDAR data can, at least partially, penetrate vegetation canopies and therefore extract the terrain surface. However, care should be exercised as 'bare-earth' products apply bespoke 'surface clutter' removal algorithms (Sithole and Vosselman, 2004). Although these might be fit-for-purpose with some applications, Smith et al. (2006) demonstrate that subtle topographic features can also be removed during this process.

The reader is referred back to Oguchi et al. (2011) for the selection of source data; however, Smith et al. (2006) provide an example of the impact of data source type on the geomorphological mapping of glacial landforms in Scotland. The authors used original field mapping for a 100 km^2 study area to validate mapping performed from six different 'off-the-shelf' DEM products.

7. SUMMARY

This chapter has outlined a method for digital geomorphological mapping, detailing the framework that is required in order to produce detailed maps of consistent quality. Prior to starting any project, it is necessary to understand the constraints that are imposed upon mapping and, more specifically, the detectability of individual landforms as a result of the experience of an individual interpreter and the source data being used. It is important to put in place procedures to ensure that interpreters are consistent in the approach they take to mapping and that, where appropriate, validation is performed through field-checking selected aspects of the mapping. Ultimately, manual mapping utilises the experience and expertise of an individual and, more generally, the ability of the human visual perspective system to identify complex patterns. This offers great benefits with the visual heuristics used to develop landform associations very hard to reproduce using automated methods. However, the weakness in this approach is in maintaining high levels of consistency in mapping not only between interpreters but also between individual mapping 'sessions'.

Interpreters have extensive control over the source data used for mapping, and the detectability of landforms will be dependent upon relative size, azimuth biasing and landform signal strength. For satellite imagery, it is necessary to review the project goals in order to determine the spatial resolution (and so *relative size*) that is best suited. For example, for moderate-scale regional mapping, it is desirable to balance high spatial resolution data with large areal coverage and low cost. Either Landsat ETM+ or Terra ASTER may be ideally suited for such prerequisites. For a detailed survey (not involving field mapping), either aerial photography or high spatial resolution commercial satellite imagery (e.g. GeoEye-1) are more suitable data sources. *Azimuth biasing* is particularly problematic as it is selective (rather than random), causing the visible breaks-of-slope to change position, with greater effects on elongate landforms. This can only be mitigated through the use of alternative data sources. *Landform signal strength* can be maximised through the acquisition of imagery with a low solar elevation and, where appropriate, the spectral manipulation of imagery.

For DEMs, data source selection will again depend upon the stipulations of the project under consideration. Spatial resolution will impact directly upon the size of landforms that are visually perceptible (Smith et al., 2006); however, the method used for data collection (e.g. InSAR, LiDAR, contours) will also have an impact upon data quality. Azimuth biasing remains a problem as relief shading is one of the most popular techniques for visually assessing a DEM. A bias-free visualisation technique (e.g. gradient, RRS) is recommended for initial mapping, which can then be supplemented with relief-shaded images.

Mapping is typically performed within a GIS using on-screen digitising. Feature types that can be mapped include points, lines and polygons (or areas). Prior to digitising, it is necessary to determine which landforms are to be mapped and the feature types that will be used to represent them. Feature types cannot be mixed within single data files and so strict adherence to an organisational schema where landform types are separated into separate layers will prevent later problems and the requirement for restructuring.

Although the outputs of digital mapping are commonly used within cartographic products (Otto et al., 2011), for example in journal publications or reports to project clients, they can also be used for quantitative assessments. Basic attributes of the digitised vector data include vertex location (point, line and polygon), orientation/line length (line), area/perimeter length (polygon). For polygons it is also possible to calculate the major and minor axes and so determine elongation and preferred orientation. DEMs also allow the derivation of other physical attributes and the reader is directed to Hengl and Reuter (2008).

Finally, all projects will incorporate some level of error. This will be composed of elements relating to *completeness*, *classification* and *locational accuracy*. The detectability of the landform (the visual presence of a feature) will depend upon the expertise of an individual interpreter and representation on an image (relative size, azimuth biasing and landform signal strength). The interpreter will also be responsible for the correct classification of a landform, whilst misdigitisation, the geolocation of the source data and the visualisation technique employed all contribute to locational accuracy.

ACKNOWLEDGEMENTS

I gratefully acknowledge the support of the Kingston University Research Development Fund for a sabbatical during which this chapter was prepared and Intermap Technologies for the supply of NextMap Britain data used in Figure 8.4.

REFERENCES

Bue, B.D., Stepinski, T.F., 2006. Automated classification of landforms on Mars. Comput. Geosci. 32, 604–614.

Clark, C.D., Evans, D.J.A., Khatwa, A., Bradwell, T., Jordan, C.J., Marsh, S.H., et al., 2004. Map and GIS database of glacial landforms and features related to the last British Ice Sheet. Boreas 33, 359–375.

Close, M.H., 1867. Notes on the general glaciation of Ireland. J. R. Geogr. Soc. London 1, 207–242.

Colwell, R.N., 1983. Manual of Photographic Intrerpretation. American Society of Photogrammetry, Falls Church, Virginia, USA.

Estes, J.E., Hajic, E.J., Tinney, L.R., 1983. Fundamentals of image analysis: analysis of visible and thermal infrared data. In: Colwell, R.N. (Ed.), Manual of Remote Sensing. American Society of Photogrammetry, Falls Church, VA.

Evans, I.S., 1972. General geomorphometry, derivatives of altitude and descriptive statistics. In: Chorley, R.J. (Ed.), Spatial Analysis in Geomorphology. Harper and Row, New York, pp. 17–90.

Ford, J.P., 1984. Mapping of glacial landforms from Seasat radar images. Quat. Res. 22, 314–327.

Hengl, T., Reuter, H.I., 2008. Geomorphometry: Concepts, Software, Applications. Elsevier, Amsterdam, 796 pp.

Hillier, J., 2007. Pacific seamount volcanism in space and time. Geophys. J. Int. 168 (2), 877–889.

Hillier, J., in press. Submarine geomorphology: quantitative methods illustrated with the Hawaiian volcanoes. In: Smith, M.J., Paron, P., Griffiths, J. (Eds.), Geomorphological Mapping: A Handbook of Techniques and Applications. Elsevier, Amsterdam.

Hillier, J., Smith, M.J., 2008. Residual relief separation: DEM enhancement for geomorphological mapping. Earth Surf. Process. Landforms 33, 2266–2276.

Hillier, J.K., 2008. Seamount detection and isolation with a modified wavelet transform. Basin Res. 20, 555–573.

Jansson, K.N., Glasser, N.F., 2005. Using Landsat 7 ETM+ imagery and digital terrain models for mapping lineations on former ice sheet beds. Int. J. Remote Sens. (26), 3931–3941.

Kirk, R.L., Squyres, S.W., Neukum, G., 2004. Topographic mapping of Mars: from hectometer to micrometer scales. Twentieth ISPRS Congress, Istanbul, Turkey.

Knight, J., Mitchell, W., Rose, J., in press. Geomorphological field mapping. In: Smith, M. J., Paron, P., Griffiths, J. (Eds.), Geomorphological Mapping: A Handbook of Techniques and Applications. Elsevier, Amsterdam.

Kraak, M.-J., Ormeling, F., 2003. Cartography: Visualisation of Spatial Data. Prentice Hall, Harlow.

Latocha, A., 2009. The geomorphological map as a tool for assessing human impact on landforms. J. Maps v2009, 103–107.

Lillesand, T.M., Kiefer, R.W., Chipman, J.W., 2008. Remote Sensing and Image Interpretation. sixth ed. John Wiley & Sons, New York, 1164 pp.

Longley, P.A., Goodchild, M.F., Maguire, D.J., Rhind, D.W., 2006. Geographic Information Systems and Science. second ed. John Wiley & Sons, London, 536 pp.

Mather, K., 2008. Drumlins in the Howgills. Unpublished M.Sc. Dissertation, Cambridge University, Cambridge, 98 pp.

Mather, P.M., 2004. Computer Processing of Remotely-Sensed Images. Wiley-Blackwell, London, 442 pp.

Oguchi, T., Hayakawa, Y., Wasklewicz, T., in press. Data sources. In: Smith, M.J., Paron, P., Griffiths, J. (Eds.), Geomorphological Mapping: A Handbook of Techniques and Applications. Elsevier, Amsterdam.

Otto, J.-C., Gustavsson, M., Geilhausen, M., in press. Cartography: design, symbolisation and visualisation of geomorphological maps. In: Smith, M.J., Paron, P., Griffiths, J. (Eds.), Geomorphological Mapping: A Handbook of Techniques and Applications. Elsevier, Amsterdam.

Philipson, W.R., 1997. Manual of Photographic Interpretation. second revised ed. American Society of Photogrammetry, Bethesda, MD, 700 pp.

Pike, R.J., Evans, I.S., Hengl, T., 2008. Geomorphometry: a brief guide. In: Hengl, T., Reuter, H.I. (Eds.), Geomorphometry: Concepts, Software, Applications. Elsevier, Amsterdam, pp. 3–30.

Punkari, M., 1982. Glacial geomorphology and dynamics in the eastern parts of the Baltic Shield interpreted using Landsat imagery. Photogramm. J. Finland 9, 77–93.

Rhind, D., 1988. Personality as a factor in the development of a discipline: the example of computer-assisted cartography. Am. Cartogr. 15 (3), 277–289.

Rose, J., Smith, M.J., 2008. Glacial geomorphological maps of the Glasgow region, western central Scotland. J. Maps v2008, 399–416.

Sahlin, E.A.U., Glasser, N.F., 2008. Geomorphological map of Cadair Idris, Wales. J. Maps v2008, 299–314.

Schmidt, J., Evans, I.S., Brinkmann, J., 2003. Comparison of polynomial models for land surface curvature calculation. Int. J. Geogr. Inf. Sci. 17 (8), 797–814.

Seijmonsbergen, A.C., Hengl, T., Anders, N.S., in press. Semi-automated identification and extraction of geomorphological features using digital elevation data. In: Smith, M.J., Paron, P., Griffiths, J. (Eds.), Geomorphological Mapping: A Handbook of Techniques and Applications. Elsevier, Amsterdam.

Short, N.M., Blair, R.W., 1986. Geomorphology from Space: A Global Overview of Regional Landforms. NASA SP-486, Washington, DC.

Siegal, B.S., 1977. Significance of operator variation and the angle of illumination in lineament analysis of synoptic images. Mod. Geol. 6, 75–85.

Sithole, G., Vosselman, G., 2004. Experimental comparison of filter algorithms for bare-Earth extraction from airborne laser scanning point clouds. ISPRS J. Photogramm. Remote Sens. 59, 85–101.

Slaney, V.R. (Ed.), 1981. Landsat images of Canada – a geological appraisal. Geological Survey of Canada Paper 80-5, 102 pp.

Smith, M.J., Clark, C.D., 2005. Methods for the visualisation of digital elevation models for landform mapping. Earth Surf. Process. Landforms 30 (7), 885–900.

Smith, M.J., Pain, C., 2009. Applications of remote sensing in geomorphology. Prog. Phys. Geogr. 33 (4), 568–582.

Smith, M.J., Wise, S.M., 2007. Mapping glacial lineaments from satellite imagery: an assessment of the problems and development of best procedure. Int. J. Appl. Earth Obs. Geoinf. 9, 65–78.

Smith, M.J., Rose, J., Booth, S., 2006. Geomorphological mapping of glacial landforms from remotely sensed data: an evaluation of the principal data sources and an assessment of their quality. Geomorphology 76, 148–165.

Smith, M.J., Rose, J., Gousie, M.B., 2009. The cookie cutter: a method for obtaining a quantitative 3D description of glacial bedforms. Geomorphology 108, 209–218.

Vencatasawmy, C.P., Clark, C.D., Martin, R.J., 1998. Landform and lineament mapping using radar remote sensing. In: Lane, S.N., Richards, K.S., Chandler, J.H. (Eds.), Landform Monitoring and Analysis. John Wiley & Sons, Chichester, pp. 165–194.

Waters, R.S., 1958. Morphological mapping. Geography 43, 10–18.

Wessell, P., 1998. An empirical method for optimal robust regional-residual separation of geophysical data. Math. Geol. 30, 391–408.

CHAPTER NINE

Cartography: Design, Symbolisation and Visualisation of Geomorphological Maps

Jan-Christoph Otto[a], Marcus Gustavsson[b] and Martin Geilhausen[a]
[a]Department of Geography and Geology, University of Salzburg, Salzburg, Austria
[b]Helsingforsgatan, Uppsala, Sweden

Contents

Developments in Earth Surface Processes, Volume 15
ISSN: 0928-2025, DOI: 10.1016/B978-0-444-53446-0.00009-4

1. INTRODUCTION

Map design is the creative act of visual communication, with the composition of the map, choice of symbols and colours and the compilation of map content requiring thoughtful consideration to transfer the message of the map. Geomorphological maps are highly complex thematic maps depicting the composition of the Earth's surface and the processes working there. To deliver this complex information, geomorphological maps commonly make full use of the various elements of cartographic design. Different kinds of symbols and colours need to be arranged and composed carefully in order to generate a readable map that clearly expresses the map content and message.

Before starting the process of map design, it is necessary to review the following questions:

- What is the purpose, message and central aspect of the map?
- Who is the map aimed at?
- Who will be using the map?
- How will the reader use the map (i.e. office, field)?

Applications of geomorphological maps range from simple descriptions of a field site, for example accompanying a journal publication or construction site report, to specialised land system analyses, for example for land management or natural hazard assessment. It is equally important to consider the production process and dissemination of the final product. Is it a paper map? Is the map produced in colour or black and white? Is the map accompanying a journal publication? Will it be published online? These issues strongly influence how you compile and arrange your data, which symbols are used, how the various map items are composed and whether colours can be used or not.

Prior to data collection, for example going into the field or digitising from aerial photographs, fundamental decisions need to be made in relation to the mapping area, scale (field scale and output scale) and in the choice of the symbols to be used. These settings influence the design, shape and final appearance of the map. When all data are collected, specifications for map composition and production need to be considered: What map sheet format shall be used? Can colour be used? What will be the size of symbols and text? Which coordinate system will be used? How will topography be represented on the map?

Besides accuracy and quality of the data, good design creates a good map. In this chapter, we will briefly review principles and elements of cartographic design and communication through maps, before we introduce common legend systems available for geomorphological mapping. Practical issues of map and symbol creation using graphic and geographical information system (GIS) software are provided, and some basic material concerning final map production are introduced. Map dissemination through the Internet is increasingly important for geomorphologists (Hake et al., 2001), and therefore, technical issues on web mapping are presented towards the end of this chapter.

2. ELEMENTS OF CARTOGRAPHIC MAP DESIGN

Geomorphological legends commonly use complex, sometimes pictorial symbols to represent landforms or landform characteristics, surface materials and processes. What differentiates geomorphological maps from other thematic maps is that qualitative information prevails over quantitative or classified data. Quantitative information in geomorphological maps is delivered by displaying proportional landform sizes (large-scale maps) or, for example, by providing data on depth, age or grain-size composition of deposits. In order to understand the differences between different symbol types and their role in map design, we will now look at the basic elements of cartographic design.

The basic representations of objects in maps are the symbol primitives of point, line and area (Figure 9.1). These are also referred to as dot, dash and patch, or termed marker, line and polygon (area) symbols in many GIS applications (Robinson et al., 1995). Whether a linear feature in nature is represented by a line symbol on the map is mainly a question of scale. For example, a river could be depicted by a blue line. On larger maps (with increasing size of the map items), the river would be depicted using an area symbol. The map scale also determines if a landform is depicted by a point symbol or if it is split up into its morphological components. Rock glaciers, for example, could be represented by a single point symbol on small-scale maps or by the assemblage of line and area symbols that differentiate the step height of the rock glacier front, furrows and ridges and the accumulation of boulders and blocks on top of the rock glacier, if the map scale increases.

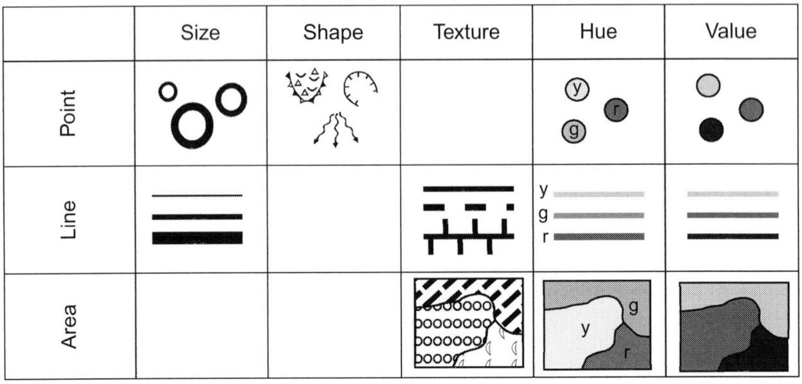

Figure 9.1 Primitives of map symbols and visual variables (y = yellow, r = red, g = green).

A differentiation of these basic representations, to express relationships among or differences between the data, can be achieved by variations of the basic visual variables: shape, size, orientation, texture or colour (Robinson et al., 1995; Kraak and Ormeling, 2002). *Shape* refers to different forms of the graphic symbol for points (marker) and lines (Figure 9.1). Shape variation demonstrates qualitative differences and is the most commonly applied visual variable in geomorphological maps because of the great number of different symbols for different landforms. Difference in symbol *size* will be apparent by changing geometric dimensions, such as area, length or width of the symbol. Size variations are typically used to represent nominal differences, for example to underline variations of importance, size or activity of a landform or process. Differences in shape and size always refer to the variations of the symbol itself and not to changes of the object shape. When using area symbols, pattern *orientation* can be altered to depict qualitative or quantitative information differences. *Texture* variations represent changes that result when the shape, orientation or the spacing of components that generates a pattern is modified. Furthermore, the spatial arrangement of the pattern, for example systematically ordered or randomly distributed, is a way to illustrate symbol differences. Patterns or hatched symbols are used in geomorphological maps, for example, to depict lithology or slope gradient.

Colour is an important visual variable, mainly used to depict qualitative differences. However, geomorphological maps are commonly produced in black and white, especially when they are part of a journal publication to keep production costs low. If colour is used, variation of colour characteristics, that is hue, value (lightness) and chroma (saturation) are the most powerful tools to emphasise certain aspects of the map (Table 9.1). As the human visual perception is adapted to colours, we strongly react to differences in colour. We can use colour variations to draw the reader's attention to specific features, or to convey information, sometimes in a subjective way (e.g. the colour red has a connotation with danger). The use of colour also demands great care because the perception of colours has physical and psychological aspects. These include the ability to differentiate contrasts between different colours, or to perceive colours in very small areas. Certain colours have conscious or unconscious connotations, for example the so-called warm (red, yellow) and cold (blue) colours. Most connotations are the result of the different wavelengths that lead to different moments when the colour reaches the eye. Long wavelengths (e.g. red) are seen 'earlier' and appear to be 'nearer', while short wavelengths (blue or green) are seen later and appear 'further away' (Rouleau, 1993). Wrong usage and composition of colours can destroy the readability of the map or lead to misunderstanding. Within cartography, some colour conventions exist that should be acknowledged to avoid confusion. For example, on topographic maps blue is used for objects related to water, for example rivers, springs or lakes; green generally represents areas covered by vegetation. A valuable assistance for colour selection is provided by the online tool 'Colorbrewer' (Brewer, 2009). The tool assists in choosing the right composition of colours by displaying different colour schemes. Colour combinations can be tested on a complex map sample that enables the user to experience the differentiation and perception of the colours used. Geomorphological maps use blue colours generally to represent features related to the hydrological processes

Table 9.1 Definitions of Hue, Value and Chroma

Hue	Refers to the colour we perceive. It describes the dominant wavelength of light (e.g. red, blue and yellow)
Value	Refers to the relative lightness or darkness of a hue. Light variations of a hue are referred to as high value, and dark changes have a low value
Chroma	Describes the colour saturation. It represents the 'colourfulness' of a hue, which can be reduced adding white or black. Chroma can range from a greyish hue with no apparent colour pigment (or proportion of light of the specific wavelength reflected) to a pure, intense hue.

and black for anthropogenic features. In many geomorphological legend systems, colours are applied to represent variations in landform genesis, process domains or lithology (see Section 3). These are either expressed by coloured area symbols (Stäblein, 1980) or by using coloured line or point symbols (Gustavsson et al., 2006).

2.1 Graphic Communication and Design Principles

Communication with maps differs significantly from other types of human communication. Maps are visual media and evoke visual stimuli that cause different reactions in people in comparison to books or conversations. In books or spoken conversation, information is delivered in a sequence, one sentence following another. In contrast, graphic communication, like maps, delivers all information at once. This means information is not perceived sequentially, but instantaneously with respect to the location and relative position on the map sheet or screen. Thus, the appearance and composition of graphical elements should be considered thoughtfully. On a map, all information is spatially related and needs to be considered holistically. The composition of map items decides if and how the reader understands the message, with perception and understanding occurring subconsciously. To allow map users to understand the meaning of the map, a visual sense to the symbols and their attributes that correspond to the intention of the cartographer needs to be assigned (Robinson et al., 1995). When looking at the graphic design of geomorphological maps, an inverse relationship commonly occurs between the ability to read the map and the amount of information expressed in colours and symbols. Thus, geomorphological maps tend to be 'overloaded' with information.

The principles of graphic design of maps include *legibility*, *visual contrast*, *figure-ground* perception and *hierarchical composition* (Robinson et al., 1995). *Legibility* is probably the most important principle and provides a challenge especially for geomorphological maps. A large number of different symbols generally are using graphic variables that bear the potential to render the map unreadable and hence not understandable (Figure 9.2). Ready-made legend systems are commonly used; however, each symbol needs to be clearly distinguishable. Legibility mainly depends upon symbol size and density, which results from the size of the map. Map space is characteristically restricted or determined by the extent and/or scale of the final map. The map maker's task is to find the right balance between

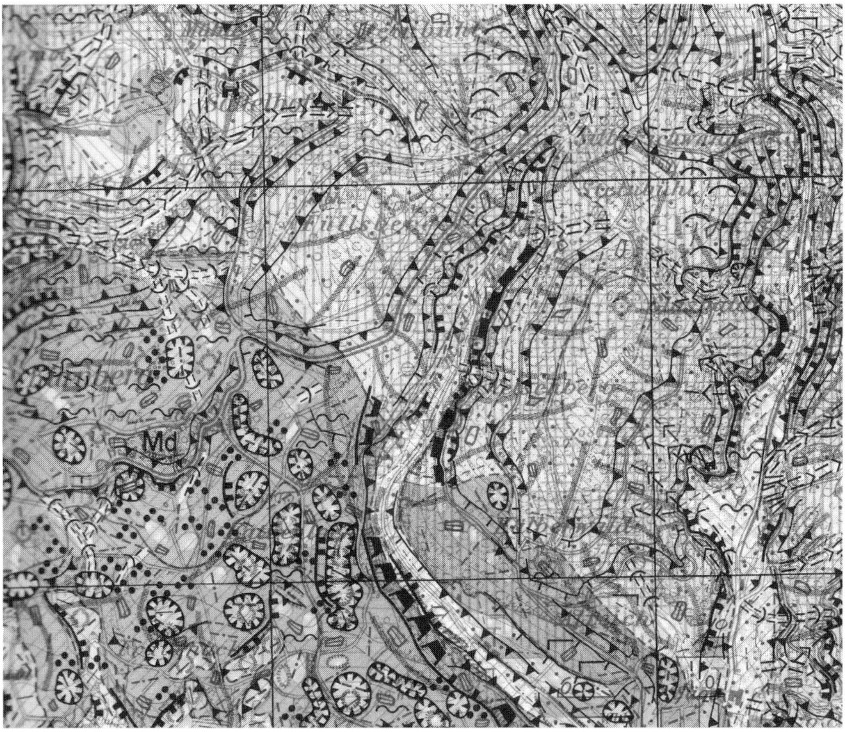

Figure 9.2 Section of the geomorphological map 1:25,000, sheet 8114 Feldberg, from the GMK 25 mapping programme in Germany. Colour intensity and the density of symbols render this map hard to read. *Extracted from Geilhausen, Otto and Dikau (2007).*

the number and size of symbols used, which includes the process of *generalisation*. Generalisation is the abstraction of map objects aiming at a simplification of the map content in order to fit the scale or purpose of the map without significantly changing the map's message (Slocum et al., 2005). In geomorphological maps, generalisation could mean that complex surface morphology is not represented by different line symbols that follow breaks in surface, but by single illustrative point symbol that depicts the landform type (Speight, 1974; see later).

Contrast is the basis of vision. Visibility of the map depends to a large extent on the right contrast between the graphic elements. Variation of contrast can be achieved by changing shape, size or colour of a symbol, or all of them. *Figure-ground* perception describes a person's ability to distinguish between an object and its surrounding. The figure, that is the object, should be clearly separated from the less distinct (back)ground. This happens automatically as a natural and fundamental characteristic of human visual perception. In relation to maps, a common example is the discrimination of land and water on a simple map of continents. The figure-ground differentiation is generated choosing different hues (brown and blue) or values (light and dark) to generate a contrast from which the continents clearly emerge from the surrounding seas (Figure 9.3). Figure-ground perception is supported when the figure is familiar. Unfamiliar objects need special effort to allow figure recognition.

Geomorphological maps require a good differentiation of map element structuring. *Hierarchical organisation* and visual layering enable separation of meaningful characteristics in order to depict differences, interrelationships or hierarchies. Different line symbols of roads on a road map, for example, are used to differentiate between different levels of

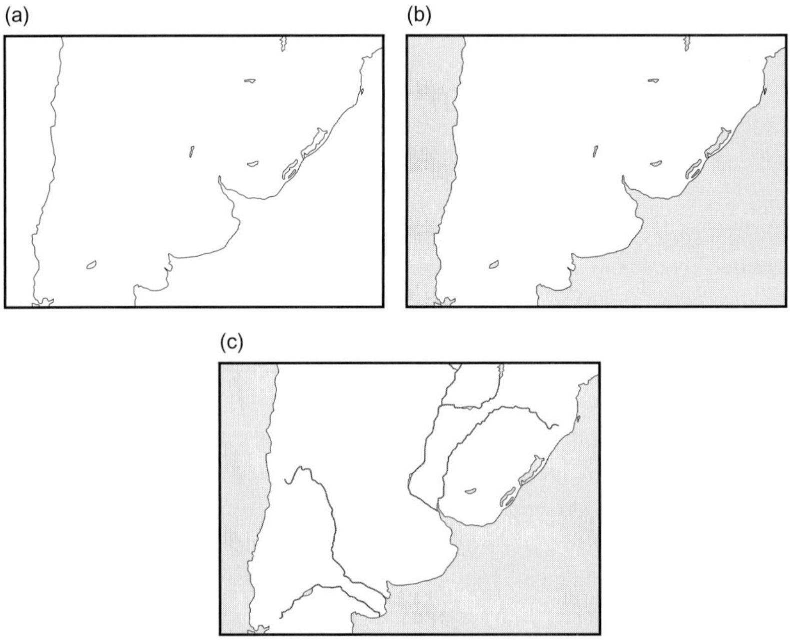

Figure 9.3 Illustrating the figure-ground relationship: (a) A simple black line on white does not help to differentiate between different levels of information. (b) The grey colour now separates the different features on the same map, but the outcome is still ambiguous. (c) By adding lines representing rivers, the separation of land and ocean becomes more obvious. *Inspired by Robinson et al. (1995).*

road types like highways, major roads and local roads. Typical rules of cartographic language only apply marginally for geomorphologic maps. These rules are related to the appropriate use of the visual variables in order to represent the level of relationship among the data types. For example, quantity relationships are depicted by varying symbol size, order relationships can be expressed using different tonal values or changing symbol size (Bertin, 1982; Rouleau, 1993). On geomorphological maps, relationships between map elements are usually expressed by the composition of the legend (see Section 3) that may put a visual focus on one set of information (e.g. morphogenesis) by altering the visual variables. The various layers of information, such as morphostructure, processes or subsurface material, can be arranged specifically to highlight one layer according to the purpose of the map (Figure 9.4). A geomorphological map created for the purpose of hazard assessment, for example, will probably highlight the active, hazardous processes. This is performed using the graphic principles mentioned above.

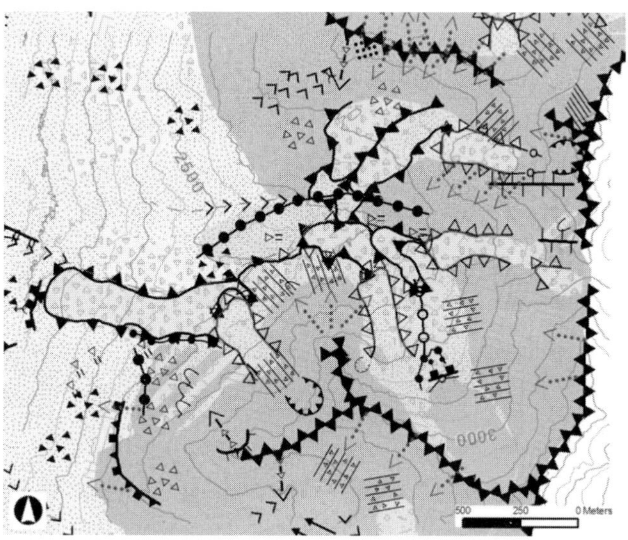

Figure 9.4 Section of the geomorphological map 1:25,000 Turtmanntal, Switzerland (Otto and Dikau, 2004). This map contains several hierarchical levels of information: coloured area symbols represent the process domains, light grey (orange in the coloured image) symbol fills show surface material information, black point and line symbols indicate landforms and processes, and point symbols in light grey depict active processes.

2.2 Map Layout and Graphic Organisation

Geomorphological maps characteristically include a great number of symbols, organised in thematic categories. This requires a large portion of the map sheet to be reserved for the legend. However, the final map is not only composed of the mapped data, its symbols and its legend, but typically includes other *map components* such as a title, scale bar, border and additional information (GITTA, 2006). These components set the mapped data into a spatial and topical context and help to identify the place, symbolisation and orientation of the map. Map components have to be systematically arranged to generate visual harmony and balance and to deliver the message of the map. Just like preparing a presentation or a publication, it can be useful to produce a basic outline of the map beforehand in the form of a sketch. This helps to get an idea where to place the title, legend, main map and other information on the map sheet. Experimenting with different layouts during the process of map making helps to find the right visual composition, which makes the map reader focus on the content and not on the layout.

Map layout consists of the arrangement of the map components into a functional composition and a meaningful and aesthetically pleasing design to facilitate the visual communication (GITTA, 2006). Geomorphological maps characteristically include the following map elements surrounding the main map (Figure 9.5): title, legend, scale, directional indicator (north

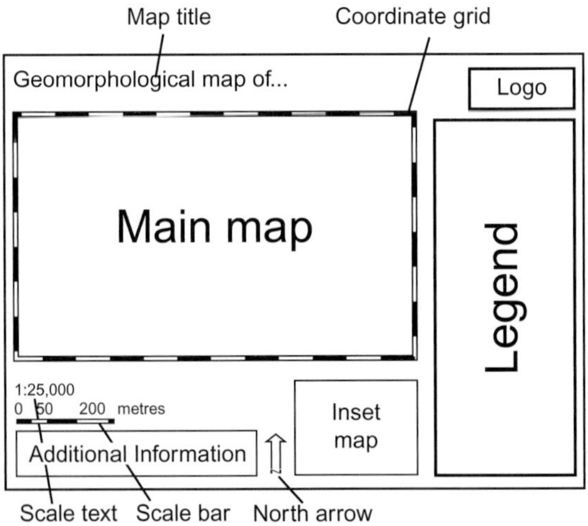

Figure 9.5 Typical items of a geomorphological map.

arrow), coordinate grid or border, information on coordinate system and map projection, and author credits. Commonly, inset maps are included to show the location of the mapped area (essential for large-scale maps), an overview of the geological situation or other additional information on the study area (e.g. a slope map). These items need to be arranged carefully to guide the viewer's eyes towards the *focus* of the map. Just like a book, a map also has a reading direction, which is usually from top-left to bottom-right. The visual centre of the map is located slightly above the actual centre (Krygier and Wood, 2005). The map reader tends to focus on the visual centre, implying that the most important information should be positioned here. This is of course not always possible on geomorphological maps, because there is probably more than just one important feature. However, the arrangement of the map elements should account for this phenomenon of human perception. Geomorphological maps generally require a coordinate grid to allow special referencing. Borders around other map elements, such as the legend or scale, should be avoided as borders separate objects and interrupt the flow of visual perception.

Between the different map components, a visual *balance* should be achieved to generate focus and keep the reader's attention on the map. Balance refers to the variable weight and direction of the map items. Lighter features are small, dully coloured or irregularly shaped, while heavier items are larger, brightly coloured and more compact in shape. Balance may be symmetrical or asymmetrical that is achieved using a central axis (vertical or horizontal). Due to the reading direction of the map, components placed in the upper part of the map and at the right side are heavier compared to objects located towards the bottom or left border of the sheet. With increasing distance to the visual centre of the map, a component's weight increases proportionally (GITTA, 2006). Using an imaginary *grid* may help to structure the positioning of map components. The grid subdivides the map sheet into horizontal and vertical spaces and generates sight-lines that create stability of the layout. Map items should be aligned along the grid to generate order and visual harmony between them (Krygier and Wood, 2005).

Colours draw the viewer's attention strongly to certain areas. The strongest colours should be used for the most important information. On many topographical maps, for example, rivers and lakes are characteristically the first features one perceives, because the dark rich blue colour contrasts strongly with more gentle colours such as green, brown and grey used for other information on the map. To verify the visual focus of the map, look at it from a distance and see what dominates the layout.

3. GEOMORPHOLOGICAL LEGEND SYSTEMS AND MAP SYMBOLS

Finding new ways to describe and visualise the landscape surrounding us has a long tradition. Even though early maps were not aimed at scientific purposes, but rather for easier orientation, military or economical uses, they did describe the landscape using simplifying symbols (later using colour) (Klimaszewski, 1982). Since the early twentieth century, the requirement for a more detailed scientific description of the landscape has been linked to a need for new symbols and cartographic designs for landscape description in geomorphological maps. Whether the symbol sets or mapping systems are used to construct thematic or comprehensive geomorphological maps, they are important both for the readability and the scientific content of the maps.

No matter what scale is chosen, depicting the physical landscape in an exact manner would be an impossible task, and thus the purpose of geomorphological mapping systems is to show an interpreted, generalised and understandable picture of the area/feature mapped. The tools available for this are the symbols and colours summarised in the legend.

When constructing a geomorphological legend, an important task is to enable the separation of descriptive and interpretative information. This is important since it opens the possibility for other map readers to draw their own conclusions or at least clarify what underlies the map maker's interpretation of the area. This also enables both the description of individual landforms, for example morphogenesis, and their relation to other forms and processes in their surroundings (St-Onge, 1981). Regarding descriptive and interpretative information, there are two commonly used models in use. The first is the *Landform Pattern Model*, which is a more interpretative model, and here the landforms are presented as repeatable, easily definable forms or patterns (e.g. hills, ridges and channels) usually not drawn at scale. The second model is the *Landform Element Model* where the landforms are described as combination of geometric elements (e.g. slope, crest and plain) and thus presents a more descriptive picture of the morphology (Speight, 1974). Depending on scale, however, the latter model often has to be complemented with the first model in various degrees. It is also an advantage if the mapping system is flexible, allowing the user to adopt the symbols most appropriate for the mapped landscape and if the mapping system is to be used at different scales (Verstappen, 1970).

3.1 Presentation of Different Legend Systems

This section outlines some of the more commonly used or recently developed detailed geomorphological mapping systems, that is mapping systems designed at scales 1:100,000 or larger (Demek et al., 1972). In addition to these there are also numerous other mapping systems or separate map sheets not connected to any mapping system published. The descriptions below outline the general characteristics of the mapping systems regarding both their scientific content and their graphical layout. More detailed descriptions of these mapping systems and their legends can be found in the references cited in each section.

The basis for most geomorphological maps is generally a base map (commonly a topographic map with reduced contrast) presenting contour lines (sometimes together with hypsometric shading) and the general layout of the hydrography. Some infrastructure may also be shown. National or global grids generally are included or indicated with 'ticks' in the margins. Also commonly found is the use of line and pattern symbols, or shadings, for illustrating information on gradient (or morphography). Many, but not all, geomorphological mapping systems also follow the guidance established by the International Geographical Union (IGU) Subcommission of Geomorphological Survey and Mapping (Gilewska, 1968) by, for example, putting the emphasis on morphogenesis and expressing this information in colour.

Even though most mapping systems share this common base for map construction, the appearance of geomorphological maps and their content varies (Table 9.2). Many of the differences in the construction of geomorphological mapping systems can be explained by the fact that the appearance of geomorphological maps is very much a result of the scientific tradition of the mapping geomorphologist and the purpose of the map and thus on what geomorphological information the emphasis is placed. These differences are reflected in the legends and consequently also in the appearance of the map sheets. Maps covering the same area but mapped by different geomorphologists using different mapping systems can

Table 9.2 Representation of Different Geomorphological Parameters in the Legend Systems Introduced

Mapping System	Morphometry/ Morphography	Hydrography	Lithology	Structure	Process/ Genesis	Age
IGU, Unified Key (1968)	Contour lines and symbols	Lines and symbols in blue	Not indicated	Not indicated	Colours, patterns, lines and symbols	Letter code
ITC, Verstappen and van Zuidam (1968)	Contour lines, symbols and lines	Hatching, lines and symbols in blue	Patterns, lines and symbols	Not indicated	Colours and symbols	Colours in separate map
The Netherlands, Maarleveld et al. (1977)	Contour lines, colour intensity and code contour lines	Lines, areas and symbols in blue	Not indicated	Partly in legend	Code, legend	Code/legend
GMK 25, Barsch et al. (1987)	Contour lines, grey shading symbols and lines	Lines, areas, symbols and patterns in blue	Red pattern and separate map	Not indicated	Colours, red and black symbols	Colour
AGRG, De Graaff et al. (1987)	Grey contour lines, symbols for breaks, etc., arrows and figures for slopes	Lines, areas, symbols and patterns in blue	Separate transparent maps, based on existing geological maps	Not indicated	Colours, symbols	Relative age according to youngest progress
Gustavsson et al. (2006)	Grey contour lines, symbols for breaks, etc., arrows and figures for slopes	Lines, areas, symbols and patterns in blue (and black)	Symbols for unconsolidated/ letter	Red lines and symbols	Coloured symbols, colours	Separate map Coloured letter code for consolidated rock

Source: Modified from Gustavsson et al. (2006).

therefore give completely different impressions, depending upon whether the emphasis is on morphometry/morphography, chronology, lithology or genesis/processes.

In order to illustrate differences between the legend systems introduced, Figure 9.6 illustrates how a moraine ridge and a fluvial terrace are represented by map symbols.

3.1.1 The IGU Unified Key

The IGU Unified Key mapping system was the result of the IGU Subcommission of Geomorphological Survey and Mapping (Gilewska, 1968) presented by Demek et al. (1972) in the Manual of Detailed Geomorphological Mapping. Another version of the mapping system designed for mapping at smaller scales was also published as the Guide to Medium-Scale Geomorphological Mapping (Demek and Embleton, 1978).

The legend of the Unified Key is comprehensive, presenting information about genesis, lithology, morphometry/morphography and age. However, since the legend is used for many different scales, the detail of this information varies. Although there is an attempt to make a comprehensive geomorphological mapping system for the whole world with an extensive legend, Demek et al. (1972) claims that it is not a ready unified legend covering all forms and processes and that the legend sometimes needs to be extended or modified to fit local or regional conditions (Demek et al., 1972; Barsch et al., 1987).

The IGU Unified Key includes 353 symbols representing different landforms, which enables a detailed inventory of the landscape. The main information in the legend is on morphogenesis, and thus this is expressed in 10 colours in combination with texture, line- and point symbols. The genesis is further divided into 3 form groups representing endogenic processes and 13 form groups representing exogenic processes. The red colour is reserved for endogenous landforms, black for biogenic/anthropogenic forms or data, grey for contour lines and slope classes and blue for water surfaces and hydrography. The rest of the colours describe different erosional and depositional exogenous forms. To describe landforms with complex genesis, two colours can be used where the first, used as a base colour, shows the original genesis, and symbols in the second colour shows the modifications of the landform. According to the IGU Commission on Geomorphological Survey and Mapping, the altitude in a detailed geomorphological mapping system should be described with contour lines while surface inclination should be described by the shade of the

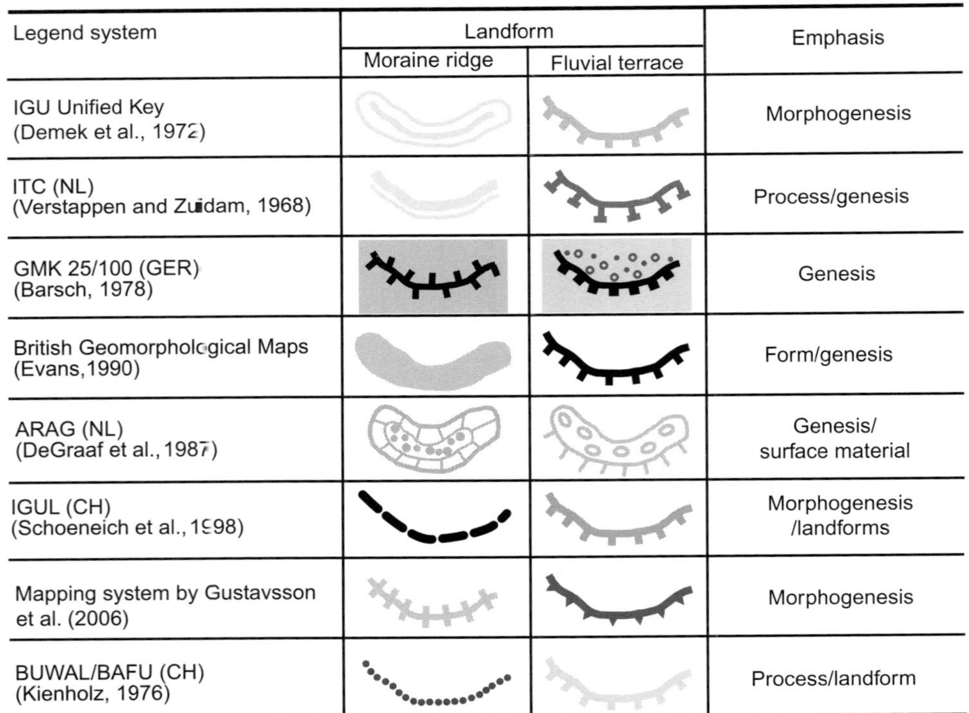

Figure 9.6 Comparing the symbols for moraine ridge and fluvial terrace of the different legend systems presented in this chapter.

genesis colour, and thus the mapping system developed uses both these ways to express information on slope. The slopes are classified into six categories according to their gradient (0–2°, 2–5°, 5–15°, 15–35°, 35–55° and >55°). The IGU Commission on Geomorphological Survey and Mapping also suggests that, in some areas, a classification based on other critical slope values may be used. Information on geological age is expressed with a letter code in black. When possible the landforms are represented by figures at scale and in other cases they are shown by symbols (Demek et al., 1972; Klimaszewski, 1982).

3.1.2 The ITC Geomorphological System (Enschede, The Netherlands)

In 1968 the Dutch International Institute for Aerial Survey and Earth Sciences (ITC) published a comprehensive geomorphological mapping system for all scales. The ITC maps are, however, divided into two groups: (1) large- and medium-scale maps and (2) small-scale maps. Depending on their content, reliability and degree of generalisation, the two map groups can also be further subdivided into several classes (Verstappen and Zuidam, 1968). The ITC geomorphological mapping system presents information about morphometry/morphography, processes/genesis, age and lithology (with particular attention to rock-type properties). Stress is placed on geomorphological processes, which determines the landscape units shown on the map.

In the ITC system, colours are used in two ways. First, shading is used to define larger landscape units based on the dominant process, which gives a good overview with pronounced geomorphological units. Second, 10 colours are used for line symbols describing processes and genesis of smaller landscape elements. The symbols in the ITC system are subdivided into 14 groups based on process/genesis, morphometry, lithology, chronology and topography. In addition to this there are also two special-purpose map legends. The use of these almost 500 unique line symbols makes the production of maps printed in greyscale possible. If presented in greyscale, the symbols describing geomorphological processes are printed in black while topography and lithology are printed in grey. There are also additional symbols available for some specialised maps connected to the system (e.g. the morpho-conservation map and the hydromorphological map). A disadvantage of this legend size is that it gets complex and hard to use for geomorphologists not familiar with the system. The age of the landforms is indicated by a letter code in black (Verstappen and Zuidam, 1968; Salomé et al., 1982).

3.1.3 The German GMK Mapping Systems

Geomorphological mapping has a long tradition in Germany with early work (Passarge, 1912) generally related to concepts of landform analysis (Kugler, 1964). In 1976 a research programme on geomorphological mapping was initiated, managed by D. Barsch; for 9 years ~40 groups from German universities mapped different landscape types typical of the Central European landscape. The research programme resulted in 27 geomorphological maps at 1:25,000 scale (GMK 25) and eight geomorphological maps at 1:100,000 scale (GMK 100). All available maps of the research programme are available online at the homepage of the German Working Group on Geomorphology (www.ak-geomorphologie.de).

In Central Europe, the GMK (GMK = Geomorphologische Karte) maps have been created with two main practical applications in mind: (1) to create a planned cultural landscape and (2) to reduce the destruction of the natural environment, in order to keep the ecology in as natural a state as possible. The GMK 25 legend system allows for the production of derivative and interpretation maps, such as the GÖK (Geoökologische Karte) 25, a geo-ecological map (Barsch and Liedtke, 1980b; Barsch et al., 1985). The development of the GMK has resulted in three versions of the legend: the red legend (1972), the green legend (1975) and the white legend (1990). The earliest legends had problems with the delineation of slope angles; this was solved by the use of mean slope angles. In 1998 a complement to the legend for mapping in alpine environments was published in the GMK Hochgebirge. This complement provided additional symbols for permafrost phenomena, slope forms and mass movement (Kneisel et al., 1998). Symbols of the GMK Hochgebirge are available for ArcGIS software and can be downloaded at http://www.geomorphology.at/ (Otto, 2008).

The information in the GMK mapping system is presented in a legend consisting of eight layers of information presenting: (1) areas of process and structure (in colours), (2) hydrography (blue), (3) actual processes (black+red), (4) subsurface material/surface rock (reddish brown), (5) curvatures (black screen), (6) steps/minor forms/valleys/roughness (black), (7) slope angles (grey raster) and (8) situation/topography (grey) (Barsch and Liedtke, 1980a,b). Bright red is used in the maps to highlight recent geomorphological processes or to give attention to active morphodynamics and areas of potential danger. Since the legend is constructed like a construction kit, individual layers can be easily modified extending the use of the mapping system to areas outside Europe where, for example, other surface forms occur.

In the GMK 25, areas larger than 50 m × 100 m are represented at accurate scale while the smallest landform presented at accurate scale in the GMK 100 is 200 m × 400 m (Barsch and Liedtke, 1980b; Barsch et al., 1987; Klimaszewski, 1990; Kuhle, 1990).

Each map sheet displays a relevant part of the complete GMK legend printed in the margin. A separate geological reconnaissance map at 1:300,000 scale printed in the margin of the GMK maps presents a good overview of the main geological conditions of the mapped area. On the main map, detailed information on lithology is presented as grain-size compositions of substrate material. When the substrate material is composed of easily weathered bedrock such as limestone, the weak resistance to weathering is also presented. In coastal areas, some submarine features are also included.

The GMK system enables a detailed and informative presentation of the geomorphology and also shows the degree of anthropogenic change in the landscape. The amount of information presented in the maps however makes them hard to read at first. In the GMK system, the symbols describing morphography and morphometry are genetically similar, and it is therefore hard to separate similar landforms originating from different genesis. Also the substrate pattern is presented in a highly differentiated symbol key inherited from a standard of pedological mapping. When this reddish pattern is printed on a similar colour describing 'areas of process and structure', it is hard to read the content. There are many colours used for describing 'areas of process and structure', and this sometimes makes the differences between them too small to differentiate. On the GMK 100, problems may arise with the placement of generalised symbols, for example by using the same symbol for deep narrow valleys and broader flatter ones (Barsch and Liedtke, 1980b; Barsch et al., 1987; Kuhle, 1990). It is also hard to get a clear picture of the shape on valleys. This is especially true for flat-floored valleys. The results of a survey in alpine environment in Switzerland also show that the information in the GMK 25 is too dense to be readable. To solve these problems, suggestions were made by Kneisel and Tressel (2000) to change the colour intensity of some features in the map legend.

3.1.4 British Geomorphological Maps

In Britain a geomorphological mapping system has been developed using the Ordnance Survey 1:25,000 as a base map. Emphasis has mostly been put on mapping form and genesis for particular groups of landforms. The tradition in Britain has been to construct geomorphological maps using the Landform Element Model (Speight, 1974) and thus the emphasis has been on morphology. Because slope gradient is an important variable for many processes and applications, classification of relevant slope class limits has been considered especially important. The maps have been shown to be useful in developing an 'eye for the landscape', and practical applications have been made in landslide areas. Depending on the purposes of the maps, materials are classed in different ways. Geomorphological maps made for geological and soil surveys classify materials based on a combination of both genesis and characteristics (till, glacial sand, gravel and so on), whereas maps constructed to describe current processes and hydrology describe the materials based on their physical properties (grain-size distribution). For the description of bedrock, special emphasis is placed on the degree of jointing (Cooke and Doornkamp, 1990; Evans, 1990).

3.1.5 The AGRG Geomorphological Mapping System (Amsterdam, The Netherlands)

The detailed geomorphological mapping system of the Alpine Geomorphology Research Group (AGRG, Amsterdam, The Netherlands) has been developed in the alpine surroundings of Vorarlberg, Austria. Although developed in alpine areas, the legend has also successfully been used in areas with less pronounced relief (with minor modifications). Due to the complex geomorphology of alpine environments, the maps are commonly made at 1:10,000 scale or larger.

The legend presents information about morphography/morphometry, lithology, process/genesis and hydrography as four different layers on a base map showing contour lines and other administrative information in grey. Because the emphasis is on the process/genesis, this information is expressed in six colours used to print the symbols. Unconsolidated materials are presented as pattern-like symbols that also can be used to indicate the direction of transport of materials. The hydrography is indicated by blue symbols with additional symbols in black for artificial drainage.

The geomorphological information is printed on a base map presenting infrastructure and contour lines in grey. Additional information about the physical and chemical properties of the materials is printed on a separate geotechnical overlay map. A natural hazard overlay map has also been developed (De Graaff et al., 1987).

Because many periglacial and nival processes working in an alpine environment are very similar to other degradational processes and thus difficult to objectively map in the field, these processes have been grouped together. Also fluvial erosion and avalanches (as far as they are geomorphologically active) are treated in the same way. Some special features (protalus ramparts, rock glaciers and stone stripes) have been separated in the legend. The AGRG mapping system does not make a distinction between active and relict processes but presents indirect information about relative age for a few features (De Graaff et al., 1987).

The materials are subdivided into four classes based on process/genesis: sediments formed in or by water, glaciogenic or related sediments, slope deposits and organic deposits. A further division of materials is made on the basis of depositional environment and/or texture (De Graaff et al., 1987).

The original legend is focused on materials (based on genesis) and processes occurring in the Alps and lacks many symbols useful elsewhere. The construction of the legend is similar to the construction of the legend of the GMK with several layers overlapping each other. The AGRG mapping system however uses an open framework, supported by contour lines, to indicate the morphography by use of the Landform Element Model (Speight, 1974). This framework and the absence of covering colours make the maps difficult to read for geomorphologists not accustomed to the system but gives the advantage of possibilities of many combinations of forms and processes. Another advantage is that the colours do not obscure other information (De Graaff et al., 1987).

3.1.6 The IGUL Mapping System (Lausanne, Switzerland)

A simple mapping system for high and middle mountain areas was developed at the Institute de Géographie de l'Université de Lausanne (IGUL), Switzerland, in the late 1980s (Schoeneich, 1993). The system has a strong morphogenetic and morphodynamic focus and only depicts landforms. It combines several principles of previously published Swiss, French and German mapping systems. According to the German system GMK 25 (see Section 3.1.3), colours are applied to express processes. However, colours are used to differentiate between the line and area systems, following the French system of Tricart (1965), to present genetic information for the landforms mapped. A differentiation of erosional and depositional dynamics is provided using white and coloured surfaces, respectively (Schoeneich et al., 1998). Morphographic information and lithology is not provided. The legend system is mainly used for educational purposes but has been applied to landform inventories and the analysis of sediment dynamics (Theler and Reynard, 2008; IGUL, 2010).

3.1.7 Mapping System by Gustavsson et al. (2006)

Using parts of the basic concept of the AGRG mapping system (De Graaff et al., 1987), the mapping system of Gustavsson et al. (2006) is constructed through a thorough study of earlier developed geomorphological mapping systems. It has tried to solve specific problems often occurring in the presentation of comprehensive geomorphological data, for example presentation of sediment of mixed composition, diagenesis, presentation of bedrock lithology and the separation between descriptive and interpretative geomorphological data. An aim has also been to enable a detailed presentation of varied and complex geomorphological environments without the use of complex legends (Demek et al., 1972). Since the scale of a geomorphological map varies due to landscape complexity and mapping purpose, the mapping system is designed to be used at different scales using the same legend (tested at 1:5000 to 1:50,000 scale) (Gustavsson and Kolstrup, 2009).

The mapping system is not aimed at being as detailed and precise in information as other more comprehensive mapping systems (Verstappen and Zuidam, 1968; Demek et al., 1972), but uses a simple structure where information is based on the combination of individual descriptive data. These data are combined in an easy-to-use legend, which enables simple conversion to a geomorphological GIS database constructed in parallel with the mapping system (Gustavsson et al., 2006). The less-extensive legend also allows for additions and improvement according to the needs of the user.

To reduce the subjectivity and to increase the possibilities for application, the mapping system presents basic descriptive geomorphological data as far as possible. Thus, the legend of the mapping system enables all geomorphological data presented to be read separately (e.g. material, process, genesis or morphography), and it is the combination of these data that enable the map reader to interpret the landscape (St-Onge, 1981). As in the AGRG mapping system, the morphography is expressed at scale (where permitted) by means of the Landform Element Model (Speight, 1974).

To enhance the readability, this mapping system avoids a saturated combination of several layers of symbols in various colours. Like the AGRG mapping system, this system instead uses an open framework that enables additional point and line symbols together with a pattern describing the materials to be more clearly presented.

As with the GMK and the AGRG systems, the system presents detailed information on anthropogenic influence, and the system also enables the description of biogenic genesis of forms and materials and point descriptions of known stratigraphy. Whereas most mapping systems include karst processes as genesis or as specific features, the legend incorporates weathering, which also includes weathering of non-calcareous rocks as a morphogenesis or origin of materials. Unconsolidated lithologies are expressed as grain-size distributions whereas bedrock types are described in letter codes printed in colour of geological age according to the Elsevier Geological time table (Haq and Eysinga, 1987). Morphography and materials, both described by the use of symbols and their genesis (11 different genesis types), are then expressed through the use of colours. Diagenesis or, for example, surface-washed materials can be expressed by combining colours. This use of coloured symbols enables the original field observations of materials and forms to be seen in the map, which allows the map reader to see what the interpreted genesis is based upon. This separation also makes the conversion to a GIS database easy. A disadvantage of this combination of data is, of course, that the maps are harder to interpret by non-geomorphologists.

3.1.8 The Swiss BUWAL Mapping System

The BUWAL mapping system (BUWAL: former Bundesamt für Umwelt, Wald und Landschaft – *Swiss Federal Agency for Environment, Forest and Landscape*, today BAFU: Bundesamt für Umwelt – *Swiss Federal Agency for the Environment*) for natural hazards has been developed for applied mapping of potentially hazardous processes and landforms (Kienholz, 1976, 1978; Kienholz and Krummenacher, 1995). Maps of natural phenomena are regarded as a prerequisite for natural hazard assessment and hazard management in Switzerland. Implemented within the procedure of hazard management, the map is considered the first step in the recognition and documentation of hazards. The final purpose of these maps is to support the hazard assessment and decision process by increasing transparency and traceability towards the engaged parties.

The legend system is compiled as a construction set to enable a greater degree of freedom and flexibility for map creation and to accommodate the purpose and requirements of the individual project. It follows three formal principles:

1. Applicability for different map scales ranging from 1:1000 to 1:50,000.

2. Applicability for specialised (restricted to one process) or general hazard maps (several sources of hazard on one sheet),
3. Map compilation generated from a combination of simple and limited basic elements (construction set).

The legend focuses on the mapping of processes and the related landforms of erosion and deposition. Three main differentiations of graphic variables are provided regarding the topical map content: (1) difference in colour (hue) depicts the various processes and (2) variations in colour intensity (value) or (3) symbol size represent changes in process intensity, activity, evidence, age or depth. Due to its origin, the symbol set concentrates on processes with hazardous potential in mountain areas and their forelands. These processes include avalanches, debris flows, rock fall, landslides and hydrological hazards (flooding). Maps generated using this mapping system contain specialised symbols for areas of process origin, transfer zones and depositional zones. What differentiates this legend from others is its potential for predictive mapping of potentially hazardous locations, for example location within small creeks that indicate the potential for blocking by woody debris during debris-flow events. Thus, these maps not only document existing phenomena but also provide an interpretation of the mapped objects with respect to hazard assessment.

4. MAP PRODUCTION AND DISSEMINATION

Traditionally, geomorphological features are mapped in the field, or at the desk using tracing paper or drawing film draped over an aerial photograph or a topographical map sheet (Evans, 1990; Lee, 2001). These field maps are then digitised or scanned to transfer the information into the computer. Alternatively, geomorphologic features are digitised directly on the screen (see Chapter 8 for further details) or by using a portable mapping device in the field (see Chapter 6 for further details). Combined with a GPS, a portable device delivers georeferenced information in a GIS format e.g. (Dykes, 2008). The final production of the map is generally performed using graphic or GIS software. Although graphic software is used for on-screen visual design, GIS software focuses on spatial data management, analysis and map creation. One advantage of map creation using a GIS is the geographical referencing of the input data so that it can be analysed and used for several applications. Although not comparable to graphic design software, the capabilities remain very good. The main difference in map creation is that within the GIS, every drawn point, line or area becomes an object in the database, whereas the graphic software enables the manual combination of graphical strokes and points to generate more complex objects and symbols.

The production of a geomorphological map requires the following steps:

- Selection of the legend system,
- Mapping of the geomorphological objects (processes, landforms, materials) in the field, or from secondary data,
- Generation of a digital symbol set (optional),
- Transfer to the mapping software which may involve the following steps:
  - scanning of the field maps,
  - georeferencing of the scanned image (only necessary for a GIS),
  - digitising of features.
- Generalisation of map data including:
  - simplification of complex objects to fit the map sheet,
  - exaggeration of features that are too small to show at the scale of the map.
- Printing or online publication of the map.

Geomorphological maps are composed of different layers of information. The base layer is generally a topographical map or simple contour lines as a reference source. As this information should not dominate or influence the geomorphologic information on the map, the base layer should be displayed using light colours (e.g. light grey). Thematic layers differ between the mapping systems, depending on the focus of the map. Typically, a geomorphological map includes layers on morphography and/or landforms, process distribution, hydrology and subsurface material. Changing the composition and the visual hierarchy of these layers allows shifting the focus of the map. Such specialised geomorphological maps, focused on, for example, process distribution or subsurface material, can be of interest for application in natural hazard management or engineering projects.

4.1 Map Creation Using Graphic Software

Graphic software can be differentiated into programs focusing on the creation of vector graphics (e.g. Adobe Illustrator, Corel Draw and Inkscape) or raster images (e.g. Adobe Photoshop, Corel Photo Paint and Gimp).

Vector graphics are made up of points (nodes) and paths (edges), whereas raster graphics are based on rectangular pixels organised on a grid. Raster graphics are typically used to edit photographical images or create artistic illustrations. Because vector graphics are not composed of a certain number of pixels, they can be scaled without losing image quality (Slocum et al., 2005). Complex graphics and sketches produced for printing are usually generated using vector graphics. The geometrical primitives, points, lines and polygons that compose a geomorphological map are best represented using vector graphics and produced with vector graphics software.

The main advantage of graphical software with respect to the generation of geomorphological maps is the great number of tools for the creation and modification of graphic objects. Generally these can be adjusted and customised to the user's needs and exceed the possibilities provided by GIS software. Common to all graphic software (as well as to GIS software) is the ability to organise the objects in different layers. This feature is particularly useful for map creation and should be utilised for the organisation and structure of items and different topical layers of the map. Using layers enables certain objects to be fixed in order to prevent unintentional modification, while working on neighbouring features. By deactivating or masking a layer, the number of objects on the screen is reduced during the process of mapping, allowing for a clear view of the object in preparation.

The greatest challenge in the process of map production is the generation of reusable symbols (see Section 4.3). Graphic software allows the definition of any drawing as symbol templates for points, lines or area fills (e.g. in Adobe Illustrator: symbol, brushes and swatches). A large number of ready-made symbols can be found on the Internet, very few however are specially made, or useful for geomorphological maps. Although point and area symbols are generally easier to apply, line symbols commonly have problems in drawing symbols at corners and curves (see Section 4.3) and thus require more effort to generate.

Due to the great number of graphic tools, graphic software offers many options for symbol creation and enables the creation of maps using very complex symbols. Graphic software is designed to produce high-quality print products and thus provide many tools for print optimisation. However, this software is often very complex and requires some expertise in order to fully handle the functionality and tools available.

4.2 Map Creation Using GIS Software

The performance of *GIS software* goes beyond making maps. Data analysis (queries, overlays and so on), data management and database storage are central features of GIS software (Longley et al., 2005). Prior to transferring field data into a GIS, the structure and design of the database should be considered. In the database, geomorphological information should be stored in a logical manner and prepared for analysis and query. Each object in the map is thereby linked to the database by its table of attributes. The database provides additional information on the object that is either gathered during the mapping campaign or generated afterwards. Thus, GIS offers the ability to combine basic information on landform/process/material type and geometry with secondary data on feature characteristics (e.g. from sampling, dating, laboratory, geophysical or GIS analyses). A simple structure for a database connected to a geomorphological map may include the following levels of information: (1) geomorphological features (landforms, processes), (2) geological/lithological data, (3) hydrological information and (4) additional data used for map construction, such as topographical maps, digital elevation models or aerial photographs. An example of a geomorphological database structure using Environmental Systems Research Institute, Inc. (ESRI) ArcGIS software is given by Gustavsson et al. (2008).

GIS analysis commonly results in the compilation of a map and consequently GIS software includes mapping facilities and graphic design capabilities. Among these are automatic tools to generate the legend, scale bar, north arrow and coordinate grid. These map elements are automatically adapted to changes, for example the scale or symbol type. Often special symbol editors are provided to compose and define the symbol set for the map (see Section 4.3). As with graphic software, GIS software offers tools to digitise vectors (points, line, polygons) with high accuracy and the ability to modify single vector nodes. As the data structure in a GIS is organised into different layers, these can be combined to form map frames. By combining several map frames, inset maps can be created, geographically referenced and created within the same GIS project.

One of the advantages of using a GIS is the geographical referencing of the data. The geomorphological map can easily be rescaled, for example, to enlarge certain areas or to fit a special sheet size. Further, the coordinate grid and direction indicator (e.g. north arrow) are automatically generated and adapted.

4.3 Creation and Utilisation of Standardised Digital Symbols in a GIS

Graphic symbols are the most fundamental element of cartographic language on geomorphological maps. They must be created to clearly express the geographic location of the feature and to display relationships between features with respect to differences, quantities or ranking (Rouleau, 1993). There are few ready-made symbol sets for standard GIS software freely available (Otto and Dikau, 2004; Otto, 2008) and therefore legend symbols commonly will need to be created. By defining the symbol type used for each data set, digitised points, lines or areas are automatically replaced by map symbols. Every graphic element on a map is a symbol that is systematically linked to the data and content of the map. In contrast to other thematic maps that commonly display numerical data, geomorphological maps depict a composition of real-world features and their interpretation, for example process activity, or genesis of landforms. Just like topographical maps, where contour lines represent elevation and therefore the shape of the land surface, geomorphological maps refine this representation of the surface using symbols, commonly with topographical maps providing base or background information. The majority of geomorphological symbols represent qualitative rather than quantitative data, because most geomorphological maps focus on the inventory and location of objects on the land surface. Land surfaces are commonly composed of a complex set of landforms and processes creating a very dense display of information. To allow good legibility and facilitate understanding of the map, symbols need to be created that are easily distinguished and understood. Understanding is closely connected to familiarity of what we see. Thus, well-chosen illustrative symbols can remind the viewer of the related feature. Abstract symbolisation requires a greater ability of spatial thinking and visual perception. However, as many users of geomorphological maps are familiar with landscapes, they will be able to perceive the map content as a whole even if some of the symbols are not familiar, as long as the map is readable and permits the perception of the land surface.

In the past, geomorphological maps and the symbols used have been drawn by hand. The transfer of these handmade symbols into a GIS often suffers from graphical restrictions produced by the computer and has to do with the composition and reproduction of vector graphics on the computer.

Symbols are generally composed of different graphic objects. Although point symbols usually consist of a single graphic, complex line and area symbols are constructed by combining different graphics to create the final symbol. For example, a ridge is commonly represented by a line symbol that consists of a solid black line in the centre and solid, black triangles on both sides indicating the directions of the two adjacent slopes (Figure 9.7). This symbol is thus composed of three different layers: (1) the black line, (2) the triangles facing upwards and (3) the triangles facing downwards.

Two important restrictions need to be considered when working with complex symbols in GIS. (1) Symbols generally do not scale automatically as do features. Thus, symbol size and line thickness have to be customised to the appropriate map scale in order to display correctly. (2) Reproduction problems commonly arise when using complex lines symbols. Line vector graphics are composed of nodes (points) and edges (lines) connecting the nodes. When a line is digitised, nodes are set by clicking the mouse and the edge is generated automatically between the nodes. Curvature of the line is a function of node density, or generated automatically by the graphics program by smoothing. If additional graphic

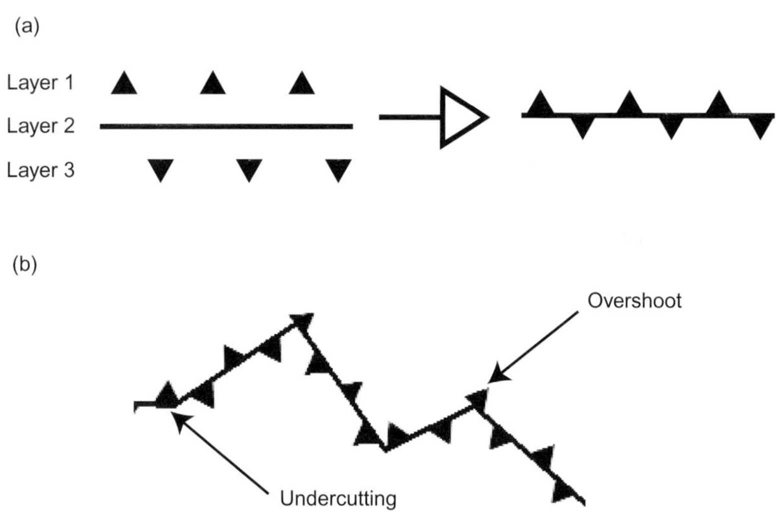

Figure 9.7 (a) A composed line symbol, constructed from three layers of symbols. (b) Typical problems of undercutting and overshoot of symbol representation in GIS.

objects are positioned along the line, for example triangles on the left and the right, overshoots and misplacements of the symbol parts can occur in GIS (Figure 9.7). Because the software automatically places these symbol parts between the nodes of the line, an exact and regular positioning is not always possible. This effect can however be removed manually by changing the node position.

4.3.1 Creation of Point Symbols

Symbols for geomorphological point features are generally used for single landforms and/or single processes that are too small to be represented at scale. Thus, point symbols are commonly applied where features have been generalised and their shape and size commonly does not represent the real extent of the object. Point symbols are the most illustrative symbols and are generally created using drawings (simple bitmap graphics) or font characters. These predefined images are created in graphic or special font character software and later imported into the GIS. Point symbols can show orientation of an object defined by an angle of rotation. If the feature direction varies between different objects, the rotation angle needs to be stored within the feature's database (e.g. attribute table).

4.3.2 Creation of Line Symbols

Line symbols are commonly applied for structural and linear features, for example ridges, moraines and rivers. Simple line symbols use solid, dotted or hashed lines. More complex symbols combine pictures or characters that are added to the line or replace the line along its length. Line symbols can have a direction, for example indicating the flow direction of a river, or the direction of valley. Direction then is indicated by a special arrangement of the symbol elements. In GIS, line direction is commonly dependent on the direction of digitising, but can also be changed by flipping the start and end node of the line.

4.3.3 Creation of Area Symbols

Not all geomorphological mapping systems make use of area symbols. However, if they are applied, area symbols mostly represent spatially continuous information, for example subsurface material or slope gradient. Area symbols are generally composed of colour or hatch (texture) fills. Variation in area symbols is therefore performed by changing colour, hatch orientation and density, or by changing the texture shape. A typical example is symbolisation for grain size, which can be depicted by increasing dot size according to different grain sizes. In contrast to line and point symbols, areas cannot indicate a feature orientation. However, this information is not relevant to most features depicted by area symbols.

4.4 Map Reproduction

Despite dissemination of maps via the Internet or within journals, many geomorphological maps are still reproduced on paper. Whichever method is chosen, the choice affects many steps of map design and production. Maps that will be printed have different requirements concerning, for example, colours or resolution than maps that are viewed on a computer screen. The map design thus has to be customised with the output method of the map in mind.

Special attention is required when preparing maps that will be printed. One common problem is that the colours of the printed map do not match the ones composed on the computer. Problems of colour management are related to the different use of colours on computer screens and printing devices. The main difference in colour representation is the process of colour combination, which can be additive or subtractive (Rouleau, 1993; Slocum et al., 2005). Computer monitors use the combination of three colours, red (R), green (G) and blue (B), to create all other colours. The RGB system is an additive method which means that when all three colours are added, white colour is generated. RGB colours are composed giving a value for each of the three colours (e.g. the combination of R: 250, G: 250, B: 0 produces a bright yellow colour). The RGB colour system should primarily be used for maps that are viewed on the computer. When a map is printed, simple computer printers generally are able to reproduce RGB colour; however, more sophisticated computer and commercial printers require a conversion into the CMYK colour system. This colour system is a subtractive process using the basic colours cyan (C), magenta (M), yellow (Y) and black (K). When combining the first three colours C, M, Y all light is absorbed or subtracted from the vision and the result is black. The same yellow given in the example above would be composed in CMYK by choosing: C 11%, M 0%, Y 91%, K 0%. Graphic software usually enables a conversion of colours from RGB into CMYK and vice versa.

Another issue for map production is the display or print resolution of the graphics. Computer monitors display at a lower resolution in comparison to printed maps. Image resolution is measured in dots per inch (DPI), which describes the density of individual points that are placed (displayed or printed) within a linear inch. Computer monitors have a resolution of 96 DPI, whereas printers generally require a minimum resolution of 300–600 DPI in order to produce sharp graphics. This needs to be considered when the map is prepared for printing.

The final step of production is the transfer of the map to the printer. Printers generally use different file formats than the standard graphic format generated by the graphic or GIS software. The digital map file needs to be converted into this printer file format, which is generally done by the application software (GIS or graphics). The most common file format used for printing is the PDF (portable document format) that contains the graphic and page description information. PDF is a standard format that can be processed by many graphic software and printers without loss of information.

A GeoPDF includes one or multiple map frames within the PDF page associated with a coordinate reference system. It enables the sharing of geospatial maps and data in PDF documents. Multiple, independent map frames with individual spatial reference systems are possible within a GeoPDF, for example, for map overlays or insets. Geospatial functionality of a GeoPDF includes scalable map display, layer visibility control, access to attribute data, coordinate queries and spatial measurements. Adobe Reader™ (starting with Version 9.0) supports geospatial functions of GeoPDFs. However, full functionality of GeoPDFs require a free and user-friendly plug-in for Adobe Reader™, the TerraGo™ toolbar (see www.terrago.com). GeoPDFs can be created either directly from GIS (e.g. ArcGIS 9.3) or using a specific software called TerraGo Publisher™ that is integrated into GIS applications such as ESRI's ArcGIS™, Intergraph's GeoMedia™ or ERDAS Imagine™. A GeoPDF enables fundamental GIS functionality turning the formerly static PDF map into an interactive, portable georeferenced PDF map. It is an interesting and valuable way of dissemination of geomorphological maps. Some geospatial data providers such as the United States Geological Survey (USGS) and the Australian Hydrographic Service (AHS), have already started publishing interactive maps using the GeoPDF format.

5. GEOMORPHOLOGICAL MAPS ON THE INTERNET

With the digital production of geomorphological maps, the dissemination of research outputs now extends beyond simple paper products. Internet technologies can contribute to both the dissemination of geomorphological maps and access to geomorphologic data and help to make geomorphological knowledge available to the general public. Indeed, many national geological surveys employ end-to-end digital workflows from data capture in the field to final map production and dissemination (e.g. USGS – see http://seamless.usgs.gov/). This section therefore deals with the potential of web mapping applications for the distribution of geomorphological information.

Mitchell (2005) mentioned two general types of Internet maps: static and dynamic maps. Static maps, scans or image exports from GIS software, are the easiest way of displaying maps on the Internet. They are simply embedded in web pages as images and detailed knowledge of web development is not required. Because static maps have been produced using GIS or graphic software, no limitations to design or symbology exists. However, web sites constrain extent and graphic resolution of the map to the capabilities of the computer screen. The term 'static' refers to the definite status of the map. Just like hard-copy maps, static maps on the web cannot be modified by the user. This implies spatial navigation and views at variable scales are impossible. There is no spatial reference so the image cannot be used by other applications, even if the map has been previously produced in a GIS.

Dynamic maps, in contrast, are characterised by interactive capabilities: the user can interact with the map by zooming, panning or adding further thematic layers, with the map refreshed after each task. Web mapping applications such as Google Maps are currently very popular and widespread and have increased the interest and access to mapping. Depending on the system components, advanced symbology, map overlays from different applications and their integration into a Desktop GIS is possible. The interoperability is achieved through the use of international open standards that include mechanisms for the integration and visualisation of information from multiple sources.

The motivation to write about the online distribution of geomorphological maps originates in the increasing number of web mapping applications available today. They indicate that the Internet has become a medium for displaying geographical information in rich forms and user-friendly interfaces. So, why not use the Internet to distribute geomorphological maps and enhance their practical application? Web mapping can play a key role in the movement towards the global dissemination of geomorphological information. We present two examples, WebGIS and Google Earth, and focus on the generation and display of complex symbols.

5.1 Principles of WebGIS

A WebGIS is a common way of presenting dynamic maps online. It links the Internet with GIS technology. The GIS processing is performed online and maps are visualised in interactive web viewers. Although there are many ways in establishing a WebGIS, depending on the software components used, most applications are based on the same principles (Figure 9.8).

The user works with a web client displayed in their Internet browser. The client contains the demanding GIS functions (e.g. zooming or panning), compiles the map requests and forwards them to the application server. The server passes the map requests to the mapserver, the central software performing the GIS processing. The mapserver, having access to the spatial data, executes the map requests and returns the maps as images to the web server, which finally sends them back to the user's web mapping client. The application acts as a web-based information system. Another way is using a web service, for example a Web Map Service (WMS), a software function that is accessible by a desktop GIS programme providing direct access to the mapserver.

WMS is a widely supported, standardised protocol for accessing maps online that contains the map request and parameters specifying GIS processing for the mapserver, for example choice of layers or spatial extent. The protocol standard is specified by the Open Geospatial Consortium (OGC), a non-profit international standards organisation with members from commercial, governmental and research organisations, including Google and Microsoft. It is leading the developments of standards to establish interoperability and ensures platform and software independent

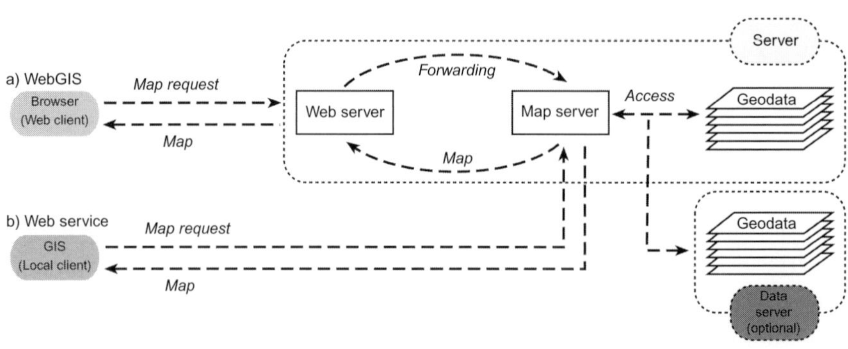

Figure 9.8 Simplified scheme of information and data transfer of a WebGIS and web service application.

Table 9.3 List of Several Open-Source (*) and Commercial Software Products Providing and Supporting the WMS Format

WMS Servers	Web Mapping Clients	Desktop Clients
UMN Mapserver*	OpenLayers*	GRASS GIS*
GeoServer*	Mapbender*	Quantum GIS*
Degree*	ka-Map!*	ArcGIS/ArcView
ArcGIS Server	Mapbuilder*	ArcGlobe
ArcIMS	Chameleon*	MapInfo
GeoMedia	ArcGIS Explorer	Global Mapper
Express Viewer	Autodesk MapGuide	Autodesk AutoCAD
ERDAS Apollo Server	Oracle Map Viewer	ERDAS Imagine
Demis Web Map Server	Worldkit	uDig*
	ERDAS Titan	OpenJUMP*
		Google Earth
		NASA World Wind
		Demis Mapper
		Gaja
		GDV Spatial Commander

usability of geospatial services and data sharing. WMS is one of the most frequently used protocols in web mapping, which is supported by many open-source and commercial software (Table 9.3).

The introduction to all available software components for WebGIS applications would go beyond the scope of this chapter. One popular package available for Windows is Maptool's 'MapServer for Windows' (www.maptools.org/ms4w/), which uses open-source components to provide a mapserver environment including libraries for data input and output. MapServer is GIS software running on a web server that enables interaction with GIS data over the Internet and generates cartographic output of geographic content. In addition, the Geospatial Data Abstraction Library (GDAL, www.gdal.org), a powerful tool for data translation and processing (which is used by several GIS programmes including GRASS, and ArcGIS) is included. An introduction to the most common WebGIS tools is given by Mitchell (2005).

Figure 9.9 shows a WebGIS that visualises the results of a geomorphological field mapping campaign in the Turtmann valley (Switzerland), which is available online at www.geomorphology.at. The application employs MapServer generating the maps as WMS, the spatial database management system PostgreSQL (www.postgresql.org) maintaining the geometries and the web mapping client Mapbender (www.mapbender.

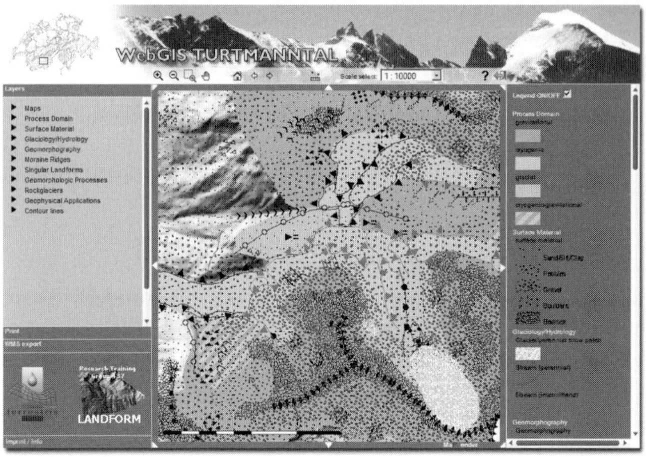

Figure 9.9 The graphical user interface (GUI) of the geomorphological WebGIS application Turtmanntal (Universities of Salzburg and Bonn, available at www.geomorphology.at).

org). Aerial images and a shaded relief map are provided as base layers and several thematic layers present information on process domains, surface materials, landforms and single processes. Due to MapServer's powerful cartographic engine, complex geomorphological symbols can be implemented and displayed. Symbols based on the legend for high mountain systems established by Kneisel et al. (1998) have been implemented. The WebGIS map thus uses the same symbology as the printed map of the same area (Otto and Dikau, 2004). The MapServer uses one symbol file that defines the composition of symbols for all types of vector geometries. Point information, such as individual landforms, is displayed using a geomorphological font (Otto and Dikau, 2004) and the spatial orientation of each character is achieved by providing the rotation angle as attribute data. Line features, for example crests and ridges, are constructed using multi-level symbols and advanced polygon symbology is supported by hatching or image fills. The Turtmanntal WebGIS offers simple functionality of a desktop GIS such as spatial navigation, coordinate queries, length and area calculations as well as selection of single layers of information. The composed image of the map frame can be exported in high-resolution PDF (300 dpi) in A4 and A3 landscape or portrait orientation. For educational purposes, a glossary delivers definitions of geomorphological terms.

The WMS online resources are accessible through an export tool and the maps can be embedded in other web or desktop GIS applications, thus the Turtmanntal WebGIS provides geospatial services as well. Figure 9.10 shows different applications of the same WMS services viewed in the original WebGIS application (a), as an overlay on Google Maps data in a web mapping application hosted on another server (b) and finally in two desktop GIS programmes, ESRI's ArcMap (c) and Quantum GIS (d), both supporting the WMS format as a data source.

The WMS protocol enables the easy implementation and integration of distributed WMSs from different servers and so the collection of huge "own data" pools becomes unnecessary. For simple visualisation of geomorphological data, a public WMS serving aerial photographs could be used as a base layer that is overlain with the WMS delivering the mapping results to produce the online geomorphological map (Figure 9.11).

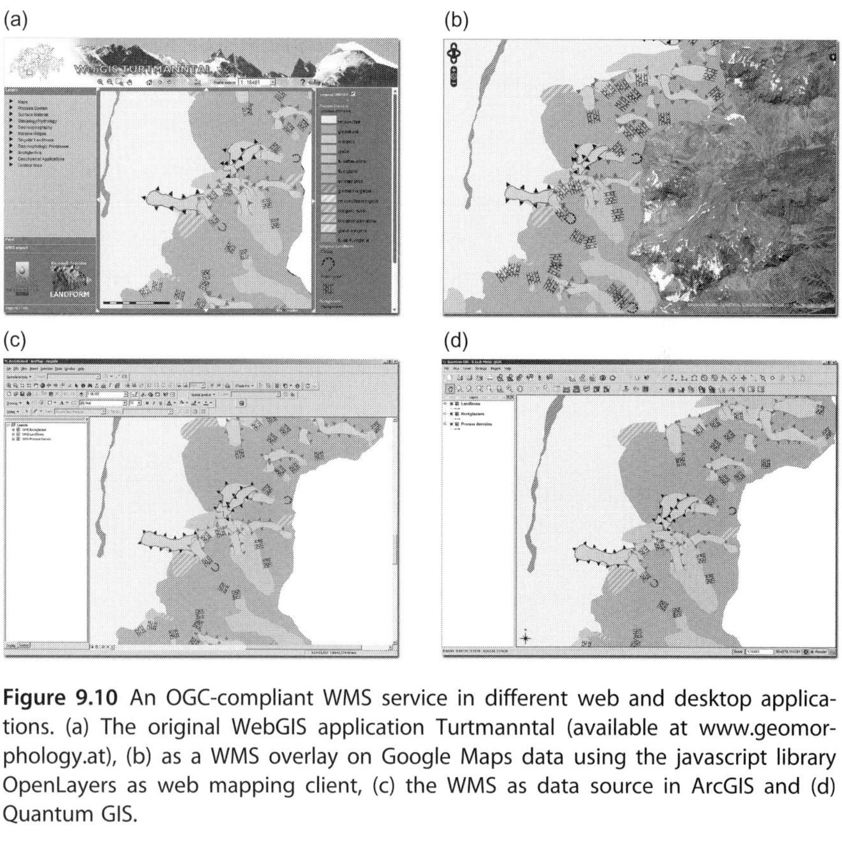

Figure 9.10 An OGC-compliant WMS service in different web and desktop applications. (a) The original WebGIS application Turtmanntal (available at www.geomorphology.at), (b) as a WMS overlay on Google Maps data using the javascript library OpenLayers as web mapping client, (c) the WMS as data source in ArcGIS and (d) Quantum GIS.

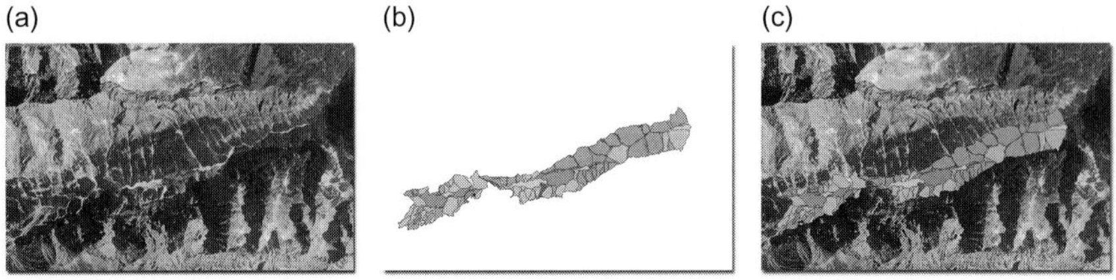

Figure 9.11 A map based on distributed WMSs from different servers (a) Orthophoto WMS of the Bavarian Survey Administration showing the Reintal basin, Bavaria, Germany (WMS available at http://www.geodaten.bayern.de/ogc/getogc.cgi?), (b) WMS displaying the spatial distribution of sediment storages in the Reintal basin (available at www.reintal-webgis.de) and (c) the final map.

We believe that the value of geomorphological data increases the more it is linked to other available information. Geomorphologists should consider the opportunity to present and share their data in a way users can easily tie to other data sources.

5.2 Maps in Google Earth

Google Earth is a free and convenient desktop application available for Windows and Mac OSX offering high performance access to global geographic data. The software provides an easy-to-use interface to a variety of data. Base data in Google Earth is the same as in the browser-based Google Maps application. A major difference lies in the way maps can be viewed and manipulated. Google Earth enables Earth image browsing in a three-dimensional view on a virtual globe (Brown, 2006). Butler (2006) noted Google Earth's popularity to a growing number of scientists is due to its excellent background imagery and the ability to place spatial data on top of them. However, Google Earth has only limited analytic functions and it is not designed to replace professional GIS software. A tool like Google Earth increases researchers' awareness to explore more powerful GIS techniques due to its easy visualisation (Butler, 2006). Google Earth uses the keyhole markup language (KML) to manage three-dimensional spatial data and also supports WMS as image overlays turning the application into a WMS client.

KML, also an OGC standard, enables the organisation and exchange of vector geometries. Numerous tools, such as GDAL, are available for data translation into KML. KML handles each type of vector geometry differently; however, advanced visualisation by complex symbology is limited. Point symbols are displayed as images, which enables a more complex symbology although data specific rotation is not possible. Line features are simply displayed with additional specifications of line width and colour, but multi-level symbols are not supported. Polygon features only support simple colour fills, with no hatching or patterned fills, and the style for polygon perimeters is the same as for line symbols (Figure 9.12). This restricts the use of KML for complex geomorphic feature visualisation and limits its suitability for the dissemination of geomorphological maps. As a rule of thumb, one should keep the symbology for a KML file as simple as possible (see Chapter 8 for further discussion on spatial data formats).

One possible method to distribute geomorphological maps for Google Earth is to display the map as an image overlay. The image is exactly positioned on Google's virtual globe by a bounding box. A single image will only be displayed at the scale the image was created and zooming will deliver blurred data. The best performance is achieved if the image is served as a network link through a WMS. The image is refreshed after each navigation task and delivers high resolution at different scales. The WMS map request can be embedded in a KML file and stored on a local hard drive. In addition, the use of Google Earth as a WMS client allows the display of additional information from any publicly available WMS.

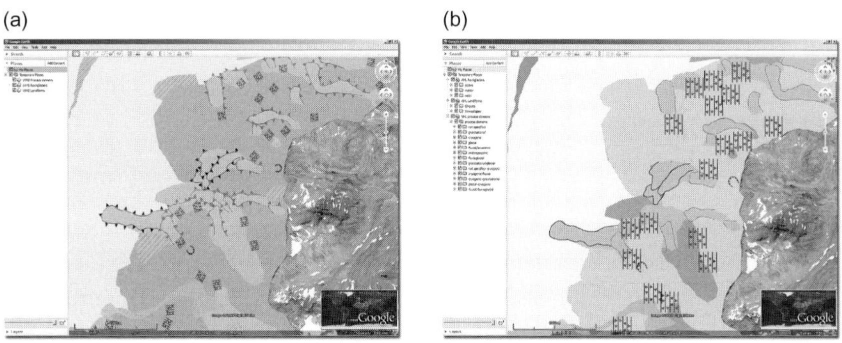

Figure 9.12 WMS overlays and the corresponding KML files in Google Earth. (a) Geomorphic features as WMS overlays in Google Earth. This lesser known feature allows the display of any publicly available WMS. The WMS appears as an image overlay that is refreshed after each navigation task. (b) The same data as a KML layer, the KML file was generated using the GDAL/OGR tool (GDAL, www.gdal.org). Compared to the WMS overlays, more sophisticated symbology like hatching, multi-level symbols or symbol rotation is not supported within the style reference of KML.

6. CONCLUSIONS

The perception and mapping of landscapes is a subjective process. The usability and quality of geomorphological maps is therefore not only dependent upon the choice and familiarity of the legend system and symbols but also to a greater extent upon good map design. Cartography provides valuable principles and techniques to focus the reader's attention to the main content of a map. To apply these principles, geomorphologists should be aware of the different cartographic elements that compose geomorphological maps and their usage when creating a map. Legibility is probably the greatest challenge to geomorphological maps. A clear hierarchical organisation, the thoughtful application of colour, contrast and symbol density and a well-balanced arrangement of map items are good preconditions for well-designed maps that deliver the intended message.

This chapter has introduced many different legend and mapping systems. The choice often depends on the author's scientific context or the purpose of the map. A legend system should be adjusted to the specific conditions of the study area and the message of the map, sometimes requiring the creation of new symbols. Due to recent advances in graphic and map design functionality, GIS software provides increasing possibilities for map creation and creative map design. However, graphics software still enables more flexible symbol generation and map creation. To make full use of GIS functionality, database structure, layer composition and data formats need to be considered prior to map creation. This enables not only the storage and distribution of data but also helps to create a clear hierarchical organisation of the map contents.

The Internet is a valuable platform for storage, exchange and dissemination of geomorphological information. Web mapping is a central part of the Internet that can be used for geomorphological maps as well. Either by static or dynamic techniques, geomorphological maps are easily published online using free software and data. Interoperability and exchange of geomorphological maps and data are provided by data and protocol standards for web mapping (e.g. WMS). Publication of geomorphological data through the Internet will contribute to the distribution and application of geomorphological maps in other scientific and non-scientific fields. In order to assure that geomorphological maps deliver the information aimed for, whether through online or printed maps, a clear and understandable design and composition of the geomorphological map is required.

REFERENCES

Barsch, D., Liedtke, H., 1980a. Methoden und Anwendbarkeit Geomorphologischer Detailkarten – Beiträge zum GMK-Schwerpunktprogramm II, Berliner Geographische Abhandlungen.

Barsch, D., Liedtke, H., 1980b. Principles, scientific value and practical applicability of the geomorphological map of the Federal Republic of Germany at the scale of 1:25,000 (GMK 25) and 1:100,000 (GMK 100). Z. Geomorphol. Suppl. 36, 296–313.

Barsch, D., Fränzle, O., Leser, H., Liedtke, H., Stäblein, G., 1985. Geomorphological mapping in the Federal Republic of Germany. First International Conference on Geomorphology, Manchester.

Barsch, D., Fischer, K., Stäblein, G., 1987. Geomorphological mapping of high mountain relief, Federal Republic of Germany (with geomorphological map Königssee, scale 1:25.000). Mt. Res. Dev. 7, 361–374.

Bertin, J., 1982. Semiology of Graphics. University of Wisconsin Press, Madison, WI.

Brewer, C.A., 2009. Colorbrewer. <http://www.ColorBrewer.org> (accessed September 2009).

Brown, M.C., 2006. Hacking Google Maps and Google Earth. Wiley, Chichester.

Butler, D., 2006. The web-wide world. Nature 439, 776–778.

Cooke, R.U., Doornkamp, J.C., 1990. Geomorphology in Environmental Management. A New Introduction. Clarendon Press, Oxford.

De Graaff, L.W.S., De Jong, M.G.G., Rupke, J., Verhofstad, J., 1987. A geomorphological mapping system at scale 1:10,000 for mountainous areas (Austria). Z. Geomorphol. 31, 229–242.

Demek, J., Embleton, C., 1978. Guide to Medium-Scale Geomorphological Mapping. E. Schweizerbart'sche Verlagsbuchhandlung, Stuttgart.

Demek, J., Embleton, C., Gellert, J.F., Verstappen, H.T., 1972. Manual of Detailed Geomorphological Mapping. International Geographical Union Commission on Geomorphological Survey and Mapping. Academia, Prague.

Dykes, A., 2008. Geomorphological maps of Irish peat landslides created using hand-held GPS. J. Maps v2008, 258–276.

Evans, I.S., 1990. Cartographic techniques in geomorphology. In: Goudie, A. (Ed.), Geomorphological Techniques. Unwin Hyman, London.

Geilhausen, M., Otto, J.-C., Dikau, R., 2007. GMK.digital – Geomorphologische Karten im Netz. <http://gidimap.giub.uni-bonn.de/gmk.digital/home.htm> (accessed September 2009).

Gilewska, S.K.M., 1968. Project of the Unified key to the geomorphological map of the World. Folia Geographica, Ser. Geographica-Physica, II. Polska Akademia Nauk, Kraków.

GITTA, 2006. Layout design settings/graphical semiology. Geographic Information Technology Training Alliance. <http://www.gitta.info/LayoutDesign/en/html/DefOrgMapEle_learningObject1.html> (accessed January 2010).

Gustavsson, M., Kolstrup, E., 2009. New geomorphological mapping system used at different scales in a Swedish glaciated area. Geomorphology 110, 37–44.

Gustavsson, M., Kolstrup, E., Seijmonsbergen, A.C., 2006. A new symbol-and-GIS based detailed geomorphological mapping system: renewal of a scientific discipline for understanding landscape development. Geomorphology 77, 90–111.

Gustavsson, M., Seijmonsbergen, A.C., Kolstrup, E., 2008. Structure and contents of a new geomorphological GIS database linked to a geomorphological map – with an example from Liden, central Sweden. Geomorphology 95, 335–349.

Hake, G., Grünreich, D., Meng, L., 2001. Kartographie. Visualisierung raum-zeitlicher Informationen. de Gruyter, Berlin.

Haq, B.U., Eysinga, F.W.B., 1987. Geological Time Table. Elsevier, Amsterdam.

IGUL, 2010. La légende géomorphologique de l'IGUL. <http://www.unil.ch/igul/page40935.html> (accessed January 2010).

Kienholz, H., 1976. Kombinierte geomorphologische Gefahrenkarte 1:10000 von Grindelwald. Catena 3, 265–294.

Kienholz, H., 1978. Maps of geomorphology and natural hazards of Grindelwald, Switzerland: scale 1:10000. Arct. Antarct. Alp. Res. 10, 169–184.

Kienholz, H., Krummenacher, B., 1995. Symbolbaukasten zur Kartierung der Phänomene. Mitteilungen des Bundesamtes für Wasser und Geologie Nr. 6. Bern, Bundesamt für Umwelt, Wald und Landschaft (BUWAL).

Klimaszewski, M., 1982. Detailed geomorphological maps. ITC J., 3, 265–271.

Klimaszewski, M., 1990. Thirty years of geomorphological mapping. Geogr. Pol. 58, 11–18.

Kneisel, C., Tressel, E., 2000. Zur Anwendung der neuen GMK Hochgebirge mit einer Beispielkartierung aus den Ostschweizer Alpen. Trierer Geographische Studien 23, 113–132.

Kneisel, C., Lehmkuhl, F., Winkler, S., 1998. Legende für geomorphologische Kartierungen in Hochgebirgen (GMK Hochgebirge). Trierer Geographische Studien 18: 24S.

Kraak, M.-J., Ormeling, F., 2002. Cartography: Visualization of Geospatial Data. Prentice Hall, Upper Saddle River, NJ.

Krygier, J., Wood, D., 2005. Making Maps: A Visual Guide to Map Design for GIS. The Guilford Press, New York.

Kugler, H., 1964. Großmaßstäbige Geomorphologische Kartierung und Geomorphologische Reliefanalyse. Universität Leipzig, Leipzig, Dissertation.

Kuhle, M., 1990. Quantificational reductionism as a risk in geography instanced by the 1:25000 Geomorphological map of the Federal Republic of Germany. Geogr. Pol. 58, 41–54.

Lee, E.M., 2001. Geomorphological mapping. In: Griffiths, J.S. (Ed.), Land Surface Evaluation for Engineering Practice. The Geological Society, London, Special Publication No. 18.

Longley, P.A., Goodchild, M.F., Maguire, D.J., Rhind, D.W., 2005. Geographic Information Systems and Science. Wiley, Chichester.

Maarleveld, G.C., ten Cate, J.A.M., de Lange, G.W., 1977. Geomorfologische kaart van Nederland schaal 1:50 000. Legenda, 20pp + Toelichting op de legenda, 91 pp. Stichting voor Bodemkartering, Wageningen/Rijks Geologische Dienst, Haarlem.

Mitchell, T., 2005. Web Mapping Illustrated. O'Reilly, Sebastopol, CA.

Otto, J.-C., 2008. Symbols for geomorphologic mapping in high mountains for ArcGIS. <http://www.geomorphology.at/index.php?option=com_content&task=view&id=133&Itemid=176> (accessed September 2009).

Otto, J.-C., Dikau, R., 2004. Geomorphic system analysis of a high mountain valley in the Swiss Alps. Z. Geomorphol. N.F. 48, 323–341.

Passarge, S., 1912. Physiologische morphologie. Mitt. Geogr. Ges. Hamburg 26, 135–337.

Robinson, A.H., Morrison, J.L., Muehrcke, P.C., Kimerling, A.J., Guptill, S.C., 1995. Elements of Cartography. Wiley, Chichester.

Rouleau, B., 1993. Theory of cartographic expression and design. In: Anson, R.W., Ormeling, F.J. (Eds.), Basic Cartography, vol. 1: Manual Series Basic Cartography for Students and Technicians. Butterworth-Heinemann, Oxford, pp. 65–91.

Salomé, A.I., Dorsser, H.J., Rieff, P.L., 1982. A comparison of geomorphological mapping systems. ITC J., 3.

Schoeneich, P., 1993. Comparison des systémes de légendes francais, allemand et suisse principes de la légende IGUL. In: Reynard, E., Schoeneich, P. (Eds.), Cartographie Géomorphologique, Cartographie des Risques, Travaux et Recherches, 9, pp. 15–24.

Schoeneich, P., Reynard, E., Pierrehumbert, G., 1998. Geomorphological mapping in the Swiss Alps and Prealps. In: Kriz, K. (Ed.), Hochgebirgskartographie Silvretta '98. Wiener Schriften zur Geographie und Kartographie, 11, pp. 145–153.

Slocum, T.A., Mcmaster, R.B., Kessler, F.C., Howard, H.H., 2005. Thematic Cartography and Geographics Visualization. Prentice Hall, Upper Saddle River, NJ.

Speight, J.G., 1974. A parametric approach to landform regions. Progress in Geomorphology, Special Publication. Institute of British Geographers, Oxford, pp. 213–229.

St-Onge, D.A., 1981. Theories, paradigms, mapping, and geomorphology. Can. Geogr. 25, 307–315.

Stäblein, G., 1980. Die Konzeption der Geomorphologischen Karten GMK25 und GMK100 im DFG-Schwerpunktprogramm. Berl. Geogr. Abhandlungen 31, 13–30.

Theler, D., Reynard, E., 2008. Geomorphological maps of a high mountain watershed: sedimentary dynamics of the Bruchi torrent, Swiss Alp. J. Maps v2008, 289.

Tricart, J., 1965. Principes et Methodes de la Geomorphologie. Masson, Paris, 465 pp.

Verstappen, H.T., 1970. Introduction to the ITC System of Geomorphological survey. K. N.A.G. Geographisch Tijdschrift IV, 87–91.

Verstappen, H.T., Zuidam, R.A., 1968. ITC Textbook of Photo-Interpretation, VII: 2-ITC System of Geomorphological Survey.

CHAPTER TEN

Semi-Automated Identification and Extraction of Geomorphological Features Using Digital Elevation Data

Arie Christoffel Seijmonsbergen, Tomislav Hengl and Niels Steven Anders
Institute for Biodiversity and Ecosystem Dynamics, University of Amsterdam, Amsterdam, the Netherlands

Contents

Developments in Earth Surface Processes, Volume 15
ISSN: 0928-2025, DOI: 10.1016/B978-0-444-53446-0.00010-0

1. INTRODUCTION

Classic geomorphological maps are slowly being replaced by geomorphological maps that are extracted from digital elevation models (DEMs). A simple visual inspection of detailed hill-shaded representations of fine elevation data reveals a wealth of information about the landscape, which often goes beyond the detail that is available in hand-drawn classical geomorphological maps. Use of DEMs for quantitative and qualitative description of landscape is the focus of the relatively new discipline of geomorphometry (Pike et al., 2008). The name 'geomorphometry' was first used by Von Humboldt in 1849 (Dikau et al., 1995), but it was the first DEMs in the 1960s and 1970s that motivated researchers to develop various methods and applications. Today, various geomorphometric algorithms are implemented in commercial and/or open-source geographical information system (GIS) software packages. Existing 'classical' geomorphological maps can be used to train and validate automatically derived landforms. In this way, the 'expert knowledge' is converted into sequences of mathematical calculation rules, which make it possible for any end-user to derive digital maps of the Earth's surface in an automated or semi-automated manner.

This chapter discusses semi-automated methods for the identification and classification of terrestrial geomorphological features, illustrated through two case studies of contrasting environments, one from the Drenthe area in the northern part of the Netherlands (low relief), and a second case study from an alpine area in Western Austria (high relief). We specifically emphasise the importance of hybrid expert-knowledge and statistical approaches to the extraction of geomorphological features. In Section 2, we first review both past and recent developments in classic and automated geomorphological mapping. The case studies are then presented and processing steps described. In the first case study (Section 3), a 5 m resolution light detection and ranging (LiDAR) DEM is used to increase the detail of an existing geomorphological map by applying multinomial logistic regression techniques in an open-source software package. In the second case study (Section 4), a 1 m resolution LiDAR data set from a high alpine mountain area is used in an object-based segmentation, combining 'topographic openness' with slope parameters at multiple scale levels using commercial software. Special attention is paid in Section 4 to the potential of 'topographic openness', which is an angular measure of enclosure of an object or pixel in the landscape (Yokoyama et al., 2002). It can be used in multi-scale landscape analysis since openness can be measured over user-specified distances. According to Prima et al. (2006), slope in combination with topographic openness can be used for genetic interpretation of topography. Combinations of slope, broad-scale openness values (measured over a search radius of 200 m) and fine-scale openness (measured over a search radius of 50 m) in a single RGB composite image enhance the recognition of geomorphological features.

All scripts, data, process trees and methods used in this chapter can be obtained from http://www.appgema.net and the www.geomorphometry.org website.

2. GEOMORPHOLOGICAL MAPPING

2.1 Classic Geomorphological Mapping – Schools and Approaches

Geometric descriptions of the Earth's surface have their roots in ancient history, but classic geomorphologic mapping systems independently developed in several, mainly, European countries. For thorough reviews of the geomorphological mapping systems, refer to, for example, Gilewska and Klimek (1968), Demek and Embleton (1978), Salomé et al. (1982), Klimaszewski (1990), Evans (1990), Gustavvson et al. (2006), Otto et al. 2011 and Verstappen (2011). Various 'mapping schools' developed and promoted different methods for representing the land surface, mostly by using symbol-based legends. Classic geomorphological mapping systems are often thought to be subjective (Carrara, 1992), non-reproducible (Van Westen et al., 1999), time consuming and designed only for scientific goals (Salomé et al., 1982). The International Geomorphological Union (IGU) mapping system (Gilewska and Klimek, 1968) was originally developed to standardise various schools and support geomorphological mapping at a global scale, but this system has not been adopted.

The skills of a trained geomorphologist permit the interpretation of polygenetic landscapes and document landscape history, former and currently active processes and materials underlying the landforms, and summarise their knowledge in a single map layer. A serious criticism of the classical approach to geomorphological mapping is that it makes no distinction of the type of boundary between the units, and the units are forced into predefined categories at specific mapping scales by the expert. In reality, three types of common geomorphological boundaries may occur in a landscape – sharp, gradational and diffuse (Batten, 2001). This illustrates a need for new models to represent geomorphological features. In addition, classic geomorphological maps are commonly not supplemented with information or an evaluation of possible map errors. It is clear that particular landforms or landform elements are open to alternative interpretations, especially if surface exposures are absent and the terrain is inaccessible or to a large extent covered by dense vegetation.

The onset of GIS-assisted mapping that started in the 1990s gave an impulse to automated geomorphological mapping and caused a paradigm shift. In parallel, new statistical techniques and GIS models evolved that allowed the enrichment of geomorphological maps. However, no standards yet exist to formalise digital geomorphological mapping in terms of unique GIS legends, map representation schemes and derivation methods (Van Westen et al., 2000; Bocco et al., 2001; Seijmonsbergen and de Graaff, 2004). Recently, Gustavsson et al. (2008) presented a standardised geomorphological GIS database, designed to be used as a basis for digital mapping projects, where geomorphological vector data, raster data and tabular data are stored in a geomorphological geodatabase. A promising initiative to document and store maps is demonstrated by the open access journal *Journal of Maps* (http://www.journalofmaps.com/) – it publishes both classic and GIS-based geomorphological maps, which allows further comparison and merging of the 'classical' and 'digital' approaches.

High-resolution DEMs, in combination with detailed orthophotos, make it possible for a surveyor to refine relative stratigraphy of deposits and events and introduce detail to geomorphological maps not known to traditional mappers (Newell and Clark, 2008). The use of new technology is also cost effective: it reduces fieldwork, speeds up the map making process and increases the use of geomorphological maps. For example, in the Netherlands, the national 1:50,000 geomorphological map (Koomen and Maas, 2004), used in combination with a LiDAR DEM (submeter altitude information of lowland areas), became crucial for flood protection at the national level. New technologies have demonstrably changed and revitalised geomorphology, however, old information should not be thrown away: it is the integration of classic and digital mapping that can significantly contribute to applied problems (van Asselen and Seijmonsbergen, 2006; Gustavsson et al., 2008). Historic photographs, information from literature, historical records and DEM-based parameters either stored in the same attribute table or in local or remote databases can be analysed in combination with digital geomorphometric data and then used to solve real-life problems. Figure 10.1 shows a classic geomorphological map fragment overlaid with digital geomorphological polygons and two examples of additional clickable information used as a basis for geoconservation in Western Austria.

The photo shows a key location for reconstruction of the Würm deglaciation history. The small map shows how individual geomorphological units translate into 'scientific relevance'. Therefore, it is important that traditional sources of information are digitised, integrated into a GIS and used in combination with digital- and remotely sensed layers.

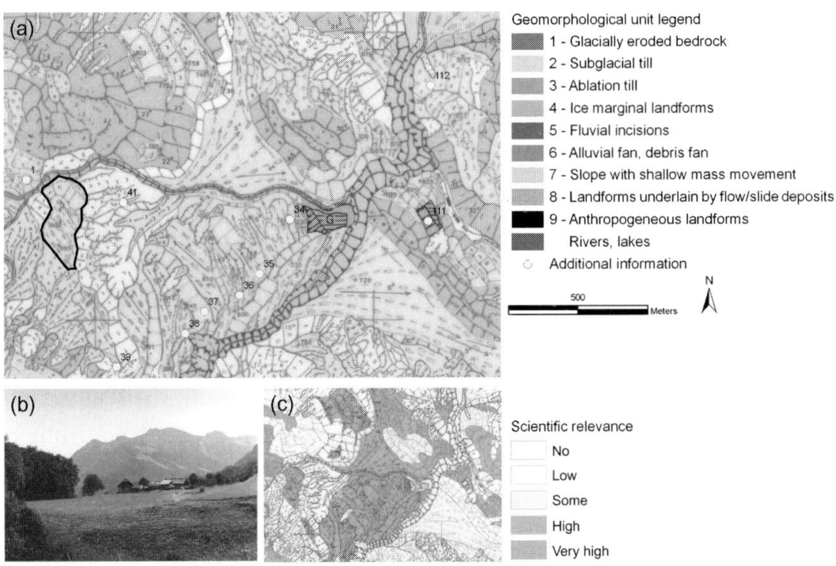

Figure 10.1 (a) Classic geomorphological map fragment of map sheet Gurtis overlaid with manually digitised geomorphological polygons and a point file linked to additional information. Two examples of linked additional information are shown: (b) a photo of an ice marginal terrace, the location indicated by a black outline in the geomorphological unit map and (c) a derived map of scientific relevance. *After Seijmonsbergen (1992).*

2.2 DEMs and Land Surface Parameter

2.2.1 Introduction to DEM Analysis

Geomorphological features can be detected, isolated, mapped and characterised using a variety of automated techniques. Some methods target a particular class of feature (Behn et al., 2004; Hiller and Smith, 2008), whereas others aim to completely divide an area into zones of different morphological characteristics. Approaches to automated analysis include:

a. mimicking the mapping method of a manual interpreter in an automated and reproducible way for a class of feature (Hillier and Watts, 2004),

b. proposing robust statistics and objective metrics to optimally isolate individual features (Wessel, 1998),

c. using algorithms that search a landscape for a class of feature using scale-invariant or multi-scale parameters (Wessel, 2001; Behn et al., 2004),

d. simultaneously using multiple land surface parameters (LSPs) in order to categorise areas within a landscape into classes with distinctive properties that relate to a type of feature.

The basis of LSPs is the DEM, a digital representation of the land surface topography (Pike, 1995; Hengl et al., 2008). DEMs may be derived from many sources (Oguchi and Hayakawa, 2011). For further information on the DEM production methods, DEM sources, accuracy, cell sizes and preparation techniques, see, for example, Maune (2001), Fisher and Tate (2006), Reuter et al. (2008) and Nelson et al. (2008).

Once created, LSPs may be derived from a DEM in order to create geomorphological information. A classic geomorphological map contains information represented in one layer – a paper or polygon-based map of geomorphological units. This layer is complex in a sense that it is an expert summary assimilating diverse information about the landscape, geology, stratigraphy and geomorphometry. An LSP extracted from a DEM on the other hand pertains to one aspect of this whole – each layer carries specific information that may be interpreted in terms of a feature (Pike et al., 2008). A selection of LSPs are shown in Figure 10.2, and refer to the same area as depicted in Figure 10.1. A common LSP is slope angle, which is the rate of change of altitude in the direction where that rate is maximised. The ice marginal landforms documented in the map in Figure 10.1 correlate well with low angle slopes presented in Figure 10.2. Closely linked to slope angle is aspect – a circular variable (0°–360°)

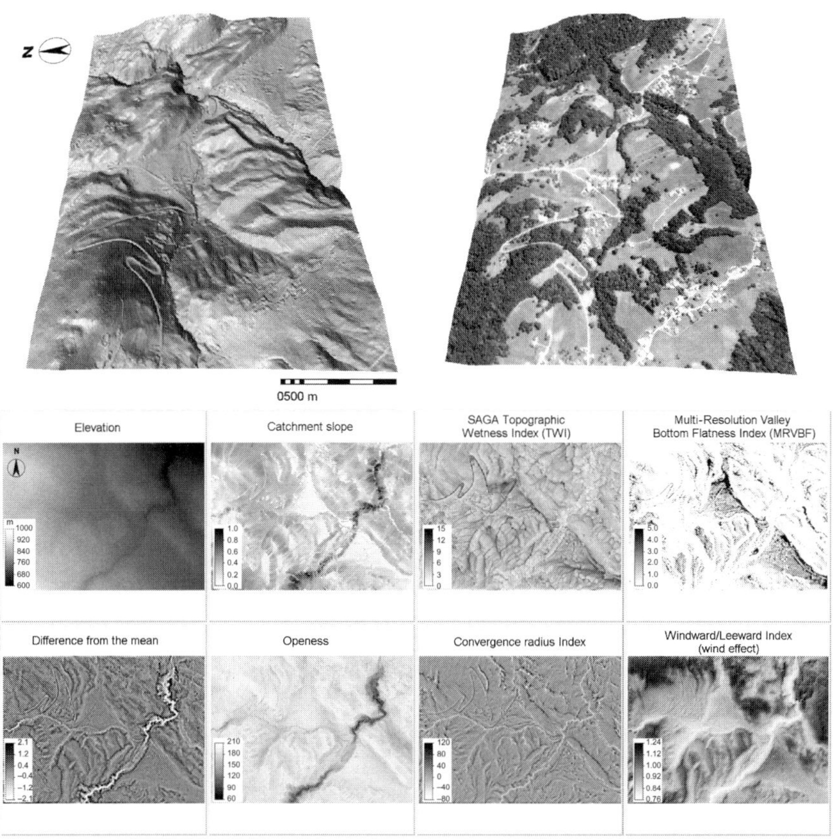

Figure 10.2 A preview of LSPs derived using 1 m LiDAR DEM for a study area in Austria (the same extent as in Figure 10.1).

describing the direction or azimuth of this true slope angle (Evans, 2004). It can be used, for example, to automatically map incisions, which are characterised by opposite slope aspect. Curvature is the second derivative of land surface with negative values representing concavity (Evans, 2004). It is often used to map foot slopes, on which colluvium may preferentially accumulate. 'Openness', explained in detail in Section 4, refers to how wide a landscape can be viewed from a certain position on a DEM. In Figure 10.2, the darker areas correlate with narrow fluvial incisions, whereas lighter areas reflect open terrain. Techniques can also be used to make features in the landscape more visible. For example, 'hill shades' are representations of a DEM created by illumination of the DEM with a virtual source. The selected LSPs shown in Figure 10.2 are only a small sample of what can be extracted from DEMs. Hengl and MacMillan (2008) argued that more than 100 basic and complex LSPs are currently available for characterising the landscape.

Computational techniques developed to extract and classify LSPs from DEMs have become integrated into commercial software packages such as ESRI ArcGIS,[1] ERDAS Imagine[2] and Definiens Developer.[3] Standard geometric calculations can be used as built in toolboxes, and special toolboxes are developed and freely distributed via the Internet (Wood, 2008). Examples of free software and open-source packages specialised for processing DEMs include SAGA GIS,[4] ILWIS GIS,[5] GRASS GIS,[6] TOPAZ,[7] TAPES,[8] Anudem,[9] LandSerf[10] and MicroDEM.[11] Recent progress in geomorphometry can be best followed via the activities of the geomorphometry[12] research group. This is possibly the best platform for exchanging new applications, development tools/scripts and experiences in the analysis of DEMs.

LSPs can be roughly divided into: (1) basic local (e.g. slope, aspect and curvature), regional (e.g. catchment area, slope length, proximity to local streams and ridges, relative relief, visual exposure) and statistical parameters (e.g. terrain roughness, complexity, anisotropy, fractal dimension), (2) LSPs connected with hydrology (e.g. topographic wetness index (TWI), height above channels) and (3) LSPs connected with climatic modelling (e.g. solar insolation, wind exposure). Basic LSPs can be derived directly from a DEM without further understanding of the area (Olaya, 2008), other LSPs require some input parameters to be set by the analyst. For overviews of LSP types, refer to Mark (1975), Wilson and Gallant (2000), Iwahashi and Pike (2007), Minár and Evans (2008) and Hengl and Reuter (2008).

[1] http://www.esri.com/
[2] http://www.erdas.com
[3] http://www.definiens.com/
[4] http://saga-gis.org
[5] http://www.ilwis.org/open_source_gis_ilwis_download.htm
[6] http://grass.itc.it
[7] http://homepage.usask.ca/~lwm885/topaz/
[8] http://uscgislab.net/incEngine/?art=software
[9] http://fennerschool.anu.edu.au/publications/software/
[10] http://www.landserf.org
[11] http://www.usna.edu/Users/oceano/pguth/website/microdem/microdemdown.htm
[12] http://geomorphometry.org/

2.2.2 Extracting Geomorphological Features

Once a variety of LSPs have been computed from a DEM, we can use them to extract geomorphological features in the same way remote sensing bands are used to extract land cover classes Lillesand et al. (2008). Here two main approaches exist: supervised and unsupervised (Figure 10.3). In the case of the supervised approach, human interpreters prepare known geomorphological features that serve as training areas

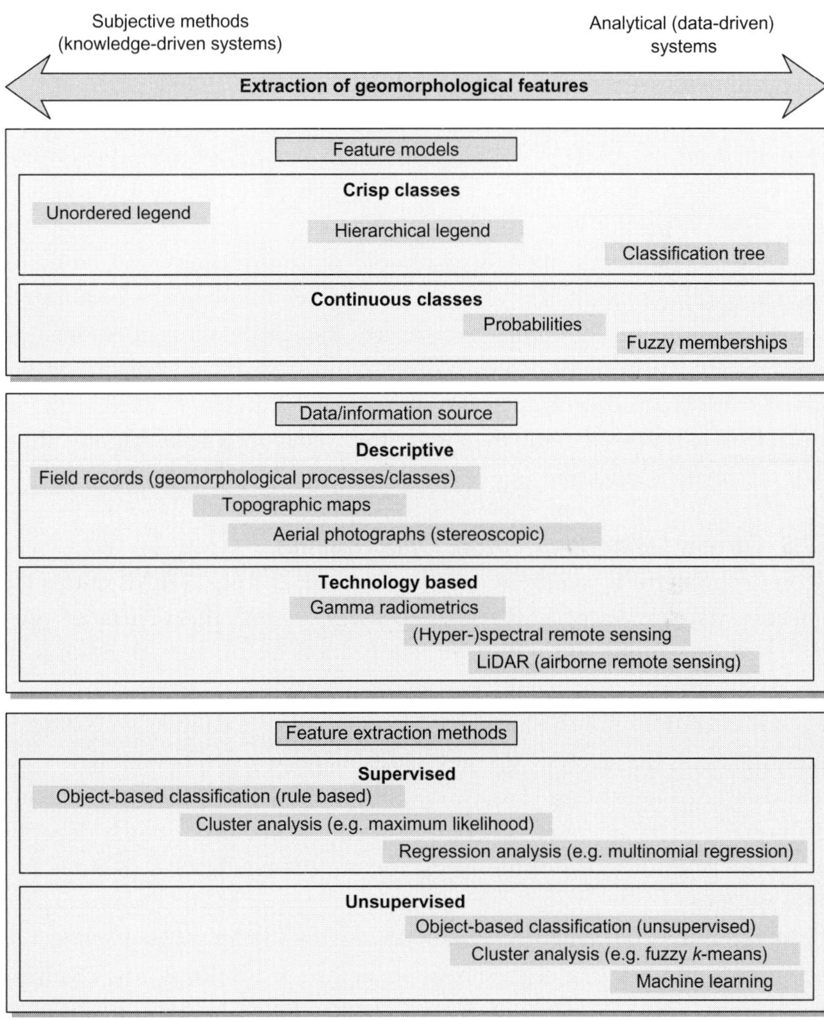

Figure 10.3 General models and approaches to extraction of geomorphological features.

from which classification rules can be developed. The unsupervised approach lets an algorithm automatically find the best fit of LSPs into a particular number of categories, which can be assigned meaning after the classification. In both approaches, challenges are similar: 'how to handle and calculate with large data sets?', 'how to filter DEMs to improve their reliability?' and 'how to design more efficient LSPs that may reflect the detail of topography at multiple scales?'.

A limitation of the pixel-based classification of LSPs is that it ignores spatial continuity. Geomorphological features can be described as groups of pixels i.e. *bodies* covering an area of the landscape, which asks for alternative approaches of digital landscape classification (Blaschke et al., 2004). Techniques such as image segmentation can be used to divide a DEM or (combinations of) extracted LSP rasters into image objects (polygons). The constructed image objects can then be classified into real-world features. Object-based classification is an alternative to pixel-based classifications and is commonly applied to remote sensing imagery and complex landscapes (Hay et al., 2003; Van Asselen and Seijmonsbergen, 2006), perhaps because it visually compares to existing fragmentation in landscapes. Much effort has been put into finding the correct image object size for subsequent classification into geomorphological features. It seems that, in general, it is more efficient to cluster the pixels to a level slightly finer than the final classification (see Section 4).

2.2.3 Current Limitations, Future Opportunities

Geomorphometrical synthesis of the landscape from DEMs aims for objective delineation of LSP data. However, despite the variety of programmed operations and statistical classification techniques, thresholds in the classifications are generally iteratively adjusted in response to subjective considerations (Iwahashi and Pike, 2007). For example, if landforms resulting from the analysis do not satisfy *a priori* expectations based on field data and/or existing classic maps (cross validation), then the classification parameters are reset. This process may be iterative, which depends on the users' knowledge of the landscape under investigation (Reuter and Nelson, 2008).

In the future, DEM-based classification should aim to go beyond the recognition of LSPs that classify basic shapes, such as hills, slopes, channels and plateau areas (Minár and Evans, 2008). DEM-based extraction of geomorphological feature should be able to distinguish landforms according to their formational process or 'morphogenetics' and even be able to discern something about the current activity level of processes. Thus, computationally derived information may come to closely resemble classic geomorphological information.

Similar to existing classifications of remote sensing imagery, any automated DEM classifications should ideally be accompanied by a methodology to assess precision and accuracy. This is necessary because it is evident that errors in DEMs will propagate to derived LSPs and modelling results in a way that is not easily predicted (Maune, 2001; Oksanen and Sarjakoski, 2005; Temme et al., 2008). As with all (semi-) automated mapping techniques (Starck et al., 2000; Wessel, 2001), it is crucial for the progress of geomorphological mapping that DEM-based digital mapping techniques all become reproducible and that standards become accepted. In this context, access to sample data sets, open-source or commercial software and relevant instruction manuals are indispensible (Neteler and Mitasova, 2008).

2.3 Contemporary Applications

From the mid-1970s, simple LSPs such as slope, aspect, hydrographical pattern and shaded relief derived from DEMs were used to improve geomorphological understanding (Adediran et al., 2004). The basic geomorphic unit to be identified and classified from LSPs was the slope (Giles and Franklin, 1998). The catena concept of Milne (1935), the nine-unit land surface concept of Dalrymple et al. (1968) and the morphological classes proposed by Speight (1990) served as early examples for automatic morphometric classifications of the landscape. The relatively new research field of pedometrics, which is the application of mathematical and statistical methods for the study of the distribution and genesis of soils (Heuvelink, 2003), still contributes to concepts and examples of soil-landscape models, which are based on terrain modelling (Hengl and Rossiter, 2003; Grunwald, 2006). Pike (1988) introduced the concept of geometric signatures in landslide terrain, which presented a further challenge for automated geomorphological feature extraction from DEMs. For example, Dikau et al. (1995) recognised five landform types – plains, tablelands and three hills and mountains types, subdivided into 24 landform classes, and based on morphometric analysis of a 200 m DEM from New Mexico.

Since then, many statistical techniques and classification procedures have been applied to DEMs of many types – all to characterise the Earth's surface shape in an efficient manner. MacMillan et al. (2000) used unsupervised neural network (UNN) analysis on slope, profile and plan curvature of a 5 m resolution DEM to produce 'Element and Landform Classifications'. Burrough et al. (2000) applied fuzzy *k*-means techniques for landform classification which resulted in classified LSP maps. Multivariate statistics were used by Adediran et al. (2004) for the classification of morphometrical parameter maps, based on various DEM sources. Drăguţ and Blaschke (2006) prepared data layers of LSPs that were segmented at several levels using object-oriented image segmentation. This resulted in nine classes of landforms, which were based on fuzzy membership relations. Prima et al. (2006) used supervised classification techniques based on topographic openness, slope and standard deviation of slope to typify seven landform classes in a volcanic mountain area in Japan. Region growing classification was used by Etzelmuller et al. (2007), based on amongst others profile and plan curvature, spatial scale and landform object, to classify 25 landform classes for Norway, which were then merged into 10 landform types. Iwahashi and Pike (2007) made an impressive effort to automate unsupervised classifications of the Earth surface based on an iterative nested-means algorithm and a three-part geometric signature (based on slope gradient, local convexity and surface texture). Bue and Stepinski (2006) used unsupervised classification based on the self-organising map technique to divide pixels into landform classes on the basis of similarity between attribute vectors, to produce a geomorphic map of part of the surface of Mars.

The production of high-resolution elevation models from LiDAR technology is a technical development that may further initiate digital landform mapping. A LiDAR scanning system employs multiple measurements of distance and the amount of energy reflected from the target. Over a vegetated surface, laser scans are generally able to penetrate through the canopy and therefore record information about both the canopy and the topographic surface below (Kraus and Pfeifer, 1998). The digital terrain and surface model combinations can be used for forestry or ecological applications (Lefsky et al., 2002), whereas the surface model below the canopy may hold geomorphometric information at greater detail than standard DEMs. LiDAR, in combination with high-resolution orthophotos, provides detailed visualisations of landscapes that show far better fit with geomorphological features than, for example, virtual globe systems such as Google Earth. Detailed monitoring of the dynamics of fine-scale land surface elements is now possible, such as riverbank erosion assessment and sediment yield calculation (Thoma et al., 2005), automated mapping of the topographic signatures of deep-seated landslides (Booth et al., 2009) and using characteristic eigenvalues and slope filter values to extract recent landslide activity (Kasai et al., 2009). McKean and Roering (2004) reported that contrasts in surface roughness can be interpreted to identify bedrock landslides, map their spatial extent and investigate the landslide internal kinematics. Dewitte et al. (2008) used (multi-temporal) DEMs from various sources to monitor and map deep-seated landslide activity in Belgium. In this respect, Arrell et al. (2007) notes that morphometric classes exhibit resolution dependency in their geographical extents (*cf.* also Schmidt and Andrew, 2005). Anders et al. (2009) used a LiDAR DEM to set initial parameters for modelling channel incisions and alpine slope development. In glaciology studies, Arnold et al. (2006) used LiDAR DEMs to derive mass balance information and detailed meltwater channel and crevasses dynamics.

MacMillan and Shary (2008) argued that automated classification of landforms almost always represent an attempt to replicate some previously conceived system of manual landform classification and mapping. Interesting in this respect is the article of Minár and Evans (2008) who proposed a concept of elementary forms (segments, units) that are defined by constant values of fundamental morphometric properties and limited by discontinuities of the properties. They further argued that geomorphological map unit boundaries in general follow morphometric boundaries. The internal homogeneity and external contrasts of segments in terms of their geometry should reflect their genesis and recent dynamics. Therefore, it is a challenge to automatically delineate and classify morphogenetic landscape units from DEMs, based on LSPs, rather than to focus only on the morphometric unit.

Several books on digital terrain analysis (i.e. geomorphometry) have been published. The following six, however, need to be emphasised. Wilson and Gallant (2000) focused on working with the TAPES-C DEM package for hydrological application of DEMs and integration with ecosystem modelling. The DEM Users Manual (Maune, 2001) summarised the sources, accuracy, user requirements, applications and analyses of DEMs. Other important sources showing the recent status of the field are the conference proceedings of the Terrain Analysis and Digital Terrain Modelling conference (Zhou et al., 2008) and the Digital Terrain Modeling book by Li et al. (2004). The most recent edition of the GRASS book (Neteler and

Mitasova, 2008) contains many illustrative examples of DEM processing. Recently, Hengl and Reuter (2008) compiled an extensive review of geomorphometry. In this book, Evans et al. (2008) listed three main automated applications of DEMs in geomorphology:

1. Automated recognition and quantification of geomorphological properties,
2. Automated extraction of hydrologic/denudation structures and
3. Automated extraction of landforms.

To reflect these main groups of applications, we have selected two case studies that focus on (1) recognition and (2) extraction, both in contrasting environments.

3. CASE STUDY BOSCHOORD – THE NETHERLANDS

3.1 Study Area and Data Sets

The case study 'Boschoord' (3024 ha) is a small area located in the province of Drenthe, in the northern part of the Netherlands (Figure 10.4a). The Boschoord area is part of the Drenthe Plateau which is underlain by till deposited by the second last (*Saalien*) ice sheet. After deglaciation, local rivers incised during low sea level stands into the plateau. In contrast, valleys were filled during high interglacial sea level stands, mostly with slope deposits derived from the surrounding plateau areas. During periglacial conditions in the last ice-age (*Würm*), several *pingos* developed in the plateau and cover sands were deposited on and along the plateau edges. During the Holocene, the remnants of the till plateau were partly overgrown by a mantle of peat. Deforestation in historic times has resulted in renewed river incision, whereas peat was stripped for fuel and the area was drained by a series of small canals, to lower groundwater tables. Local 'plaggen' farming during medieval and recent periods disturbed the local heath vegetation on top of the cover sand, after which the formation of irregular dune topography began. This rather complex genesis created a fragmented landscape in which hydrological differences are strongly linked to this polycyclic landscape development (see Figure 10.7a).

What makes this data set especially interesting is that it is an area of low relief but with distinct geomorphological classes that have been

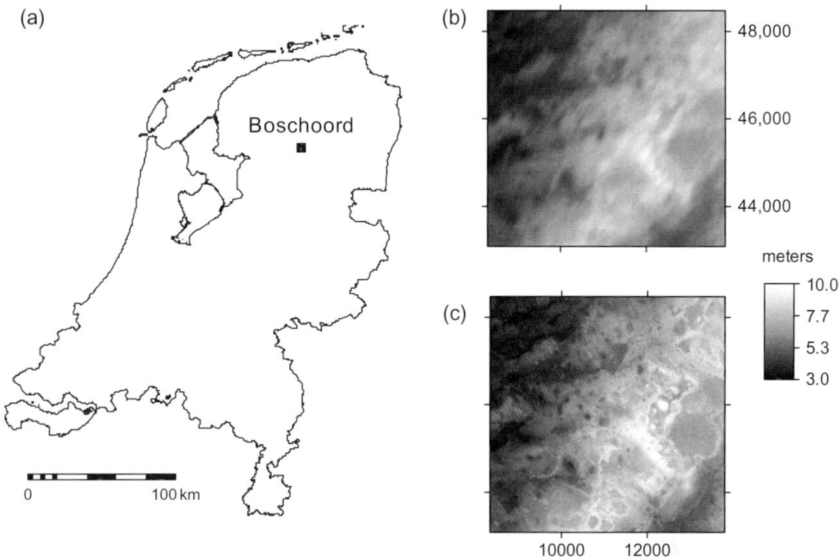

Figure 10.4 Location of the study area (a) and the two main DEM data sources used for analysis: DEM25TOPO – generated using ordinary kriging (b) and DEM25LIDAR (c).

mapped with relatively high accuracy (Koomen and Maas, 2004). The elevations range from 3 to 10 m above the sea level, with a standard deviation of 1.54 m; changes in topography are difficult to identify even in the field. 'Boschoord' is specifically selected to highlight the DEM-based extraction of geomorphological features in areas of low relief. Another reason why this area has been selected is because it has been surveyed and mapped in high detail. DEMs of various resolution and vertical accuracy are available, as well as numerous land cover and topographic maps. Additionally, we compiled several transect surveys in order to validate the quality of the geomorphological map and cross-check suspicious features in the LiDAR DEM.

The specific objective of this exercise was to suggest a way to improve the existing geomorphological map of the Netherlands (Koomen and Maas, 2004) by using various sources of DEMs and statistical techniques. In particular, we wanted to see if the differences in the accuracy between maps generated using the LiDAR-based DEM and traditional DEMs are significant. Additionally, we compared the results of supervised and unsupervised classifications using the same set of DEM parameters. The data set consists of three groups of layers:

- Elevation – This includes the 5 m LiDAR DEM (surveyed in 2004) and a point data set with 5010 measurements of heights (surveyed in 1960–1969). Both data sets show elevations measured with a high precision (±10–20 cm),
- Geomorphological map (GKN50) – The map contains 12 classes: ground moraines (3L1), low plains with ridges (3N3), peat bog depressions (2R4), cover sand undulated (3L5), low plains/depressions without ridges (3N4), low dunes+plains (3L8), cover sand undulated (3K14), ground moraines (high) (3L2a), low dunes+plains (3L9), areas partially covered with cover sand (2M14), low dunes (4K19) and cover sand areas (2M13),
- Topographic data – Includes all roads and infrastructure, land use classes and similar features from the TOP10VECTOR basic topographic map of the Netherlands (1:5000 scale). This data is used only for orientation purposes.

The original data set can be downloaded from http://www.appgema.net and the geomorphometry.org website.[13]

3.2 Data Processing and Analysis Steps

3.2.1 Supervised Extraction of Geomorphological Units

Statistical prediction of geomorphological classes follows the computational framework shown in Figure 10.5. The heart of this framework is the multinomial logistic regression algorithm, as implemented in the multinom method of the *nnet* package (Venables and Ripley, 2002) within the R Statistical Environment (http://www.r-project.org); this method iteratively fits logistic models for a number of classes given a set of training pixels. The output predictions can then be evaluated against the complete geomorphological map to see how well the two maps match and where the most problematic areas are. There are two inputs to the supervised classification scheme in Figure 10.5: (1) raw elevation measurements (either points or un-filtered rasters); (2) existing map. The raw elevations are used to generate the initial DEM, which is filtered for artefacts. After that, the expert needs to define a set of suitable LSPs that can be used to parameterise the features of interest. For example, we can derive DEM parameters that describe shape (curvature, wetness index), hydrologic

[13] http://geomorphometry.org/content/boschoord-case-study

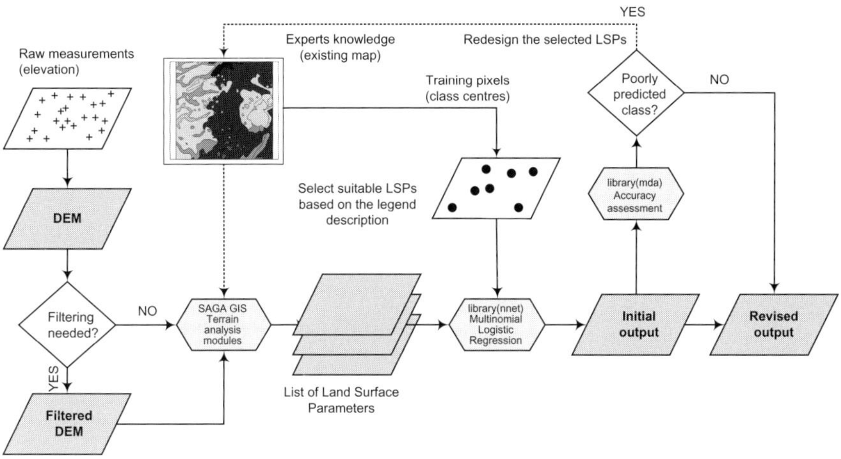

Figure 10.5 Data analysis scheme: supervised extraction of geomorphological classes using the existing geomorphological map (a hybrid expert/statistical-based approach). Software used to run different DEM and statistical analysis steps (SAGA GIS, R libraries nnet and mda) are also indicated.

context (distance from streams, height above the drainage network) or climatic conditions (incoming solar radiation). In practice, however, many geomorphological features will relate to both land surface and sub-surface parameters that are difficult to obtain and/or are not possible to derive from the existing DEM: the derived model will therefore have problems predicting the spatial location of geomorphological features accurately. We will possibly never be able to model such features with only DEM data, but we can at least iteratively adjust the initial list of LSPs until the prediction accuracy is satisfactory for all classes.

Because the objective was to refine the existing geomorphological map, a selection of pixels from the existing map was used to train the model. A simple approach would be to randomly sample points from the existing maps and then use them to train the model, but this has a disadvantage of (wrongly) assuming that the map is the same quality across the entire area covered. Instead, we use an algorithm which selects training pixels from the centre of classified areas. This comprises two steps: a map of medial axes for polygons (geomorphological units) is first derived to avoid selecting transitional pixels that might well be in either of the two neighbouring classes. Medial axes are locations that are most distant from the edges of polygons. Once the medial axes have been determined, points can be selected using the rpoint function of the *spatstat* package (see Figure 10.7a; and the R script[14] on the http://www.appgema.net for details). This will randomly allocate N points given a mask map. In this case study, we have considered that N=1000 is enough to build a model; higher sampling densities are also possible but could significantly increase the time needed to fit the model. The advantage in using medial axes to locate the training pixels is that relatively small polygons will be represented in the training pixels set. Or in other words, with this technique, large polygons will typically be proportionally under-sampled; it is important to have a balanced representation of features regardless of the spatial extent.

3.2.2 Unsupervised Extraction of Landforms

An alternative approach to extract geomorphological classes is the cluster analysis approach, i.e. different versions of unsupervised classification. In this case study, we considered only the fuzzy k-means clustering approach as implemented in that *stats* package (Venables and Ripley, 2002) and the results of supervised extraction of memberships as explained in Hengl et al. (2004). For this purpose, we use the same list of LSPs previously selected for the supervised classification and also the same number of classes as found on the geomorphological map. Refer to the R script on the http://www.appgema.net website for more details.

3.2.3 Software and Scripting

The computational framework described above is implemented using a combination of the R software for statistical computing (R Development Core Team, 2009) and the open-source desktop GIS packages SAGA GIS and ILWIS GIS. This combination is referred to as 'R+GIS'. SAGA GIS (Conrad, 2007) is used to extract DEMs, reproject and rescale maps and run various types of filters. ILWIS GIS is used to visualise the data and to run additional processing on the maps. The complete data set and the scripts used to extract geomorphological features shown in this section are available on the http://www.appgema.net website. Users who would like to repeat this analysis will need to obtain and install (in chronological order): R and necessary packages (RSAGA, maptools, rgdal, gstat), SAGA GIS and ILWIS GIS. In principle, R has full control over SAGA and ILWIS GIS, hence the complete processing can be run from a single R script.

[14] Boschoord. R available at http://geomorphometry.org/content/geomorphological-mapping

3.2.4 DEM Data Sources

The supervised extraction of geomorphological units is repeated using a DEM of the same study area derived from two different sources (Figure 10.4b and c):

1. hoogte_16ef.shp – the 5020 field measurements of elevation (land survey) collected in the 1960s by the 'Meetkundige Dienst Rijkswaterstaat'. This was used to generate the 25 m DEM25TOPO.
2. ahn5m.img – the 5 m LiDAR-based DEM distributed by the Ministry of Transportation and Water Management (measurements in centimetre). This data set is also known as 'Actueel Hoogtebestand Nederland' (AHN[15]) (van Heerd et al., 2008).

The LiDAR DEM shows much higher detail and depicts small depressions and elevations not visible from the DEM25TOPO (Figure 10.4). There are also considerable differences in elevation (up to ~2–3 m) between the measurements in 1960 and the LiDAR DEM in areas with peat soils (northwest part of the area) due to oxidation of peat and resultant lowering of the land surface.

Although the original LiDAR product has already been filtered for forest canopy and human-built objects, we identified several artificial spikes by isolating pixels with much higher elevation values than the neighbouring pixels (Figure 10.6). We visited these areas in the field (GPS–PDA system with a map overlay) and found that these are all areas of densely planted pine trees. Such dense parts of forest are obviously difficult for LiDAR to penetrate, hence only the upper object surface model has been generated.

Spikes, roads and similar linear features are not really connected with the geomorphology and need to be filtered before we can use the DEM for geomorphological mapping (Milledge et al., 2009). The unusual spikes and linear features can be detected (in SAGA GIS) using two parameters: 'difference from the mean (DFM) value given a search radius' and 'representativeness index' (Conrad, 2007). Where the value of either of the two LSPs exceeds a threshold value, we can remove the LiDAR values and then re-interpolate them from the neighbouring pixels using the 'close gap' operation in SAGA GIS (Figure 10.6). By visually inspecting the results of the analysis and the search radius/smoothing parameters, optimal parameters were manually set, which allowed us to mask out >90% of 'suspicious' pixels.

[15] http://www.ahn.nl

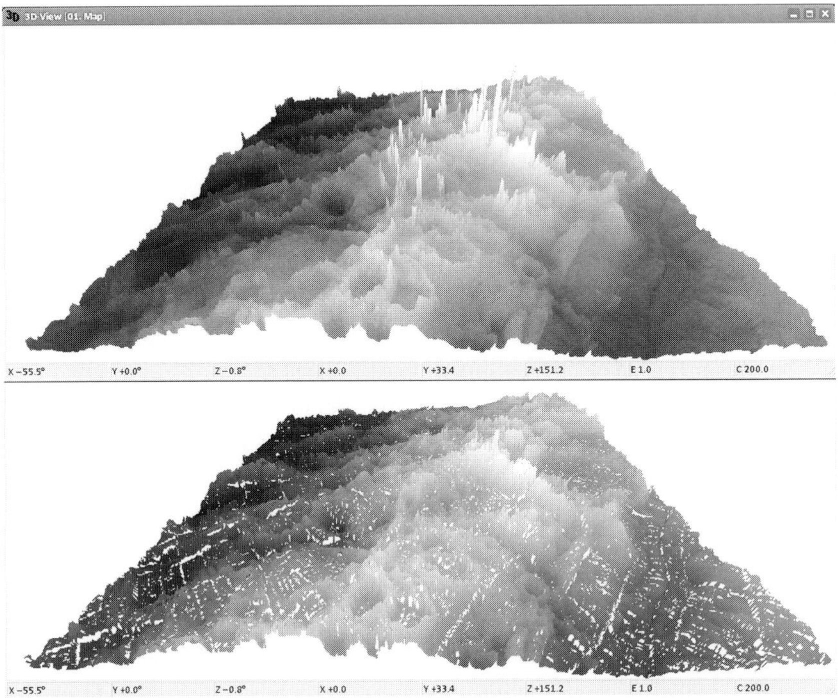

Figure 10.6 Spikes and similar artefacts on the LiDAR DEM, as seen from the south (above). Artefacts (below) masked using two LSPs derived in SAGA GIS: DFM value and representativeness. Exaggeration factor: $\times 10$.

3.3 Results

3.3.1 DEM Filtering and Extraction of LSPs

A variogram was derived using the field-measured elevations (data set 'hoogte_16ef.shp') in the gstat package (Pebesma, 2008) and showed that the features of interest vary smoothly in the study area, which is typical for elevation data. Information about the smoothness of terrain can help to determine the amount of filtering needed to decrease the effects of man-made objects and artefacts in the LiDAR DEM. The anisotropy is significant and therefore needs to be incorporated. Ordinary kriging was used to produce the output DEM and is shown in Figure 10.4b.

After the DEMs had been filtered for artefacts, it was used to generate a list of LSPs that are able to explain the distribution of geomorphological classes. Although SAGA GIS can be used to derive over 100 LSPs given the input DEM, only LSPs that are relevant to the mapping objectives, the study area characteristics and the scale of application were utilised. After several iterations, the following list of LSPs was produced: (1) elevation, (2) SAGA TWI, (3) Valley depth (VDEPTH), (4) Multi Resolution Valley Bottom Flatness Index (MRVBF), (5) DFM, (6) Residual Percentage Index (PERC) and (7) Convergence Index (CONI). These LSPs can be used to depict small changes in morphology and surface roughness, which would possibly not be visible using other LSPs. Note also that a wide search radius was used to derive the LSPs, whilst for residual analysis we use a search radius of 80 pixels. For TWI, we use a floating point of 120. We need to emphasise that these were heuristic settings determined by visually comparing the overlaid geomorphological map boundaries and the intermediate LSPs until the matching was satisfactory.

3.3.2 Extraction of Geomorphological Classes

The results of Kappa statistics show that both DEM25LIDAR-based and DEM25TOPO match the original map relatively well ($\kappa = 56\%$ and $\kappa = 57\%$). The most problematic classes are 3N3 (low plains with ridges), 3N4 (low plains/depressions without ridges) and 3L2 (ground moraines). These are classes that are determined not only by relief but also by the sub-surface composition (rock fragments) and specific shape (ridges). Multinomial logistic regression is shown to be an unbiased estimator – none of the classes have been reduced or omitted from the map (Figure 10.7c). The method is able to reconstruct the geomorphological map (especially the dominant units 3L5 and 3L2a), but the spatial location of some classes differs (*cf.* Figure 10.7a and c). A relatively low kappa is typical for soil and/or geomorphological mapping (Kempen et al., 2009 for a discussion). The advantage of using the DEM25LIDAR is that it depicts small depressions (3N3, 3N4) and ridges (3K14) more accurately than the DEM25TOPO. Because the surveyors likely had problems mapping all small polygons manually, the result of kappa statistics do not show that the map derived using the DEM25LIDAR is more accurate than with using the DEM25TOPO.

The results of unsupervised classification show that the original legend can be refined (Figure 10.7d). The optimal number of classes we estimated using the k-means method as described in Bivand et al. (2008) exceeds the original 14 classes. There are certainly more unique geomorphological features than shown on the GKN50. The question remains which of the two approaches would be more beneficial for geomorphological

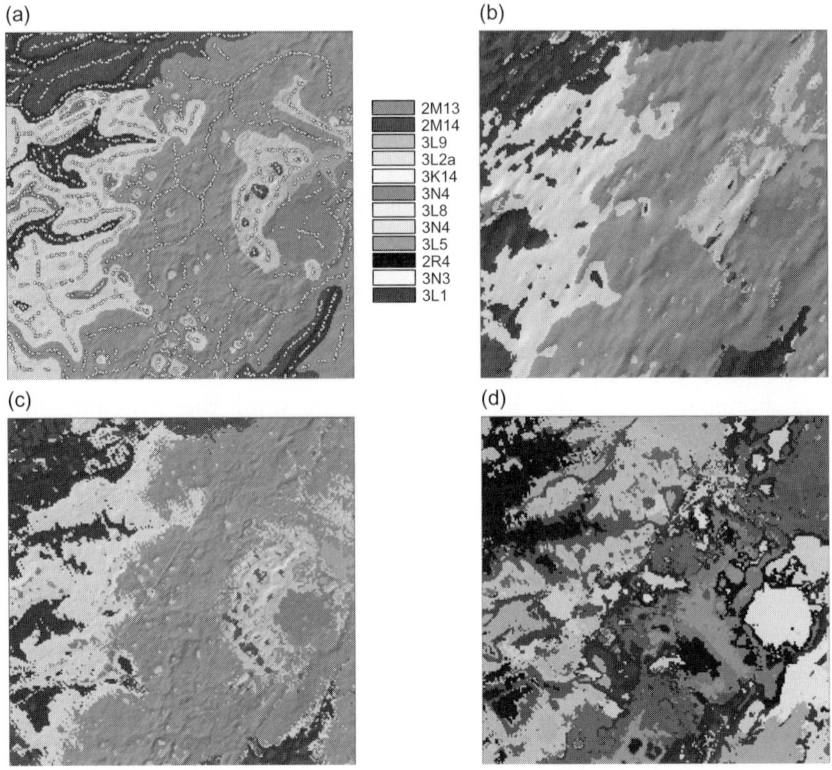

Figure 10.7 Results of supervised classification for Section 3: (a) the original geomorphological map and the training pixels (along medial axes); (b) classes predicted using the multinomial logistic regression and DEM25TOPO; (c) classes predicted using multinomial logistic regression and DEM25LIDAR; (d) results of unsupervised classification using the same number of classes (no legend). See text for description of classes in the legend.

mapping: a completely supervised approach so that the classes fit expert knowledge, or an unsupervised approach and then assignment of geomorphological meaning to the extracted units.

For a comparison, we also present the results of extracting memberships (0–1 values) following the fuzzy *k*-means algorithm outlined in Hengl et al. (2004). For this purpose we use the same training pixel set, but then associate the pixels to classes just by standardising the distances in feature space determined by the LSPs. Figure 10.8 shows the results of mapping classes 3L9 and 4K19. Note that the algorithm finds a much higher number of small patches of class 4K19 (depressions), which were

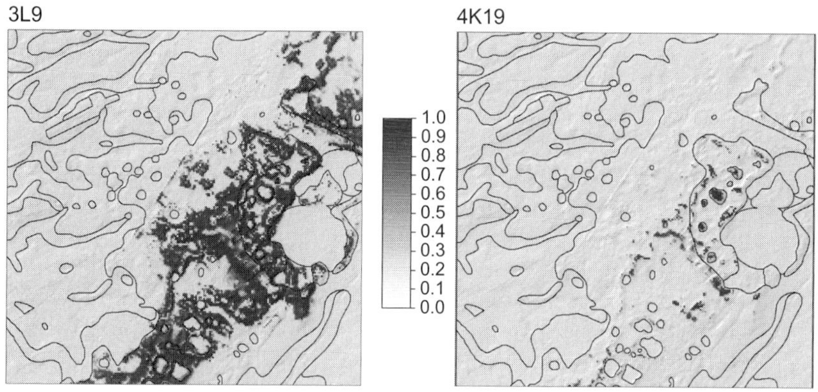

Figure 10.8 Membership maps for geomorphological classes 3L9 (low dunes+plains) and 4K19 (low dunes/depressions); both based on the DEM25LIDAR. Visualised in SAGA GIS.

indicated at only few locations on the geomorphological map (*cf.* with Figure 10.7a). This demonstrates that the LIDAR DEM is particularly suitable at improving the spatial detail of small patchy classes.

The advantage of using membership is that one can observe how crisp transitions between certain classes are, and where the confusion of classes is high (Hengl et al., 2004). This way the analyst has an opportunity to focus on mapping a single geomorphological unit, adjust training pixels where necessary and improve the quality of resulting maps.

3.4 Discussion and Conclusions

The results of this case study indicate that the LiDAR DEMs can be used to improve geomorphological mapping in areas of low relief. For instance, we were able to map many small features (depressions and ridges) that have been overlooked by previous surveyors (e.g. class 4K19 in Figure 10.8). This demonstrates that multinomial logistic regression can be used to increase the detail of existing geomorphological maps, without a need for manually delineating such features. The results of this case study show that the predictions are unbiased and the main features match the existing map moderately well (Figure 10.7). The number of spatial features (polygons) in the new map has increased by 50–100%. Further field validation, however, is needed to determine if these small patches represent the landscape more accurately than the classical geomorphological map.

There are several remaining issues about LiDAR data processing. For example, we are still not certain if the DEM filtering could be completely automated, and therefore should all artificial objects be filtered out or included as separate classes in the legend? Furthermore, how should an optimal set of LSPs for a given study area be selected? Our experience is that the LSPs of interest for geomorphological mapping need to be iteratively fine-tuned in order to allow optimal information extraction. We can foresee that, in the near future, automated optimisation algorithms will be developed that iteratively compare the LSP settings until an optimal product is reached (maximisation of the classification accuracy). This could, however, become computationally intensive as the number of combinations is rather high. For example, LSPs such as MRVBF or TWI require numerous initial parameters to be set by the user (e.g. initial slope, number of iterations and search radius). To test which combination is the best, one would need to rerun the analysis on hundreds and hundreds of variants of LSPs.

4. CASE STUDY LECH – AUSTRIA

4.1 Study Area and Data Sets

The Lech area is a high alpine area in the province of Vorarlberg, Western Austria. The elevation ranges between 1650 m in the valleys and 2450 m at the highest summit (Figure 10.9). The area is underlain by the 'Lechtal Decke', a tectonic nappe composed mainly of limestone, marl and evaporatic formations. The geomorphology reflects glacial, fluvial, mass movement and karst landforms. The Lech area has been subject to severe glacial erosion and subsequent postglacial mass wasting, which includes rock fall, slide and flow-type mass movement (Cammeraat, 1986; Ruff and Czurda, 2008). Landforms related to bare and covered gypsum karst are common. Recently, successful automated geomorphological mapping of alpine areas from DEM data has been performed using object-based classification (Drăguţ and Blaschke, 2006) and the morphometric parameterisation through self-organising maps (Ehsani and Quiel, 2008), despite the mountains' morphometric complexity (Rasemann et al., 2004). However, geomorphologists are also interested in the morphogenetic background (Minár and Evans, 2008). The specific objective of this case study is to suggest a semi-automatic and object-based method

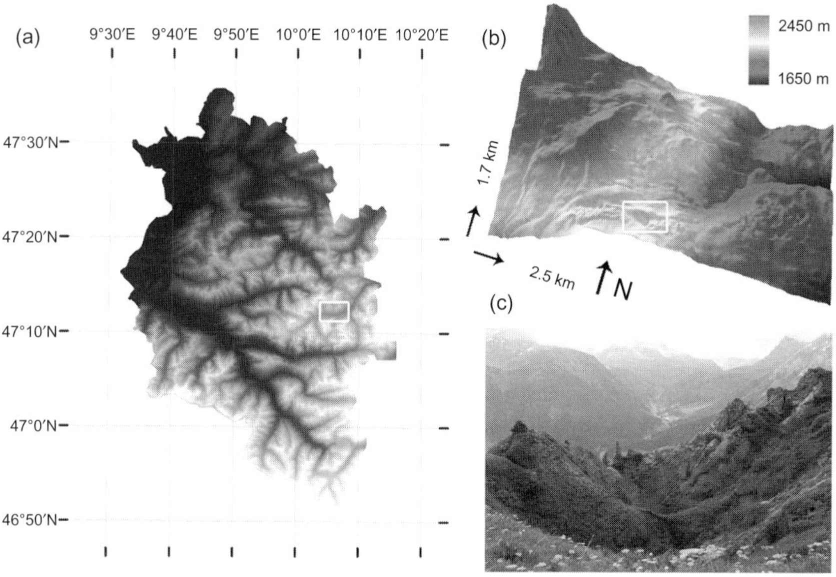

Figure 10.9 (a) White box indicates the location of the 'Lech' study area (DEM in (b)) in Vorarlberg, Western Austria. (b) DEM of study area (vertical exaggeration of 1.5). (c) Bare gypsum karst geomorphology near Lech, location photo indicated by the white box in (b).

of image analysis for the classification of geomorphological landforms in complex alpine terrain. In this method, a geomorphological feature is likely described by a set of pixels (Blaschke et al., 2004). Specific classification rules for each geomorphological feature are applied to the DEM to identify and categorise the various geomorphological classes.

For this study, we used parameters derived from a 1 m LiDAR DEM, kindly provided by the Land Vorarlberg.[16] In addition, 0.25 m resolution false-colour ortho-rectified air photos were used as a reference for a field campaign, during which the pre-field constructed objects were classified using a mobile GIS device. A classic 1:25,000 scale geomorphological map of Cammeraat (1986) was used for validation of image object boundaries. The classification method has been tested in the study area and will be applied to other mountain areas (not shown here). The data set and process tree used in Definiens Developer is available on the http://www.appgema.net website.

[16] http://www.vorarlberg.at

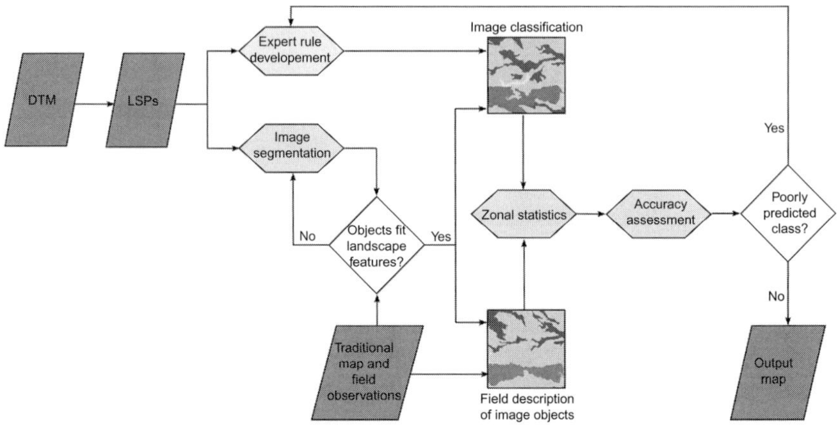

Figure 10.10 Data analysis scheme illustrating how field-based and automated mapping are combined for the classification of geomorphological features. See text for detailed explanation.

4.2 Mapping Scheme

The general data analysis framework is given in Figure 10.10, with LSPs extracted from the LiDAR DEM. These parameters serve as input for a multiresolution image segmentation procedure (Baatz and Schäpe, 2000) that calculates image objects with internal homogeneous conditions of the user-specified LSP layers – at multiple scale levels. After comparison with field observations and a geomorphological map, the expert decides which scale levels are used, and the classification type. Poorly segmented image objects can be adjusted by choosing different sets of LSPs for the segmentation procedure. Subsequently, the expert designs specific classification rules to describe a particular geomorphological feature, which is based on internal image object statistics and spatial relations between image objects at the target scale level or between upper and/or lower scale levels. The final step is the actual image classification using the developed expert rules.

The classification results are iteratively compared with field observations and – if available – a classic geomorphological map. The accuracy assessment procedure uses ESRI ArcGIS Zonal Statistics to evaluate confusion between individual classes and to improve classification rules of each specific landform.

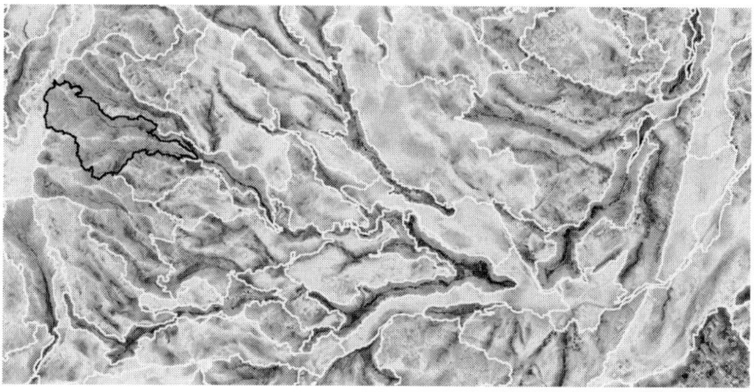

Figure 10.11 Fragment of segmented LiDAR DEM. The segments are based on the underlying three layer composite image that includes slope, openness R50 and openness R200.

4.2.1 Extraction of LSPs

Seven LSPs were used for classification: elevation, curvature, slope, elevation percentile (EPC), upstream area, topographic openness – measured over a radius of 50 and 200 m (R50 and R200, approximately) – and 'Filled Area'. These LSPs are calculated with ArcGIS Desktop tools and a MATLAB script (EPC and openness). Image segmentation, classification rule design and rule implementation for classification were carried out in the Definiens Developer software. Curvature, slope and topographic openness maps were combined in a single RGB composite (Figure 10.11), for visualisation purposes, which proved useful during the field campaign. EPC maps were used to determine topographic position (e.g. relatively low/high) of image objects in the landscape. Upstream area values were used to identify fluvial incisions and alluvial/debris fans. Dissolution of gypsum results in sharp dolines in the landscape that show as sinks in the corresponding LiDAR DEM. Sinks in DEMs are often considered artificial and are filled to create a hydrological-corrected DEM. Jenkins and McCauley (2006) studied the effect of this 'correction' tool in wetland areas and found that real sinks are also filled. In this gypsum karst area, this tool also removes the sinks.

The difference between the filled and original DEM ('Filled Area' parameter) is used to identify the location and size of karst features. Other landforms and processes are identified on the basis of the combined statistical properties of the slope and openness maps and the spatial relations that exist between image objects.

4.2.2 Image Segmentation and Rule Sets for Classification

A hierarchical structure of image objects is used as the result of multi-level segmentation. This means that relatively large image objects contain smaller, fine-scale image objects. These fine-scale image objects can only belong to one single broad-scale object. Each scale level of image objects is processed in (semi-) automated image analyses in which relations with objects from other scale levels can be used. The number and scale parameter of image object levels are controlled by the user and depend on the purpose of analysis. The process tree that is used for identification of geomorphological features from LSPs can contain pixel-based values (min, max and so on), object-based internal statistics (mean, standard deviation and so on), shape (length/width ratio, area and so on) and relations to sub, super or neighbouring image objects (bordering to, existence of and so on).

In our method, image classification follows a step-by-step procedure: easily recognised geomorphological features with sharp boundaries are classified first. These are erosion channels or gypsum dolines which can be identified from relatively small image objects. Smooth geomorphological features (e.g. glacial erosion or depositional landforms) are more efficiently extracted using relatively large image objects. After the extraction and classification of fluvial incision and gypsum dolines, the unclassified image objects are aggregated into larger objects before further classifications are made. Since individual geomorphic units, such as a fluvial incision, often consist of several image objects, the extraction needed several steps before a combination into the desired geomorphic unit was made. This means that each geomorphic unit is extracted by applying a unique rule set, which is based on feature-specific parameter criteria (Table 10.1). Such classification rules are set up by the expert, based on comparison between field knowledge or observations of landscape features and LSP values.

4.2.3 Field Observations

During a field campaign – prior to the final classification – we validated the image objects using a Trimble Mobile GIS device in combination with digital ortho-rectified air photos and the RGB composite of slope and topographic openness parameters. We used the classes 'glacially eroded bedrock', 'fluvial incision', 'alluvial/debris fan', 'landforms underlain by fall deposits' and 'karst'. For each image object, we evaluated the primary geomorphological process responsible for that landform, along with the

Table 10.1 Overview of the LSPs and Criteria Used in the Step-By-Step Feature Extraction

Step	Action	Scale	Feature	LSP	Criteria
1	Define position in the landscape	10	Low–high	EPC	0–1
2	Classify active erosion features	20–50	Gypsum dolines	Filled area	>50 m^3
			Slope subject to karst	Mean curvature	>1
				Adjacent to gypsum doline[a]	1
			Fluvial incision	Upstream area	>10,000 m^2
				Mean curvature	<−3
				Mean openness (R200)	110°–170°
				Existence of low/ medium features[a]	1
				Length/width ratio	>2
				Mean slope	>25°
3	Classify fossil erosion or deposition features	100	Landforms underlain by fall deposits	Brightness (defined by elevation, EPS, slope and openness)	640–700
				Mean slope	25°–40°
			Alluvial/debris fan	Upstream area	>75,000 m^2
				Mean curvature	0–0.5
				Mean slope	0°–15°
				Bordering to classified fluvial features[a]	1
			Glacially eroded Bedrock (summits)	Mean EPC	>0.4
				Standard deviation openness	>5
				Standard deviation slope	>8
4	Classify unclassified objects	100	Karst	Filled area	>5 m^3
				Bordering to karst area (step 2) [a]	1
			Fluvial incision	Filled area	<10 m^3
				Mean openness (R200)	<170°
				Mean slope	>22°
			Alluvial/debris fan	Mean curvature	<−0.5
				Mean EPC	<0.4
			Glacially eroded bedrock	Mean openness (R200)	>140°
			Landforms underlain by fall deposits	Mean openness (R200)	<140°

[a]Value acts as a Boolean number: 0, no; 1, yes.

current activity weight of this process. For example, a glacially eroded bedrock slope has no current activity since glaciers are no longer present. In addition, secondary processes acting on such a slope, e.g. solifluction, are evaluated in a similar way. The derived data set covers 100% of the study area and serves as a reference to determine the final classification's accuracy, based on percentages of classified geomorphological features within the reference image objects (Van Asselen and Seijmonsbergen, 2006).

4.3 Results

The final extracted geomorphological map is shown in Figure 10.12. The five major legend categories in the landscape occur in clear patterns and visually match the classic geomorphological map quite well. The confusion matrix in Table 10.2 shows a comparison between the field observations and the classified map and reveals an overall accuracy of 76.5%. Glacially eroded bedrock (84.1%) and karst features (76.6%) show relatively high classification scores.

Fluvial incisions (52.9%), alluvial/debris fans (49.7%) and fall deposits (62.5%) are sometimes confused with other classes. Fluvial incisions are often confused with glacially eroded bedrock.

4.3.1 Discussion and Conclusions

During the field inspection, the composite RGB of slope and openness values, in combination with LiDAR-derived contour lines, proved useful for recognition of the image objects. Although human interference in this landscape is relatively high, our experience was that the occurrence of roads, houses, ski-runs and other infrastructure did not greatly affect the segmentation shape. Within larger image objects, often smaller patterns of openness reflected 'secondary' geomorphological processes, such as shallow incisions into glacially eroded bedrock. Field observations confirm these assumptions. This may represent a gradual transition between glacially eroded bedrock under influence of postglacial fluvial erosion and may result in overlap between two classes in the confusion matrix. A fuzzy approach to landscape classification (see also MacMillan et al., 2000; Schmidt and Hewitt, 2004; Arrell et al., 2007), rather than crisp geomorphological units, might help to overcome this problem. This understanding is promising for further rule-set optimisation. The relatively low accuracy values of the classes 'alluvial/debris fan' and 'fall deposits' are also caused by confusion between the classes since their morphology and internal object statistics are relatively comparable. This is especially true if

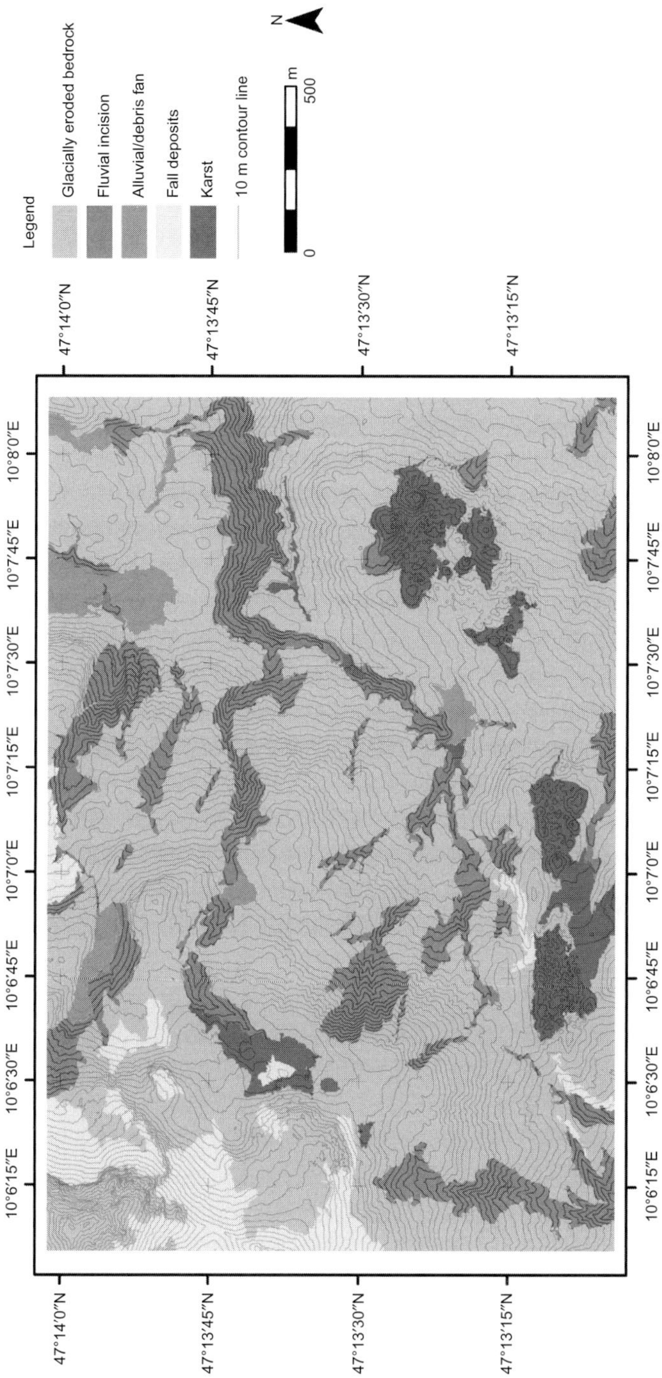

Figure 10.12 Fragment of the classified geomorphological map.

Table 10.2 Confusion Matrix Showing the Number of Pixels of Classified Geomorphological Features within the Reference Data Set

Geomorphological Unit	**Classification**						
	Glacially Eroded Bedrock	**Fluvial Incision**	**Alluvial or Debris Fan**	**Fall Deposits**	**Karst**	**Total**	**Correctly Classified**
Glacially eroded bedrock	2,638,920	325,362	51,924	70,673	63,348	3,150,227	84.1
Fluvial incision	114,939	393,894	225	7,033	871	516,962	52.9
Alluvial/debris fan	0	1,848	51,527	0	0	53,375	49.7
Fall deposits	139,729	9,085	23	157,926	2,306	309,069	62.5
Karst	186,013	14,832	0	17,213	218,061	436,119	76.6
Total	3,079,601	745,021	103,699	252,845	284,586	4,465,752	
Overall accuracy	76.5	Congalton (1991)					
Average user's accuracy	65.2	Story and Congalton (1986)					
Average producer's accuracy	61.6	Story and Congalton (1986)					
Kappa coefficient	0.52	Congalton and Green (1999)					

the original landforms are transformed by secondary processes, such as solifluction. Further rule-set optimisation and additional LSPs are necessary to improve final accuracies.

5. CLOSING REMARKS

Two contrasting landscapes, LiDAR and DEMs, have been analysed to illustrate the variety and possibilities in the use of LSPs for geomorphological feature extraction. Both case studies demonstrate that existing classic geomorphological information is a valuable source for fine-tuning, selection and classification of the relevant LSPs. Landscape management will certainly profit from the improvements that are made to existing information sources by automated classification of fine-scale DEMs. The importance and added value of DEMs for geomorphological mapping will increase significantly, especially for countries/regions with limited budgets and limited thematic information.

The future of automated mapping using technologies such as LiDAR is in combination with other optical, radar and hyperspectral sensors. This will enable an analyst to work with surface and sub-surface parameters that describe all aspects of a terrain/surface material so that important geomorphological properties are not overlooked. In this respect, we welcome further developments within open-source modelling environments, GIS and morphometrical analysis software. Further refinement of existing statistical models could also improve the mapping of landform categories; these include regression trees or machine learning algorithms.

ACKNOWLEDGEMENTS

This research was carried out in the context of the Virtual Laboratory for e-Science project supported by a BSIK grant from the Dutch Ministry of Education, Culture and Science (OC&W) and is part of the ICT innovation programme of the Ministry of Economic Affairs (EZ). We are grateful to the 'Land Vorarlberg' in Austria for allowing us to use the 1 m resolution LiDAR data. We also thank 'inatura, Naturerlebnis Dornbirn' for their continuous support. Our colleague Erik Cammeraat of IBED is thanked for allowing us to use his classic geomorphological map of the Northern Lech Quellengebirge.

REFERENCES

Adediran, A.O., Parcharidid, I., Poscolieri, M., Pavlopoulos, K., 2004. Computer-assisted discrimination of morphological units on north-central Grete (Greece) by applying multi-variate statistics to local relief gradients. Geomorphology 58, 357–370.

Anders, N.S., Seijmonsbergen, A.C., Bouten, W., 2009. Modelling channel incision and alpine hillslope development using laser altimetry data. Geomorphology 113, 35–46.

Arnold, N.S., Rees, W.G., Devereux, B.J., Amable, G.S., 2006. Evaluating the potential of high-resolution airborne LIDAR data in glaciology. Int. J. Remote Sens. 6, 1233–1251.

Arrell, K.E., Fisher, P.F., Tate, N., Bastin, L., 2007. A fuzzy *c*-means classification of elevation derivatives to extract the morphometric classification of landforms in Snowdonia, Wales. Comput. Geosci. 33, 1366–1381. <http://dx.doi.org/10.1016/j.cageo.2007.05.005>

Baatz, M., Schäpe, A., 2000. Multiresolution segmentation – an optimization approach for high quality multi-scale image segmentation. In: Strobl, J., Blaschke, T., Griesebner, G. (Eds.), Angewandte Geographische Informationsverarbeitung, vol. XII. Wichmann, Heidelberg, pp. 12–23.

Batten, P., 2001. A new approach for landscape mapping. Proceedings of the Sixth International Conference on Geocomputation. University of Queensland, Brisbane, Australia, 24–26 September.

Behn, M.D., Sinton, J.M., Detrick, R.S., 2004. Effect of the Galápagos hotspot on seamount volcanism along the Galápagos Spreading Center. Earth Planet. Sci. Lett. 217, 331–347.

Bivand, R., Pebesma, E., Rubio, V., 2008. Applied Spatial Data Analysis with R. Use R Series. Springer, Heidelberg.

Blaschke, T, Burnett, C., Pekkarinen, A., 2004. Image segmentation methods for object-based analysis and classification. In: De Jong, S.M., Van der Meer, F.D. (Eds.), Remote Sensing Image Analysis: Including the Spatial Domain. Kluwer Academic, Dordrecht.

Bocco, G., Mendoza, M., Velásquez, A., 2001. Remote sensing and GIS-based regional geomorphological mapping – a tool for land use planning in developing countries. Geomorphology 39, 211–219.

Booth, A.M., Roering, J.J., Perron, J.T., 2009. Automated landslide mapping using spectral analysis and high resolution topographic data: Pudget Sound lowlands, Washington, and Portland Hills, Oregon. Geomorphology 109, 132–147.

Bue, B.D., Stepinski, T.F., 2006. Automated classification of landforms on Mars. Comput. Geosci. 32, 604–614.

Burrough, P.A., van Gaans, P.F.M., MacMillan, R.A., 2000. High resolution landform classification using fuzzy *k*-means. Fuzzy Sets Syst. 113, 37–52.

Cammeraat, L.H., 1986. A Geomorphological Investigation of the Northern Lechquellengebirge, Vorarlberg, Austria. Unpublished M.Sc. Thesis. Laboratory for Physical Geography and Soil Science, Universiteit van Amsterdam, Amsterdam, 174 pp.

Carrara, A., 1992. Landslide hazard assessment. Proceedings of 1er Simposio Internacional sobre Sensores Remotos y Sistemas de Informacion Geografica (SIG) Para el Estudio de Riesgos Naturales. Bogota, pp. 329–356.

Congalton, R.G., 1991. A review of assessing the accuracy of classification of remotely sensed data. Remote Sens. Environ. 37, 45–46.

Congalton, R.G., Green, K., 1999. Assessing the Accuracy of Remotely Sensed Data: Principles and Practices. Lewis Publishers, Boca Raton, FL, 137 pp.

Conrad, O., 2007. SAGA – Entwurf, Funktionsumfang und Anwendung eines Systems für Automatisierte Geowissenschaftliche Analysen. Ph.D. Thesis. University of Göttingen, Göttingen, 221 pp.

Dalrymple, J.B., Blong, R.J., Conacher, A.J., 1968. A hypothetical nine unit land surface model. Z. Geomorphol. 12, 60–76.

Demek, J., Embleton, C. (Eds.), 1978. Guide to Medium-Scale Geomorphological Mapping. IGU Commission on Geomorphological Survey and Mapping. E. Schweizerbart'sche Verlagsbuchhandlung, Stuttgart.

Dewitte, O., Jaselette, J.-C., Cornet, Y., Van den Eeckhout, M., Collignon, A., Poesen, J., et al., 2008. Tracking landslide displacements by multi-temporal DTMs: a combined aerial stereophotogrammetric and LIDAR approach in western Belgium. Eng. Geol. 99, 11–22.

Dikau, R., Brabb, E.E., Mark, R.K., Pike, R.J., 1995. Morphometric landform analysis of New Mexico. Z. Geomorphol. Suppl. 101, 109–126.

Drăguţ, L., Blaschke, T., 2006. Automated classification of landform elements using object-based image analysis. Geomorphology 81, 330–344.

Ehsani, A.H., Quiel, F., 2008. Geometric feature analysis using morphometric parameterization and artificial neural networks. Geomorphology 99, 1–12.

Etzelmuller, B., Romstad, B., Fjellanger, J., 2007. Automated regional classification of topography in Norway. Norw. J. Geol. 87, 167–180.

Evans, I.S., 1990. Cartographic techniques in geomorphology. In: Goudie, A., Anderson, M., Burt, T., Lewin, J., Richards, K., Whalley, B., et al. (Eds.), Geomorphological Techniques, second ed. Routledge, London, pp. 97–108.

Evans, I.S., 2004. Geomorphometry. In: Goudie, A.S. (Ed.), Encyclopedia of Geomorphology. International Association of Geomorphology. Routledge, London, pp. 435–439.

Evans, I.S., Hengl, T., Gorsevski, P., 2008. Applications in geomorphology. In: Hengl, T., Reuter, H.I. (Eds.), Geomorphometry: Concepts, Software, Applications. Developments in Soil Science, vol. 33. Elsevier, Amsterdam, pp. 497–525.

Fisher, P.F., Tate, N.J., 2006. Causes and consequences of error in digital elevation models. Prog. Phys. Geogr. 30, 467–489.

Giles, P.T., Franklin, S.E., 1998. An automated approach to the classification of slope units using digital data. Geomorphology 21, 251–264.

Gilewska, S., Klimek, M., 1968. Project of the Unified key to the detailed geomorphological map of the World, Folia Geographica, Series Geographica-Physica, vol. II. Polska Akademia Nauk, Kraków.

Grunwald, S. (Ed.), 2006. Environmental Soil–Landscape Modeling: Geographic Information Technologies and Pedometrics. CRC Press, Taylor & Francis, New York, 488 pp.

Gustavsson, M., Seijmonsbergen, A.C., Kolstrup, E., 2008. Structure and contents of a new geomorphological GIS database linked to a geomorphological map – with an example from Liden, central Sweden. Geomorphology 95, 335–349.

Gustavvson, M., Kolstrup, E., Seijmonsbergen, A.C., 2006. A new symbol-and-GIS based detailed geomorphological mapping system: renewal of a scientific discipline for understanding landscape development. Geomorphology 77, 90–111.

Hay, H.J., Blascke, T., Marceau, D.J., Bouchard, A., 2003. A comparison of three methods for the multiscale analysis of landscape structure. ISPRS J. Photogramm. Remote Sens. 57, 327–345.

Hengl, T., MacMillan, R.A., 2008. Geomorphometry – a key to landscape mapping and modelling. In: Hengl, T., Reuter, H.I. (Eds.), Geomorphometry: Concepts, Software,

Applications. Developments in Soil Science, vol. 33. Elsevier, Amsterdam, pp. 433–460.

Hengl, T., Reuter, H.I. (Eds.), 2008. Geomorphometry: Concepts, Software, Applications. Developments in Soil Science, vol. 33. Elsevier, Amsterdam, , 525 pp.

Hengl, T., Rossiter, D.G., 2003. Supervised landform classification to enhance and replace photointerpretation in semi-detailed soil survey. Soil Sci. Soc. Am. J. 67, 1810–1822.

Hengl, T., Walvoort, D.J.J., Brown, A., 2004. A double continuous approach to visualization and analysis of categorical maps. Int. J. Geogr. Inf. Sci. 18 (2), 183–202.

Hengl, T., Bajat, B., Reuter, H.I., Blagojevic, D., 2008. Geostatistical modelling of topography using auxiliary maps. Comput. Geosci. 34, 1886–1899.

Heuvelink, G., 2003. 'The Definition of Pedometrics'. Pedometron (International Working Group on Pedometrics – Provisional Commission on Pedometrics of the International Union of Soil Sciences, 15, pp. 11–12. <http://www.pedometrics.org/pedometron/pedometron15.pdf>

Hiller, J.K., Smith, M.J., 2008. Residual relief separation: digital elevation model enhancement for geomorphological mapping. Earth Surf. Process. Landforms 33, 2266–2276, doi:10.1002/esp.1659.

Hillier, J.K., Watts, A.B., 2004. 'Plate-like' subsidence of the East Pacific Rise – South Pacific superswell system. J. Geophys. Res. 109, B10102, doi:10.1029/2004JB003041.

Iwahashi, J., Pike, R.J., 2007. Automated classifications of topography from DEMs by an unsupervised nested-means algorithm and a three-part geometric signature. Geomorphology 86, 409–440.

Jenkins, D.G., McCauley, L.A., 2006. GIS, SINKS, FILL, and disappearing wetlands: unintended consequences in algorithm development and use. Twenty-First Annual ACM Symposium on Applied Computing, Dijon, France, pp. 277–282.

Kasai, M., Ikeda, M., Asahina, T., Fujisawa, K., 2009. LiDAR-derived DEM evaluation of deep-seated landslides in a steep and rocky region of Japan. Geomorphology 113, 57–69.

Kempen, B., Brus, D.J., Heuvelink, G.B.M., Stoorvogel, J.J., 2009. Updating the 1:50,000 Dutch soil map using legacy soil data: a multinomial logistic regression approach. Geoderma 151, 311–326.

Klimaszewski, M., 1990. Thirty years of geomorphological mapping. Geogr. Pol. 58, 1–18.

Koomen, A.J., Maas, G.J., 2004. Geomorfologische Kaart Nederland (GKN). Achtergrond document bij het landsdekkende digitale bestand. Alterra rapport 1039, Wageningen.

Kraus, K., Pfeifer, N., 1998. Determination of terrain models in wooded areas with airborne laser scanner data. ISPRS J. Photogramm. Remote Sens. 53, 193–203.

Lefsky, M.A., Cohen, W.B., Parker, G.G., Harding, D.J., 2002. LiDAR remote sensing for ecosystem studies. BioScience 52, 9–30.

Li, Z., Zhu, Q., Gold, C., 2004. Digital Terrain Modeling: Principles and Methodology. CRC Press, London, 323 pp.

Lillesand, T.M., Kiefer, R.W., Chipman, J.W., 2008. Remote sensing and image interpretation. John Wiley and Sons, New York.

MacMillan, R.A., Shary, P.A., 2008. Landforms and landform elements in geomorphometry. In: Hengl, T., Reuter, H.I. (Eds.), Geomorphometry: Concepts, Software, Applications. Developments in Soil Science, vol. 33. Elsevier, Amsterdam, pp. 227–354.

MacMillan, R.A., Pettapiece, W.W., Nolan, S.C., Goddard, T.W., 2000. A generic procedure for automatically segmenting landforms into landform elements using DEMs, heuristic rules and fuzzy logic. Fuzzy Sets Syst. 113, 81–109.

Mark, D.M., 1975. Geomorphometric parameters: a review and evaluation. Geogr. Ann. 57A, 165–177.

Maune, D.F. (Ed.), 2001. Maryland, MD. American Society for Photogrammetry and Remote Sensing, , 539 pp.

McKean, J., Roering, J., 2004. Objective landslide detection and surface morphology mapping using high-resolution airborne laser altimetry. Geomorphology 57, 331–351.

Milledge, D.G., Lane, S.N., Warburton, J., 2009. The potential of digital filtering of generic topographic data for geomorphological research. Earth Surf. Process. Landforms 34, 63–71.

Milne, G., 1935. Some suggested units of classification and mapping particularly for East African soils. Soil Res. 4, 183–198.

Minár, J., Evans, I.S., 2008. Elementary forms for land surface segmentation: the theoretical basis of terrain analysis and geomorphological mapping. Geomorphology 95, 236–259.

Nelson, A., Reuter, H.I., Gessler, P., 2008. DEM production methods and sources. In: Hengl, T., Reuter, H.I. (Eds.), Geomorphometry: Concepts, Software, Applications. Developments in Soil Science, vol. 33. Elsevier, Amsterdam, pp. 65–85.

Neteler, M., Mitasova, H., 2008. Open Source GIS: A GRASS GIS Approach, third ed.. The International Series in Engineering and Computer Science, vol. 773. Springer, New York, 406 pp.

Newell, W., Clark, I., 2008. Geomorphic map of Worcester County, Maryland, interpreted from a LiDAR-based, digital elevation model. Open-File Report 2008–1005. U.S. Geological Survey, 34 pp. <http://pubs.usgs.gov/of/2008/1005/>

Oguchi, T., Hayakawa, Y., 2011. Data sources. In: Smith, M.J., Paron, P., Griffiths, J. (Eds.), Geomorphological Mapping: A Handbook of Techniques and Applications. Elsevier, Amsterdam, pp. 189–224.

Oksanen, J., Sarjakoski, T., 2005. Error propagation of DEM-based surface derivatives. Comput. Geosci. 31, 1015–1027.

Olaya, V., 2008. Basic land-surface parameters. In: Hengl, T., Reuter, H.I. (Eds.), Geomorphometry: Concepts, Software, Applications. Developments in Soil Science, vol. 33. Elsevier, Amsterdam, pp. 141–169.

Otto, J.-C., Gustavsson, M., Geilhausen, M., 2011. Cartography: design, symbolisation and visualisation of geomorphological maps. In: Smith, M.J., Paron, P., Griffiths, J. (Eds.), Geomorphological Mapping: A Handbook of Techniques and Applications. Elsevier, Amsterdam.

Pebesma, E., 2008. Interpolation and geostatistics. In: Bivand, R., Pebesma, E., Gómez-Rubio, V. (Eds.), Applied Spatial Data Analysis with R. Use R Series. Springer, Heidelberg, pp. 191–236.

Pike, R.J., 1988. The Geometric Signature: Quantifying Landslide-Terrain Types from Digital Elevation Models. Math. Geol. 20 (5), 491–511.

Pike, R.J., 1995. Geomorphometry – progress, practice, and prospect. Z. Geomorphol. Suppl. 101, 221–238.

Pike, R.J., Evans, I.S., Hengl, T., 2008. Geomorphometry: a brief guide. In: Hengl, T., Reuter, H.I. (Eds.), Geomorphometry: Concepts, Software, Applications. Developments in Soil Science, vol. 33. Elsevier, Amsterdam, pp. 3–30.

Prima, O.D.A., Echigo, A., Yokoyama, R., Yoshida, T., 2006. Supervised landform classification of Northeast Honshu from DEM-derived thematic maps. Geomorphology 78, 373–386.

R Development Core Team, 2009. R: A Language and Environment for Statistical Computing. R Foundation for Statistical Computing, Vienna, Austria. ISBN 3-900051-07-0. <http://www.R-project.org>

Rasemann, S., Schmidt, J., Schrott, L., Dikau, R., 2004. Geomorphometry in mountain terrain. In: Bishop, N., Schroder, J.F. (Eds.), Geographic Information Science in Mountain Geomorphology, Springer-Praxis. pp. 101–145.

Reuter, H.I., Nelson, A., 2008. Geomorphometry in ESRI packages. In: Hengl, T., Reuter, H.I. (Eds.), Geomorphometry: Concepts, Software, Applications. Developments in Soil Science, vol. 33. Elsevier, Amsterdam, pp. 269–291.

Reuter, H.I., Hengl, T., Gessler, P., Soille, P., 2008. Preparation of DEMs for geomorphometric analysis. In: Hengl, T., Reuter, H.I. (Eds.), Geomorphometry: Concepts, Software, Applications. Developments in Soil Science, vol. 33. Elsevier, Amsterdam, pp. 87–120.

Ruff, M., Czurda, K., 2008. Landslide susceptibility analysis with a heuristic approach in the Eastern Alps (Vorarlberg, Austria). Geomorphology 94, 314–324.

Salomé, A.I., van Dorsser, H.J., Rieff, Ph.L., 1982. A comparison of geomorphological mapping systems. ITC J. 3, 272–274.

Schmidt, J., Andrew, R., 2005. Multi-scale landform characterization. Area 37, 341–350.

Schmidt, J., Hewitt, A., 2004. Fuzzy land element classification from DTMs based on geometry and terrain position. Geoderma 121, 243–256.

Seijmonsbergen, A.C., 1992. Geomorphological Evolution of an Alpine Area and its Application to Geotechnical and Natural Hazard Appraisal in the NW. Rätikon Mountains and S. Walgau (Vorarlberg, Austria). Ph.D. Thesis. University of Amsterdam, Amsterdam, 109 pp.

Seijmonsbergen, A.C., de Graaff, L.W.S., 2004. Geomorphological mapping and geophysical profiling for the evaluation of natural hazards in an alpine catchment. Nat. Hazards Earth Syst. Sci. 6, 185–193.

Speight, J.G., 1990. Landform. In: McDonald, R.C., Isbell, R.F., Speight, J.G., Walker, J., Hopkins, M.S. (Eds.), Australian Soil and Land Survey Field Handbook, second ed. Inkarta Press, Melbourne, pp. 9–57.

Starck, J.L., Bijaoui, A., Valtchanov, I., Murtagh, F., 2000. A combined approach for object detection and deconvolution. Astron. Astrophys. Suppl. Ser. 147, 139–149.

Story, M., Congalton, R.G., 1986. Accuracy assessment: a user's perspective. Photogramm. Eng. Remote Sens. 52, 397–399.

Temme, A.J.A.M., Heuvelink, G.B.M., Schoorl, J., Claessens, L., 2008. Geostatistical simulation and error propagation in geomorphometry. In: Hengl, T., Reuter, H.I. (Eds.), Geomorphometry: Concepts, Software, Applications. Developments in Soil Science, vol. 33. Elsevier, Amsterdam, pp. 121–139.

Thoma, D.P., Gupta, S.C., Bauer, M.E., Kirchhoff, C.E., 2005. Airborne laser scanning for riverbank erosion assessment. Remote Sens. Environ. 95, 493–501.

Van Asselen, S., Seijmonsbergen, A.C., 2006. Expert-driven semi-automated geomorphological mapping using a laser DTM. Geomorphology 78, 309–320.

Van Heerd, R., Kuijlaars, E., Teeuw, M., Van 't Zand, R., 2008. Productspecificatie AHN 2000. Rijkswaterstaat, Adviesdienst Geo-informatie en ICT, Delft, 23 pp.

Van Westen, C.J., Seijmonsbergen, A.C., Mantovani, F., 1999. Comparing landslide hazard maps. Nat. Hazards 20, 137–158.

Van Westen, C.J., Soeters, R., Sijmons, K., 2000. Digital geomorphological landslide hazard mapping of the Alpago area, Italy. Int. J. Appl. Earth Obs. Geoinf. 2 (1), 51–60.

Venables, W.N., Ripley, B.D., 2002. Modern Applied Statistics with S. fourth ed. Springer, Heidelberg.

Verstappen, H., 2011. Old and new trends in geomorphological and landform mapping. In: Smith, M.J., Paron, P., Griffiths, J. (Eds.), Geomorphological Mapping: A Handbook of Techniques and Applications. Elsevier, Amsterdam, pp. 13–38.

Wessel, P., 1998. An empirical method for optimal robust regional–residual separation of geophysical data. Math. Geol. 30, 391–408.

Wessel, P., 2001. Global distribution of seamounts inferred from gridded Geosat/ERS-1 altimetry. J. Geophys. Res. 106, 19431–19441.

Wilson, J.P., Gallant, J.C. (Eds.), 2000. Terrain Analysis: Principles and Applications. Wiley, New York, 303 pp.

Wood, J., 2008. Overview of software packages used in geomorphometry. In: Hengl, T., Reuter, H.I. (Eds.), Geomorphometry: Concepts, Software, Applications. Developments in Soil Science, vol. 33. Elsevier, Amsterdam, pp. 257–267.

Yokoyama, R., Shirasawa, M., Pike, R.J., 2002. Visualizing topography by openness: a new application of image processing to digital elevation models. Photogramm. Eng. Remote Sens. 68, 257–265.

Zhou, Q., Lees, B., Tang, G., 2008. Advances in Digital Terrain Analysis. Lecture Notes in Geoinformation and Cartography. Springer-Verlag, Berlin, 456 pp.

SECTION 3

Case Studies

CHAPTER ELEVEN

Mapping Ireland's Glaciated Continental Margin Using Marine Geophysical Data

Paul Dunlop[a], Fabio Sacchetti[a], Sara Benetti[a] and Colm O'Cofaigh[b]
[a]School of Environmental Sciences, University of Ulster, Coleraine, Northern Ireland
[b]Department of Geography, Durham University, Durham, UK

Contents

1. INTRODUCTION

The continental shelf northwest of Ireland is located in the southern regions of the glaciated continental margin of Northwest Europe where glacial processes are known to have heavily influenced the evolution of the shelf (Weaver et al., 2000; Sejrup et al., 2005). Since ice sheets are climatically controlled systems (Imbrie et al., 1993), research on former ice sheet margins, which are sensitive to climatic forcing, can provide important insights into the nature and timing of regional climatic events

Developments in Earth Surface Processes, Volume 15
ISSN: 0928-2025, DOI: 10.1016/B978-0-444-53446-0.00011-2

(McCabe and Clark, 1998). Research conducted on the margins of the former British Irish Ice Sheet (BIIS) has shown that it extended onto the continental shelf at various stages during the Pleistocene (Belderson et al., 1973; Bailey et al., 1974; Fyfe et al., 1993; Stoker et al., 1993; Gordon et al., 1997) meaning important information on how the ice sheet responded to climate forcing remains concealed beneath present sea level. Unravelling the submerged glacial record on the continental shelf is therefore a major research challenge that is essential for developing a spatially consistent understanding of the climatically sensitive BIIS.

1.1 Advances in Marine Remote Sensing Techniques

Although traditional marine surveys using single-beam sonar, seismic and sedimentary coring techniques have been critical for developing our understanding of glaciated continental margins (Stoker et al., 1993; Sejrup et al., 2005), the spatial distribution of the collected data is not always suitable for geomorphological mapping. For example, single-beam sonar or seismic lines taken across the seabed are excellent for showing the cross-sectional morphology of submarine landforms (King et al., 1998). However, geomorphological mapping is best achieved when imagery showing the shape of the entire landform is available. This allows more detailed morphological analysis to be conducted, which aids the production of detailed geomorphological maps (Bradwell et al., 2008; Spagnolo and Clark, 2009). Terrestrial remote sensing data sets, which have national coverage, are routinely used for detailed geomorphological mapping onshore, where large sectors of former ice sheets, or entire ice sheet beds, have been mapped (Boulton and Clark, 1990; Dunlop and Clark, 2006a,b; McCabe and Dunlop, 2006; Clark et al., 2009; Greenwood and Clark, 2009a,b). Here, landforms, such as drumlins, which provide a proxy record of former ice flow, are mapped to help identify the former ice dispersal centres and trajectory pathways of the ice sheet. Moraine systems are mapped in order to help locate the maximum extent of glaciation and the retreat pattern of the ice sheet during deglaciation. This approach has been less developed offshore where the limiting factor is the lack of data suitable for geomorphological mapping. However, the recent development of multibeam swath bathymetry systems allows the seabed to be imaged at resolutions previously unobtainable, and multibeam data are now being used to address this shortfall. Multibeam swath bathymetry systems retrieve depth measurements of the seabed in a large swath that radiates outwards beneath the survey vessel. These data points are then used to create seamless high-resolution

Digital Elevation Models (DEMs) of the seafloor that provide detailed three-dimensional views of the seabed that are suitable for geomorphological mapping. A large range of glacial landforms have now been identified using multibeam data which has led to significant breakthroughs in our understanding of the extent and dynamics of ice sheet activity along formerly glaciated continental margins and the Antarctic shelf (O'Cofaigh et al., 2002; Ottesen et al., 2005; Bradwell et al., 2007). Multibeam surveys therefore offer real potential for mapping the glacial geomorphology of the continental shelf. This potential, however, has been limited by the fact that multibeam surveys are very time consuming to collect in shallow water (due to the restricted swath width) and, as a result, most research programmes are forced to target areas of specific interest rather than undertaking systematic regional-scale surveys of large areas of the continental shelf. As a result, multibeam coverage is patchy and this has limited our ability to perform ice sheet reconstructions over large parts of the continental shelf.

Where multibeam coverage does not exist, other large data sets have been used to great effect. An example is the Olex echo-sounder bathymetry database, which provides imagery of the seabed over large parts of the North Atlantic, Greenland and Canadian margins. Olex is a compilation of bathymetry data collected mainly from fishing vessels and is managed by the Norwegian company Olex AS. The Olex database is a bathymetric DEM with a cell size of 5 m, positional accuracy generally better than 10 m and vertical accuracy of 1 m in water depths >100 m and 0.1 m at depths <100 m which makes it suitable for glacial geomorphological mapping. Recently, it has been used to reconstruct ice sheet events on 80% of the Icelandic shelf (Spagnolo and Clark, 2009) and on the shelf around northern Britain where it has been used to show how the former British and Fennoscandian Ice Sheets were coalescent at the last glacial maximum (*ca.* 30 and 25 ka BP) (Bradwell et al., 2008).

Although large-scale studies such as these are improving our understanding of the morphology of the Northwest European margin, the precise extent and timing of shelf glaciation(s) offshore of Britain and Ireland remains controversial. This is largely due to the fragmentary nature of the evidence in many areas (Bowen et al., 2002; Boulton and Hagdorn, 2006; O'Cofaigh and Evans, 2007; Bradwell et al., 2008) and historically, a lack of data on the shelf around Ireland where surveys investigating glacial processes were traditionally of an exploratory nature (King et al., 1998). This changed in 1999 when under the auspices of the Irish National Seabed Survey (INSS) and Integrated Mapping for the Sustainable

Development of Ireland's Marine Resource (INFOMAR) programmes, the Geological Survey of Ireland and the Irish Marine Institute initiated the first systematic multibeam survey of the seabed within Ireland's territorial waters. To date, the INSS/INFOMAR programmes have acquired and processed over 850,000 km^2 of multibeam bathymetry and backscatter data of the Irish seabed. As one of the most comprehensive surveys of its kind, the INSS/INFOMAR programmes have provided an unprecedented opportunity to conduct detailed mapping of the submerged glacial record in Irish territorial waters.

2. CASE STUDY: MAPPING IRELAND'S GLACIATED CONTINENTAL MARGIN

This chapter presents results from the first systematic investigation of the submerged glacial record on the northwest Irish shelf using data from the INSS/INFOMAR programmes (Dunlop et al., 2010; Benetti et al., 2010; O'Cofaigh et al., in press) (Figure 11.1). This work has been important in establishing direct evidence for the expansion and retreat of the former BIIS across the northwest shelf off the coast of Donegal. The primary acoustic data used for mapping in this study was multibeam echo-sounder-derived bathymetric and backscatter data. The main techniques used to acquire, process and map glacial and glacially related geomorphological features on the Irish shelf are summarised below.

2.1 Acquisition and Processing of Data from the Shelf

The primary acoustic data collected by INSS/INFOMAR[1] on the Irish shelf and in deep water (>200 m water depth) is multibeam echo-sounder bathymetric and backscatter data. The continental shelf off the north and northwest coast of Ireland was surveyed in the period 2002–2008 by the *RV Celtic Explorer* and *RV Celtic Voyager* using a hull-mounted Simrad EM1002S and EM3002D on the *RV Celtic Voyager* and an EM1002 on the *RV Celtic Explorer* with decimetric vertical and horizontal accuracy from 10 to 50 cm according to water depth. The deep water regions were surveyed between 2000 and 2002 by the *RV Bligh* using a hull-mounted Kongsberg-Simrad EM120 multibeam echo-sounder with horizontal and

[1] Bathymetry data from the INSS and INFOMAR projects are freely available for download from the Geological Survey of Ireland archives via their Interactive Web Data Delivery System: https://jetstream.gsi.ie/iwdds/index.html

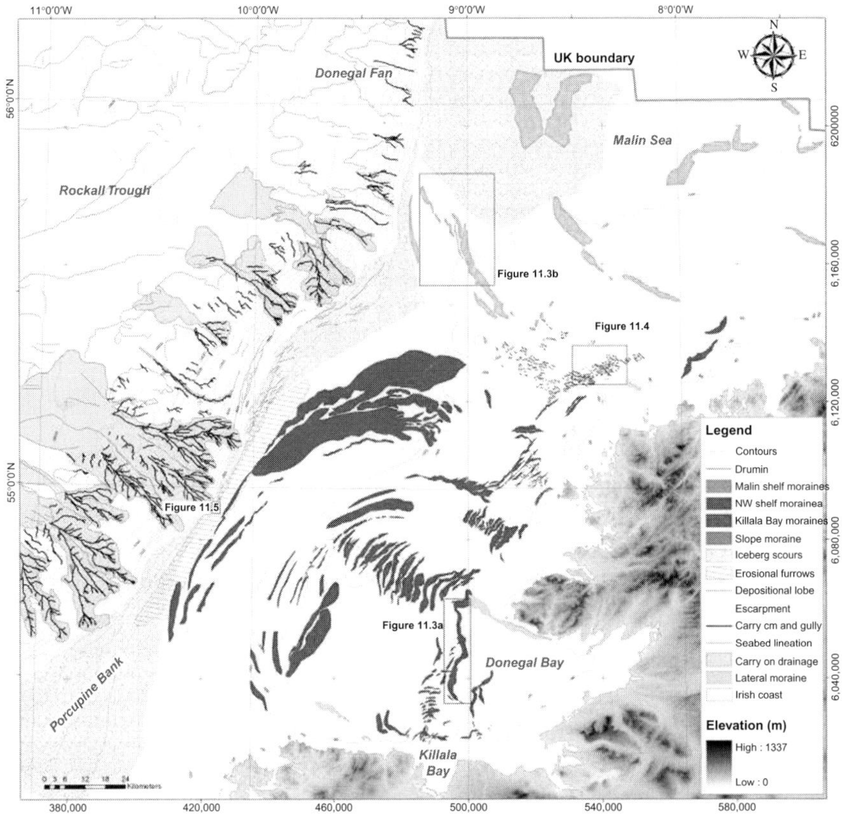

Figure 11.1 Geomorphological interpretation of the continental shelf off northwest Ireland showing all the glacial and glacially related features identified on the INSS/INFOMAR multibeam swath bathymetry data. The location of Figures 11.3, 11.4 and 11.5 are shown by grey boxes on the map.

vertical resolutions of a few metres. The frequency used by these systems varies between 11.75 and 12.75 kHz for the EM120 to 95 kHz for the EM1002S and up to 300 kHz for the EM3002D. On average they can achieve an angular coverage sector of up to six or seven times the water depth. Each collected beam is accurately positioned and fully compensated for pitch, roll and yaw using high-accuracy, differential global positioning system (DGPS) and real-time heading and attitude systems. Bathymetric and backscatter data were processed using CARIS/Hydrographic Information Processing System (HIPS) and CARIS/Sonar Image Processing Software (SIPS) v6.1 and v7.0 from Universal Systems Limited.

Multibeam data collected on the continental shelf (between 10 and 230 m water depth) was tidally corrected using a combination of predicted tides from *Polpred* (Proudman Oceanographic Institute, Continental Shelf Model CS3-30HC) and tide gauge data. Each collected sounding was reduced to Chart Datum (lowest astronomical tide datum) and then gridded at a cell size of 10, 15 or 20 m depending on water depth. Data collected on the continental slope and in deep water was processed to remove erroneous soundings and corrected for vessel motion, navigational jumps and sound velocity refraction artefacts caused by inappropriate sound velocity profiles and then gridded at 30 m.

2.2 Geomorphological Mapping Using Multibeam Bathymetry and Backscatter Data

The main products used for geomorphological mapping are DEMs derived from the multibeam data. These were imported into IVS Fledermaus v6.7 and Erdas Imagine 9.2 and ArcGIS 9.3 for visual analysis and interpretation using shaded renditions of the DEMs. To minimise the problem of azimuth biasing (Clark and Meehan, 2001; Smith et al., 2001), northwest, northeast and non-azimuth-biased renditions were consulted. Geomorphological mapping was achieved by on-screen digitising directly into a GIS using ArcGIS 9.3. Landforms identified on the DEMs were mapped by either digitising polygons along the break of slope for large geomorphic features or digitising the crest line of linear features. ArcInfo tools and IVS 3D Fledermaus profiles were used to derive morphometric attributes of the interpreted bedforms.

Initial mapping of the slope using 50 m and 100 m resolution DEMs identified a series of glacially related features, including canyon systems, depositional lobes and scarps on the northwest Irish continental margin (Benetti et al., 2010; O'Cofaigh et al., in press). For this study, the 30 m resolution DEM was used as it provided a more detailed image of these systems. Shaded relief images of the slope and deep water area were produced as well as gradient, aspect and contour plots. The contour plot of the slope was particularly useful for aiding the interpretation of escarpments related to downslope erosion and mass transport. Escarpments could be easily traced following the 10° slope contour or higher. Escarpment interpretation was validated by taking profiles on the DEM in IVS 3D Fledermaus.

To aid the identification of gullies and canyon systems, an automated approach was adopted using ArcHydro, which is an ArcGIS-based hydrological extension. This was used to perform a hydrological analysis of the

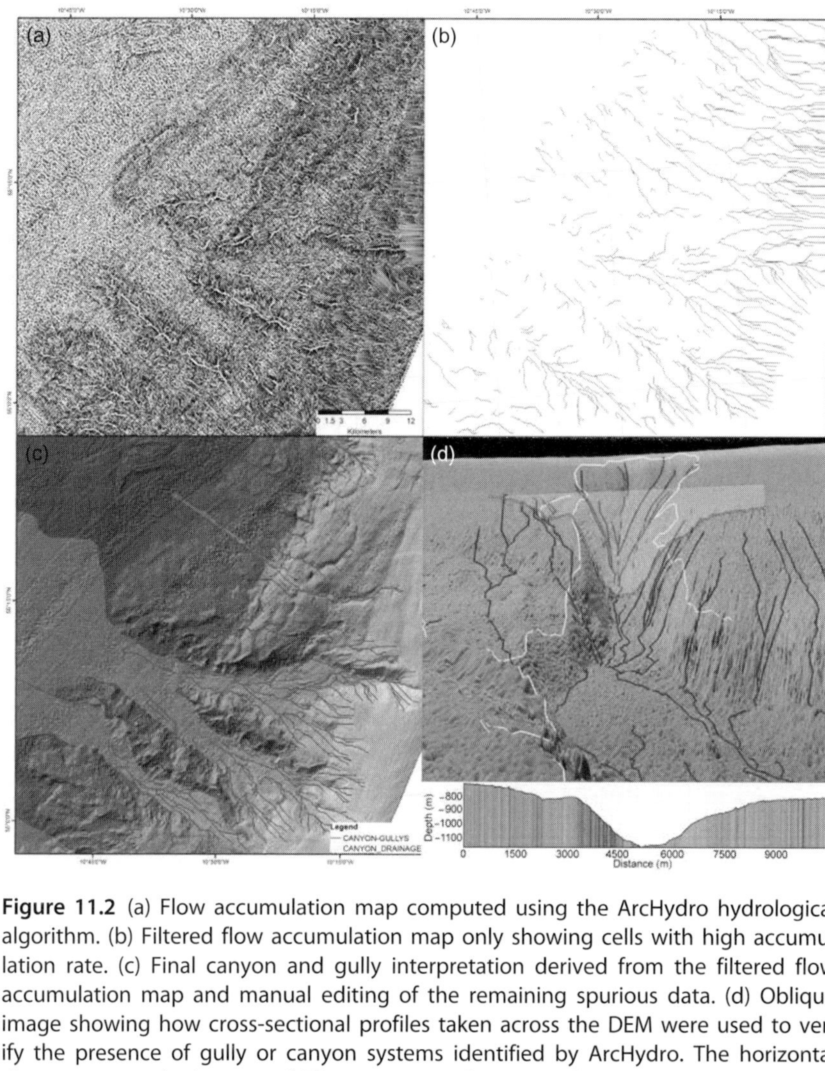

Figure 11.2 (a) Flow accumulation map computed using the ArcHydro hydrological algorithm. (b) Filtered flow accumulation map only showing cells with high accumulation rate. (c) Final canyon and gully interpretation derived from the filtered flow accumulation map and manual editing of the remaining spurious data. (d) Oblique image showing how cross-sectional profiles taken across the DEM were used to verify the presence of gully or canyon systems identified by ArcHydro. The horizontal distance across the bottom of (d) measures 8.5 km. Vertical exaggeration 8.5 ×.

raster grids. Various hydrological surfaces were generated including flow direction, flow accumulation and stream orders, which can aid the identification and classification of gully and canyon systems (Strahler, 1957). The flow accumulation cells were first filtered to remove spurious data and then converted into an ArcGIS shape file that showed the canyon and gully systems identified by the analysis. The shape file was then imported in IVS 3D Fledermaus and draped onto the DEM where its accuracy was assessed (Figure 11.2).

2.3 Backscatter Data

A second important data set that is derived by multibeam systems is acoustic backscatter data. The acoustic backscatter returned to the multibeam sonar is the result of interactions of the acoustic wave front with the seafloor. The acoustic energy returned to the multibeam transducer carries important information about the seafloor morphology and physical properties that can aid geomorphological interpretations of the seabed (de Moustier and Matsumoto, 1993; Hughes-Clarke et al., 1997). Backscatter data were analysed using the Geocoder algorithm, which is a recently developed method implemented in IVS 3D Fledermaus v7.0 and Caris Hips and Sips v7.0 (Fonseca and Calder, 2005; Fonseca et al., 2009). On the continental rise and abyssal plain, seabed features, such as depositional lobes from turbidite or mass transport deposition, tend to lose their topographic expression and so backscatter was used to aid their interpretation.

3. THE GLACIAL GEOMORPHOLOGY OF THE NORTH AND NORTHWEST IRISH SHELF – DESCRIPTION AND INTERPRETATION

A range of glacial and glacially related landforms have now been identified on the North and Northwest Irish shelf. These provide a proxy record for a number of ice sheet processes and events from former ice sheet extent, dynamic ice flow, sediment delivery pathways and ice sheet break-up and retreat during deglaciation. The following sections discuss the geomorphology of the various features that have been mapped on the shelf and used to reconstruct this sector of the BIIS (Dunlop et al., 2010; Benetti et al., 2010; O'Cofaigh et al., in press).

3.1 Submarine Ridges

The most conspicuous geomorphological features in the study area are sequences of large submarine ridges that are oriented in three distinct arrangements across the shelf. By far the most striking sequence is a set of closely nested northeast–southwest-oriented ridges that stretch from the centre of Donegal Bay to the shelf break ~90 km off the northwest coast of Donegal (Northwest Shelf Moraines in Figure 11.1). The two outermost ridges in the sequence extend discontinuously for ~125 km across the shelf break forming a large prominent arc on the outer shelf (Figure 11.1). The outermost ridge is widest in its northwestern section being up to 11 km wide and 14 m high. In the central and southeastern parts it is much narrower and lower in height, being ~0.5 km wide and 6 m high. The ridge behind is similar in size. At its widest part it measures 11 km and is up to 6 m in height. At its western end, it splinters into finger-like ridges that are narrower than the main ridge and range from 0.5–2 km in width and up to 7 km in length (Figure 11.1). Immediately to the south (54° 52′ N 10° 03′ W), the ridge becomes narrower and more fragmented. Individual sections are straight to arcuate in planform and are 2.5–25 km in length, 1–1.5 km in width and 1–2 m in height. The ridges form a closely nested sequence across the shelf towards outer Donegal Bay, 18 km to the southeast. The eastern-most ridges are lobate in form and can be traced for 40–70 km across the shelf. The nested ridges are much narrower than the ridges at shelf break and vary in width from 0.2 to 3.5 km. They are also generally lower in relief, ranging from 1 to 4 m in height. The most prominent ridge in the sequence is a ~35 km long ridge that stretches across the mouth of Donegal Bay between 54° 44′ and 54° 27′ N, and along 9° 2′ W (Figure 11.3a). This ridge ranges from 1 to 2.5 km in width and is up to 15 m in height. Generally in the central and southern parts of the bay, the ridge is sharp crested with a steeper eastward (landward) face and gentler westward side however; further to the north the ridge has a broader, flatter profile. The final ridge in this sequence is located ~5 km to the east of this ridge in the centre of Donegal Bay (54° 28′ N 8° 57′ W). It is 16 km in length, 1 km in width and 3–5 m in height.

To the north in the Malin Sea, a sequence of northeast–southwest-aligned ridges step eastwards across the Malin Shelf from the shelf break (Malin Shelf Moraines in Figure 11.1). The ridges are similar in morphology to the ridges on the northwest shelf; however, they are shorter in length and occur less frequently. The most prominent examples are located in the western sector of the shelf where they form two large discontinuous arcs on the seabed (Figure 11.3b). The most westerly ridges are the most prominent on the seafloor and form a sequence of coalescing sinuous ridges ~45 km long (ridge centre located at 55° 34′ N 8° 59′ W). The most continuous section of the ridge measures 30 km in length, 2.5 km wide and is up to 5 m in height. The smaller sections vary in length

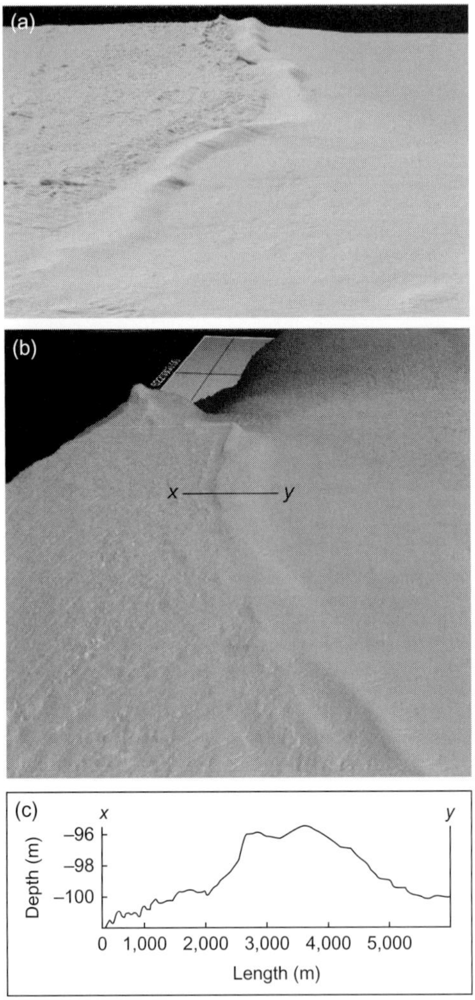

Figure 11.3 Oblique views of the large end moraine ridges located on the shelf northwest of Ireland (see Figure 11.1 for their location). (a) The prominent ridge that runs across the outer reaches of Donegal Bay. The image measures 8.5 km across the bottom. (b) The outermost ridge positioned near the shelf edge in the Malin Sea. The image measures 12 km across the bottom. (c) Cross-sectional profile taken across the ridge shows it has an asymmetric profile that is typical of many of the moraines on the shelf.

and range between 0.4–2.5 km wide and 4–8 m in height. Cross-sectional profiles taken across the ridge show it has an asymmetric profile that is much steeper on its western side (Figure 11.3c).

The third set of ridges extend southwards in a closely nested fashion from the centre of Donegal Bay towards the mouth of Killala Bay ~8 km off the north Mayo coast (Killala Bay Moraines in Figure 11.1). They are straight to arcuate in planform and are 0.7–3 km in length, 0.1–0.7 km in width and 2.5–5 m in height. Most have a strong east–west alignment except a few in the central parts of Donegal Bay which have a slight southeast–northwest orientation. Here they are superimposed on top of the prominent ridge at the mouth of Donegal Bay and three other northeast–southwest-aligned sinuous ridges that occur at the mouth of Killala Bay (54° 21′ N 9° 12′ W) (Figure 11.1) indicating they were deposited after formation of the northeast–southwest-aligned ridges.

Based on their morphology, scale and nested, arcuate pattern, the submarine ridges are interpreted as being recessional moraines that record former ice sheet presence at the shelf edge and then subsequent retreat inshore (O'Cofaigh et al., in press). The pattern of retreat in all cases appears to have been in the form of lobate ice masses that occupied different parts of the continental shelf off the coast of Donegal and North Mayo.

3.2 Streamlined Mounds

At two locations in the study area, swarms of closely nested elongate mounds are clearly visible on the seafloor northwest of Donegal (Figure 11.4 and Figure 11.1). The mounds tend to have a blunt steep face on their eastern side and a tapering lee on the western side, and the long axis tends to be aligned in northwest–southeast orientation which gives the seabed a streamlined appearance (Figure 11.4). The landforms range in height from 1 to 5 m, and statistical measures show they have a mean length of 657 m, mean width of 350 m and mean elongation ratio of 2.8. Statistical measures of their length, width and elongation ratio is consistent with those taken from a large sample of 58,983 of drumlins, which found mean drumlin length to be 629 m, mean width to be 209 m and mean elongation ratio to be 2.9 (Clark et al., 2009). Based on their morphological and morphometric properties, the streamlined mounds are interpreted as being drumlins which are formed subglacially by actively flowing ice and are found commonly in glaciated environments (Fader et al., 1997; O'Cofaigh et al., 2002; McCabe and Dunlop, 2006; Greenwood and Clark, 2009a,b). The drumlins here record northwesterly directed ice flow across the shelf.

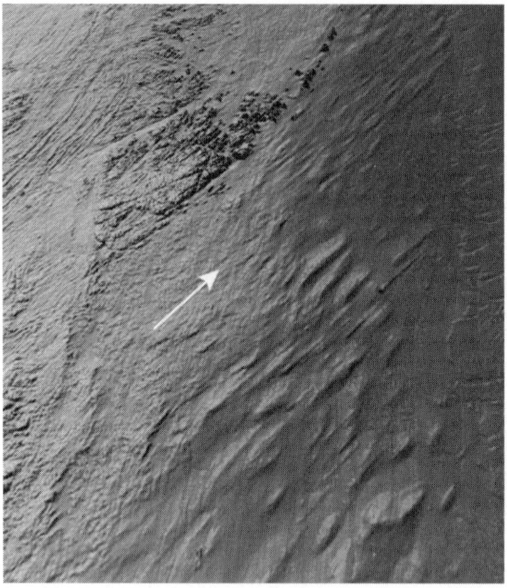

Figure 11.4 Oblique image of a swarm of drumlins 22 km northwest off the coast of Donegal give the seabed a streamlined appearance (see Figure 11.1 for location). They provide a record of northwesterly ice flow across the shelf. The image measures 11 km across the bottom.

3.3 Furrowed Seabed

The outer part of continental shelf and upper slope down to water depths of 500 m are characterised by large furrows that are incised into the seabed (Figures 11.1 and 11.5). The furrows are commonly cross-cutting; however, in the southern part of the study area, they exhibit an overall northeast to southwest orientation. In dimensions, they range from a few hundred metres to 7 km in length, 15 m wide and are 3–4 m deep. Cross-sectional profiles show the furrows have a predominant V-shaped profile with steep slopes that are often flanked by either single or double lateral berms (Figure 11.5). On the basis of their dimensions, form (grooves flanked by lateral berms) and cross-cutting pattern, the furrows on the outer shelf area are interpreted as iceberg scours created by the movement of icebergs grounded on the seafloor during iceberg calving events (*cf.* Belderson et al., 1973; Dowdeswell et al., 1993; Long and Praeg, 1997; O'Cofaigh et al., 2002).

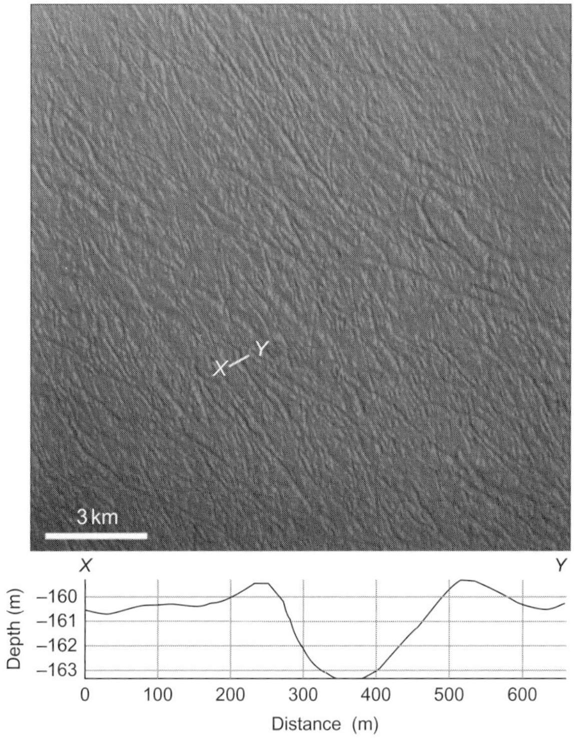

Figure 11.5 Iceberg scours on the outer shelf northwest of Donegal Bay (see Figure 11.1 for their location). In cross section, many iceberg furrows have troughs several metres deep that are flanked by pronounced lateral berms.

4. THE GLACIALLY RELATED GEOMORPHOLOGY OF THE NORTHWEST IRISH CONTINENTAL MARGIN

It is now well established that the delivery of sediments and meltwater from ice sheet margins that were situated near the shelf edge had a controlling influence on the geomorphology of the continental slope, rise and abyssal plain along glaciated continental margins. Here, frequent sediment failures, large trough mouth fans and the development of submarine canyons have been related to ice-marginal processes (Klaucke and Hesse, 1996; Ó Cofaigh et al., 2003; Piper, 2005; Sejrup et al., 2005; Skene and Piper, 2006; Noormets et al., 2009). Since the moraine systems on the northwest Irish shelf record an ice margin at the shelf break (O'Cofaigh

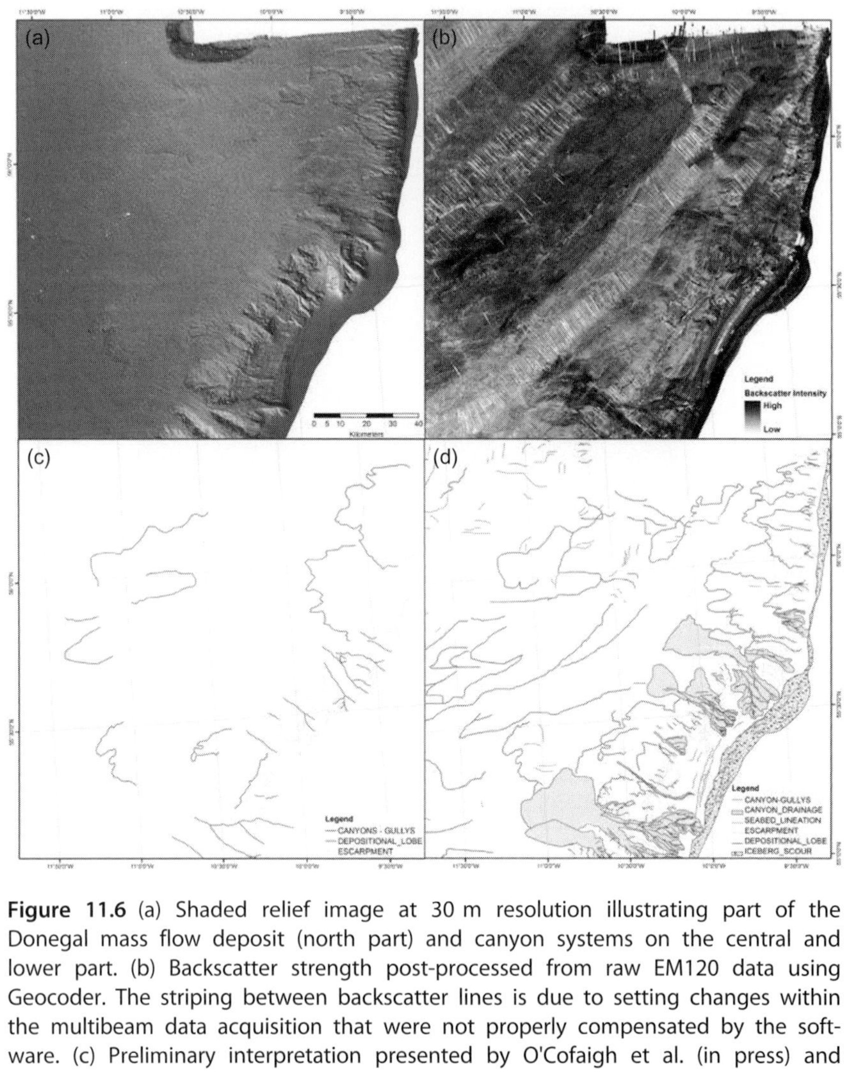

Figure 11.6 (a) Shaded relief image at 30 m resolution illustrating part of the Donegal mass flow deposit (north part) and canyon systems on the central and lower part. (b) Backscatter strength post-processed from raw EM120 data using Geocoder. The striping between backscatter lines is due to setting changes within the multibeam data acquisition that were not properly compensated by the software. (c) Preliminary interpretation presented by O'Cofaigh et al. (in press) and Benetti et al. (2010) based on visual interpretation of multibeam data gridded at 100 m resolution. (d) Improved interpretation of the same study area using the same multibeam raw data set reprocessed with advanced tools and gridded at higher resolution. Both the images and maps shown in (a)–(d) are of the same area and scale.

et al., in press), geomorphological mapping conducted on the outer shelf, slope and abyssal plane is essential for providing information on former ice-marginal processes in this region.

A series of gully and canyon systems are visible on the multibeam data along the continental slope in water depths below 250 m (Figure 11.6;

Elliott et al., 2006; O'Reilly, 2007; O'Cofaigh et al., in press). The gullies have a U-shaped profile and are typically 1–20 km long, 1–3 km wide and are up to 200 m deep. They merge downslope into larger and deeper V-shaped canyons up to 6 km wide and up to 40 m deep. The canyons extend for 10–35 km downslope to water depths of ~2700 m. Improved cleaning and processing of INSS multibeam data (see Sections 2.1 and 2.2) has revealed extensive depositional lobes throughout the study area, frequent escarpments in the upper and mid-slope and well-developed canyon systems south of 55°30′ N (Figures 11.1 and 11.2).

These canyon systems have been related to the incision of the slope by downslope transport during the Pliocene (5.3–2.6 Ma; Elliott et al., 2006), and it is likely that they acted as pathways for ice-sourced meltwater and sediment when the ice margin was sitting near the shelf edge. This downslope redistribution of glacially derived sediment may have contributed to further deepening the large canyon systems observed south of 55°30′ N (*cf.* Cronin et al., 2005; O'Reilly et al., 2007). In other places, glaciomarine debris has prograded onto the slope, as in the upper reaches of the Donegal Fan, and failed downslope as large mass wasting events that resulted in the progradation of large depositional lobes onto the Rockall Trough basin (Figure 11.6). The escarpments situated between canyon systems on the upper and mid-slope are a result of this style of mass transport. O'Cofaigh et al. (in press) have argued that the difference in these slope depositional processes is related to variations in sediment flux related to palaeoglaciology, with a main zone of ice-streaming focused in the area of the Donegal Fan.

5. DISCUSSION AND CONCLUSIONS

Glacial geomorphological mapping using recently acquired multibeam data of the Irish shelf has provided direct evidence that northwest of Ireland the former BIIS was grounded as far as the shelf break. Evidence of ice sheet extension to the shelf edge is indicated by the presence of the large arcuate moraine systems, whose configuration indicates different phases of lobate ice sheet activity on the shelf. The sequence of northeast–southwest moraines offshore of Donegal Bay records the former presence of a large ice lobe, which extended over 80 km from the mouth of Donegal Bay to the shelf edge and was about 120 km across at its widest point. This lobe would have been fed by ice from dispersal centres in the Donegal mountains and the Omagh Basin (McCabe, 2008). The extensive zones of iceberg scours distal to the outermost moraine indicate that the initial retreat from the shelf edge was associated with an episode of ice sheet break-up and calving (O'Cofaigh et al., in press). The pattern of continuous, closely spaced nested arcuate moraines implies episodic, possibly slow, ice-marginal recession across much of the shelf, punctuated by occasional minor readvances or oscillations (*cf.* Shipp et al., 2002; Nygård et al., 2004; O'Cofaigh et al., 2008). The orientation of the moraines in the Malin Sea suggests the ice lobe here was formed by ice that advanced from western Scotland onto the shelf and then subsequently retreated eastwards across the Malin Shelf during deglaciation. The fact that the Killala Bay Moraines are superimposed on top of the large moraine that cuts across the mouth of Donegal Bay implies they are younger than the Donegal Bay moraine system. The most likely explanation is that the Killala Bay Moraines represent a late-stage readvance into the southern part of Donegal Bay from ice centred in County Mayo (O'Cofaigh et al., in press). The well-developed system of gullies and canyons on the slope have been interpreted as recording a former line-sourced sediment supply related to an ice sheet margin that was positioned at or close to the shelf edge (as indicated by the distribution of the arcuate moraines) and would have acted as pathways for the downslope transfer of sediment via turbidity current activity and mass wasting (O'Cofaigh et al., in press).

Currently, no dated sediment cores from the moraines are available to age constrain these ice sheet events. However, O'Cofaigh et al. (in press) have used marine stratigraphic records from the wider northwest margin to argue that ice sheet extension to the shelf edge most likely occurred around 29–27 cal ka BP, prior to the globally defined last glacial maximum and that retreat from this shelf edge occurred after 24 cal ka BP. Although dated sediment cores are required to resolve this key issue, this study clearly demonstrates the contribution that geomorphological mapping can make towards unravelling former ice sheet events on the continental shelf.

ACKNOWLEDGEMENTS

The authors would like to thank the Irish Marine Institute and the Geological Survey of Ireland for providing access to the entire INSS and INFOMAR data sets and for help and support throughout the project.

REFERENCES

Bailey, R.J., Grzywacz, J.M., Buckley, J.S., 1974. Seismic reflection profiles of the continental margin bordering the Rockfall Trough. J. Geol. Soc. Lond. 130, 55–69.

Belderson, R.H., Kenyon, N.H., Wilson, J.B., 1973. Iceberg plough marks in the northeast Atlantic. Palaeogeogr. Palaeoclimatol. Palaeoecol. 13, 215–224.

Benetti, S., Dunlop, P., O'Cofaigh, C., 2010. Glacial and glacially-related features on the continental margin of northwest Ireland mapped from marine geophysical data. J. Maps 2010, 14–29.

Bradwell, T., Stoker, M.S., Golledge, N.R., Wilson, C.K., Merritt, J.W., Long, D., et al., 2008. The northern sector of the last British Ice Sheet: maximum extent and demise. Earth Sci. Rev. 88, 207–226.

Bradwell, T., Stoker, M.S., Larter, R., 2007. Geomorphological signature and flow dynamics of the Mich palaeo-ice stream, northwest Scotland. J. Quat. Sci. 22, 609–617.

Boulton, G.S., Clark, C.D., 1990. A highly mobile Laurentide Ice Sheet revealed by satellite images of glacial lineations. Nature 346, 813–817, doi:10.1038/346813a0.

Boulton, G.S., Hagdorn, M., 2006. Glaciology of the British Isles Ice Sheet during the last glacial cycle: form, flow, streams and lobes. Quat. Sci. Rev. 25, 3359–3390.

Bowen, D.Q., Phillips, F.M., McCabe, A.M., Knutz, P.C., Sykes, G.A., 2002. New data on the Last Glacial Maximum in Great Britain and Ireland. Quat. Sci. Rev. 21, 89–101.

Clark, C.D., Hughes, A.L.C., Greenwood, S.L., Spagnolo, M., Ng, F.S.L., 2009. Size and shape characteristics of drumlins, derived from a large sample, and associated scaling laws. Quat. Sci. Rev. 28, 677–692.

Clark, C.D., Meehan, R.T., 2001. Subglacial bedform geomorphology of the Irish Ice Sheet reveals major configuration changes during growth and decay. J. Quat. Sci. 16, 483–496.

Cronin, B.T., Akhmetzhanov, A.M., Mazzini, A., Akhmanov, G., Ivanov, M., Kenyon, N. H., 2005. Morphology, evolution and fill: implications for sand and mud distribution in filling deep-water canyons and slope channel complexes. Sediment. Geol. 179, 71–97.

de Moustier, C., Matsumoto, H., 1993. Seafloor acoustic remote sensing with multibeam echo-sounders and bathymetric sidescan sonar systems. Mar. Geophys. Res. 15 (1), 27–42.

Dowdeswell, J.A., Villinger, H., Whittington, R.J., Marienfeld, P., 1993. Iceberg scouring in Scoresby Sund and on the East Greenland continental shelf. Mar. Geol. 111, 37–53.

Dunlop, P., Clark, C.D., 2006a. The morphological characteristics of ribbed moraine. Quat. Sci. Rev. 25, 1668–1691.

Dunlop, P., Clark, C.D., 2006b. Distribution of Ribbed Moraine in the Lac Naococane Region, Central Québec, Canada. J. Maps v2006, 59–70.

Dunlop, P., Shannon, R., McCabe, M.A., Quinn, R., Doyle, E., 2010. Marine geophysical evidence for ice sheet extension on the Malin Shelf: new evidence for the western limits of the British Irish Ice Sheet. Mar. Geol. 276, 86–99.

Elliott, G.M., Shannon, P.M., Haughton, P.D.W., Praeg, D., O'Reilly, B., 2006. Mid- to Late Cenozoic canyon development on the eastern margin of the Rockall trough, offshore Ireland. Mar. Geol. 229, 113–132.

Fader, G.B.J., Stea, R.R., Courtney, R.C., 1997. A seabed drumlin field on the inner Scotian Shelf, Canada. In: Davies, T.A., Bell, T., Cooper, A.G., Josenhans, H., Polyak, L., Solheim, A., et al.,Glaciated Continental Margins: An Atlas of Acoustic Images. Chapman and Hall, London, pp. 50–51.

Fonseca, L., Brown, C.J., Calder, B., Mayer, L., Rzhanov, Y., 2009. Angular range analysis of acoustic themes from Stanton Banks Ireland: a link between visual interpretation and multibeam echosounder angular signatures. Appl. Acoust. 70, 1298–1304.

Fonseca, L., Calder, B., 2005. Geocoder: an efficient backscatter map constructor. Proceedings of the U.S. Hydrographic Conference 2005, San Diego.

Fyfe, J.A., Long, D., Evans, D., 1993. The Geology of the Malin-Hebrides Sea Area, United Kingdom Offshore Regional Report. British Geological Survey, HMSO, London.

Gordon, B.J., Rudolph, R.S., Courtney, R.C., 1997. A seabed drumlin field on the inner Scotian shelf, Canada. In: Davies, T.A., Bell, T., Cooper, A.G., Josenhans, H., Polyak, L., Solheim, A., et al. (Eds.), Glaciated Continental Margins: An Atlas of Acoustic Images. Chapman and Hall, London, pp. 50–51.

Greenwood, S.L., Clark, C.D., 2009a. Reconstructing the last Irish Ice Sheet 1: changing flow geometries and ice flow dynamics deciphered from the glacial landform record. Quat. Sci. Rev. 28, 3085–3100.

Greenwood, S.L., Clark, C.D., 2009b. Reconstructing the last Irish Ice Sheet 2: a geomorphologically-driven model of ice sheet growth, retreat and dynamics. Quat. Sci. Rev. 28, 3101–3123.

Hughes-Clarke, J., Danforth, B.W., Valentine, P.C., 1997. Areal seabed classification using backscatter angular response at 95kHz. SACLANTCEN Conference Proceeding CP-45, Lerici.

Imbrie, J., Berger, A., Boyle, W.A., Clemens, S.C., Duffy, A., Howard, W.A., et al., 1993. On the structure and origins of major glaciation cycles 2. The 100,000 year cycle. Palaeoceanography 8, 699–735.

King, E.L., Haflidason, H., Sejrup, H.P., Austin, W.E.N., Duffey, M., Helland, H., et al., 1998. End moraines on the northwest Irish continental shelf. Third ENAM II Workshop, Edinburgh, 1998 (abstract volume).

Klaucke, I., Hesse, R., 1996. Fluvial features in the deep-sea: new insights from the glacigenic submarine drainage system of the Northwest Atlantic Mid-Ocean Channel in the Labrador Sea. Sediment. Geol. 106 (3–4), 223–234.

Long, D., Praeg, D., 1997. Buried ice scours: 2D vs. 3D-seismic geomorphology. In: Davies, T.A., Bell, T., Cooper, A.K., Josenhans, H., Polyak, L., Solheim, A., Stoker, M.S., Stravers, J.A. (Eds.), Glaciated Continental Margins: An Atlas of Acoustic Images. Chapman and Hall, London, pp. 142–143.

McCabe, A.M., 2008. Glacial Geology and Geomorphology: The Landscapes of Ireland. Dunedin Academic Press, Edinburgh, 274 pp.

McCabe, A.M., Clark, P.U., 1998. Ice-sheet variability around the North Atlantic Ocean during the last deglaciation. Nature 392, 373–377.

McCabe, A.M, Dunlop, P., 2006. The Last Glacial Termination in Northern Ireland. Geological Survey of Northern Ireland, Belfast, 93 pp.

Noormets, R., Dowdeswell, J.A., Larter, R.D., Ó Cofaigh, C., Evans, J., 2009. Morphology of the upper continental slope in the Bellingshausen and Amundsen Seas – implications for sedimentary processes at the shelf edge of West Antarctica. Mar. Geol. 258, 100–114.

Nygård, A., Sejrup, H.P., Haflidason, H., Cecchi, M., Ottesen, D., 2004. Deglaciation history of the southwestern Fennoscandian Ice Sheet between 15 and 13 14C ka. Boreas 33, 1–17.

O'Cofaigh, C., Dunlop, P., Benetti, S., in press. Marine geophysical evidence for Late Pleistocene ice sheet extent and recession off northwest Ireland. Quat. Sci. Rev. doi:10.1016/j.quascirev.2010.02.005.

O'Cofaigh, C., Dowdeswell, J.A., Evans, J., Larter, R.D., 2008. Geological constraints on Antarctic palaeo-ice stream retreat. Earth Surf. Process. Landforms 33, 513–525.

O'Cofaigh, C., Evans, D.J.A., 2007. Radiocarbon constraints on the age of the maximum advance of the British–Irish Ice Sheet in the Celtic Sea. Quat. Sci. Rev. 26, 1197–1203.

O'Cofaigh, C., Pudsey, C.J., Dowdeswell, J.A., Morris, P., 2002. Evolution of subglacial bedforms along a palaeo ice stream, Antarctic Peninsula continental shelf. Geophys. Res. Lett. 29, 1119–1136.

O'Cofaigh, C., Taylor, J., Dowdeswell, J.A., Pudsey, C.J., 2003. Palaeo-ice streams, trough mouth fans and high-latitude continental slope sedimentation. Boreas 32, 37–55.

O'Reilly, P.M., Shannon, B.M., Readman, P.W., 2007. Shelf to slope sedimentation processes and the impact of Plio–Pleistocene glaciations in the northeast Atlantic, west of Ireland. Mar. Geol. 238, 21–44.

Ottesen, D., Dowdeswell, J.A., Rise, L., 2005. Submarine landforms and the reconstruction of fast-flowing ice streams within a large Quaternary ice sheet: the 2500-km-long Norwegian-Svalbard margin (57°–80°N). Geol. Soc. Am. Bull. 117 (7–8), 1033–1050, doi:10.1130/B25577.1.

Piper, D.J.W., 2005. Late Cenozoic evolution of the continental margin of eastern Canada. Norw. J. Geol. 85 (4), 305–318.

Sejrup, H.P., Hjelstuen, B.O., Dahlgren, K.I.T., Haflidason, H., Kuijpers, A., Nygård, A., et al., 2005. Pleistocene glacial history of the NW European continental margin. Mar. Pet. Geol. 22, 1111–1129.

Shipp, S.S., Wellner, J.S., Anderson, J.B., 2002. Retreat signature of a polar ice stream: sub-glacial geomorphic features and sediments from the Ross Sea, Antarctica. In: Dowdeswell, J.A., Ó Cofaigh, C. (Eds.), Glacier-Influenced Sedimentation on High-Latitude Continental Margins. Geological Society, London, Special Publication 203, pp. 277–304.

Skene, K.I., Piper, D.J.W., 2006. Late Cenozoic evolution of Laurentian Fan: development of a glacially-fed submarine fan. Mar. Geol. 227 (1–2), 67–92.

Smith, M.J., Clark, C.D., Wise, S.M., 2001. Mapping glacial lineaments from satellite imagery: an assessment of the problems and development of best procedure. Slovak Geol. Mag. 7, 263–271.

Spagnolo, M., Clark, C.D., 2009. A geomorphological overview of glacial landforms on the Icelandic continental shelf. J. Maps v2009, 37–52.

Stoker, M.S., Hitchen, K., Graham, C.C., Abraham, D.A., Ritchie, J.D., 1993. The Geology of the Hebrides and West Shetland Shelves, and Adjacent Deep-Water Areas. United Kingdom Offshore Regional Report, British Geological Survey, HMSO, London.

Strahler, A.N., 1957. Quantitative analysis of watershed geomorphology. Trans. Am. Geophys. Union 8 (6), 913–920.

Weaver, P.P.E., Wynn, R.B., Kenyon, N.H., Evans, J., 2000. Continental margin sedimentation, with special reference to the north-east Atlantic margin. Sedimentology 47, 239–256.

CHAPTER TWELVE

Submarine Geomorphology: Quantitative Methods Illustrated with the Hawaiian Volcanoes

John K. Hillier
Dept. Geography, Loughborough University, Leicestershire, UK, LE11 3TU

Contents

1. INTRODUCTION

Humans cannot directly observe the vast majority of the seafloor as seawater absorbs nearly all light by a depth of a few hundred metres, which is in direct contrast to the sub-aerial landscape; contrast, for example, a pilot looking down from an aircraft and a sailor looking over the side of his craft. Electronic geophysical equipment is therefore necessary to measure the deep ocean at any level beyond the most cursory and crude, which opened a divide; on land there was no reason to strive in the same way for automated measuring equipment capable of giving relatively sparse height measurements.

Analysis of the seafloor has been strongly linked to geophysics. Since the early 1970s, digital elevation models (DEMs), although they were not called that at the time, and quantitative analyses have predominated. First, Section 1.1 describes the historical developments leading to this situation. Section 1.2 then documents the development of digital bathymetric data,

Developments in Earth Surface Processes, Volume 15
ISSN: 0928-2025, DOI: 10.1016/B978-0-444-53446-0.00012-4

culminating in the recent convergence of both submarine and sub-aerial topography data towards high-resolution DEMs. Section 1.3 introduces *Regional–Residual Separation*, leading into the Hawaii case study. Submarine volcanoes are difficult to isolate accurately, so the Hawaiian volcanoes are ideal to illustrate quantitative techniques and indeed have driven their development. Equally, the Hawaiian region is chosen because the mapping and geomorphological quantification of that region's features has been critical to improving our understanding of volcanism and its causes. To conclude, the final section draws together wider opportunities and lessons about quantitative analysis in submarine geomorphology.

1.1 Historical Development

Knowledge of water depth has long been critical for navigation, with depth measured by 'soundings' using a variety of techniques. Initially, sounding used a long pole; from the 1870s hemp rope with a lead weight attached was used (e.g. the steamer *Albatross*) until the early twentieth century when a wire cable method began to be used. These soundings were, however, neither quick nor cheap to obtain, so only a few thousand were available in the deep oceans. Further, substantial progress was not made until 1935 when *USS Ramapo* of the US Navy took some of the first 'sonic soundings', measuring the return time of a sound pulse, at relatively closely spaced intervals (Menard, 1964). So, although many local, detailed sub-aerial observations were made, submarine geomorphology resolved features at scales of hundreds or even thousands of kilometres. Depth data, however, were electronically collected, if not yet electronically stored.

Using sonic soundings, Betz and Hess (1942) produced one of the earliest bathymetry maps of the Pacific, or indeed any ocean, with 'many' soundings (Menard, 1964) i.e. $\gg$10,000. They likened its resolution to a map of North America created from an airplane flown at 200 mph across the continent 100 times taking soundings every 10 mins. Broad features were, however, represented 'with a fair degree of accuracy'; for instance, they noted and named the 'Hawaiian Swell' (e.g. Figures 12.1a and 12.5c). Subsequently, this feature's morphology has been the inspiration for, and is still a critical constraint upon, many theories about the processes driving the vigorous flows within the planet, how the Earth melts and volcanism (Crough, 1978; McKenzie et al., 1980; Wessel, 1993; Ribe and Christensen, 1999; Watts and Zhong, 2002). Similarly, using

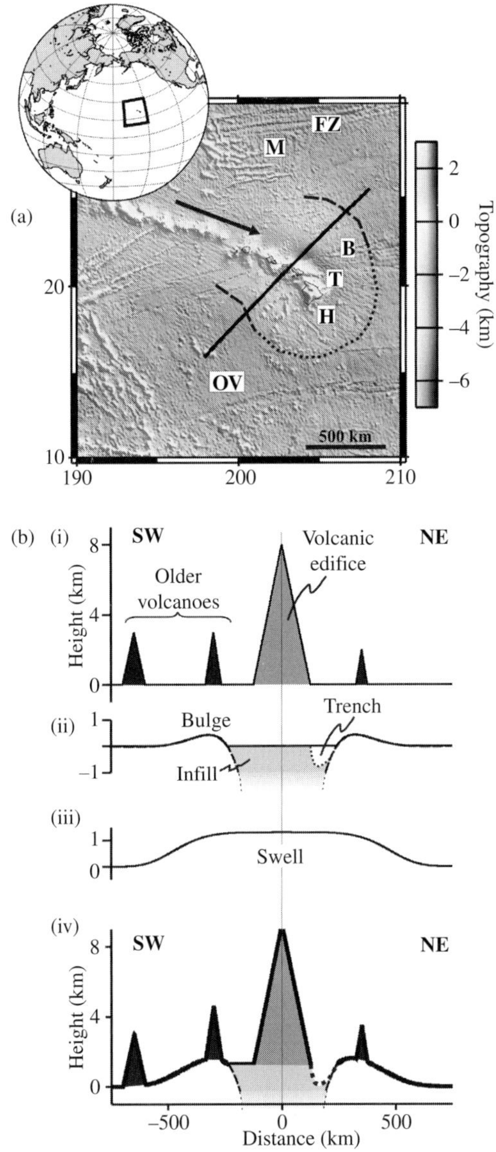

Figure 12.1 (a) 2′ × 2′ relief-shaded topography (Smith and Sandwell, 1997) of the Hawaiian Region as Hillier (2008) located on inset. Thin lines are coastlines. H, Hawaii; T, Trench; B, Bulge; FZ, fracture zone; M, Musicians Seamounts; OV, older volcanoes. Dashed and dotted lines illustrate limit of southeast end of the 'Hawaiian Swell' (Betz and Hess, 1942; Wessel, 1993). Solid line locates profile (Figure 12.4), selected proximal to those of Watts (1978). (b) Schematic illustration of interaction between (i) a volcanic edifice, (ii) seafloor warping due to the volcano's weight and (iii) an ~1000 km wide swell. (i)–(iii) are components of, and sum to, the total bathymetry in (iv). Colour version is available at http://www.appgema.net/.

echo-sound data Menard and Smith (1966) were able to verify the hypsometric curve (i.e. area versus depth) of Murray and Hjort (1912) and discuss modes of formation of the deep oceans. This is a form of analysis also used for sub-aerial (Willgoose and Hancock, 1998; Montgomery et al., 2001) and planetary (Rosenblatt et al., 1994) geomorphology.

Perhaps because of the electronic nature of data collection, a digital global seafloor depth map (or DEM) called SYNBAPS was created in the 1970s (Van Wyckhouse, 1973). Quantitative, and even computational, analyses of the shapes of different classes of feature on the seafloor became common from this time (Menard, 1969; Sclater et al., 1971; Watts, 1976, 1978; Parsons and Sclater, 1977; McKenzie et al., 1980).

1.2 Data Development

Ocean-wide digital bathymetry maps (GEBCO, 2003, ETOPO-5 (NOAA, 1988)) from soundings now contain $\sim 40 \times 10^6$ km of measurements along ship tracks (Hillier and Watts, 2007). Interpolation between these sparse soundings has been improved by using bathymetry predicted, via a derived gravity field, from satellite altimetry (Smith and Sandwell, 1997). Additionally, high-resolution data have been collected (RMBS, 2009). Even though all publicly available data only cover a small fraction (i.e. few percent) of the deep ocean (Smith and Sandwell, 2004), swath bathymetry has allowed the morphology of many other types of smaller feature to be examined; for example seamounts (Smith and Cann, 1992; Scheirer and Macdonald, 1995), abyssal hills (Mitchell and Searle, 1998; Behn et al., 2002) and lava flows (White et al., 2000). Concurrently, this increase in data resolution (commonly to $\ll 100$ m grid cells) has started to permit quantitative geomorphological investigation of features more commonly associated with sub-aerial processes (e.g. landsliding, Mitchell, 2001; Watts and Masson, 2001, and canyon formation). Equally sub-aerially, satellite-derived DEMs (e.g. NASA's SRTM data with 90 m $\times$ 90 m grid spacing) have recently become available, are at similar resolutions and permit similar quantitative analyses of these same processes.

1.3 Regional–Residual Relief Separation

To better understand processes that affect the Earth's surface, landforms (or 'features') are divided into classes or categories, based on some similarity of form, or *a priori* knowledge that they relate to the physical process under investigation. Implicit in this categorisation is the notion that landscapes are composed of, and can be divided into, meaningful classes, or 'components', of features related to a process. If H is height, $H\_DEM = H_1 + H_2 + \cdots + H_n$, where n is the number of components. Regional–residual separation methods (Sclater et al., 1975; Wessel, 1998) aim to divide the landscape, so that all features of interest are completely and uniquely in one component. The features in, and the process represented by, the component may then be more easily quantified, analysed and investigated. Although regional–residual separation allows features to be more clearly seen, by whatever visualisation technique is chosen (see Chapter 8 by Smith), it must be emphasised that such separation is conceptually very different from visualisation alone and has the benefit that numerical analyses can be separately performed for each class of feature even where they are superimposed.

Nature is complex, so regional–residual separation is often difficult. Features within components may be separate or overlapping and may vary in shape and size. For example, submarine volcanoes range from <100 m to 6.7 km in height (Hillier and Watts, 2007). Features in different components may be superimposed on each other, for example 200 m wide drumlins superimposed on 10 km wide rolling hills. As the last example suggests, the skill is identifying a distinctive difference such as 'width-scale' (Hillier and Smith, 2008) between components. Most simply, but effectively, features of interest can be identified manually and a trend extrapolated underneath them (Smith et al., 2009). Usually, the aim is automated and objective determination of the larger scale regional from which the smaller residual features can be derived (McKenzie et al., 1980; Hillier and Smith, 2008), or the direct isolation (i.e. identifying and determining accurate spatial limits for) of all individual features of a certain type within landscapes (Hillier and Watts, 2004) leaving the regional trend.

In short, geophysics has a long history, using various techniques, of dividing the seafloor into components. Techniques vary with the task in hand. For more regional–residual separation techniques, references and background, the reader is referred to Wessel (1998). The case study below illustrates the development of geomorphological techniques as applied to Hawaii and geological insights gained. The older techniques are good first approximations here and remain excellent solutions in some situations (Hillier and Smith, 2008). The desire, however, for more accurate observations (e.g. of height and volume), preferably of many volcanoes and without the possibility of influencing the result through subjective choice separation parameters (e.g. width-scale), has driven the development of methods. In a wider context, these methods have application in many analogous situations.

2. CASE STUDY: HAWAII

Figure 12.1a shows Hawaii, a large volcano of diameter ~250 km in the Pacific Ocean, at the southeast end of the Hawaiian Island Chain. The volcanoes in the chain get progressively younger towards Hawaii [H] (Clouard and Bonneville, 2005), a current 'hot-spot' (Christofferson, 1968) of volcanic activity. Figure 12.1b(iv) is a simplified profile across the chain: Hawaii is dark grey and older, smaller volcanoes are black. Hawaii sits atop an ~1000 km wide 'super-Gaussian' shaped (Wessel, 1993) bulge or swell, also shown in Figure 12.1b(iii). Hawaii is directly surrounded by a trench then, farther away, a bulge. The scars of old fracture zones (large inactive fault systems) are also present (Figure 12.1a).

The paragraph above describes the geomorphology of the seafloor in the region of Hawaii, but what can this usefully tell us about our planet? What do the shapes mean? To interpret this window into the Earth's interior, geomorphological techniques and mapping must be used. The paragraph below illustrates the spectrum of what geomorphology can tell us, and then the case study focuses on the mechanics of isolating volcanoes.

Through isolating the various features and using them as constraints in physical models, we now much better understand the main processes acting to cause this fascinating array of geomorphological features. The swell is probably caused (directly or indirectly) by an upward flowing jet of hot material also feeding the volcanism (Wilson, 1963; Morgan, 1971). The volume of the volcanoes constrains, for example, the temperature of the jet, which limits how vigorously it can push up and how much it can reheat the tectonic plate; both of these processes could cause the swell (Crough, 1978; Ribe and Christensen, 1999). The volcano volumes also record the waxing and waning of this jet through time: its dynamics (White, 1993; Ark and Lin, 2004). Seafloor warping (i.e. bulge and trench) is due to Hawaii's weight (Vening-Meinesz, 1941; Watts, 2001: equation 3.25), so these shapes are indicative of the strength of the tectonic plates. Lastly, the shape of the Hawaiian Swell is a further constraint upon possible origins of both itself and the volcanoes (Wessel, 1993). It is therefore important to isolate geomorphic components in the DEM.

Volcanoes are the most striking component of the DEM but, since key to successful regional–residual separation is identifying a distinctive difference between the component of interest and others, there are a number of difficulties in isolating them. This, however, makes them ideal to illustrate the progression of techniques. The main complications are as follows:

- There are 40±10 thousand of them taller than 1 km (Hillier and Watts, 2007), which strongly favours the use of automated techniques,
- Heights range from <0.1 km to 10 km, and diameters from <1 km to 250 km (Wessel, 2001; Behn et al., 2004), which limits the use of a 'width-scale' or a size-related distinction,
- Approximately conical, volcanoes' shapes vary enormously (Jordan et al., 1983), so there is no distinctive defining shape for pattern matching,
- Volcanoes overlap, are superimposed, and some have been tilted,
- Volcanoes occur on slopes, in trenches, and sometimes in very high spatial densities, which complicates any universal attempt to directly determine a regional.

Initial processes for regional–residual separation involved the definition of a regional larger scale trend from which the "residual" containing small features (i.e. volcanoes) could be calculated. The earliest efforts were manual (Menard, 1973; Sclater et al., 1975). Computationally, early methods used efficient 'frequency domain' filters (Watts and Daly, 1981; Cazenave et al., 1986) to retain long wavelengths (λ), for example $400<\lambda<4000$ km, but these 'are not a good quantitative translation of "regional" in a seamount province' (Smith, 1990). Alternatively, regional depth can be estimated from a 'linear combination' of data within an area, the simplest of which is the mean (Watts, 1976). A mean computed within a window and output to the window's central point is a sliding mean or 'boxcar' filter. Figure 12.2a illustrates the (equal) weights assigned to the data within the window for a mean; variants include using a Gaussian weighting (McKenzie et al., 1980). These work well with an appropriately selected filter width where residual topography is both positive and negative around the regional, and work reasonably where 'normal' flat seafloor overwhelmingly dominates. Figures 12.3a and 12.4a, however, show the computed output on simplified and observed volcano topographies, respectively, which highlight Smith's (1990) point. Because

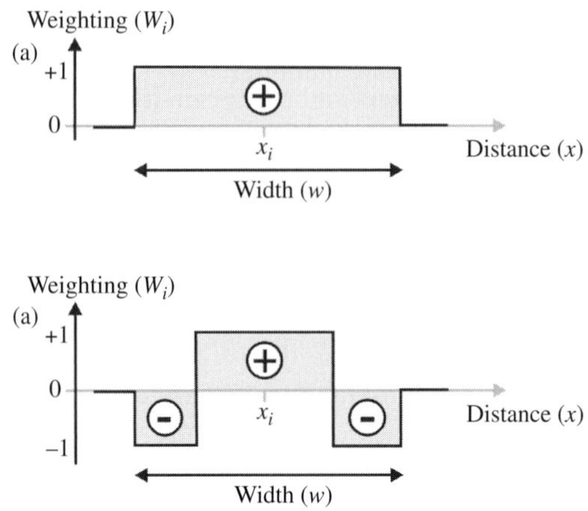

Figure 12.2 Weightings creating (a) sliding mean filter (e.g. GMT; Wessel and Smith, 1998) and (b) SWT (Hillier, 2008).

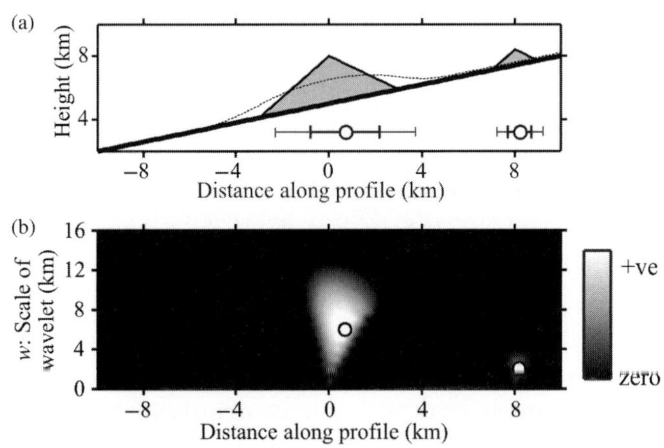

Figure 12.3 Wavelet transform of two synthetic seamounts, one small and one large on a sloping regional bathymetry. (a) Bathymetry profile (thin line) overlies the seamounts (grey shades). White circles outlined in black locate the highest amplitude coefficients in (b); the associated bold horizontal bars indicate the span of the central part (i.e. $x_i \pm w/4$) of the best-fitting wavelets, and the thin bar the whole width (Figure 12.2b). Thick black line is the regional bathymetry (i.e. pre-existing seafloor before seamount was added) as estimated by the SWT method (Hillier, 2008) (see text for details). Thin dotted line is a 6 km wide mean filter. (b) WT of the profile. Coefficients $C_{x,w}$ at each scale w centred on distances x_i along the profile are grey-shaded with large $C_{x,w}$ light coloured. White circles outlined in black are the highest amplitude best-fitting coefficients. SWT, spatial wavelet transform.

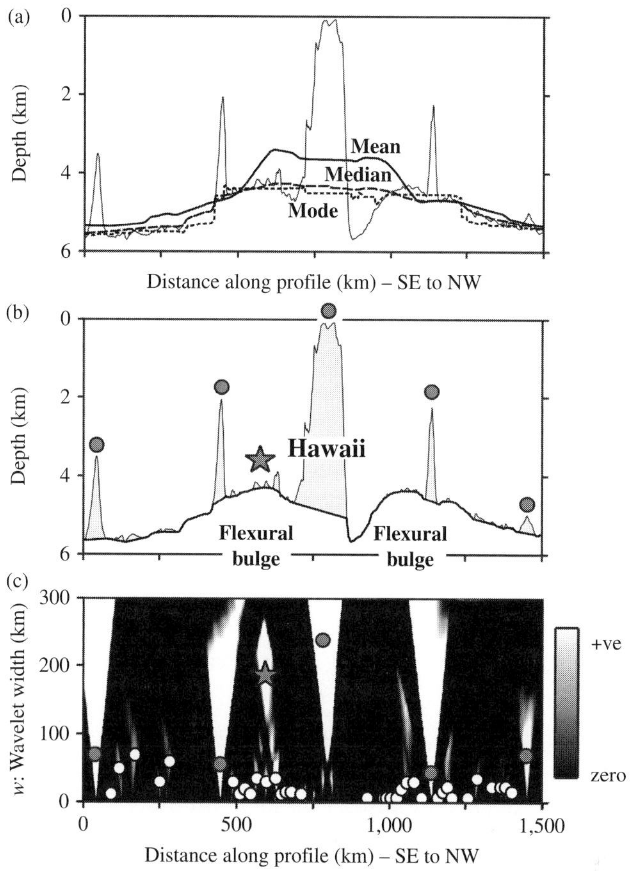

Figure 12.4 Comparison of windowed filters and the SWT method on a bathymetric profile across the Hawaiian Chain, as in Hillier (2008). Profile located on Figure 12.1. (a) Bathymetry profile (thin line) and regional bathymetries estimated by optimal (Wessel, 1998) 480 km wide mean (thick line), median (dashed line) and mode (dotted line) filters. (b) Regional bathymetry estimated by the SWT method (bold line) by extrapolating under-detected seamounts (light grey). WT of the bathymetry profile. Circles are the coefficients best-fitting the seamounts, $w > 20$ km only for clarity. Illustratively, grey circles are linked to seamounts in (b). Another, scale-invariant, MiMIC technique produces very similar results (Hillier and Watts, 2004). Star indicates coefficients relating to the flexural bulge; eliminated and not used to create regional in (b). (c) WT of the profile. Colour version is available at http://www.appgema.net/.

all volcanoes protrude upwards, height is never removed, just spread around. In both cases the width, height and volume of edifices are underestimated, and smaller volcanoes isolated in Figure 12.4b are completely missed. 'Robust' statistics can mitigate this (Smith, 1990).

Robust statistics such as the median and mode ignore 'outliers' (e.g. volcanoes if flat seafloor is the statistical norm) when computed in windows along profiles or in areas. So, such statistics better estimate regional depth in the presence of volcanoes (Smith, 1990; Levitt and Sandwell, 1996) (Figure 12.3a). Consequently, the number and dimensions of the volcanoes themselves are better estimated. Techniques such as identifying the deepest of multiple modes (Crosby et al., 2006), or using directional medians on gridded data (Kim and Wessel, 2008), have improved results further. Problems, however, remain; mainly (i) 'spectral overlap' or the sharing of width-scales with other seafloor features (Wessel, 1998), which prevents simple approaches that can succeed in isolating some superimposed landforms (Hillier and Smith, 2008) and (ii) width-scale selection, which has not been automated as would be necessary for optimally separating each individual edifice in a large population. Additionally, conceptually proceeding down this route, one must consider how *a priori* the existence and location of every edifice is known for a separation to be optimised.

An alternative approach is to directly isolate (i.e. identify and determine accurate spatial limits for), by their own descriptive characteristics, all individual features of a certain type within a landscape. For multi-scale seamounts, the challenge is to isolate them computationally whatever their shape with no *a priori* knowledge of where they are or how big they are. Hillier (2008) details the history of such attempts, criteria for evaluating methods, and proposes a method based on a 'spatial wavelet transform' (SWT).

A 'wavelet', as its name suggests, is a small wave that grows and decays over a limited range of distance (Percival and Walden, 2000). The wave consists of weights to create a weighted average of another function (e.g. a depth profile such as Figure 12.4a) *at a given scale and location*, and can be designed (Figure 12.2b) to be a maximum at the scale and location of a volcano (Figure 12.3b) (Hillier, 2008). Conceptually, the wavelet in Figure 12.2b is simply the difference in mean heights on the central and edge sections of the wavelet. Once wavelet coefficients are computed across a grid of distances and scales (Stage 1), the largest 'best-fitting' one can be found giving a location and scale for each feature (Figure 12.4c). Then, completing Stage 2, other features such as the flexural bulge (star) are distinguished in size-scale space as they are much less tall than volcanoes at that width-scale. Finally, in Stage 3, using the derived widths and locations and in conjunction with the original bathymetry, the limits of the volcanoes are determined. Figure 12.4c clearly shows that the SWT method works accurately for the whole size-range of volcanoes, even in the presence of several other topographic components.

Figure 12.5 shows the application of the SWT method to gridded data, isolating Hawaii from its swell. Currently, the SWT method is only coded to work along profiles, so this result is produced using a mesh of profiles across the grid. Scatter, produced by differences between profiles, is random about the desired regional, so a 100 km wide median filter was applied before computing components for display. The components in Figure 12.5b and c sum together to make the DEM in Figure 12.1a, and the success of the separation can be judged by the absence of seamounts in Figure 12.5c and the absence of swell in Figure 12.5b.

The three stages used by Hillier (2008) provide an analytical framework and are entirely interchangeable with any equivalent preferred by a future investigator. For instance, the 'spatial wavelet' does not conform to some mathematical strictures usually taken to define wavelets and so has great flexibility. For example, to eliminate large artefacts at the edge of trenches, the coefficient was reduced to zero if either edge was above the centre; this is not easily possible with formal wavelets; however, depending upon the task, one may prefer a readily available wavelet such as a 2D 'Mexican Hat' (Gonzalez-Nuevo et al., 2006). Likewise, rules to distinguish features of interest in the size-scale space of the WT can be customised, as can the translation of this information into outlines of features.

The SWT is also relatively computationally efficient. For the 750 data spaced at 2 km intervals across the Hawaiian chain (Figure 12.4a and b), a 1.33 GHz ibook with 512 MB RAM took 0.8 s to find the 38 features. Analysis of gridded data is considerably slower: the 844 profiles took 22 min 43 s to analyse, compared to 13 min 57 s for the traditional 480 km wide median filter (grdfilter of GMT).

Automated detection and isolation of seamounts has permitted work such as the geophysical dating of ~2000 Pacific volcanoes (Hillier, 2007), and the refinement of the estimate of the global volcano population taller than 1 km from 10–100 thousand to 40±10 thousand (Hillier and Watts, 2007). Additionally, however, the automated isolation permitted a large, self-consistent and morphologically parameterised catalogue of seamounts to be created, which has generated interest in diverse communities such as those studying fisheries and habitats for corals (Tittensor et al., 2009). Similarly, large (e.g. 100,000 entries), self-consistent and parameterised catalogues of features could assist in sub-aerial geomorphology such as the

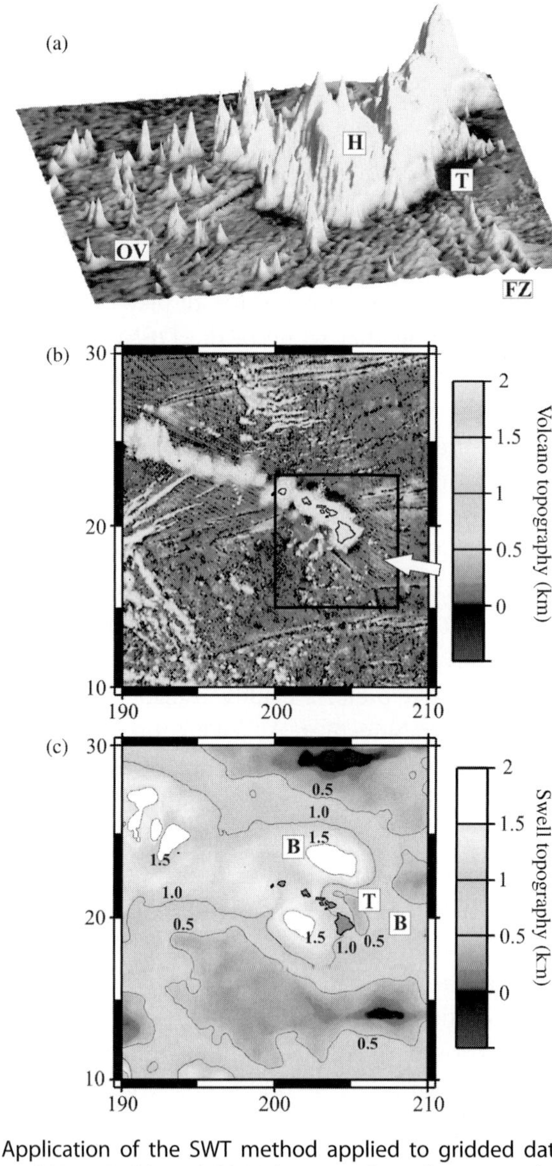

Figure 12.5 Application of the SWT method applied to gridded data (Figure 12.1a) in the region of Hawaii. (b) and (c) Volcano and swell topography above a 6 km deep baseline, respectively. Letters as in Figure 12.1. Coastline shown for reference in (b) and (c), land shaded dark grey in (c). (a) 3D relief-shaded view of estimated volcano component of bathymetry near Hawaii i.e. in box in (b). View from 100/25, white arrow. Relief is coloured as in (b). Note that features within both components are much more evident than in Figure 12.1, and that any desired visualisation technique may now be used on the components. Colour version is available at http://www.appgema.net/.

study of drumlins and their formation processes. These opportunities exist as part of a wider context and are summarised below.

3. DISCUSSION AND CONCLUSIONS

Quantitative analysis of DEMs of the seafloor has existed since the early 1970s, and this case study illustrates some of the *regional–residual separation* techniques used by focusing on the important region around Hawaii. The techniques divide the submarine landscape into 'components' containing features representing processes. Such division is extremely valuable. First, it permits better visualisation and mapping of features, but more importantly the features and thus the processes can be quantitatively analysed (e.g. via volcano volume). The mapping and subsequent analysis of the Hawaiian region's morphology has been a major contributor to our understanding of how, why and when the Earth melts, rises and erupts to the surface.

Always, the skill in performing a successful regional–residual separation is determining a property that makes the features you wish to isolate distinctively different. If, for instance, features of interest are $\sim$100 m wide and have a complex morphology but lie on a gently undulating regional topography with undulations of 1 km in width, then determine the shape of the simple component, the regional: a sliding mean may be appropriate. If, however, both regional and residual topographies have complex and varied morphologies but the features of interest are known to be constructional (i.e. *always* build upwards): a wavelet detecting areas that rise above their surroundings may be appropriate.

By dint of how submarine and sub-aerial topography has to be measured, the available data has differed. Now, however, as data in both environments tend to high-resolution DEMs, a re-merging of submarine and sub-aerial geomorphology is occurring. This presents a great opportunity for exchange of knowledge about processes, technical expertise and thus opportunities for improving our understanding of our planet.

Drawing on work presented in the session on 'Seafloor expression of tectonic and geomorphic processes' at the 2007 EGU General Assembly, Hillier et al. (2008) identify key future opportunities in submarine geomorphology. They promote a vision where submarine geomorphology (i) unites processes typically studied in sub-aerial geomorphology (e.g. landsliding and channel erosion) and marine geophysics (e.g. volcanism, tectonics and geodynamics), (ii) strives to progress beyond qualitative methods and embrace quantitative approaches and (iii) integrates bathymetry with standard marine techniques that readily image sub-surface structures.

Methodological progress will have a significant impact. Several areas for progress are as follows:

- Increasingly quantitative analysis,
- Time-lapse studies, for instance repeated multibeam studies (Smith et al., 2005), to reveal the kinematics and link a feature with its formational process. Possible targets: Coastal erosion by wave cutting; how does material disperse after a sea-cliff fails? How do sedimentary dunes migrate?
- Further integration of bathymetry information with seismic and other techniques that image the sub-surface in a similar way to combining data from multiple satellites.

Four prospective or emergent areas of study highlighted to have great potential are as follows:

- The interaction between submarine erosional or mass wasting processes (e.g. landsliding and turbidity flows) and tectonics, which has proved fruitful on land (Montgomery et al., 2001),
- Quantitative comparison between sub-aerial and submarine analogues, such as slope-area scaling relations and channel morphology (Ramsey et al., 2006), using the same techniques,
- Insight into the way earthquakes create landscapes by examining the evolution of a surface expression through time as recorded in the morphology of surface and sub-surface layers. Reflection seismic data is routinely recorded at sea, whilst fully excavating a structure on land using trenches is rarely done,
- Understanding the processes of landform creation by resolving their interior structures and relationship to bedrock, for example drumlins formed during the last ice age that are now drowned and are visible on OLEX data (Spangolo and Clark, 2009), or possible signatures of catastrophic floods (Gupta et al., 2007).

4. SOFTWARE AND DATA

GMT (Generic Mapping Tools) is an open-source collection of ~60 tools for manipulating Cartesian and geographic data sets, and producing simple $x-y$ scatter plots to artificially illuminated perspective views. GMT is downloaded from http://gmt.soest.hawaii.edu/ (Windows, UNIX, Linux and Mac OSX). Code, c-shell scripts and compilation checks for MiMIC (Hillier and Watts, 2004) and SWT (Hillier, 2008) as used in the production of the papers are, without any guarantee whatsoever. If used or altered, the appropriate reference should be cited. (Available at http://www.appgema.net/).

REFERENCES

Ark, E.V., Lin, J., 2004. Time variation in igneous volume flux of the Hawaii-Emperor hot spot seamount chain. J. Geophys. Res. 109, B11401, doi:10.1029/2003JB002949.

Behn, M.D., Lin, J., Zuber, M.T., 2002. Mechanisms of normal fault development at mid-ocean ridges. J. Geophys. Res. 107 (B4), 2083, doi:10710.1029/2001JB000503.

Behn, M.D., Sinton, J.M., Deitrick, R.S., 2004. Effect of the Galapagos hotspot on seamount volcanism along the Galapagos Spreading Center. Earth Planet. Sci. Lett. 217, 331–347.

Betz, F., Hess, H.H., 1942. The floor of the North Pacific Ocean. Geogr. Rev. 32, 99–116.

Cazenave, A., Dominh, K., Allègre, C.J., Marsh, J.G., 1986. Global relationship between oceanic geoid and topography. J. Geophys. Res. 91, 11439–11450.

Christofferson, E., 1968. The relationship between sea-floor spreading in the Pacific to the origin of the Emperor Seamounts and the Hawaiian Island chain (Abs.). Eos Trans. AGU 49, 214.

Clouard, V., Bonneville, A., 2005. Ages of seamounts, islands, and plateaus on the Pacific plate. In: Foulger, G.R., Natland, J.H., Presnall, D.C., Anderson, D.L. (Eds.), Plates, Plumes and Paradigms. Geological Society of America Special Paper 388, pp. 71–90.

Crosby, A., McKenzie, D., Sclater, J.G., 2006. The relationship between depth, age and gravity in the oceans. Geophys. J. Int. 166, 553–573.

Crough, T.S., 1978. Thermal origin of mid-plate hot-spot swells. Geophys. J. R. Astron. Soc. 55, 451–469.

GEBCO, 2003. 1-Minute grid. <http://www.gebco.net/>.

Gonzalez-Nuevo, J., Argueso, F., Lopez-Caniego, M., Toffoatti, L., Snaz, J.L., Viela, P., et al., 2006. The Mexican hat wavelet family: application to point. Mon. Not. R. Astron. Soc. 369, 1603–1610.

Gupta, S., Collier, J., Palmer-Felgate, A., Potter, G., 2007. Catastrophic flooding origin of shelf valley systems in the English Channel. Nature 448, 342–345.

Hillier, J.K., 2007. Pacific seamount volcanism in space and time. Geophys. J. Int. 168, 877–889, doi:10.1111/j.1365-246X.2006.03250.x.

Hillier, J.K., 2008. Seamount detection and isolation with a modified wavelet transform. Basin Res. 20, 555–573.

Hillier, J.K., Smith, M.J., 2008. Residual relief separation: digital elevation model enhancement for geomorphological mapping. Earth Surf. Process. Landforms 33, 2266–2276, doi:10.1002/esp.1659.

Hillier, J.K., Watts, A.B., 2004. 'Plate-like' subsidence of the East Pacific Rise – South Pacific superswell system. J. Geophys. Res. 109, B10102, doi:10.1029/2004JB003041.

Hillier, J.K., Watts, A.B., 2007. Global distribution of seamounts from ship-track bathymetry data. Geophys. Res. Lett. 34, 1–5, doi:10.1029/2007GL029874.

Hillier, J.K., Tilmann, F., Hovius, N., 2008. Submarine geomorphology: new views on an 'unseen' landscape. Basin Res. 20, 467–472.

Jordan, T.H., Menard, H.W., Smith, D.K., 1983. Density and size distribution of seamounts in the Eastern Pacific inferred from wide-beam sounding data. J. Geophys. Res. 88, 10508–10518.
Kim, S., Wessel, P., 2008. Directional median filtering for the regional–residual separation of bathymetry. Geochem. Geophys. Geosyst. 9, 1–11, doi:10.1029/2007GC001850.
Levitt, D.A., Sandwell, D.T., 1996. Modal depth anomalies from multibeam bathymetry: is there a South Pacific superswell? 139, 1–16Earth Planet. Sci. Lett. 139, 1–16.
McKenzie, D.P., Watts, A.B., Parsons, B., Roufosse, M., 1980. Planform of mantle convection beneath the Pacific Ocean. Nature 288, 442–446.
Menard, H.W., 1964. Marine Geology of the Pacific. McGraw-Hill, New York.
Menard, H.W., 1969. Elevation and subsidence of oceanic crust. Earth Planet. Sci. Lett. 6, 275–284.
Menard, H.W., 1973. Depth anomalies and the bobbing motion of drifting islands. J. Geophys. Res. 78, 5128–5137.
Menard, H.W., Smith, S.M., 1966. Hypsometry of ocean basin provinces. J. Geophys. Res. 71, 4305–4325.
Mitchell, N.C., 2001. Transition from circular to stellate forms of submarine volcanoes. J. Geophys. Res. 106, 1987–2003.
Mitchell, N., Searle, R.C., 1998. Fault scarp statistics and the Galapagos spreading centre from deep tow data. Mar. Geophys. Res. 20, 183–193.
Montgomery, D.R., Balco, G., Willett, S.D., 2001. Climate tectonic and the morphology of the Andes. Geology 29, 579–582.
Morgan, W., 1971. Convection plumes in the lower mantle. Nature 230, 42–43.
Murray, J., Hjort, J., 1912. The Depths of the Ocean: A General Account of the Modern Science of Oceanography Based Largely on Scientific Researches of the Norweigian Steamer Michael Sars in the North Atlantic. Macmillan, London, 821 pp.
NOAA, 1988. Data Announcement 88-MGG-02, Digital Relief of the Surface of the Earth. National Geophysical Data Center, Boulder, CO.
Parsons, B., Sclater, J., 1977. An analysis of the variation of ocean floor bathymetry and heat flow with age. J. Geophys. Res. 82, 803–827.
Percival, D.B., Walden, A.T., 2000. Wavelet Methods for Time Series Analysis. Cambridge University Press, Cambridge, pp. 1–19.
Ramsey, L., Hovius, N., Lauge, D., Liu, C.S., 2006. Topographic characteristic of the submarine Taiwan orogen.. J. Geophys. Res. 1–21, doi:11110.1029/2005JF000314.
Ribe, N.M., Christensen, U.R., 1999. The dynamical origin of Hawaiian volcanism. Earth Planet. Sci. Lett. 171 (4), 517–531.
RMBS, 2009. RIDGE Multibeam Synthesis Project. <http://ocean-ridge.ldeo.columbia.edu>.
Rosenblatt, P., Pinet, P.C., Thouvenot, E., 1994. Comparative hypsometric analysis of Earth and Venus. Geophys. Res. Lett. 21, 465–468.
Scheirer, D.S., Macdonald, K.C., 1995. Near-axis seamounts on the flanks of the East Pacific Rise, 8N to 17N. Mar. Geophys. Res. 100, 2239–2259.
Sclater, J.G., Anderson, R.N., Bell, L.M., 1971. Elevation of ridges and evolution of the Central Eastern Pacific. J. Geophys. Res. 76, 7888–7915.
Sclater, J.G., Lawver, L.A., Parsons, B., 1975. Comparison of long-wavelength residual elevation and free air gravity anomalies in the North Atlantic and possible implications for the thickness of the lithospheric plate. J. Geophys. Res. 80, 1031–1042.
Smith, D.K., Cann, J.R., 1992. The role of seamount volcanism in crustal construction at the Mid-Atlantic Ridge. J. Geophys. Res. 97, 1645–1658.
Smith, D.P., Ruiz, G., Kvitek, R., Iampietro, P.J., 2005. Semiannual patterns of erosion and deposition in the upper Monterey Canyon from serial multibeam bathymetry. Geol. Soc. Am. Bull. 117, 1123–1133.

Smith, W.H.F., 1990. Marine Geophysical Studies of Seamounts in the Pacific Ocean Basin. Columbia University, New York, 216 pp.

Smith, W.H.F., Sandwell, D.T., 1997. Global sea floor topography from satellite altimetry and ship depth soundings. Science 277, 1956–1962.

Smith, W.H.F., Sandwell, D.T., 2004. Conventional bathymetry, bathymetry from space, and geodetic altimetry. Oceanography 17, 8–23.

Smith, M.J., Rose, J., Gousie, M.B., 2009. The cookie cutter: a method for obtaining a quantitative 3D description of glacial bedforms. Geomorphology 108, 209–218.

Spangolo, M., Clark, C., 2009. A geomorphological overview of glacial landforms on the islandic continental shelf. J. Maps 37–52.

Tittensor, D.P., et al., 2009. Predicting global habitat suitability for stony corals on seamounts. J. Biogeogr. 1–18.

Van Wyckhouse, R., 1973. SYNBAPS Technical Report TR-233. National Oceanographic Office, Washington, DC.

Vening-Meinesz, F.A., 1941. Gravity over the Hawaiian Archipelago and over the Madeira area. Proc. K. Ned. Akad. Wet. 44, 1–14.

Watts, A.B., 1976. Gravity and bathymetry in the Central Pacific Ocean. J. Geophys. Res. 81, 1533–1548.

Watts, A.B., 1978. An analysis of isostasy in the world's oceans 1. Hawaiian-Emperor seamount chain. J. Geophys. Res. 83 (B12), 5989–6004.

Watts, A.B., 2001. Isostasy and Flexure of the Lithosphere. Cambridge University Press, Cambridge, 458 pp.

Watts, A.B., Daly, S.F., 1981. Long wavelength gravity and topography. Anomalies Annu. Rev. Earth Planet. Sci. 9, 415–448.

Watts, A.B., Masson, D., 2001. New sonar evidence for recent catastrophic collapses of the north flank of Tenerife, Canary Islands. Bull. Volcanol. 63, 8–19.

Watts, A.B., Zhong, S., 2002. Constraints on the dynamics of mantle plumes from uplift of the Hawaiian Islands. Earth Planet. Sci. Lett. 203, 105–116.

Wessel, P., 1993. Observational constraints on models of the Hawaiian hot spot swell. J. Geophys. Res. 98, 16095-16104.

Wessel, P., 1998. An empirical method for optimal robust regional–residual separation of geophysical data. Math. Geol. 30, 391–408.

Wessel, P., 2001. Global distribution of seamounts inferred from gridded Geosat/ERS-1 altimetry. J. Geophys. Res. 106, 19431–19441.

Wessel, P., Smith, W.H.F., 1998. New, improved version of generic mapping tools released. Eos Trans. Am. Geophys. Union 79, 579.

White, R.S., 1993. Melt production-rates in mantle plumes. Philos. Trans. R. Soc. Lond. A 342, 137–153.

White, S.M., Macdonald, K.C., Haymon, R.M., 2000. Basaltic lava domes, lava lakes, and volcanic segmentation on the East Pacific Rise. J. Geophys. Res. 105, 23519–23536.

Willgoose, G., Hancock, G., 1998. Revisiting the hypsometric curve as an indicator of form and process in transport limited catchments. Earth Surf. Process. Landforms 23, 611–638.

Wilson, J.T.A, 1963. Possible origin of the Hawaiian Islands. Can. J. Phys. 41, 863–870.

CHAPTER THIRTEEN

Marine Geomorphology: Geomorphological Mapping and the Study of Submarine Landslides

Aaron Micallef
University of Malta, Msida, Malta

Contents

1. INTRODUCTION

In the 1960s, the growing interest of the oil industry in the ocean basins triggered an active programme of drilling and surveying of the ocean floor. Since then, a remarkable effort has been made by both

Developments in Earth Surface Processes, Volume 15
ISSN: 0928-2025, DOI: 10.1016/B978-0-444-53446-0.00013-6

industrial and academic institutions to generate a wealth of information about submarine landscapes and to better understand the geologic processes that shaped them. The more geologists have learnt about the ocean floor, the clearer it became that submarine landslides are ubiquitous features of submarine slopes (Canals et al., 2004; Hühnerbach and Masson, 2004). During the last decade, the effort made to characterise submarine landslides along the European margins has significantly increased our understanding of these geological phenomena (Mienert and Weaver, 2002 and references therein; Locat and Mienert, 2003). Submarine landslides have been shown to be common on both active and passive margins, and occur in a wide range of geological settings and depths (Hampton et al., 1996; Mienert et al., 2002). From a geological point of view, submarine landslides are important processes because they are the most effective agents through which sediments are transferred across the continental slope to the abyssal plains (Prior et al., 1982; De Blasio et al., 2004; Masson et al., 2006). Sediment and rock transport in submarine landslides is mainly gravity-driven and occurs on very low slopes (Mulder and Cochonat, 1996). A range of submarine mass movement processes have been recognised, although slides, debris flows and turbidity currents are the most widespread (Weaver et al., 2000; Masson et al., 2006). Submarine landslides can be complex events that vary largely in size and features, depending on the amount of disintegration that the mobilised sediment or rock undergoes. They can be up to three orders of magnitude bigger than the largest known terrestrial landslide (Haflidason et al., 2004; Guthrie and Evans, 2007). Slide deposits can measure hundreds of metres in thickness, and displacement of sediment can be in the range of hundreds of kilometres (Masson et al., 1998; Lykousis et al., 2002; Canals et al., 2004). The triggers of submarine landslides are both intrinsic (e.g. loading and underconsolidation of slope sediments, and the associated development of excess pore pressure; presence of 'weak layers' due to the alternating deposition of sediments with different geotechnical characteristics) and extrinsic (e.g. earthquakes; eustatic changes in sea level; storms; volcanic activity; gas hydrate phase changes) (Locat and Lee, 2002; Masson et al., 2006).

In the last decade, geomorphological mapping has become a fundamental tool in the study of submarine landslides (Lastras et al., 2002; Haflidason et al., 2004). Geomorphological maps constitute the best explanatory presentation of landforms and landscape development in marine geomorphology today. The main uses of geomorphological maps in the study of submarine landslides are (i) to display collected data and their interpretation, (ii) to summarise geological information, (iii) to estimate landslide dimensions, (iv) to identify the main geomorphological processes and reconstruct the evolution of the landscape and (v) to plan future data collection surveys and sampling programmes. The mapping procedure of marine geomorphology varies from that used in the production of terrestrial geomorphological maps for two reasons: (i) submarine landslides are inaccessible, and the data collection and interpretation techniques are different, and (ii) generally, the mapping has been carried out by scientists who are not trained geomorphologists. The objectives of this chapter are (i) to review the methodology adopted by marine geomorphologists to generate geomorphological maps, (ii) to assess the characteristics of marine geomorphological maps, (iii) to demonstrate how geomorphological mapping has improved our understanding of submarine landslide processes and (iv) to suggest ways in which marine geomorphological mapping can be improved. Although the focus of this chapter is on submarine landslides, most of what will be discussed is equally applicable to the study of other marine geological processes.

2. MARINE GEOMORPHOLOGICAL MAPPING METHODOLOGY

2.1 Data Collection

Since direct observation and fieldwork are impossible to undertake when studying submarine landslides, the marine geomorphologist has had to rely on a number of acoustic data acquisition systems and seabed sampling techniques to survey the seafloor. The following are amongst the most popular systems used.

2.1.1 Multibeam Bathymetry

Bathymetry is the study of underwater depth and it constitutes the main source of information on the morphology of the seafloor. Depth is estimated by measuring the time it takes for a beam of sound to travel from a sounder to the seafloor and be reflected back to the sounder. Bathymetric maps are typically compiled using a multibeam echo sounder mounted beneath or over the side of a research vessel. The echo sounder features tens of narrow adjacent beams arranged in a fan-like swath 90°–180°

across. The beams update many times a second, allowing a wide coverage of the seafloor and generating bathymetric maps and digital elevation models (DEMs) that can have a horizontal resolution of just a few centimetres. Nowadays, such high-resolution bathymetric data are collected using remotely operated vehicles (ROVs), which are unoccupied, highly manoeuvrable underwater robots operated by a person aboard a surface vessel, or autonomous underwater vehicles (AUVs), which are robotic devices designed to perform a pre-programmed set of manoeuvres. The state-of-the-art in the use of multibeam bathymetry to map submarine landslides has been discussed at the recent International Conference on Seafloor Mapping for Geohazard Assessment in Ischia, Italy, in May 2009.

2.1.2 Side-Scan Sonar

The side-scan sonar is a category of sonar system that is used to create an image of large areas of the seafloor. The system consists of a sonar device that is towed from a research vessel and emits fan-shaped pulses down towards the seafloor across a wide angle perpendicular to the path of a sensor. The intensity of the acoustic reflections from the seafloor is recorded as a series of cross-track slices that have a nominal resolution of tens of centimetres. Acoustic soundings have been principally used to obtain information about seafloor morphology, although recent studies have shown that they can reveal additional information on surface sediment properties, such as density, water/sediment density ratios, texture, compaction, porosity and benthic vegetation cover (Urick, 1975; Mitchell and Clarke, 1994; Medialdea et al., 2008).

2.1.3 Seismic Reflection Surveying

Seismic reflection is a method of exploration geophysics that provides information about the sub-surface structure of the seafloor. The general principle involves sending artificially generated acoustic waves down the water column and into the seafloor, where the different structures and objects within the Earth's crust reflect this energy back according to their acoustic impedance. These reflected energy waves are recorded by hydrophones, and the data are processed to produce a visual representation of the seabed sub-surface. 2D seismic reflection methods, used for high-resolution sub-bottom profiling, produce individual vertical cross sections, whereas 3D seismic reflection methods reveal the three-dimensional geometry of geological structures down to kilometres depth with a resolution of tens of metres. Ultra-high resolutions for shallower depths can be achieved using Chirp profiling systems.

2.1.4 Geotechnical, Sedimentological and In Situ *Techniques*

Samples of the surface and sub-surface rocks and sediments can be obtained via the use of grabs, cores and geological drillings. The collected rocks and sediment are then analysed in the laboratory using a wide range of techniques to determine their geotechnical and sedimentological properties (e.g. texture, composition, density, water content and shear strength), to ground-truth geophysical interpretations and to perform age analyses. Additional sediment properties and processes, such as excess pore water pressure and seabed displacement, can be measured using *in situ* direct monitoring systems (Strout and Tjelta, 2005).

2.2 Interpretation

2.2.1 Qualitative Interpretation

The approach to analysing the acoustic data from multibeam bathymetry, side-scan sonar and seismic reflection systems has predominantly involved the visual interpretation of charts of contour bathymetry or shaded relief maps, of grey-scale sonographs and of acoustic profiles, respectively (Laberg and Vorren, 2000; Clouard and Bonneville, 2001; Lastras et al., 2002; Lykousis et al., 2002; Imbo et al., 2003). In geohazard risk assessment exercises, the best results have been obtained when different types of data were integrated (Evans et al., 2007; Kvalstad, 2007; Moore et al., 2007). Data are nowadays integrated in a geographic information system (GIS), and the interpretation procedure involves the identification of forms or patterns that may be indicative of a landform or a geological process. This procedure is carried out across several dimensions and depends on the experience of the interpreter. It is therefore both time consuming and subjective.

2.2.2 Quantitative Interpretation

2.2.2.1 Multibeam Bathymetric Data

General geomorphometry refers to the measurement and analysis of those characteristics of landforms that are applicable to any continuous rough surface (Evans, 1980). When general geomorphometric techniques were first applied to terrestrial environments in the 1960s, attempts were made to transfer the techniques directly to the submarine environments.

The results of such studies were, however, limited by the one-dimensionality and low resolution of the bathymetric data available at the time (Krause and Menard, 1965; Neidell, 1966). In the last two decades, significant improvements in acoustic data acquisition techniques have resulted in a renewed interest in geomorphometry and its application in the study of seafloor morphology. General geomorphometric techniques used in the study of submarine landscapes are, however, less numerous and varied than those used in terrestrial geomorphology. The majority of studies have involved either spectral analyses of the bathymetric data or the statistical analysis of morphometric attributes (Fox and Hayes, 1985; Mitchell et al., 2000). This is due to the fact that submarine landscapes and bathymetric data have a number of characteristics that make the application of terrestrial general geomorphometric techniques problematic (Micallef et al., 2007a):

(i) In comparison to terrestrial landscapes, submarine topographies are generally smoother and changes in elevation occur over more extensive areas (Shepard, 1963), which means that the range of morphometric attributes and their statistics, over which changes in the landscape can be observed, are in general much narrower than for terrestrial terrains.

(ii) Submarine DEMs cover larger areas than terrestrial DEMs; they include data sets with different resolutions and they are generally difficult to ground-truth. As a result, general geomorphometric techniques for the study of submarine landscapes need to be more robust than their terrestrial counterparts.

A novel methodology for general geomorphometry for the improved quantitative analysis of submarine elevation data has been recently proposed by Micallef et al. (2007a). The method integrates three main geomorphometric techniques: (1) morphometric attributes and their statistical analyses, (2) feature-based quantitative representation and (3) automated topographic classification. The application of these techniques allows the user to characterise morphology objectively, extract and map morphological features automatically and reveal patterns in the spatial organisation and distribution of these features. The capabilities significantly enhance marine geomorphological mapping based on visual interpretation, as shown by the application of the methodology in the investigation of two aspects of submarine slope instability: spreading (Micallef et al., 2007b) and slide development (Micallef et al., 2009).

2.2.2.2 Side-Scan Sonar Data

The quantitative characterisation of the seabed using acoustic backscatter is an emerging field, and one of the goals of current research is to extend the quantitative interpretation of side-scan sonar data to enable direct extraction of specific seafloor properties such as mean grain size and lithology. Data produced by sonar are, however, nonlinear and the extraction of quantitative information is therefore a difficult process. Several methods have been applied to classify side-scan sonar data and detect features. Some of them are as follows:

(i) Textural analysis (Huvenne et al., 2002),
(ii) Neural networks (Stewart et al., 1994),
(iii) Edge-detecting algorithms (Pratt, 1978),
(iv) Power spectra (Pace and Gao, 1988),
(v) Basic image processing and unsupervised classification techniques (Reed and Hussong, 1989).

In textural analysis, the grey-level distribution of a side-scan sonar image is modelled using second-order statistical techniques, which have been shown to be the most adaptable tools to capture the textural content of an image (Haralick, 1979). A popular way to describe the average spatial relationship between grey levels is by calculating grey-level co-occurrence matrices (GLCMs). GLCMs quantify the frequency of occurrence of two grey levels at specified distances and angles from each other. GLCMs are difficult to interpret, so statistical measures or indices are used to describe them. Out of 25 textural indices available in the current literature, only two indices have been found to be useful for explaining all the variation in texture of high-resolution side-scan sonar images: entropy and textural homogeneity (Blondel, 1996). General textural analysis of side-scan sonar imagery has been incorporated into software called TexAn (Blondel et al., 1998).

Neural networks have also been particularly successful in classifying side-scan sonar images. In comparison to statistical methods, neural networks make no *a priori* assumption about the distribution of input data. Neural networks offer a high tolerance to noise, integrate information from multiple sources and are more efficient than statistical classifiers in terms of parallel processing (Stewart et al., 1994).

2.2.2.3 Seismic Data

Apart from travel time, amplitude, phase and frequency, seismic data can provide additional quantitative information in the form of seismic attributes. Since the 1970s, numerous seismic attributes have emerged from a variety of computational methods in order to predict and map lithology, facies, porosity or fluid type, amongst others. Geometrical attributes, for example, depict spatial and temporal patterns related to bedding geometries, discontinuities such as faults and fracture swarms, bedding similarity, bedding dips and depositional patterns (Taner, 2001).

2.3 Representation

A geomorphological map should ideally contain information about morphometry, lithology, structure, age and process/genesis (Gustavsson et al., 2006). In marine geomorphology this goal has been hard to achieve, and geomorphological maps generally tend to represent a scientific understanding of landscapes and their development that is dictated by mapping scale and the geological characteristics given the highest priority. Combined with the fact that many of the scientists who compile marine geomorphological maps have not been trained in geomorphology, this has resulted in the majority of marine geomorphological maps sharing the following characteristics:

a. The large majority of marine geomorphological maps are thematic and interpretational rather than holistic, scientific maps. The representation of processes and landform genesis has been given the highest priority (Masson, 1996; Laberg et al., 2002; Lastras et al., 2002) (Figure 13.1). The focus is generally on the most recent processes that acted upon the seafloor. The number and type of processes included tends to vary according to selective expert criteria, the scale of the map and the geology of the area,

b. There is no standard theoretical base for the definition and delineation of mapping units, and the segmentation process of the seafloor into landforms or landform elements is subjective,

c. Although the primary information layer of most marine geomorphological maps is bathymetry, morphological information is generally absent or poorly represented. Rarely, bathymetric contours may be provided. When morphological information is included, it is morphographic rather than morphometric (Tricart, 1965), and the description of landform shape is predominantly qualitative and intuitive,

d. Information about lithology and age is generally absent. Obtaining a complete coverage of lithological and chronological data in the submarine environment is challenging because of the limited number of

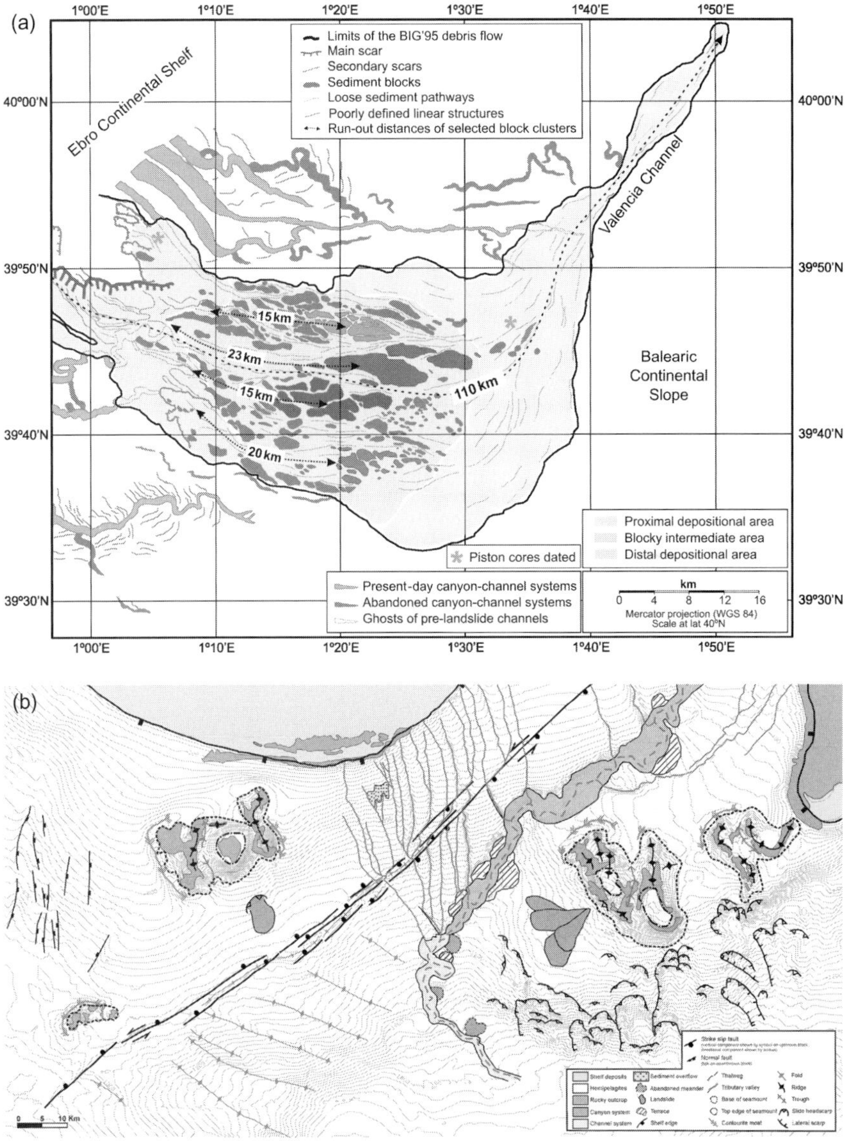

Figure 13.1 (a) Geomorphological map of the BIG'95 debris flow, Ebro continental slope, offshore Spain. (b) Geomorphological map of the Almerian margin, offshore Spain. This is a particularly good example of marine geomorphological map because it combines process interpretation with morphological and structural information. *Part (a) reprinted from Lastras et al. (2002), with permission of The Geological Society of America, and Part (b) reprinted from Lo Iacono et al. (2008), with permission of Elsevier.*

chronological techniques that can be applied, and the fact that the sampling strategy adopted when utilising grabs and cores is sporadic and irregular. Advances being made in correlating side-scan sonar data and seismic attributes with surface sediment properties should provide additional lithological information for marine geomorphological maps,

e. In comparison to terrestrial geomorphological maps, marine geomorphological maps are, in general, less detailed and simpler in their structure. This is attributed to two factors: (i) the scale of marine geomorphological maps varies tremendously (in large-scale maps, for example, certain elements such as slope gradient or small morphological features are difficult to represent) and (ii) the interpretation of acoustic data sets has principally been qualitative, and certain changes in morphology have been difficult to detect,

f. Original data on which the geomorphological interpretation is based are not always shown,

g. The use of landslide terminology is not standardised and is sometimes imprecise,

h. No homogeneous, standardised symbology exists for marine geomorphological maps.

3. EXAMPLE: GEOMORPHOLOGICAL MAPPING AND THE STUDY OF THE STOREGGA SLIDE

3.1 The Storegga Slide

A number of submarine landslides are known to occur along the Norwegian continental margin (Evans et al., 2005). The largest of these is the Storegga Slide, located 120 km offshore Norway (Figure 13.2). The Storegga Slide is 770 km long and has a maximum width of 255 km. It has a total area of 95,000 km^2, with the slide scar comprising 30% of this area (Haflidason et al., 2004). The most recent studies suggest a maximum estimated slide volume of 2400–3200 km^3 (Bryn et al., 2005). The main failure event is dated to *ca.* 7250 $\pm$ 250 ^{14}C year BP (8100 $\pm$ 250 cal. year BP) (Haflidason et al., 2005).

Most of the interest in the Storegga Slide, both industrial and academic, has been linked to the Ormen Lange gas field, Norway's second-largest gas reservoir. Discovered in 1997, Ormen Lange is located within

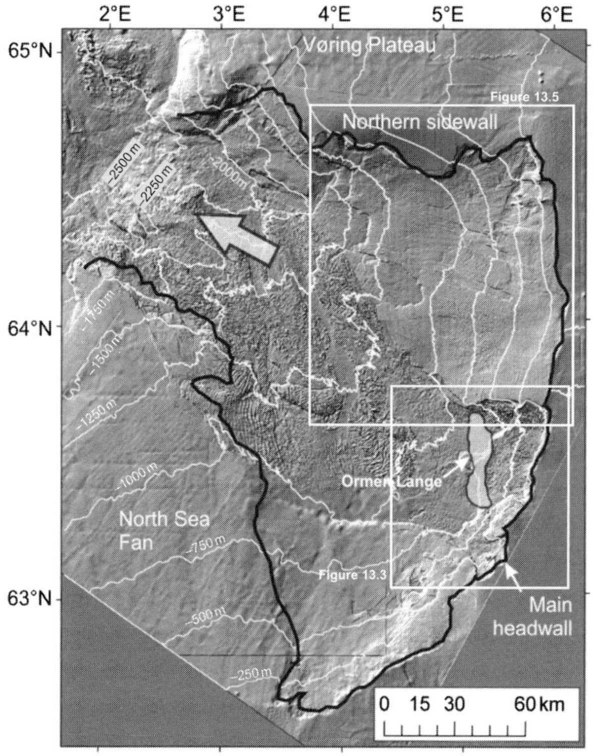

Figure 13.2 Shaded relief map of the Storegga Slide scar. The solid black line indicates the boundary of the Storegga Slide scar. The white lines represent bathymetric contours at 250 m intervals. The block arrow denotes the direction of sediment movement. The location of Figures 13.3 and 13.5 is shown. *From: Norsk Hydro ASA.*

the scar created by the Storegga Slide, some 3000 m below sea level (Solheim et al., 2005). Since the first survey was carried out, state-of-the-art acoustic data acquisition systems and geotechnical techniques have been employed to survey the slide scar seafloor. Today, the database includes multibeam bathymetry images, 2D and 3D seismic data, side-scan sonar imagery, sub-bottom profiles, ROV images, piston cores, gravity cores, geotechnical/geological drillings and *in situ* measurements of excess pore water pressure. The analyses of these data have resulted in numerous academic publications and industrial reports that have addressed various geological aspects of the Storegga Slide (Berg et al., 2005; Kvalstad et al., 2005).

3.2 Application of Geomorphological Mapping

Geomorphological mapping has been employed in the investigation of two important aspects of the Storegga Slide.

3.2.1 *The Assessment of the Risk Associated with the Ormen Lange Gas Field Development and Subsequent Hydrocarbon Extraction (Industrial Application)*

The 'Ormen Lange Project' was initiated by the license partners to secure a safe development of the Ormen Lange gas field. The aim of the project was to understand whether small slides, triggered naturally or induced by human activities, could threaten field installations and generate tsunamis. The approach adopted was a multidisciplinary one, and the role of geomorphological mapping was to classify the slide scar into morphological provinces and present the boundaries of the 63 identified submarine landslides (Haflidason et al., 2004). The geomorphological map of the Ormen Lange region, shown in Figure 13.3, was based on a thorough visual

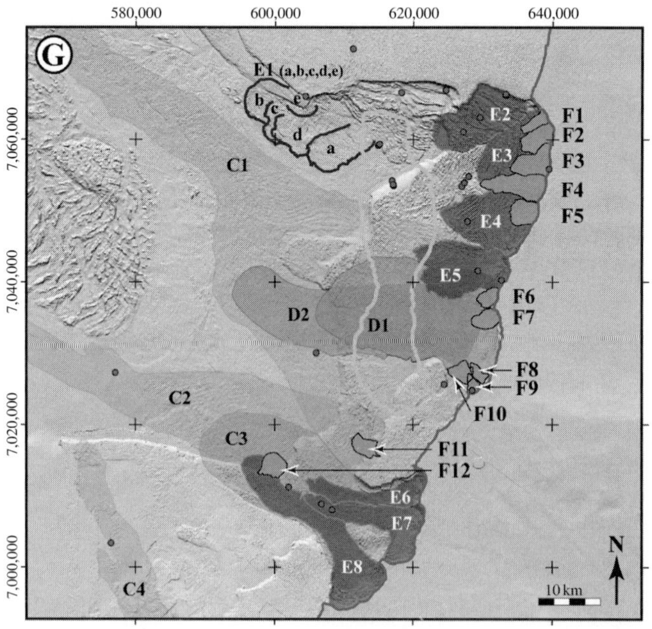

Figure 13.3 Geomorphological map of the Ormen Lange region, Storegga Slide. The zones labelled C, D, E and F correspond to slide lobes, whereas the orange line delimits the Ormen Lange gas field. *Reprinted from Haflidason et al. (2004), with permission of Elsevier.*

interpretation of 3D seismic blocks, 2D seismic profiles, side-scan sonar and multibeam bathymetric data. The map was utilised to estimate the area and volume of individual landslides, plan additional data collection surveys and guide geotechnical sampling. Morphological, geological and geotechnical information provided an input for numerical modelling (e.g. deterministic and probabilistic slope stability modelling) to evaluate slope stability and assess the risk associated with the field development. The analyses showed that only a severe earthquake with a very low probability of occurrence has the potential to renew slide activity within the Ormen Lange region (Nadim et al., 2005). Another important application of geomorphological maps was in identifying the best locations for positioning anchors, pipeline routes and other infrastructure related to the development of the gas field (Figure 13.4).

3.2.2 Improving Our Understanding of the Dynamics of Submarine Landslides (Academic Application)

In Micallef et al. (2009), a suite of geomorphometric techniques have been used to quantitatively analyse bathymetric data from the Storegga Slide. The results were integrated with side-scan sonar and seismic data to generate a detailed geomorphological map of the north-eastern Storegga Slide (Figure 13.5). The geomorphological map was used to characterise submarine landslide processes and to understand their dynamics with the north-eastern Storegga Slide scar. The study demonstrates that the development of this part of the Storegga Slide has not occurred by simple retrogression, but involved a complex interaction of various types of

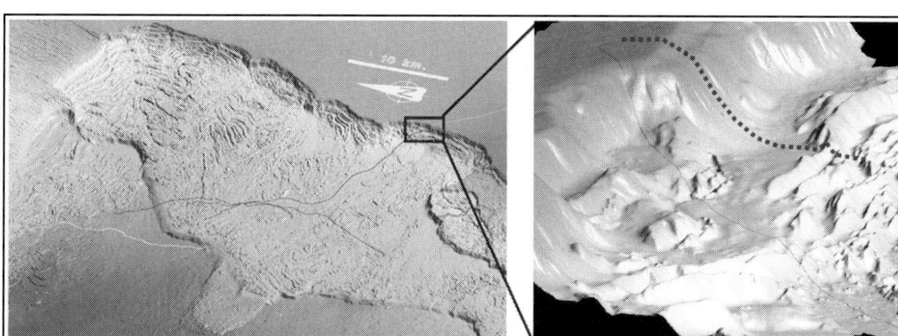

Figure 13.4 Routing of pipelines across the upper headwall of the Storegga Slide, shown on a 3D bathymetric view from the north-west. *Reprinted from Kvalstad et al. (2005), with permission of Elsevier.*

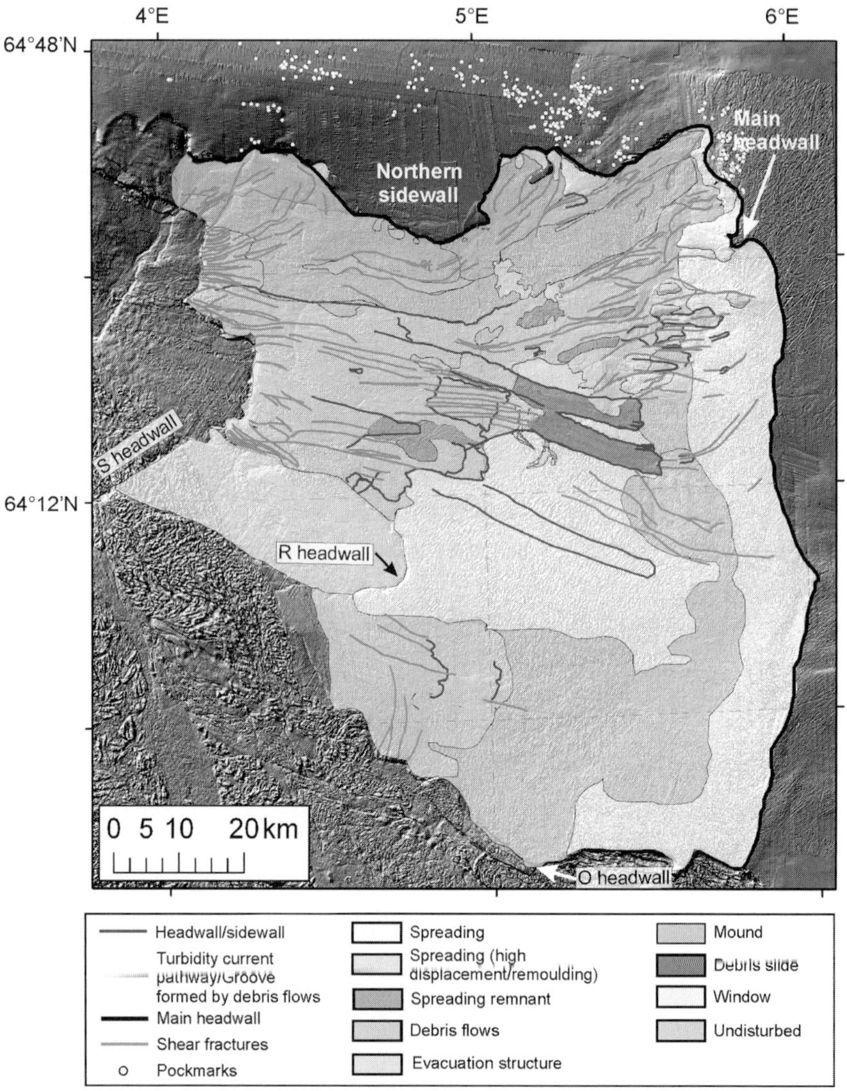

Figure 13.5 Geomorphological map of the mass movements and geological processes that have shaped the north-eastern Storegga Slide scar. *Reprinted from Micallef et al. (2009) with permission of Elsevier.*

submarine landslide processes. In the new development model proposed in the study, spreading, sediment evacuation, debris sliding and compression are shown to play a prominent role (Micallef et al., 2009). The development model also demonstrates how submarine landslides have the potential to evolve between different types, and that the same type of sediment can fail in different ways, according to the nature of sub-surface structures.

4. CONCLUSIONS

Submarine landslides are significant from a geological point of view – they are important processes in terms of their widespread occurrence, their size and the volume of sediment they displace (Canals et al., 2004). Submarine landslides also constitute a geohazard to humans and their infrastructure, not just in terms of rock/sediment mobilisation that can damage seabed infrastructure, but also in the generation of tsunamis (Gracia et al., 2003) and the dissolution/dissociation of gas hydrates, which may contribute to climate change (Kennett et al., 2003).

Despite their geological importance, submarine landslides are still not very well understood, and there are still many gaps in our understanding of submarine mass movement processes. For example, the causes and triggers of slope failures, and the mode of failure and mechanisms that characterise submarine landslides, are poorly constrained (Locat, 2001; Bünz et al., 2005; Masson et al., 2006). This is mainly due to the fact that submarine landslides are largely inaccessible. Since the main type of information available on the seabed is morphological, geomorphological mapping constitutes a fundamental research tool that provides many promising applications in the study of submarine landslides, as demonstrated by its employment in the study of the Storegga Slide. Marine geomorphological maps are, however, characterised by a number of limitations. The following are some suggestions to improve the compilation and representation of geomorphological maps of submarine landslides and other marine geological processes:

a. The geomorphological map should be built into a GIS. Apart from its holistic scientific potential, a GIS allows greater precision and easy transfer into a digital database or mathematical model,

b. The mapping system should be constructed in a way that allows the reader to see the original data upon which the map is based. In this way, the geomorphological map is open to alternative interpretations,

c. The morphologic element of the marine geomorphological map must be of primary importance because the defensible representation of the genetic element depends on an accurate appreciation of morphology. Geomorphological map unit boundaries should follow morphological boundaries (Lee, 2001). In this respect, general geomorphometric techniques are extremely useful because, in comparison to visual interpretation, they allow an easier and more precise identification of natural geomorphic boundaries. Also, an integrated geomorphometric methodology can compensate for the difficulty in validating map interpretations and landscape classification in the field,

d. Geomorphological legends should be standardised and basic mapping units should be strictly defined.

REFERENCES

Berg, K., Solheim, A., Bryn, P., 2005. The Pleistocene to recent geological development of the Ormen Lange area. Mar. Pet. Geol. 22, 45–56.

Blondel, P., 1996. Segmentation of the Mid-Atlantic Ridge south of the Azores, based on acoustic classification of TOBI data. In: MacLeod, C.J., Tyler, P.A., Walker, C.L. (Eds.), Tectonic, Magmatic, Hydrothermal and Biological Segmentation of Mid-Ocean Ridges. Geological Society, London, pp. 17–28. Special Publication 118.

Blondel, P., Parson, L.M., Robigou, V., 1998. TexAn: textural analysis of sidescan sonar imagery and generic seafloor characterisation. Paper Presented at Oceans '98 IEEE/OES Conference Proceedings, Nice, France.

Bryn, P., Berg, K., Forberg, C.F., Solheim, A., Kvalstad, T.J., 2005. Explaining the Storegga Slide. Mar. Pet. Geol. 22, 11–19.

Bünz, S., Mienert, J., Bryn, P., Berg, K., 2005. Fluid flow impact on slope failure from 3D seismic data: a case study in the Storegga Slide. Basin Res. 17, 109–122.

Canals, M., Lastras, G., Urgeles, R., Casamor, J.L., Mienert, J., Cattaneo, A., et al., 2004. Slope failure dynamics and impacts from seafloor and shallow sub-seafloor geophysical data: case studies from the COSTA project. Mar. Geol. 213, 9–72.

Clouard, V., Bonneville, A., 2001. A giant landslide on the southern flank of Tahiti Island, French Polynesia. Geophys. Res. Lett. 29, 2253–2256.

De Blasio, F.V., Elverhøi, A., Issler, D., Harbitz, C.B., Bryn, P., Lien, R., 2004. Flow models of natural debris flows originating from overconsolidated clay materials. Mar. Geol. 213, 439–455.

Evans, I.S., 1980. An integrated system of terrain analysis and slope mapping. Z. Geomorphol. N.F. Suppl. 36, 274–295.

Evans, D., Harrison, Z., Shannon, P.M., Laberg, J.S., Nielsen, T., Ayers, S., et al., 2005. Palaeoslides and other mass failures of Pliocene to Pleistocene age along the glaciated European margin. Mar. Pet. Geol. 22, 1131–1148.

Evans, T., Moore, R., Usher, N., 2007. Management of Geotechnical and Geohazard Risks in the West Nile Delta. Sixth International Conference, Offshore Site Investigation and Geotechniques: Confronting New Challenges and Sharing Knowledge, Society for Underwater Technology, London.

Fox, C.G., Hayes, D.E., 1985. Quantitative methods for analyzing the roughness of the seafloor. Rev. Geophys. 23, 1–48.

Gracia, E., Dañobeita, J.J., PARSIFAL Team, 2003. Mapping active faults offshore Portugal (36°N-38°N): implications for seismic hazard assessment along the southwest Iberian margin. Geology 31, 83–86.

Gustavsson, M., Kolstrup, E., Seijmonsbergens, A.C., 2006. A new symbol-and-GIS based detailed geomorphological mapping system: renewal of a scientific discipline for understanding landscape development. Geomorphology 77, 90–111.

Guthrie, R.H., Evans, S.G., 2007. Work, persistence, and formative events: the geomorphic impact of landslides. Geomorphology 88, 266–275.

Haflidason, H., Sejrup, H.P., Nygård, A., Bryn, P., Lien, R., Forsberg, C.F., et al., 2004. The Storegga Slide: architecture, geometry and slide-development. Mar. Geol. 231, 201–234.

Haflidason, H., Lien, R., Sejrup, H.P., Forsberg, C.F., Bryn, P., 2005. The dating and morphometry of the Storegga Slide. Mar. Pet. Geol. 22, 123–136.

Hampton, M.A., Lee, H.J., Locat, J., 1996. Submarine landslides. Rev. Geophys. 34, 33–59.

Haralick, R.M., 1979. Statistical and structural approaches to texture. Proc. Inst. Electr. Electron. Eng. 67, 786–804.

Hühnerbach, V., Masson, D.G., 2004. Landslides in the North Atlantic and its adjacent seas: an analysis of their morphology, setting and behaviour. Mar. Geol. 213, 343–362.

Huvenne, V.A.I., Blondel, P., Henriet, J.P., 2002. Textural analyses of sidescan sonar imagery from two mound provinces in the Porcupine Seabight. Mar. Geol. 189, 323–341.

Imbo, Y., De Batist, M., Canals, M., Prieto, M.J., Baraza, J., 2003. The Gebra Slide: a submarine slide on the Trinity Peninsula Margin, Antarctica. Mar. Geol. 193, 235–252.

Kennett, J.P., Cannariato, K.G., Hendy, I.L., Behl, R.J., 2003. Methane hydrates in quaternary climate change: the clathrate gun hypothesis. Am. Geophys. Union Spec. Publ. 54, 216.

Krause, D.C., Menard, H.W., 1965. Depth distribution and bathymetric classification of some seafloor profiles. Mar. Geol. 3, 169–193.

Kvalstad, T.J., 2007. What is the current 'best practice' in offshore geohazard investigations? A state-of-the-art review, Offshore Technology Conference, Houston.

Kvalstad, T.J., Nadim, F., Kaynia, A.M., Mokkelbost, K.H., Bryn, P., 2005. Soil conditions and slope stability in the Ormen Lange area. Mar. Pet. Geol. 22, 299–310.

Laberg, J.S., Vorren, T.O., 2000. The Trænadjupet Slide, offshore Norway – morphology, evacuation and triggering mechanisms. Mar. Geol. 171, 95–114.

Laberg, J.S., Vorren, T.O., Mienert, J., Evans, D., Lindberg, B., Ottesen, D., et al., 2002. Late Quaternary palaeoenvironment and chronology in the Trænadjupet Slide area offshore Norway. Mar. Geol. 188, 35–60.

Lastras, G., Canals, M., Hughes-Clarke, J.E., Moreno, A., De Batist, M., Masson, D.G., et al., 2002. Seafloor imagery from the BIG'95 debris flow, western Mediterranean. Geology 30, 871–874.

Lee, E.M., 2001. Geomorphological mapping. In: Griffiths, J.S. (Ed.), Land Surface Evaluation for Engineering Practice. Geological Society, London, pp. 53–56. Special Publication 18.

Locat, J., 2001. Instabilities along ocean margins: a geomorphological and geotechnical perspective. Mar. Pet. Geol. 18, 503–512.

Locat, J., Lee, H.J., 2002. Submarine landslides: advances and challenges. Can. Geotech. J. 39, 191–212.

Locat, J., Mienert, J. (Eds.), 2003. Dordrecht. Kluwer Academic Publishers, 540 pp.

Lo Iacono, C., Gràcia, E., Diez, S., Bozzano, G., Moreno, X., Dañobeitia, J., et al., 2008. Seafloor characterization and backscatter variability of the Almería Margin (Alboran Sea, SW Mediterranean) based on high-resolution acoustic data. Mar. Geol. 250, 1–18.

Lykousis, V., Roussakis, G., Alexandri, M., Pavlakis, P., Papoulia, I., 2002. Sliding and regional slope stability in active margins: North Aegean Trough (Mediterranean). Mar. Geol. 18, 281–298.

Masson, D.G., 1996. Catastrophic collapse of the volcanic island of Hierro 15 ka ago and the history of landslides in the Canary Islands. Geology 24 (3), 231–234.

Masson, D.G., Canals, M., Alonso, B., Urgeles, R., Huhnerbach, V., 1998. The Canary Debris Flow: source area morphology and failure mechanisms. Sedimentology 45, 411–432.

Masson, D.G., Harbitz, C.B., Wynn, R.B., Pedersen, G., Løvholt, F., 2006. Submarine landslides: processes, triggers and hazard prediction. Philos. Trans. R. Soc. 364, 2009–2039.

Medialdea, T., Somoza, L., León, R., Farrán, M., Ercilla, G., Maestro, A., et al., 2008. Multibeam backscatter as a tool for sea-floor characterization and identification of oil spills in the Galicia Bank. Mar. Geol. 249, 93–107.

Micallef, A., Berndt, C., Masson, D.G., Stow, D.A.V., 2007a. A technique for the morphological characterization of submarine landscapes as exemplified by debris flows of the Storegga Slide. J. Geophys. Res. 112, F02001.

Micallef, A., Masson, D.G., Berndt, C., Stow, D.A.V., 2007b. Morphology and mechanics of submarine spreading: a case study from the Storegga Slide. J. Geophys. Res. 112, F03023.

Micallef, A., Masson, D.G., Berndt, C., Stow, D.A.V., 2009. Development and mass movement processes of the north-eastern Storegga Slide. Quat. Sci. Rev. 28 (5–6), 433–448.

Mienert, J., Weaver, P.P.E. (Eds.), 2002. European Margin Sediment Dynamics: Side-Scan Sonar and Seismic Images. Springer Verlag, Berlin, , 309 pp.

Mienert, J., Berndt, C., Laberg, J.S., Vorren, T.O., 2002. Submarine landslides on continental margins. In: Wefer, G., Billet, D., Hebbeln, D., Jørgensen, B.B., Schlüter, M., van Veering, T. (Eds.), Ocean Margin Systems. Springer Verlag, Heidelberg, pp. 179–193.

Mitchell, N.C., Clarke, J.E.H., 1994. Classification of seafloor geology using multibeam sonar data from the Scotian Shelf. Mar. Geol. 121, 143–160.

Mitchell, N.C., Tivey, M.A., Gente, P., 2000. Seafloor slopes at mid-ocean ridges from submersible observations and implications for interpreting geology from seafloor topography. Earth Planet. Sci. Lett. 183, 543–555.

Moore, R., Usher, N., Evans, T., 2007. Integrated multidisciplinary assessment of West Nile Delta geohazards. Sixth International Conference, Offshore Site Investigation and Geotechniques: Confronting New Challenges and Sharing Knowledge, Society for Underwater Technology, London.

Mulder, R., Cochonat, P., 1996. Classification of offshore mass movements. J. Sediment. Res. 66, 43–57.

Nadim, F., Kvalstad, T.J., Guttormsen, T., 2005. Quantification of risks associated with seabed instability at Ormen Lange. Mar. Pet. Geol. 22 (1–2), 299–310.

Neidell, N.S., 1966. Spectral studies of marine geophysical profiles. Geophysics 31, 122–134.

Pace, N.G., Gao, H., 1988. Swathe seabed classification. IEEE J. Oceanic Eng. 13, 83–90.

Pratt, W.K., 1978. Digital Image Processing. Wiley, Chichester.

Prior, D.B., Coleman, J.M., Bornhold, B.D., 1982. Results of a known seafloor instability event. Geo. Mar. Lett. 2, 117–122.

Reed, T.B., Hussong, D., 1989. Digital image processing techniques for enhancement and classification of SeaMARC I1 side scan sonar imagery. J. Geophys. Res. 94 (B6), 7469–7490.

Shepard, F.P., 1963. Submarine Geology. second ed. Harper and Row, London.
Solheim, A., Bryn, P., Sejrup, H.P., Mienert, J., Berg, K., 2005. Ormen Lange – an integrated study for the safe development of a deep-water gas field within the Storegga Slide Complex, NE Atlantic continental margin; executive summary. Mar. Pet. Geol. 22, 1–9.
Stewart, W.K., Jiang, M., Marran, M., 1994. A neural network approach to classification of sidescan sonar imagery from a mid-ocean ridge. IEEE J. Ocean Eng. 19 (2), 214–224.
Strout, J.M., Tjelta, T.I., 2005. In situ pore pressures: what is their significance and how can they be reliably measured? Mar. Pet. Geol. 22, 275–285.
Taner, M.T., 2001. Seismic attributes, CSEG Recorder, pp. 48–56, September Issue.
Tricart, J., 1965. Principes et Méthodes de la Géomorphologie. Masson & Cie, Paris.
Urick, R.J., 1975. Principles of Underwater Sound. McGraw Hill, New York.
Weaver, P.P.E., Wynn, R.B., Kenyon, N.H., Evans, J., 2000. Continental margin sedimentation, with special reference to the north-east Atlantic margin. Sedimentology 47, 239–256.

CHAPTER FOURTEEN

The Cherry Garden Landslide, Etchinghill Escarpment, Southeast England

James S. Griffiths[a], E. Mark Lee[b], Denys Brunsden[c] and David K.C. Jones[d]

[a]SoGEES, University of Plymouth, Plymouth, UK
[b]York, UK
[c]Chideock near Bridport, Dorset, UK
[d]Beckley near Rye, Sussex, UK

Contents

1. INTRODUCTION

The Channel Tunnel between England and France was investigated, designed and built between 1986 and 1990 (Varley and Warren, 1995). As part of the investigations for the project, geomorphological surveys were undertaken (Birch and Griffiths, 1995), notably at the location of the main UK terminal near Folkestone in Kent (Griffiths et al., 1995). Previous investigations in this area had indicated there were extensive areas of landsliding associated with both large-scale failures from the Etchinghill Escarpment, a scarp developed in Upper Cretaceous rocks and engineering soils (Lower Chalk overlying Gault Clay), and smaller scale solifluction features possibly including cambering and coombe rock

Developments in Earth Surface Processes, Volume 15
ISSN: 0928-2025, DOI: 10.1016/B978-0-444-53446-0.00014-8

deposits (Kerney et al., 1980), relicts of the intensive periglacial activity that occurred in this part of England during the Devensian glacial period (80,000–10,000 year BP). Large-scale (1:500) geomorphological mapping was carried out across the whole of the terminal footprint, the main tunnel portals and in an area of cut-and-cover tunnelling. Of particular interest was the Cherry Garden Coombe area (an embayment in the scarp occupied by a reservoir) where all the rail lines from the terminal came together to form two mainlines (Figures 14.1 and 14.2). At this junction, the rail lines crossed a landslide complex that had previously been identified on 1:10,560 scale British Geological Survey (BGS, 1967) maps and, subsequently, investigated during a previous attempt to get the Channel Tunnel project underway (Aarons et al., 1977). In this chapter, the results from the 1987 large-scale geomorphological mapping of the site are described with comments on the value of the technique when carried out at such large scales. The main aims of the mapping were to define the nature and extent of the landslide and periglacial features in the area and to provide a framework for designing the subsequent detailed ground investigations.

2. SITE TOPOGRAPHY

The site is dominated by the Etchinghill escarpment that reaches a height of 160 m AOD (above Ordnance datum) and is aligned roughly east–west through the area (Figures 14.1 and 14.2). This can be divided into the steep southward facing scarp slope and the undulating plateau of the north-facing dip slope. Indented into the escarpment are several embayments known as coombes. The scarp slope is concave, flattening downslope to form a flat lowland at around 90 m AOD.

3. SITE GEOLOGY

The overall site geology of the area is presented in Table 14.1 and shown in Figure 14.1. The details of the surface geology at the Cherry Garden landslide complex are shown in Figure 14.3. Neither the bedrock formations of Lower Cretaceous Folkestone Beds and Upper Cretaceous

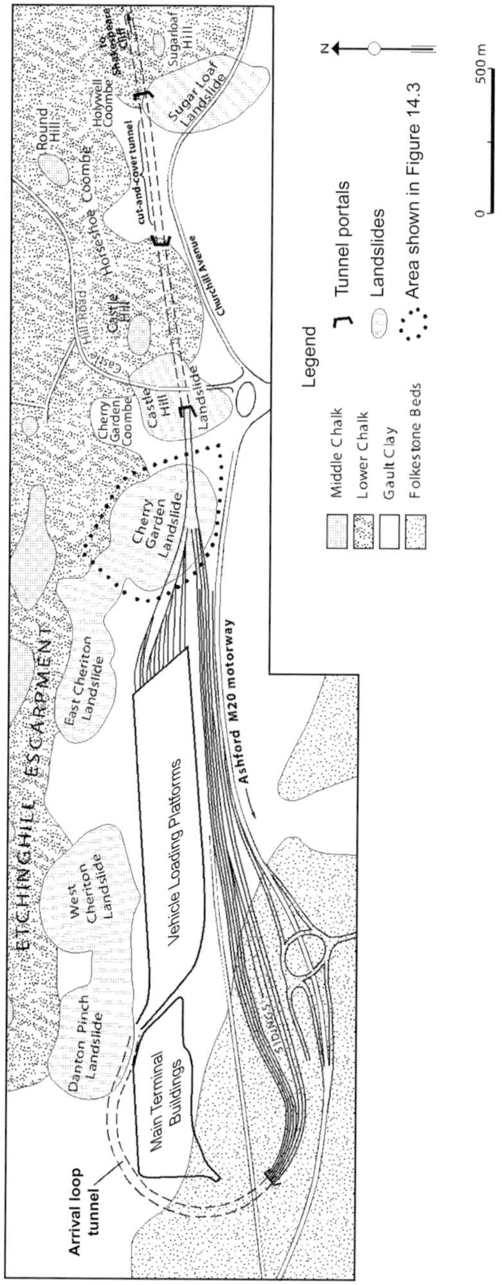

Figure 14.1 General layout and geology of the Channel Tunnel Folkestone Terminal area.

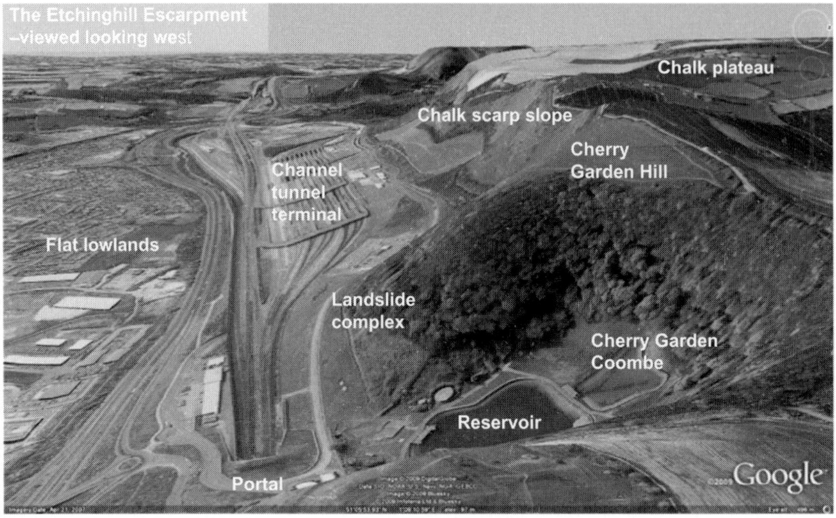

Figure 14.2 Synthetic oblique aerial image of the Channel Tunnel Terminal looking westward. Cherry Garden Coombe and the reservoir are visible on the right-hand side of the image. The Cherry Garden landslide complex occupies the centre of the image from the top of the Etchinghill escarpment down to, and beyond, where the rail lines converge before entering the Castle Hill tunnel portal. *Google Earth copyright.*

Middle Chalk nor the unconsolidated Tertiary and Quaternary deposits had an influence on the development of the main landslide complex, although the stratigraphical relationship between the landslide debris and the Late Glacial coombe rock deposits is important in helping to ascribe a date to the landslide movements.

From the engineering geological and geomorphological perspective, the most effective subdivision of the Cenomanian, Lower Chalk deposits was into the White Chalk and the Chalk Marl devised by the BGS (Smart et al., 1966). Within these broad categories, there was a useful further subdivision of the Lower Chalk Marl into the grey chalk, upper and lower chalk marl, glauconitic marl and a basal transitional bed with the underlying Gault Clay, possibly also of Albian age and known as Zone 6A. Each unit of this subdivision has distinctive geotechnical, structural and lithological characteristics which have had a significant bearing on the development of the landslide complex (Table 14.2).

Table 14.1 General Geology[a] of the Study Area

Geological Period	Deposit	Description
Recent	Colluvium	Chalky silt and clay (hillwash accumulation), generally <1 m
Late- and Early Post-Glacial	Landslides	Chalk and Gault Clay blocks and debris
	Coombe rock	Soliflucted chalky silt and clay debris
	Allerød soil	Thin brown to black organic silt
		Whole sequence generally <3 m thick
Tertiary (undifferentiated)	Clay-with-flints	Sub-angular to sub-rounded flint cobbles and gravel in a matrix of silt and clay; variable thickness but normally <5 m
Upper Cretaceous	Turonian Middle Chalk[a]	Strong, nodular yellowish white Chalk (Melbourn Rock) *ca.* 15 m thick
	Cenomanian Lower Chalk[a]	Top of the sequence identified by the occurrence of the Plenus Marl overlying moderately weak white chalk (the White Chalk), over weak clayey chalk (the Grey Chalk of the upper Chalk Marl) and weak clayey carbonate mudstones (the chalk marl of the lower chalk marl). Overall thickness *ca.* 80 m, with a basal horizon of moderately strong to weak glauconitic sandy marl (Glauconitic Marl) 1–2 m thick. The Glauconitic Marl overlies a transition zone (possibly of Albian Age) of moderately weak calcareous, glauconitic mudstone (labelled Zone 6A)
	Albian Gault Clay	Stiff blue, blue-grey or grey-brown fissured, heavily overconsolidated silty clay becoming a weak calcareous clayey mudstone, *ca.* 40–50 m thick
Lower Cretaceous	Aptian Folkestone Beds	Weakly cemented fine to medium sandstone with lenses (doggers) of strong sandy limestone

[a]The geological chronology and stratigraphy of the Upper Cretaceous Period has undergone significant refinement and updating since the site work was carried out (Harris et al., 1995b; Mortimore et al., 2001; Rawson, 2006). However, this was the system used during the mapping as it was the preferred system in place at the time and had been set up by the BGS (Smart et al., 1966).

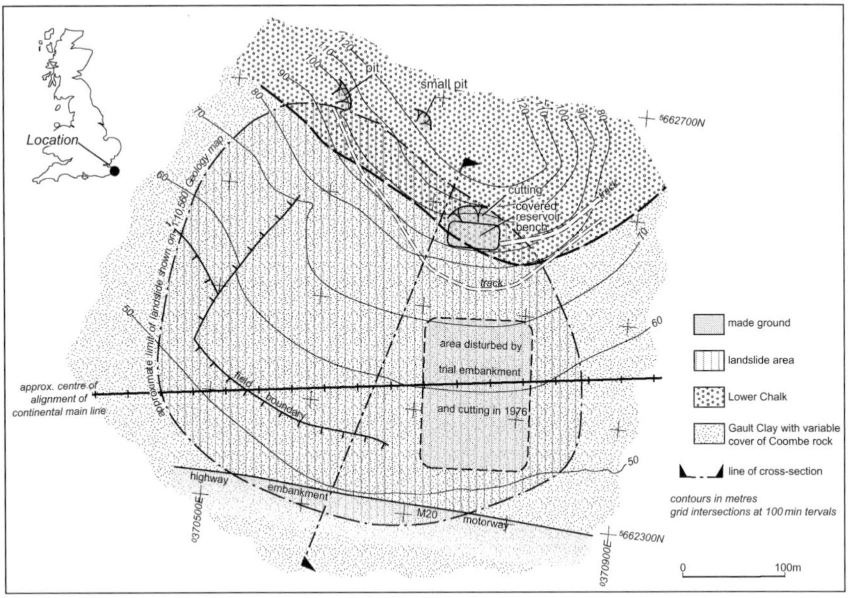

Figure 14.3 Geology of the Cherry Garden landslide.

4. MAPPING METHODOLOGY

Following desk studies of the available literature, interpretation of the limited aerial photographs of the area was undertaken. The aerial photographs comprised 1:5000 scale colour visible spectrum and false colour near-infrared images flown during the mid-1980s. Previous site investigation data were also reviewed before a field reconnaissance was carried out and a mapping programme for the geomorphology devised. The mapping used 1:500 scale engineering plans as base map field sheets and was undertaken by a two-man team as a tape, compass and clinometer survey initially using the Savigear (1965) methodology of morphological mapping. These base maps were subject to geomorphological interpretation for engineering purposes as described by Brunsden et al. (1976) and Griffiths and Marsh (1986). The mapping was performed in a systematic manner by covering the ground in a series of transects ~30 m apart, with each transect aligned normal to the main slope direction. All changes and breaks of slope, along with the slope angles within the planar slope facets, were identified on the field map. The field map was also extensively annotated with observations on the geology and geomorphology.

Table 14.2 Average Strata Thickness and Some Typical Intact Properties

Material	Average Thickness (m)	Intact Undrained Strength (Mpa)	Peak Effective Strength	Calcium Carbonate (%)	Coefficient of Mass Permeability (m/s)
Grey Chalk	5	3.75	c', 675 kPa; ϕ', 39°		2×10^{-6}
Upper Chalk Marl	20	4.25	c', 675 kPa; ϕ', 39°	75	2×10^{-6} to 4×10^{-7}
Lower and Basal Chalk Marl	12	1.75–3.75	c', 590 kPa; ϕ', 36°	50–60	1×10^{-7} to 7×10^{-8}
Glauconitic Marl	2.5	6.5	c', 1 MPa; ϕ', 40°	15	5×10^{-5} to 5×10^{-10}
Transitional bed (Zone 6A)	8	1.75–3.0		30	5×10^{-5} to 5×10^{-10}
Gault Clay	17	1.0	c', 520 kPa; ϕ', 24°	10	10^{-9}

Source: Summarised from Harris et al. (1995a,b).

5. MAPPING RESULTS: MAIN GEOMORPHOLOGICAL UNITS

The site can be divided into six significant geomorphological units as presented in Table 14.3 and Figures 14.1 and 14.2 (based on Griffiths et al., 1995).

6. MAPPING RESULTS: THE CHERRY GARDEN LANDSLIDE

The Cherry Garden landslide complex (Figures 14.3, 14.4 and 14.5) has developed from the scarp slope associated with the Upper Cretaceous Lower Chalk and involved disruption to the Grey Chalk, Chalk Marl, Glauconitic Marl, Zone 6A and the Gault Clay (Figure 14.3). The landslide area can be subdivided into four major identifiable landslide units (labelled A, B, C and D in Figure 14.4) plus a number of smaller features produced by more superficial movements (notably the feature labelled landslide E in Figure 14.4). Each of the main landslides has a backscar located in the Chalk scarp slope, although none reach to the top of the escarpment. There are a series of dislocated landslide blocks in midslope, and the movements extend downslope to form part of the lower footslopes that grade into the lowlands formed in the Lower Cretaceous Folkestone Beds (Figures 14.1 and 14.2; Table 14.3).

Landslide A is located on the eastern flanks of Cherry Garden Hill adjacent to Cherry Garden Coombe and has a south-easterly movement vector. The arcuate backscar is clearly defined with slopes in excess of 40°. There are two separate landslide blocks beneath the backscar (benches A1 and A2 in Figure 14.4) separated by a 2 m bluff. The scarp in front of bench A2 has a slope of 14°, which is too high for Gault Clay, suggesting the block comprises Lower Chalk Marl. The concavity at the base of the lower footslopes is taken to be the limit of the main landslide movement.

The morphology of landslide B suggests it was truncated on both flanks by later landslide movement (landslides A and C), indicating it is probably the oldest of the landslide movements. The extent of the landslide was difficult to define as it has been heavily disturbed by earthworks. The backscar lies in the vicinity of a bench cut for a small, now disused, covered reservoir. The lower limits have been obliterated by the excavations associated

Table 14.3 General Geomorphology of the Study Area

Geomorphological Unit and Description	**Geomorphological History**
Chalk plateau Gently inclined (<5°) northward dipping undulating surface developed in the Middle and Upper Chalk sub-parallel to the chalk bedding. Contains a number of dry valleys and is covered by a variable thickness of solifluction deposits, clay-with-flints and plateau drift with the greatest thicknesses lying in the valleys	Long period of sub-aerial denudation, probably since the Palaeogene (see Jones, 1980). Major joint sets developed during the warping associated with the formation of the Wealden anticlinorium (Alpine orogeny). These joint sets control the distribution and pattern of the dry valleys that in part formed during periods of higher groundwater levels during the Tertiary and Quaternary. Frost-shattered chalk, fossil ice wedges, solution pipes and infilled dolines are also present as relict periglacial features formed during the Pleistocene
Chalk scarp slope A steeply inclined (>20°) slope developed in the Lower Chalk that approximately follows the line of strike. Disrupted in many places by major landslides and subject to soil creep and soil erosion	Scarp probably reached its present location as a result of scarp retreat by sub-aerial processes by the end of the Tertiary Period (Jones, 1980). Meso-scale morphology modified by repeated periglacial activity throughout the Pleistocene and subject to Late and Early Post-Glacial landsliding
Chalk Marl footslope Low-angle (<10°) hillside slopes developed on the low permeability Chalk Marl. Formed primarily by higher rates of surface flow resulting in sheetwash. Surface has thin cover (<5 m) of periglacial coombe rock and colluvium and has been subject to cryoturbation. Has been disrupted by the large-scale landsliding *Lower valleyside slopes developed in the Gault Clay* Low-angle slopes (<5°) developed in the low permeability Gault Clay, located below the Chalk Marl footslope but formed in a similar way. Surface has the same thin cover of periglacial coombe rock and colluvial deposits and has also been subject to cryoturbation. In the area	Formation of both the Chalk Marl footslope and the lower valleyside slopes developed in the Gault Clay are associated with sub-aerial processes that occurred during the long-term retreat of the Chalk scarp slope throughout the Tertiary Period. Sheetwash, gelifluction and cryoturbation shaped these lower hillside slopes during the Pleistocene before the occurrence of Late and Early Post-Glacial landsliding

(continued)

Table 14.3 (continued)

Geomorphological Unit and Description	Geomorphological History
closest to the scarp slope, the deposit has been disrupted by large-scale landsliding, with the main shear surfaces located within clayey zones in the Gault Clay sequence	
Flat lowlands	
General flat (slopes <5°) lowland area south on the Etchinghill Escarpment developed in the Lower Greensand Folkestone Beds	Vale created by the long-term erosion of the Wealden anticlinorium initially through fluvial downcutting and lateral erosion in the Tertiary Period and subsequently modified by periglacial processes during cold phases of the Pleistocene
Coombes	
In a number of locations, the scarp slope has been eroded into embayments or coombes. These were formed by a combination of gelifluction (including debris flows) under periglacial conditions (Kerney et al., 1980) and subsequent erosion by spring-head sapping. These coombes are the major source of the coombe rock deposits that cover the lower valley side and lowlands	The present coombes developed under periglacial conditions during the Devensian (Kerney et al., 1980), although may have been initiated during earlier cold phases during the Pleistocene. Their location is related to the major joint sets that are associated with warping during the Alpine orogeny during the Tertiary. During the Post-Glacial period, these coombes will have been subject to further retreat through spring-head sapping and general hillwash
Landslides	
All the landslides in the vicinity of the Channel Tunnel terminal have backscars in the chalk scarp slope, but also disrupted the lower hillside slopes and involved the Chalk Marl and Gault Clay	The landslides are likely to be Late- to Early Post-Glacial in age, probably either Older or Younger Dryas. The landslide toes appear to interdigitate with different levels of the coombe rock deposits suggesting they are in part contemporaneous.

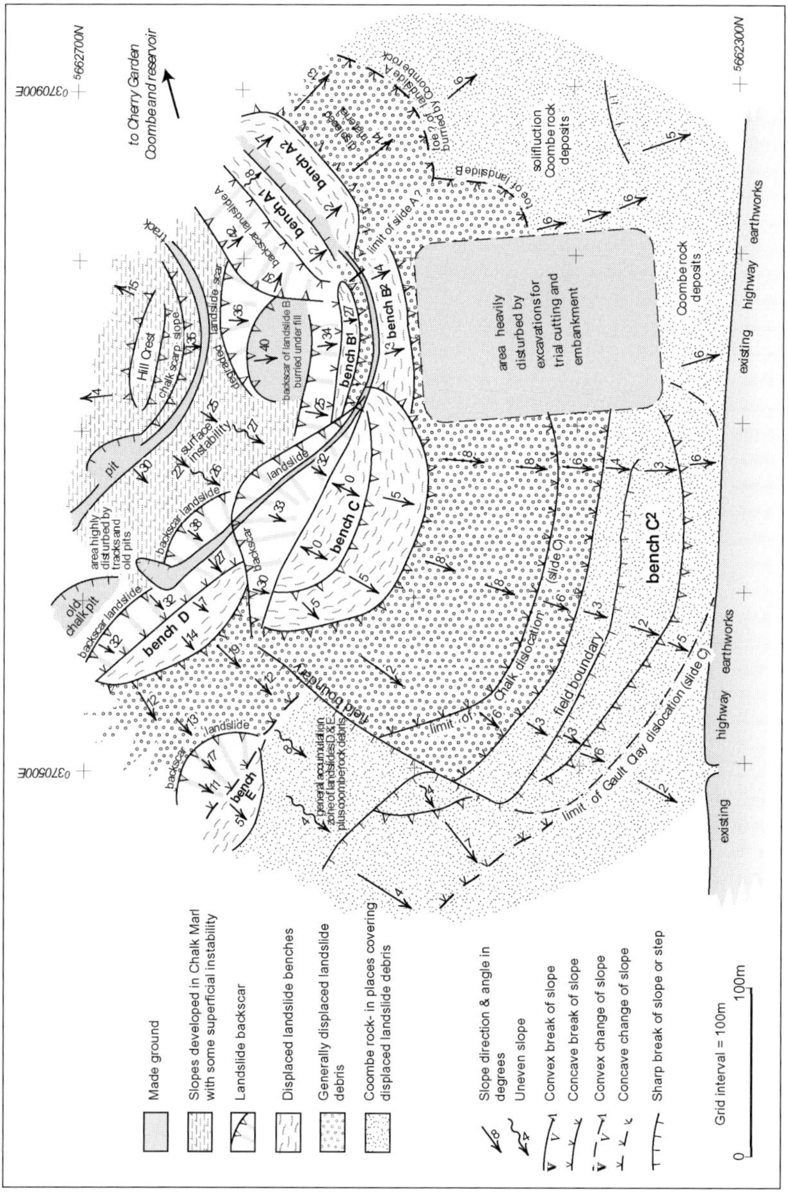

Figure 14.4 Geomorphological map of the Cherry Garden Landslide.

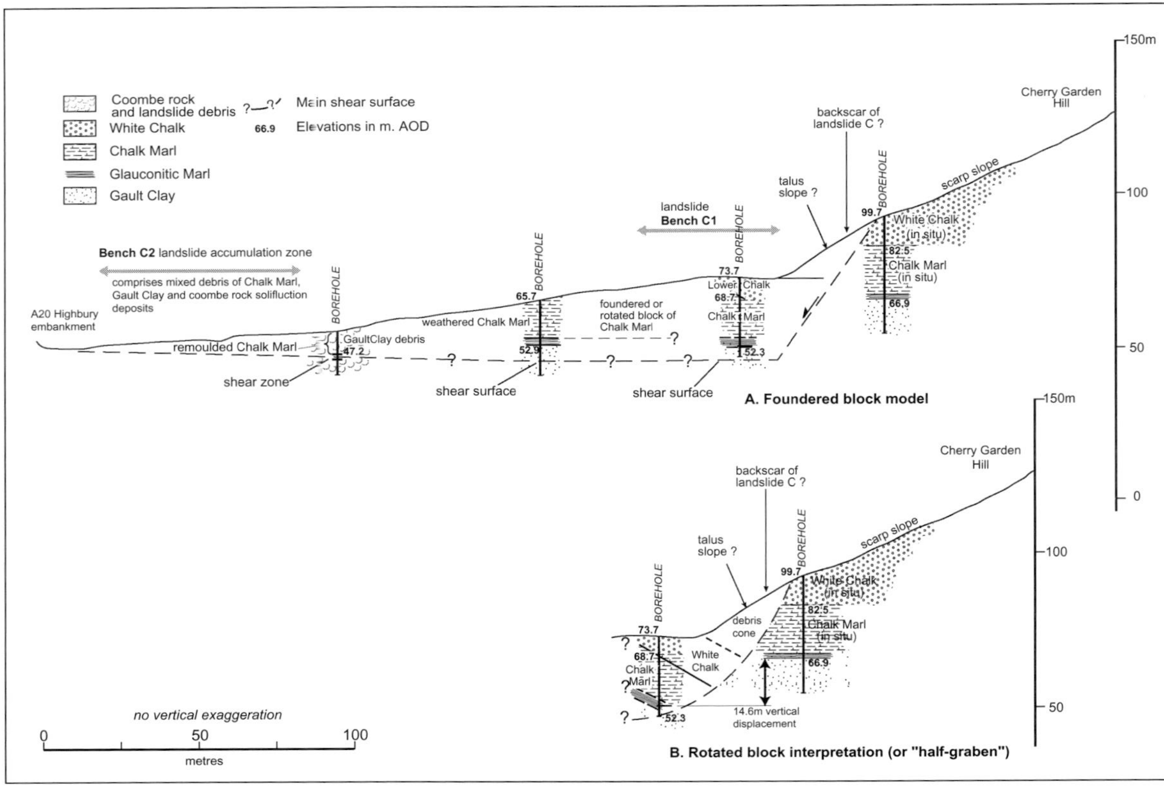

Figure 14.5 Hypothsised cross section through the Cherry Garden landslide C based on limited sub-surface data and surface mapping.

with a trail cutting and embankment during a previous investigation for the Channel Tunnel (Aarons et al., 1977). Two landslide benches (benches B1 and B2 in Figure 14.4) are visible with bench B2 being marked lower in elevation than bench A2 to the east and slightly lower than the extensive bench C1 to the west. The general landslide movement direction appears to have been due south.

Landslide C appears to be the largest landslide within the complex. The backscar is well defined and suggests a movement in the south–southwest direction. There is a marked concavity between the 33° slopes of the backscar and the relatively flat surface of the main disrupted landslide block (bench C1). This block has a maximum width of 30 m when the shallow front scarp is included. An earlier site investigation survey put a borehole through this bench (shown in the cross section in Figure 14.5), which showed it was composed of intact but displaced White Chalk and Chalk Marl. In front of bench C1, marked by a convex change of slope, lies a long convexo-concave slope (8°–3°) that leads down to the narrow, elongate bench C2. This lower bench and the slightly steeper 5°–6° scarp at the front of it are taken to be downslope limits of the landslide C movements and likely to comprise Lower Chalk and Gault Clay landslide debris mixed with coombe rock deposits (Figure 14.5).

Landslide D is a more complex unit on the western side of the Cherry Garden Hill with one clear backscar (slope 32°–38°), which is located upslope of a well-defined landslide bench (bench D). Downslope of the bench, the landslide grades into a general area of undulating ground that also contains a shallow landslide, landslide E. This is probably a more recent degradational failure. The direction of movement is west–southwest. The boundary between landslides D and C is obscured by a hedgerow.

7. GEOMORPHOLOGICAL INTERPRETATION

The surface morphology of landslides A–E suggests the forms of movements are generally planar in the lower zones of the landslides, which produced the benches A2, B2 and possibly C2. However, there appears to have been upslope retrogression possibly on quasi-rotational failure surfaces that gave rise to benches A1, B1 and possibly C1. The limited borehole data available along the direction of movement for landslide C1 (Figure 14.5) allow the use of the Glauconitic Marl as a marker horizon to show that bench C1 had been subject to nearly 15 m of vertical displacement. However, whether bench C1 failed as a single 'foundered' block or if there were several rotated blocks within the bench could not be determined during the mapping. The main planar movements appear to have taken place along a shear surface in the Gault Clay, and it is possible that the failure was a non-rotational block slide with a now-infilled graben. The lower components of the slide movements in the vicinity of bench C2 could not be disentangled from the coombe rock solifluction debris. This suggests that the main movements of landslide C were contemporaneous with the deposition of the coombe rock, indicating a Late Glacial age. Landslides A and B are likely to be similar in age with landslide B probably slightly older than A, but the indications are that landslides D and E occurred later, and indeed landslide D shows signs of relatively recent movements.

8. CONCLUSION

The engineering geomorphological mapping clarified the nature and extent of the landslide area shown on the geological maps of the area and formed the basis for more detailed ground investigations of the site. These were to be primarily aimed at clarifying the nature of the failure mechanisms of the landslides A–D. There was also concern that there might have been retrogressive block disruption or cambering (see Horswill and Horton, 1976) upslope of the main backscars. The downslope limits of landslides movements would also need determining to establish how the rail construction would affect the overall stability of the landslides. Although the maps were not updated when the new ground information became available, they did form the basis for the engineering geological and geomorphological ground model (Fookes, 1997; Brunsden, 2002) against which new findings were checked and the interpretation of ground conditions re-examined.

REFERENCES

Aarons, A., Weeks, A.G., Parkes, R.D., 1977. Site investigation for the Channel Tunnel British ferry terminal. Ground Eng. May, 43–47.

Birch, G.P., Griffiths, J.S., 1995. Engineering geomorphology. In: Harris, C.S., Hart, M.B., Varley, P., Warren, C.D. (Eds.), Engineering Geology of the Channel Tunnel. Thomas Telford, London, pp. 64–75. (Chapter 6).

British Geological Survey, 1967. Geological Sheet TR 23 NW. Published by the British Geological Survey, scale 1:10,560 (6 inches to the mile).
Brunsden, D., 2002. Geomorphological roulette for engineers and planners: some insights into an old game. Q. J. Eng. Geol. Hydrogeol. 35, 101–142.
Brunsden, D., Doornkamp, J.C., Fookes, P.G., Jones, D.K.C., Kelly, J.M.H., 1976. Large scale geomorphological mapping for highway engineering. Q. J. Eng. Geol. 8, 227–253.
Fookes, P.G., 1997. Geology for engineers: the geological model, prediction and performance. Q. J. Eng. Geol. 30, 290–424.
Griffiths, J.S., Marsh, A.H., 1986. BS 5930: the role of geomorphological and geological techniques in a preliminary site investigation. In: Hawkins, A.B. (Ed.), Site Investigation Practice: Assessing BS 5930. Geological Society, London. Engineering Geology Special Publication 2, pp. 261–267.
Griffiths, J.S., Brunsden, D., Lee, E.M., Jones, D.K.C., 1995. Geomorphological investigations for the Channel Tunnel Terminal and Portal. Geogr. J. 161, 275–284.
Harris, C.S., Hart, M.B., Varley, P., Warren, C.D. (Eds.), 1995a. Engineering Geology of the Channel Tunnel. Thomas Telford, London, 526 pp.
Harris, C.S., Hart, M.B., Wood, C.J., 1995b. A revised stratigraphy. In: Harris, C.S., Hart, M.B., Varley, P., Warren, C.D. (Eds.), Engineering Geology of the Channel Tunnel. Thomas Telford, London, pp. 398–420. (Chapter 26).
Horswill, P., Horton, A., 1976. Cambering and valley bulging in the Gwash Valley at Empingham. Philos. Trans. R. Soc. Lond. Ser. A 283, 427–462.
Jones, D.K.C., 1980. The Tertiary evolution of south-east England with particular reference to the Weald. In: Jones, D.K.C. (Ed.), The Shaping of Southern England. Academic Press, London.
Kerney, M.P., Preece, R.C., Turner, C., 1980. Molluscan and plant biostratigraphy of some Late Devensian and Flandrian deposits, Kent. Philos. Trans. R. Soc. Lond. Ser. B 291, 1–43.
Mortimore, R.N., Wood, C.J., Gallois, R.W., 2001. British Upper Cretaceous Stratigraphy. Joint Nature Conservation Committee, Geological Conservation Review Series, 23.
Rawson, P.F., 2006. Cretaceous: sea levels peak as the North Atlantic opens. In: Brenchley, P.J., Rawson, P.F. (Eds.), The Geology of England and Wales. Geological Society, London, pp. 365–394. (Chapter 15).
Savigear, R.A.G., 1965. A technique of morphological mapping. Ann. Assoc. Am. Geogr. 53, 514–538.
Smart, J.G.O., Bissom, G., Worssam, B.C., 1966. The Geology of the Country Around Canterbury and Folkestone. Memoir of the British Geological Survey.
Varley, P.M., Warren, C.D., 1995. History of the geological investigations for the Channel Tunnel. In: Harris, C.S., Hart, M.B., Varley, P., Warren, C.D. (Eds.), Engineering Geology of the Channel Tunnel. Thomas Telford, London, pp. 5–18. (Chapter 2).

CHAPTER FIFTEEN

The Application of Geomorphological Mapping in the Assessment of Landslide Hazard in Hong Kong

Steve Parry
GeoRisk Solutions Ltd, Hollywood Centre, Sheung Wan, Hong Kong, China

Contents

Developments in Earth Surface Processes, Volume 15
ISSN: 0928-2025, DOI: 10.1016/B978-0-444-53446-0.00015-X

1. HONG KONG AND LANDSLIDE HAZARDS

Hong Kong is a Special Administrative Region of China with a population of over 7 million and an area of ~1100 km^2. It is situated on the South China Sea and has a subtropical monsoonal climate, with a mean annual rainfall of 2800 mm. Some 80% of the rainfall is between May and September and commonly associated with tropical cyclones. The terrain typically rises steeply from sea level to heights in excess of 900 m over relatively short distances. As a result of the high rainfall and steep slopes, much of the terrain is susceptible to landsliding. Historically, the main population centres were located on the coastal fringe of Hong Kong Island and the Kowloon Peninsula away from the steeper terrain. However, population increases have meant developments that commonly infringe onto the steeper mountain slopes with a consequential increase in landslide risk (Figure 15.1). The term 'landslide' in this chapter follows the definition of Varnes (1984) and includes all varieties of mass movement on slopes including rockfall, rock toppling and debris flows that involve no true sliding. This definition also conforms with the concept of 'natural terrain hazards' as defined in Hong Kong (Ng et al., 2003).

2. NATURAL TERRAIN LANDSLIDES IN HONG KONG

Natural terrain is defined in Hong Kong as terrain that has not been substantially modified by human activities but includes areas where grazing, hill fires and deforestation may have occurred (Ng et al., 2003).

Figure 15.1 Western Hong Kong Island. Mount Davis (269 mPD) in foreground with High West (494 mPD) and Victoria Peak (552 mPD) behind.

Natural terrain covers over 60% of the total land area of Hong Kong, with almost 50% of the natural terrain sloping at 30° or more. Based on a review of high-level (flight height >2400 m) aerial photographs from 1945 to 2003, over 16,000 landslides have occurred on natural terrain, an average of 275 landslides per year (MFJV, 2007). However, during intense single storms, hundreds of landslides can occur in a few hours. For example, during a severe rainstorm on 7 June 2008, a rain gauge on Lantau Island recorded a peak rainfall of 384 mm in a 4 h period, whereas the total rainfall for the 24 h period was 622 mm. This storm triggered over 2400 landslides on Lantau Island resulting in numerous road links being severed and many homes being temporarily evacuated (Wong, 2009). Figure 15.2a shows a landslide swarm resulting in the closure of both lanes of Tai O Road, the only access road to south-west Lantau Island. Figure 15.2b shows a 3000 m^3 channelised debris flow (CDF) which resulted in the closure of both lanes of the Yu Tung Road dual carriageway. This landslide was filmed and can be viewed at http://www.youtube.com/watch?v=R2uTKyK1c9k.

The consequences of natural terrain landslides to date have been considerably less than those from man-made slope failures. This is mainly because of the proximity of man-made slopes to developments with the result that even relatively small landslides can have potentially serious consequences. However, with the improvements to man-made slope safety in Hong Kong, the balance of landslide risk is changing (Chan and Mak, 2007). Furthermore, whilst the evaluation of high-frequency, low-magnitude landslides can be undertaken with a reasonable amount of

Figure 15.2 Landslides following a severe rainstorm on 7 June 2008, Lantau Island, Hong Kong. Left: Landslide swarm resulting in closure of both lanes of the only road access to SW Lantau Island. Right: A 3000 m^3 CDF closed both lanes of a dual carriageway.

certainty, natural terrain has the potential for relatively high-magnitude, low-frequency events, and the evaluation of such failures is a key focus of any engineering assessment.

3. GEOLOGICAL AND GEOMORPHOLOGICAL SETTING

The geology of Hong Kong is well documented in Sewell et al. (2000) and Fyfe et al. (2000). The majority of the land area (~80%) comprises Mesozoic plutonic and volcanic rocks with the remaining area underlain by sedimentary rocks of Devonian to Tertiary age (Figure 15.3). The sedimentary units are generally restricted to country parks with limited development in the eastern side of Hong Kong or as buried marble beneath alluvial floodplains in the north-west New Territories. As a result, the Mesozoic plutonic and volcanic rocks form the vast majority of the natural terrain from which landslides have the potential for significant consequences.

Much of the present-day landscape of Hong Kong is commonly assumed to have been formed during the Quaternary. However, the subtropical weathering that affects Hong Kong may have commenced as early as the Tertiary (Fyfe et al., 2000). The weathering results in the decomposition of rock and the development of thick saprolite (rock weathered to a soil in which the original structure and fabric are preserved; Fookes, 1997a). Natural terrain landslides are typically triggered by the transient elevation of pore water pressures during and after rainstorms, commonly at geological interfaces such as colluvium/saprolite or saprolite/rock interfaces. In addition to tropical weathering, significant climatic variations have also influenced the rate of hillslope denudation. Of particular importance are episodic changes in sea level promoting fluvial downcutting during low sea level stands. The influence of these changes will have varied depending upon location and geology, and consequently, landscape responses are commonly complex and site specific. Hansen (1984) presented a landscape evolutionary model for Hong Kong based on geomorphological principles. He proposed a simple three-form model with three ages of landform assembly. The upper, older assembly containing deep weathering profiles, a middle assembly containing the oldest colluvial deposits and the lowest, younger assembly, which was a product of stream rejuvenation associated with Pleistocene sea level regression. All three

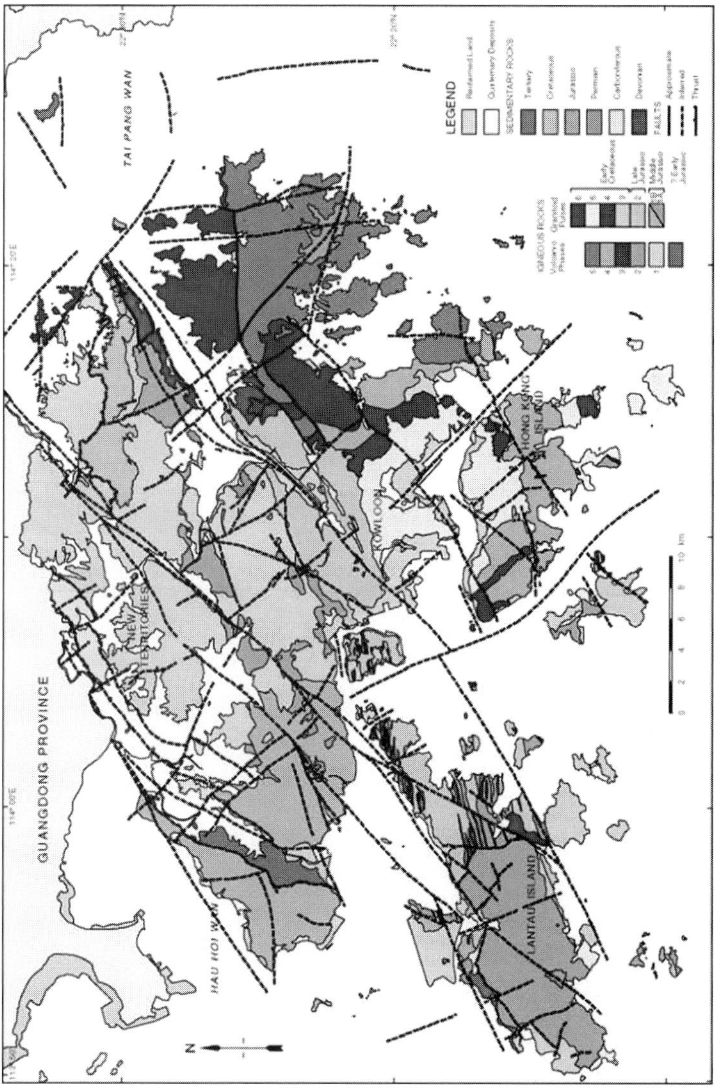

Figure 15.3 Simplified geological map of Hong Kong (Sewell et al., 2000).

assemblages are subject to different types and rates of processes with the greatest potential for erosion at the assemblage boundaries. In particular, streams above the over-steep lower assemblage are likely to be incising into thicker superficial deposits (Figure 15.4).

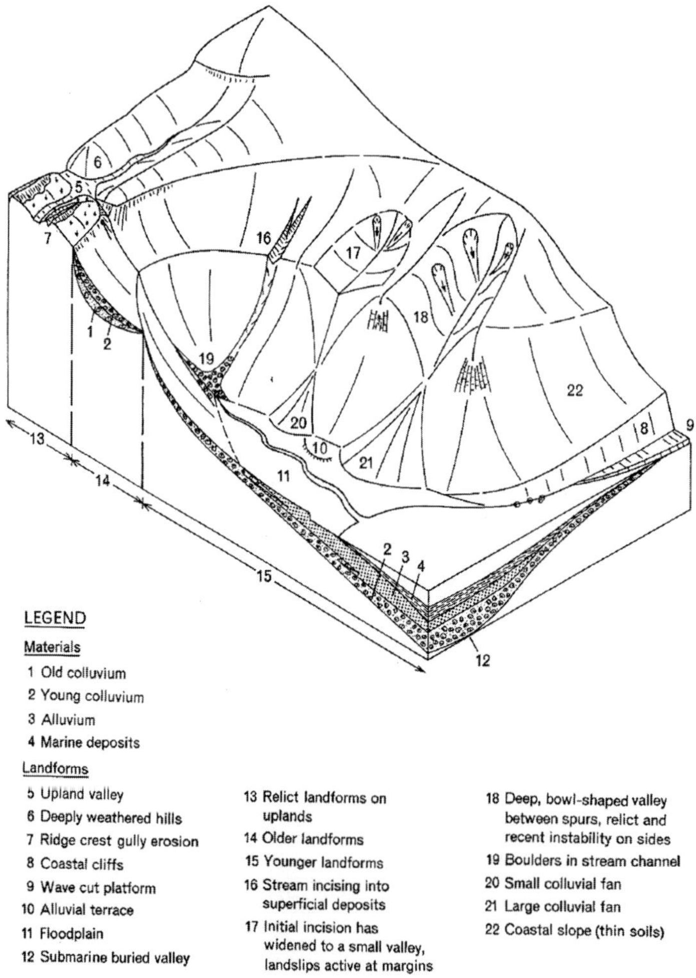

Figure 15.4 Hillslope model for Hong Kong (Hansen, 1984). This is a simple three-form model with three ages of landform assembly. The upper, older assembly containing deep weathering profiles, a middle assembly containing the oldest colluvial deposits and the lowest, younger assembly, which was a product of stream rejuvenation associated with Pleistocene sea level regression. All three assemblages are subject to different types and rates of processes with the greatest potential for erosion at the assemblage boundaries.

As a result of this complexity, whereas the solid geology of Hong Kong is well documented, the superficial deposits and in particular the age and process involved in the evolution of the present-day landscape are less well understood. Consequently, in order to understand and evaluate landslide risk, each assessment requires a site-specific, geomorphological assessment.

4. APPROACH AND METHODOLOGY FOR LANDSLIDE ASSESSMENTS IN HONG KONG

Unlike many regions in the world, where landslide assessments on natural slopes are carried out at a regional scale and relatively large degrees of uncertainty may be acceptable, the majority of Hong Kong's landslide assessments are facility driven, resulting in the study of relatively small areas, often comprising small first-order drainage line catchments, cut by man-made slopes associated with residential developments at their toe. Where larger scale studies have been undertaken, these have been related to the generation of landslide swarms, i.e. the 2000 landslide swarm at Tsing Shan, western New Territories (Parry et al., 2002) and the 2007 landslide swarm in SW Lantau (Parry et al., 2010). Consequently, even relatively small natural landslides have the potential to impact on the facilities. As a result, uncertainties in the assessment could have potentially serious consequences.

The engineering geomorphological input to landslide assessments in Hong Kong is typically undertaken in the following three inter-related and iterative stages:

Stage 1: The preliminary assessment, based on a desk study with considerable emphasis on aerial photograph interpretation (API) and the development of a conceptual ground model,

Stage 2: Verification/amendment of the conceptual model by site-specific, detailed engineering geomorphological and engineering geological mapping,

Stage 3: The determination of appropriate design events for each hazard identified.

In Hong Kong the majority of natural terrain assessments are carried out following the 'Design Event Approach'. Hong Kong Guidelines (Ng et al., 2003) present a framework for assessment based on notional susceptibility of the terrain, the type of facility and the steepness of the terrain. Figure 15.5 shows the derivation of the design event based on facility

Group No.	Facilities
1	(a) Buildings - any residential building, commercial office, store and shop, hotel, factory, school, power station, ambulance depot, market, hospital/polyclinic/ clinic, welfare centre
	(b) Others - bus shelter, railway platform and other sheltered public waiting area - cottage, licensed and squatter area - dangerous goods storage site (e.g. petrol station) - road with very heavy vehicular or pedestrian traffic density
2	(a) Buildings - built-up area (e.g. indoor car park, building within barracks, abattoir, incinerator, indoor games' sport hall, sewage treatment plant, refuse transfer station, church, temple, monastery, civic centre, [illegible] substation)
	(b) Others - road with heavy vehicular or pedestrian traffic density - major infrastructure facility (e.g. railway, tramway, flyover, subway, tunnel portal, service reservoir)
3	- densely-used open space and public waiting area (e.g. densely-used playground, open car park, densely-used sitting out area, horticultural garden) - quarry - road with moderate vehicular or pedestrian traffic density
4	- lightly-used open-air recreation area (e.g. district open space, lightly-used playground, cemetery, columbarium) - non-dangerous goods storage site - road with low vehicular or pedestrian traffic density
5	- remote area (e.g. country park, undeveloped green belt, abandoned quarry) - road with very low vehicular or pedestrian traffic density
Note:	For roads, the Facility Group should be based on Figure A13 (Wong, 1998) taking into account the actual Annual Average Daily Traffic and the number of road lanes. For footpaths alongside roads, it may be assumed that footpaths are within the same group as the adjoining roads, except for Expressways (EX), Urban Trunk Roads (UT) and Rural Trunk Roads (RT). Footpaths alongside EX, UT and RT roads may be taken, by default, as a Group 5 facility, unless dictated otherwise by site-specific conditions.

Table 1a: Facility group

Proximity	Facility Group	
	1 & 2	3
Very Close (e.g. if angular elevation from the site is ≥ 30°)	I	II
Moderately Close (e.g. if angular elevation from the site is ≥ 25°)	II	III
Far (e.g. if angular elevation from the site is < 25°)	III	IV

Notes: (1) Facility groups are given in Table 1. Facilities of Groups 4 and 5 normally do not require natural terrain hazard study and hence they are not included in Table 3.
(2) For channelized debris flow, if the worst credible event affecting the site is judged to have a volume exceeding 2000 m³, the angular elevation given in the above examples should be reduced by 5°.
(3) The examples given above are for general guidance only. Other factors, such as credible debris path, topographical conditions and site-specific historical data, should also be taken into account in assessing the "proximity" of the natural terrain to the site.
(4) This Table is applicable only to sites that require natural terrain hazard study.

Table 1b: Consequence class

Susceptibility Class	Description
A	The natural terrain is extremely susceptible to the type of failure under consideration, with a notional annual probability of occurrence in the order of 1:10 or higher. For example, there are signs of instability, continued movement, or records of repeated recent failures (say over the past 50 to 100 years as observed from aerial photographs) in the catchment and its relevant vicinity.
B	The natural terrain is highly susceptible to the type of failure under consideration, with a notional annual probability of occurrence within the order of 1:10 to 1:100. For example, there are records of occasional recent failures in the catchment and its relevant vicinity.
C	The natural terrain is moderately susceptible to the type of failure under consideration, with a notional annual probability of occurrence within the order of 1:100 to 1:1000. For example, there are few records of recent failures, but there are indications of relict failures, or geomorphological evidence of potential problems in the catchment and its relevant vicinity, or any other evidence from similar terrain in Hong Kong.
D	The natural terrain is of low susceptibility to the type of failure under consideration, with a notional annual probability of occurrence less than 1:1000. For example, there are no records of recent and relict failures, little geomorphological and other evidence of potential problems in the catchment and its relevant vicinity, and little other evidence from similar terrain in Hong Kong.
Note:	In assessing the susceptibility of the hillside to failure, consideration should be taken of the potential effects of changes in environmental factors, e.g. any changes to the overall setting of the terrain such as fill fires and construction [illegible], and the relevance of the available historical landslide records. Most of the information on environmental changes can be obtained from a review of past aerial photographs.

Table 1c: Susceptibility class

Susceptibility Class (see Table 4)	Consequence Class (see Table 3)			
	I	II	III	IV
A	WCE	WCE	WCE	CE
B	WCE	WCE	CE	CE
C	WCE	CE	CE	N
D	N	N	N	N

Notes: (1) The recommended minimum design requirements are given in this Table. The designer may adopt a more conservative design or provision of other precautionary/warning measures if considered necessary.
(2) This Table will be applied to each sub-catchment and normally the predominant type of hazard will control the design requirements.
(3) WCE = Adopt a "worst credible event" as the design event.
CE = Adopt a "conservative event" as the design event
N = Further study is not required

Table 1d: Design requirement for design event approach

Figure 15.5 The Derivation of the Design Event Landslide (Ng et al., 2003). The Hong Kong Government has produced guidelines for the selection of an initial estimation of landslide source volume that may affect a site. The type of facility is classified based on use and the consequence of a landslide is estimated based on the angle of the terrain. The suitability is initially selected based on published landslide databases and the combination of these factors indicates whether a 'worst credible event' or a 'conservative event' should be selected.

group, slope angle and an approximation of susceptibility. Based on this, either a 'conservative event' (generally corresponding to a reasonably conservative estimate based on fresh landslides evident from API in the last 50–100 years, i.e. within the available aerial photograph record for Hong Kong) or a 'worst credible event' (WCE, generally corresponding to the largest landform interpreted from API as potentiality resulting from a single landslide event) is adopted. However, the WCE is nominally defined as being approximate to a 1 in 1000 year event, and therefore should exclude 'events that occurred in the geological past' (Ng et al., 2003). Although the term 'geological past' is not defined in the guidelines, it is considered as implying different climatic conditions compared with the present day.

The Design Event Approach was developed to enable a rapid evaluation of the possible magnitudes of hazards a site may face and therefore allows potential cost implications and alternative layouts to be considered at the feasibility stage. The guidelines note that the Design Event should be reviewed and modified as the study progresses. However, it is commonly assumed that the largest landslides occurring within the time frame of aerial photograph coverage (defined as 'recent') is a 'conservative event' and the largest landform interpreted from the aerial photographs as a landslide (defined as 'relict') is the 'WCE' with little or no consideration of the site setting or with minimal field mapping. Such a simplistic approach, particularly with respect to the WCE, can seriously under- or overestimate the landslide hazard. The importance of placing a site within its geomorphological context so that the complex interrelationships between susceptibility, magnitude, frequency, entrainability and run-out can be evaluated in terms of past, present and potential future geomorphological processes cannot be over emphasised.

5. CONCEPTUAL GROUND MODELS

The conceptual ground model developed from the desk study, in particular, the API is, in the author's opinion, crucial to a successful assessment. The approach adopted follows that discussed by Fookes (1997b), whereby the site is the product of a sum of its geological and geomorphological history. The development of a conceptual ground model allows for the following:

- focused mapping, which is especially important given the generally restricted access as a result of the dense subtropical vegetation present over much of Hong Kong's natural landscape,
- the development of alternative hypotheses,
- the consideration and documentation of uncertainty.

As with any model, however, continuous verification and openness to new ideas are critical to the success of this approach.

Aerial photographs of Hong Kong date back to 1924, but the first virtually territory-wide coverage dates from 1963. This set of aerial photographs is commonly used as a baseline because of their clarity, scale (1:5400 to 1:7800) and the relatively low vegetation cover. In addition, annual high-level aerial photograph coverage has been available since 1978 (1:20,000 to 1:25,000).

Despite the comprehensive aerial photograph coverage in Hong Kong, the certainty with which an individual morphological feature can be interpreted and identified as being related to landslide processes will vary depending on a number of factors. These include the following:

- Original source volume: Larger failure scars tend to be preserved longer in the landscape. Sewell and Campbell (2005) reported that the upper bound age for the large relict landslides they examined was ~34,000 years BP, i.e. landsides within the API record can be in excess of 1000 years old. With respect to a lower bound age, Evans et al. (1997) suggested that a landslide source would typically be 90% re-vegetated after 20 years,
- The presence of associated debris below the feature: This provides the clearest evidence that a landslide has occurred. However, unless the landslide is relatively large in size or the debris is deposited as levees outside of a drainage line, fluvial erosion commonly results in the removal/reworking of the debris with time,
- The sharpness of the feature: Landslide scarps tend to degrade with time, although this may be affected by subsequent minor failures of the oversteepened scarp area,
- The position of the feature within the landscape: Landslides are commonly interpreted as such when a depression is inconsistent with the adjacent landform.

In addition to the identification of landslides, the API should also place the site in the context of its geomorphological setting, i.e. in a framework that integrates surface features such as form, materials, processes and age as well as the underlying geological controls such as lithology and geological structure on past and future landform development, i.e. a conceptual model.

A Hong Kong landslide inventory (the Enhanced Natural Terrain Landslide Inventory or ENTLI) exists which provides a valuable starting point for hazard assessments (MFJV, 2007). However, there are three main limitations to this data set:

1. Although guidelines were produced to improve consistency with respect to the interpretation of relict landslides during the ENTLI (Parry et al., 2006), the interpreters' classification varied both with experience and training,
2. There was no field verification,
3. Given the project constraints (over 105,000 aerial photographs were reviewed in 16 months), the interpreters could only examine for relatively clear evidence of landslides, e.g. obvious scarps, with very little time to consider geomorphological setting, often critical with

respect to the identification of older degraded and possibly landslide related features.

In addition, a separate Large Landslide Database also exists (Scott Wilson (Hong Kong) Ltd., 1991), which is based on a more detailed API of possible landslide features >20 m in width with very limited field verification. However, to date there has been no systematic comparison of the two data sets. Consequently, there is concern that:

- the existing landslide inventories, which have been developed without field verification, may be used mechanically and without appreciation of their compilation methodology and limitations,
- degraded landforms that may represent high-magnitude, low-frequency landslide events and that are not identified in the data sets will be overlooked.

Figure 15.6 shows the development of a conceptual model and initial design event for a hillside, predominantly based on API (Parry and Ng, 2010). The hillside of concern rises from 90 to 295 mPD and comprises two catchments with a total area of about 56,000 m^2. Figure 15.6a is an engineering geomorphological map derived from API which shows considerable areas of rock and intermittent rock outcrop associated with incising drainage lines, saprolite on the spurlines and thicker saprolite in the upper terrain. Superficial deposits, predominantly comprising taluvium (defined below in Section 7.3.3), are associated with the drainage lines. Figure 15.6a also shows the location of ENTLI features which are predominantly located in the incising terrain. Figure 15.6b shows the conceptual model for the site focusing on the likely location and magnitude of landslides. Based on the Design Event approach, a WCE is applicable. In order to generate a WCE, rather than simply using the largest ENTLI feature, an evaluation was undertaken of the largest section of steep terrain adjacent to the boundary of the upper terrain unit which could potentially fail based on the interpreted morphology selected, whereby a source landslide would result in the debris impacting upon the greatest extent of taluvium within the incising terrain unit below. Where the taluvium is considered to be underlain by rock or intermittent rock outcrop (typically steeper terrain), it has been assumed that all the taluvium, over the width of the source landslide, would mobilise under the impact of the debris. The thickness of taluvium was conservatively assigned as 4 m at this time. Such a WCE corresponds to an initial landslide source volume of 1500 m^3 in the upper terrain saprolite combined with a secondary failure of 1400 m^3 in the taluvium, resulting in a total volume of 2900 m^3.

(a)

(b)

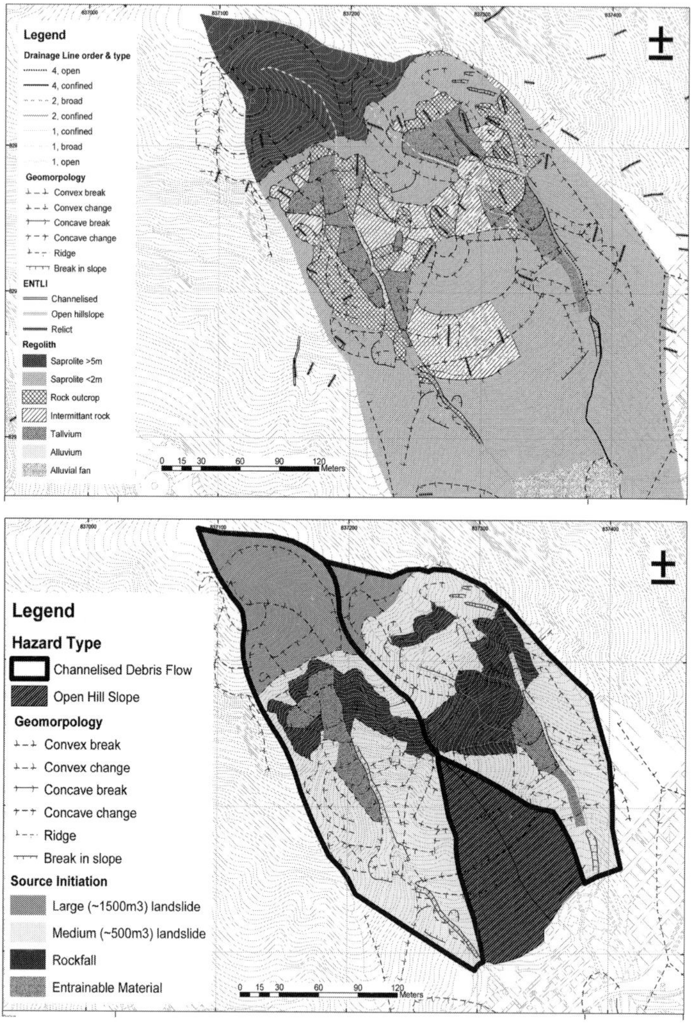

Figure 15.6 Initial Design Event derivation based on engineering geomorphological mapping. (a) Engineering geomorphological map. (b) Conceptual model used to generate the Design Event at review stage. Both are based on API and were re-evaluated during subsequent field mapping. Incising drainage lines form two adjacent catchments. Within both catchments, extensive areas of rock outcrop are present. Also shown are ENTLI landslides. The Upper Terrain above the incision was interpreted as potentially containing thicker saprolite. Part (b) shows the conceptual model based on the engineering geomorphology with potential design events varying with setting. The largest initial design event was considered to be a failure of deeper saprolite in the Upper Terrain (1500 m^3) with the landslide entraining a further 1400 m^3 of colluvium, resulting in a total volume of 2900 m^3.

The subdivision of the landscape into Terrain Units is a key component of the conceptual model and is extremely beneficial. Terrain Units are components of the landscape comprising similar material, age and processes. As a result, they are commonly associated with similar magnitude/frequency relationships with respect to landslide source volumes. In this example, the Upper Terrain was interpreted as potentially older and hence probably comprises thicker saprolite and appeared less active. Below this is Middle Terrain also comprising saprolite but interpreted as being thinner. Fluvial incision is related to the formation of Incising Terrain which has affected a large proportion of the Middle Terrain. The conceptual hazard model suggests a WCE comprising a low frequency but high-magnitude source landslide originating in the Upper Terrain with significant secondary entrainment within the Incising Terrain.

Obviously, there are uncertainties associated with such conceptual ground models, e.g. the potential source volumes and the amount of entrainment, but critically such models provide focus for subsequent field mapping and ground investigation. They also allow a preliminary evaluation of the types of mitigation works potentially required and allow factors such as cost and environmental impact to be considered at an early stage. In the example given, an initial evaluation of the maximum potential landslide magnitude was derived and key areas delineated for targeted ground investigation, i.e. thickness and stability of the upper saprolite and depth and entrainability of the taluvium.

6. SITE-SPECIFIC FIELD MAPPING

A combination of steep and densely vegetated terrain and high temperatures/humidity from May to September makes field mapping in Hong Kong both difficult and physically demanding. However, such mapping is an essential component for assessing the potential hazards affecting a site. The purpose of the mapping is threefold:

- to evaluate observations and validate interpretation from API,
- to evaluate the conceptual model,
- to record any additional evidence with respect to the geomorphological processes that control the location, magnitude and frequency of landsliding. One key area is in estimating potential entrainment with

respect to channelised debris flows. The evaluation of such potential from API alone is very problematic.

Given the limitations of API, during the field mapping geomorphological processes are re-evaluated and possible landslide features are inspected and verified. This results in a more accurate determination of previous source volumes, the examination of the failure mechanism and assessment of the mobility of the past failures (Figure 15.7). The field mapping also allows the identification of landslides that are not evident from API, e.g. where complete detachment of a previous landslide has not occurred or where landslides are obscured by vegetation. The field work should be carried out by the same personnel who undertook the API.

It should be noted that the Design Event is continuously evaluated and revised during this process, particularly during field mapping that includes field mapping of API-identified landslides, evaluation of entrainable material and determination of potential hazard not identified from API, e.g. first-time failures.

Field mapping is undertaken at scales of 1:500 to 1:1000 and utilises the 1:1000 scale topographical maps and global positioning system (GPS). For Hong Kong Island, high-quality airborne, LiDAR-derived data exist and the contour maps generated from the LiDAR data are a considerable improvement on the existing 1:1000 topographical maps, with respect to evaluating landslide run-out. However, relatively large-scale topographical variations that can affect the mobility analysis may not be evident from LiDAR and require detailed evaluation of channel morphology (see also Parry and Jonas, 2007) (Figure 15.8).

7. CASE STUDY

The case study discusses the methodology and results of a hazard assessment of natural terrain in Hong Kong. The site covers an area of $\sim 90{,}000\ m^2$ and rises from 160 to 495 mPD. At the toe of the study area, a number of permanently occupied facilities are present. The Design Event approach indicates that a WCE is applicable. The case study serves to illustrate the application of the use of geomorphological mapping in assessing design events for different types of natural terrain hazards including landslides, rockfalls and boulder falls. As discussed below, focused

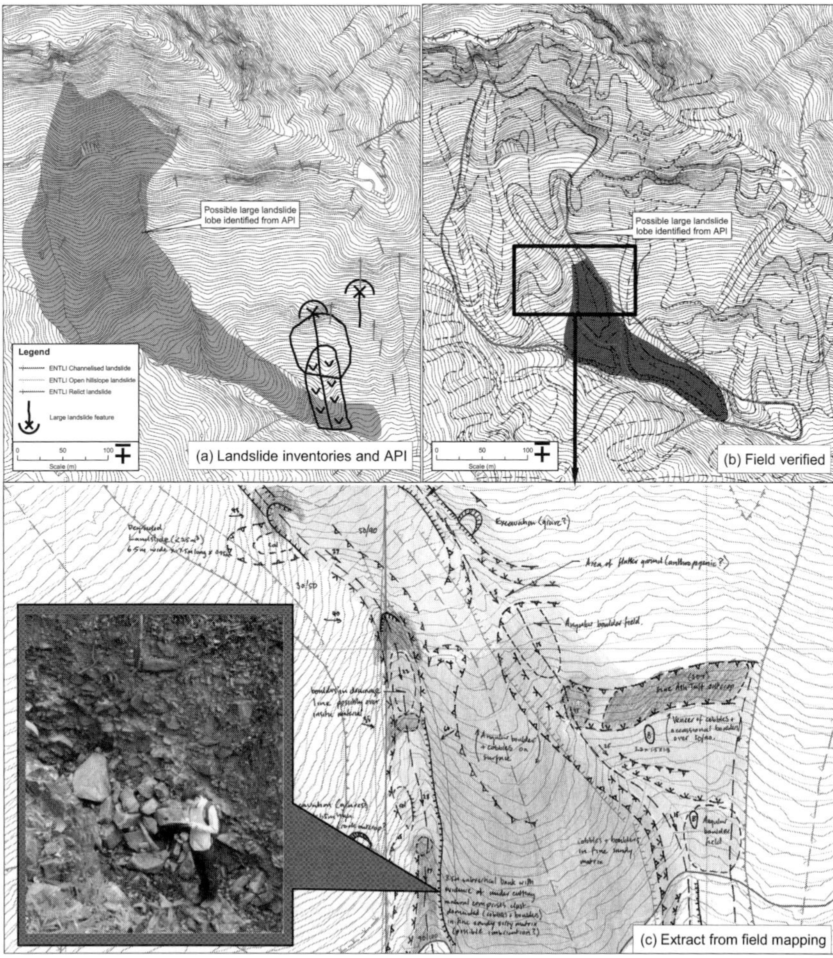

Figure 15.7 (a) Landslides recorded in the various existing inventories and an additional possible large degraded landslide debris lobe identified from site-specific API. (b) Field mapping at 1:500 scale indicated that the lobate landforms identified from API can be subdivided and have separate origins. For example, the feature identified in dark grey consists of a distinct lobate deposit comprising angular to sub-angular, slightly to moderately decomposed, clast-supported cobbles and boulders. A depression is evident above this lobe. The field evidence suggests that the lobe may represent debris from a large rock avalanche. Although the decomposition of the clasts suggests the feature occurred 'within the geological past', absolute age dating from carefully selected material is however necessary to confirm this.

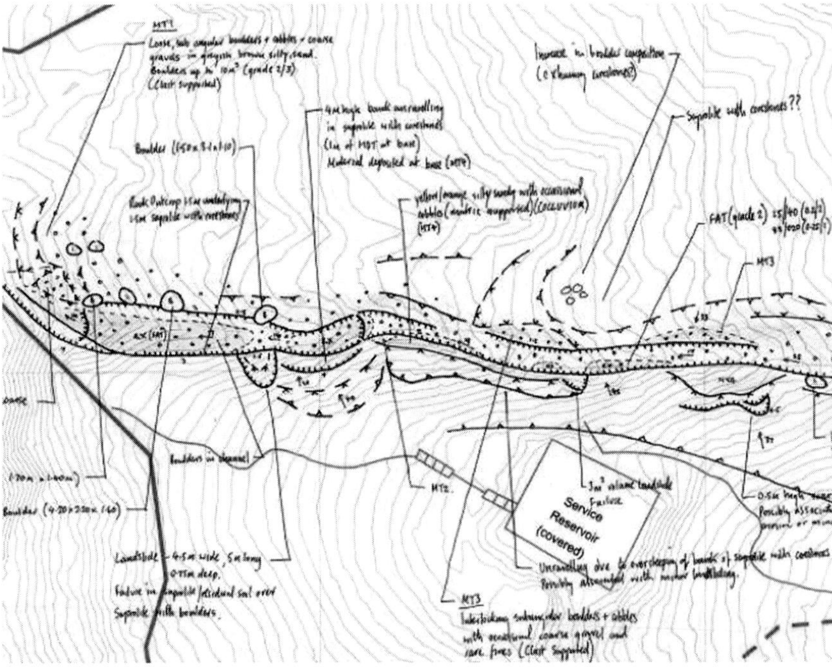

Figure 15.8 Engineering geological mapping at 1:500 scale on LiDAR-generated contours. The mapping identified an incised drainage line with vertical banks up to 4 m in height that are not evident from LiDAR. Such information is critical for mobility modelling. Also shown is over-steep terrain resulting from fluvial incision with associated evidence of instability. The initial hazard models generated from API and an existing data review were re-evaluated based on these field observations. The potential bed load of the drainage line is also recorded, as is any evidence for bank collapse, both of which can substantially influence entrainment potential.

ground investigation and absolute age dating are considered necessary to confirm the values adopted.

7.1 Methodology

The API was undertaken using a Sokkisha stereoscope with ×3 and ×8 binocular attachments. Based on the API, a conceptual model was developed for the site. This suggested an Upper Terrain unit possibly associated with extensive areas of thick (>5 m) saprolite. Consequently, potential landslide source volumes of up to 1000 m^3 were adopted. A Lower Spurline Terrain unit was considered potentially associated with relatively thick (>4 m) saprolite but this was more limited in extent.

Consequently, potential landslide source volumes of up to 500 m^3 were considered applicable in this terrain. An Upper Depositional Terrain was associated with source volumes up to 300 m^3 and a Lower Depositional Terrain considered to possess the potential for 400 m^3 landslide source volumes.

Based on a failure within the Upper Terrain, it was considered that debris could potential impact and mobilise deposits of colluvium farther downslope. It was assumed that mobilisation (either associated with secondary failure or with entrainment) would occur with a width similar to that of the landslide source area. Additional entrainment was estimated for each regolith unit over which the debris may flow. This resulted in initial estimates of the WCE varying between 1000 and 2700 m^3 depending on the location of initial failure. In addition, open hillside slope failures of 500 and 400 m^3 were considered and based on an area of potentially unstable rock identified during site reconnaissance, a possible rockfall/rock avalanche with an estimated volume of 800 m^3. Although it was noted that secondary entrainment from a large rockfall could occur, this was not estimated due to the uncertainty at that time.

Engineering geomorphological and geological mapping was carried out based on the approaches given in Anonymous (1972), Anonymous (1982) and Brunsden and Griffiths (2001). These approaches are also in accordance with Geoguide 2 (GCO, 1987) and Geoguide 3 (GCO, 1988). The field mapping was focused on evaluating the observations and initial models developed from API. Where accessible, potential landslide features identified during the API were inspected as were geomorphological boundaries. In accordance with Geoguide 3 (GCO, 1988), the underlying volcanic rock with decomposition Grade I to III was mapped as 'rock', whereas decomposition Grades of IV to V were mapped as 'saprolite'.

Much of the site is covered with dense vegetation and so a combination of GPS positioning using a hand-held Garmin 60CSx and tape and bearing were used for location purposes. In addition to the dense vegetation, a large proportion of the site comprises steep (>45°) rock outcrops also restricting access. All field observations were transferred on to a series of 1:500 scale working maps to ensure that all data were systematically recorded and evaluated. These working maps were subsequently interpreted to generate the Engineering Geomorphological Map (originally at a scale of 1:2000) shown in Figure 15.9.

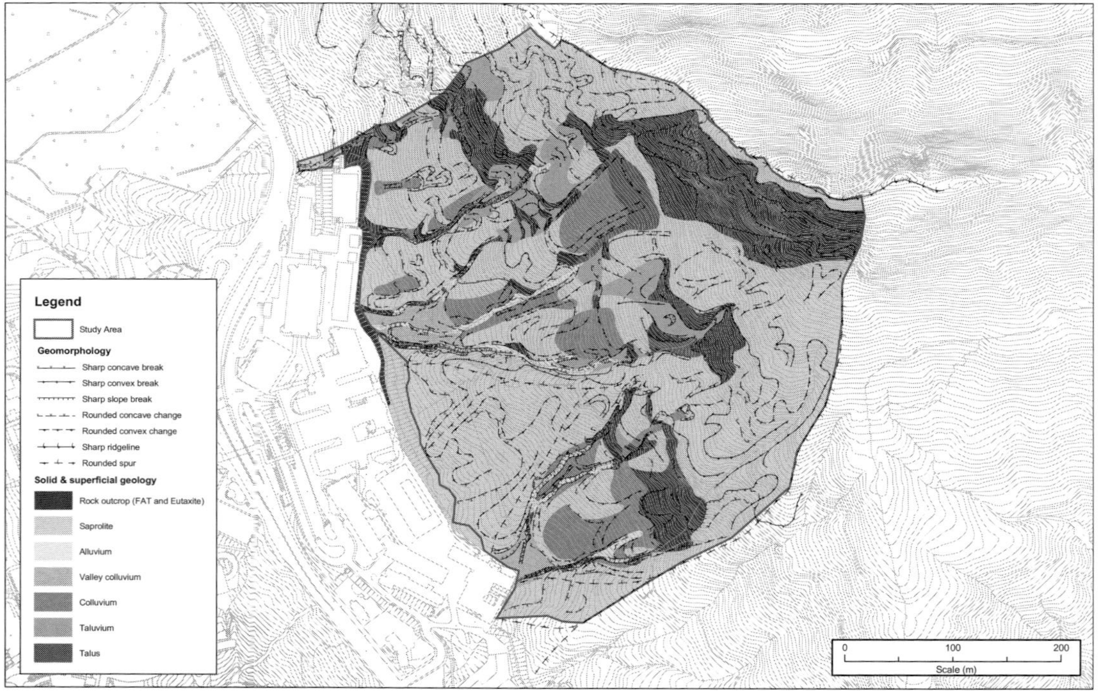

Figure 15.9 Engineering geomorphological map. Note that colluvium is mapped where it is >1 m in thickness. Thinner colluvial deposits may be more widespread. The Valley Colluvium may in fact represent the heads of fan deposits; however, considerable early anthropogenic modification has removed all evidence of these fans below the site.

7.2 Landslides

The ENTLI records seven recent landsides and 62 relict landslides within the study area. The Large Landslide Inventory records no features within the study area. Where accessible, morphological depressions identified from API were inspected, and based on the geomorphological setting, morphology and material characteristics, an interpretation was made as to whether the origin of the morphological feature involved landsliding. Where, based on the balance of evidence, the feature was considered to have involved a landslide it was mapped as such on the engineering geological map and an estimate made of the source volume adopting the formula $1/6\pi \times D \times L \times W$ (IAEG, 1990).

A total of 10 recent natural terrain landslides were confirmed during detailed field mapping and/or API. All recent landslides except two have estimated source volumes <50 m^3. These small events typically represent bank collapse along the deeply incised drainage lines. Of the two exceptions one could not be accessed due to dense vegetation and a source volume (140 m^3) and debris run-out of 70 m (with a further 150 m run-out due to fluvial reworking) was estimated from the 1976 aerial photographs. The second landslide has an estimated source volume based on field mapping of 150 m^3.

Thirty-seven relict landslides were observed during detailed field mapping and/or API. The relict landslides were degraded to varying degrees and, consequently, there is a degree of uncertainty with respect to the original source volume. It is considered that regression of some of the original steep main relict scarps have probably resulted in enlargement of the features into that visible today. As a result, the source volumes estimated have an inherent degree of conservatism and as such represent maximum volumes. Due to the degradation, it was not possible to classify the type of landslide. However, based on the geomorphological setting, it is considered that the majority were debris avalanches. Landslide debris was not evident below the main scarp of any relict landslide, so the actual travel distance of these landslides is unknown. The difference in numbers between the ENTLI relict landslides and site-specific relict landslide inventory is considered to be due to a number of joint controlled rock erosional features being misclassified as landslides during the ENTLI.

7.3 Superficial Deposits

The superficial deposits were interpreted from API and verified and described during field mapping and the following material types identified.

7.3.1 Alluvium

Alluvium was observed in all drainage lines in the study area and is probably derived from the channel banks by fluvial erosion. The alluvium typically comprises slightly gravelly, angular to sub-rounded cobbles and boulders. The thickness of the alluvium varies substantially along each drainage line. In some locations, the alluvium comprises a few isolated cobbles over rock outcrop, whereas in other locations it can reach up to 1 m thick. Locally the alluvium contains a matrix of sand and silt forming banks up to 0.5 m high.

7.3.2 Colluvium

Colluvium is defined as 'a superficial deposit transported predominantly by gravity containing <50% of material of >60 mm in size' (i.e. cobbles). Colluvium comprises dense, silty sand with many cobbles and boulders and is generally located in the lower and middle portions of the study area. The colluvium appears to be intermittent in extent, separated by extensive zones of saprolite and rock outcrop. This suggests that the present-day occurrence of colluvium may reflect the remnants of a much more extensive colluvium drape which has been subsequently removed by fluvial incision exposing the underlying saprolite and rock. Exposures along drainage lines suggest that the colluvium may be up to 3 m thick.

Valley colluvium, a subdivision of colluvium, was mapped in the lower portions of study area where the terrain generally becomes shallower (10°–20°). The material is typically matrix supported and comprises dense, sandy, gravelly silt with much rounded to sub-angular cobbles and sporadic boulders. The valley colluvium has been severely incised by drainage lines in the study area and exposures of valley colluvium up to 2.5 m high were recorded. Within valley colluvium, a lobate deposit of boulder dominant colluvium is present. The boulders within this material are atypical, reaching a maximum size of 58 m^3. Although no obvious source area(s) were mapped, it is possible that this deposit represents a large debris lobe comprising one or more events. Based on its mapped extent and assuming a thickness of 4.5 m, a conservative total volume of about 3500 m^3 was estimated. This was not evident in the API due to its masking by fluvial incision and dense vegetation.

7.3.3 Taluvium

Taluvium (transported by mass movement, i.e. landslides and screes; Fookes et al., 2007) is defined as 'a superficial deposit transported predominantly by gravity containing >50% of material, typically angular in nature, >60 mm in size'. The taluvium comprises dense, light greyish brown, angular to sub-rounded, cobbles and boulders with a matrix of finer material (gravelly, sandy silt). The taluvium is generally limited to the upper and middle portions of the study area and is typically deposited on either rock steps (~10°) or on the moderately steep terrain (25°–35°) downslope of rock outcrop. The taluvium observed on the rock steps is typically <1 m thick and has been completely removed by fluvial incision within the drainage lines. Taluvium on the moderately steep terrain is generally poorly exposed. However, a shallow recent landslide scar suggests that the taluvium deposits may be <1 m thick.

7.3.4 Talus

Talus was observed on the steep terrain (>35°) at the foot of the upper cliffs and typically comprises interlocking angular bounders <2 m^3 in volume with occasional larger boulders of <8 m^3. The talus is considered to be generally <2 m thick, although limited exposures of this material gives a large degree of uncertainty to this estimate. Talus is generally restricted to within 30 m of the foot of a cliff located above 300 mPD. However, talus was also recorded ~90 m downslope of the upper cliff (i.e. above 260 mPD). This increased distance from the cliffs may be the result of fluvial reworking.

7.3.5 Boulders

A total of 118 individual boulders (excluding boulders within talus deposits), with volume >1 m^3, were mapped where encountered in the field. The largest boulder has an estimated volume of 58 m^3, although many of the boulders mapped (~65%) had volumes <3 m^3. Field evidence suggests that they may have been transported by gravity either by mass movement or by rockfall events and are not exhumed corestones. The recorded boulders do not represent a complete boulder inventory for the study area as large portions of the terrain was not examined during field mapping due to dense vegetation and steep terrain. However, they were considered to provide a reasonable sample size to evaluate boulder size and distribution.

7.4 Terrain Units

On the basis of the API and field mapping, the study area was subdivided into five Terrain Units based on the location, morphology, geomorphological processes and solid and superficial geology (Figure 15.10). The terrain

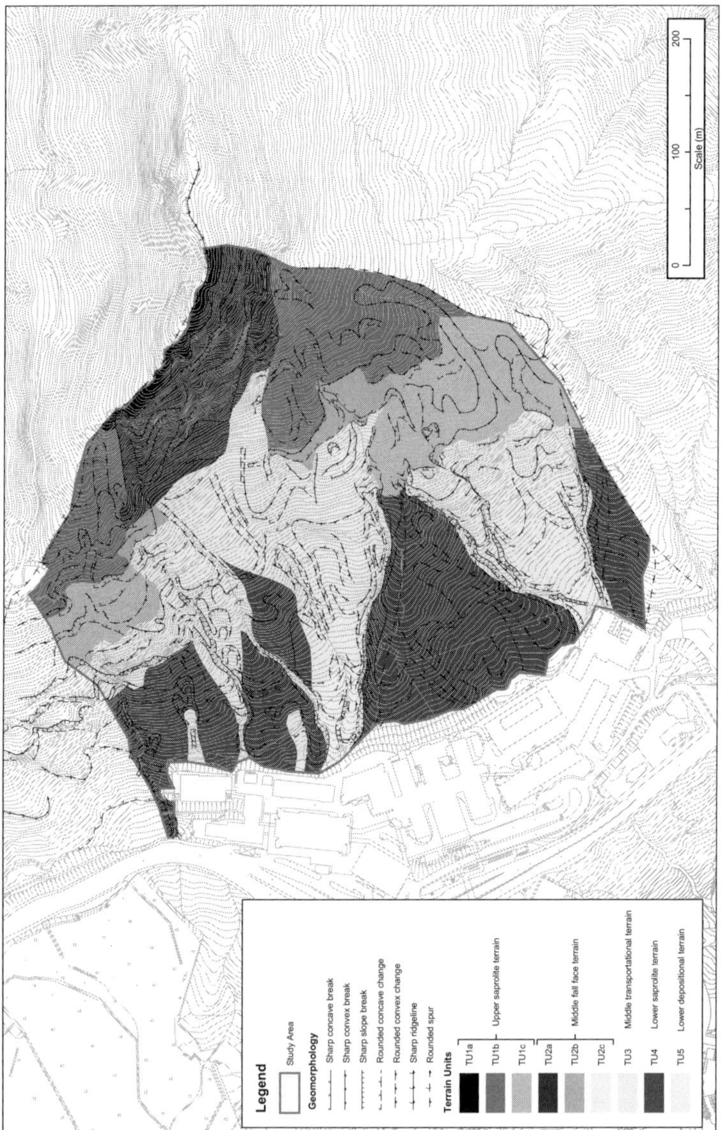

Figure 15.10 Terrain units.

units shown are broadly similar to those derived during the generation of the conceptual model based predominantly on API.

7.4.1 Upper Saprolite Terrain (TU1)

The Upper Saprolite Terrain consists of rounded topography with localised areas of intermittent rock outcrop and numerous depressions of varying size and sharpness. The lower boundary of the Upper Saprolite Terrain is generally marked by a distinct, sharp convex break in slope and is typically associated with surface or near-surface rock material, commonly forming rock cliffs. Drainage lines within this terrain unit are generally broad, although deeper incision in the lower portions of the terrain was observed in one drainage line. Given the topographical position, the Upper Saprolite Terrain may represent the oldest landforms in the study area. Based on the relative location of the rock cliffs that form the toe of this terrain unit, the Upper Saprolite Terrain has been tentatively divided into three subunits, possibly reflecting differing ages.

The main active process associated with the Upper Saprolite Terrain is considered to be mass wasting. The Upper Saprolite Terrain contains 23 relict landslides and two recent landslides. The landslides in the Upper Saprolite Terrain include the largest landslide (630 m^3) in the study areas.

7.4.2 Middle Fall Face Terrain (TU2)

A series of three main cliffs/areas of steep intermittent rock outcrop form the Middle Fall Face Terrain at the base of the Upper Saprolite Terrain. The rock cliffs are between 40 and 100 m high. From API the rock cliffs exhibit numerous sharp breaks which are associated with two sub-vertical and one sub-horizontal joint sets. Overhanging rock blocks and dilated joints were observed within the rock mass.

As with the Upper Saprolite Terrain, based on the relative location of the rock cliffs, the Middle Fall Face Terrain has been tentatively divided into three subunits, possibly reflecting differing ages.

The main active process associated with the Middle Fall Face Terrain is considered to be rockfall, with talus and taluvium commonly present below much of this terrain. Given the lack of access, these were assessed using photogrammetry techniques and the largest block is estimated to be 100 m^3 and is comparable to the 58 m^3 boulder observed during field mapping.

7.4.3 Middle Transportational Terrain (TU3)

Below the Middle Fall Face Terrain, a series of superficial deposits are present which grade from talus to taluvium to colluvium, reflecting material sorting downslope and the change from gravity to fluvial processes.

Given the distribution of the superficial deposits, it was considered that these were probably more extensive in the geological past and have been gradually removed by fluvial incision exposing the underlying saprolite and rock.

The main active processes associated with the Middle Transportational Terrain are considered to be small magnitude mass wasting and fluvial erosion and transportation along drainage lines. The Middle Transportational Terrain contains five relict landslides (<200 m^3) and two recent landslides (<50 m^3). These landslides are associated with shallow failures within the taluvium deposits which are estimated to be <2 m thick.

7.4.4 Lower Saprolite Terrain (TU4)

Lower Saprolite Terrain is present in the study area typically below 280 mPD and comprises a series of spurlines that extend downslope from the convex break in slope that forms the boundary with the Middle Fall Face Terrain to the toe of the study area. The spurlines are generally rounded and appear to be predominantly formed in saprolite, although some occasional intermittent rock outcrop was also observed.

The main active process associated with this terrain is mass wasting. The Lower Saprolite Terrain contains nine relict landslides and two recent landslides (<10 m^3). Three of the relict landslides, including the largest landslide (420 m^3), are probably associated with a deeper weathering profile associated with the largest extent of Lower Terrain saprolite. The remaining six relicts within the terrain unit are <250 m^3 and are probably associated with shallower rock head.

7.4.5 Lower Depositional Terrain (TU5)

Lower Depositional Terrain is present in part of the study area and was probably evident further downslope elsewhere. However, these deposits are no longer evident due to anthropogenic modification. The terrain is associated with gently inclined (<10°) slope gradients and is predominantly associated with valley colluvium. The main active process associated within the lower depositional terrain is considered to be fluvial reworking of mass wasting deposits. The possible large debris lobe is located within this terrain unit.

7.5 Hazard Types

Given the geomorphological setting, landslides will travel downslope and intersect drainage lines. Consequently, most of the landslide hazards associated with the study area can be categorised as CDF. Open hillslope landslide (OHL) hazards do exist; however, these are restricted to areas located at the toe of the study area. Rockfall hazards also exist.

Based on the results of the site-specific engineering geological mapping, the largest credible worst landslide source volume is estimated to be 630 $m^3$and is located within the Upper Saprolite Terrain. The landslides from such an event would enter into drainage lines and transform into CDF. Entrainment potential of these CDF is related to material type and geomorphological location. Based on the site-specific mapping, entrainment was considered to be generally restricted to areas of loose alluvium located within the drainage lines. The exception to this is where steep incised channels are present and it is considered that bank failure could occur resulting in additional entrainment.

In addition to this scenario, it was considered that a large source volume rockfall could trigger secondary failure in the underlying talus in the central portion of the study area. Finally, OHLs with a source volume of 420 m^3 were adopted as the WCE in the Lower Saprolite Terrain.

7.6 Design Event Assessment

Based on the site-specific mapping, the source volume for the WCE within the Upper Terrain was reduced from 1000 to 630 m^3. In addition, specific entrainment values in drainage lines were then derived to estimate the maximum credible volume for each individual drainage line.

In the central portion of the study area, the Upper Terrain is largely absent. Consequently, a design event comprising 100 m^3 rockfall from the Middle Fall Face Terrain and secondary failure of the talus below was adopted. Assuming that the entire thickness of talus over this width was mobilised, it was conservatively estimated that the secondary entrainment due to the rockfall impact would be 1100 m^3. Such a hazard and design-event magnitude is not apparent from the evaluation of the landslide scars alone.

With respect to entrainment within the channels, the vertical geometry of the debris run-out paths was derived from LiDAR data, whereas the dimensions of the source area and the horizontal width of the debris run-out path (channel width) were measured during detailed field mapping. In order to model entrainment from bank collapse for sub-vertical

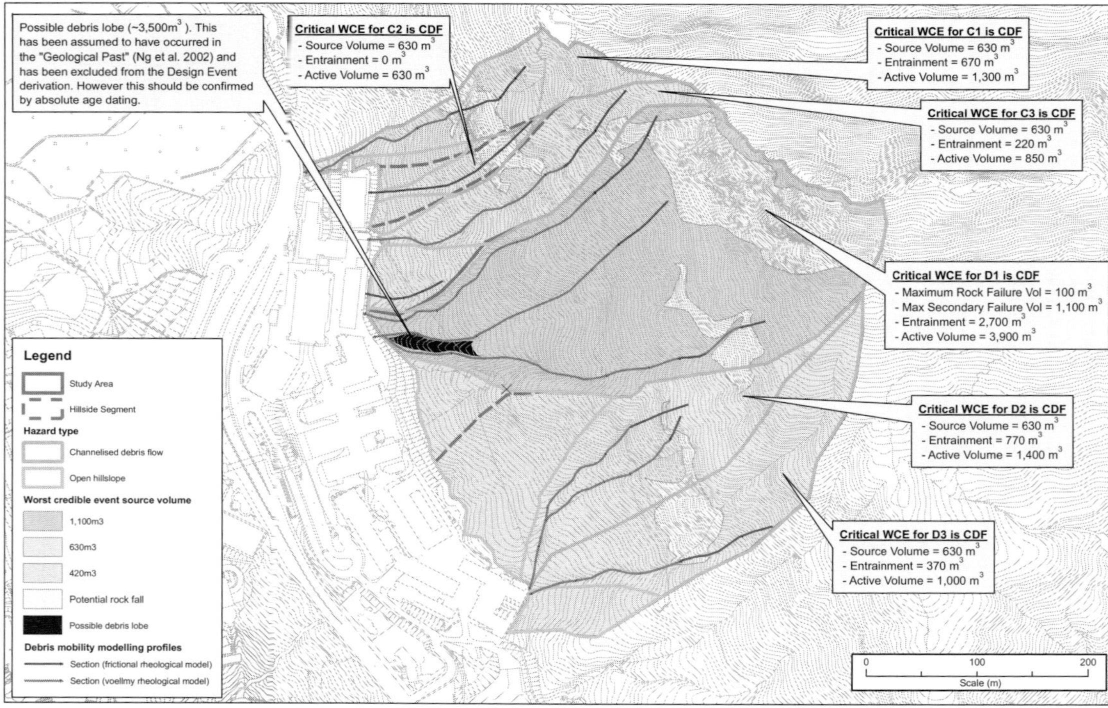

Figure 15.11 Adopted Design Events by catchments.

channel banks >2 m high, it was conservatively assumed that the current banks would upon failure, reduce back from a current slope angle of about 80°–60°. This results in an additional 1.8 m^3 of entrainable material per metre length of channel bank. With respect to an OHL hazard, a WCE of 420 m^3 was adopted based on an evaluation of relict landsides within similar terrain, i.e. the Lower Saprolite Terrain (TU4).

The resulting design event for each catchment is shown in Figure 15.11 with the design event for mitigation purposes varying from a CDF with volumes ranging from 630 to 1400 m^3 in the NW and SE catchments (compared with 1000–2700 m^3 estimated from the API alone), a rock avalanche event of 3900 m^3 in the central catchments and open hillslope failures of 420 m^3 within the Lower Saprolite Terrain.

8. CONCLUSIONS

A significant degree of engineering geological input is crucial to the successful assessment of landslide hazards from natural slopes in Hong Kong. Specifically, engineering geomorphological knowledge and experience provides the basis for the identification of geological and geomorphological controls on the location, type, magnitude, frequency and run-out characteristics of the hazards concerned. This allows the evaluation of design events beyond those evident from API alone. Such mapping skills are fundamental to the evaluation of landslide risk. Of particular importance is the use of conceptual hazard models, in particular, evaluating similar geomorphological settings or terrain units and evaluating the magnitude and frequency relationships with respect to landslide sourced volumes. Field mapping is of critical importance with respect to the identification of processes not evident from API and for the evaluation of entrainment and the derivation of parameters for mobility modelling.

ACKNOWLEDGEMENTS

This study is published with the permission of the Head of the Geotechnical Engineering Office and the Director of Civil Engineering and Development, Government of the Hong Kong Special Administrative Region, the client for the case study. The case study was carried out as part of a larger study undertaken by Fugro Hong Kong Ltd.

REFERENCES

Anonymous, 1972. The preparation of maps and plans in terms of engineering geology. Quarterly Journal of Engineering Geology 5, 295–382.

Anonymous, 1982. Land surface evaluation for engineering practice. Q. J. Eng. Geol. 15, 265–316.

Brunsden, D., Griffiths, J.S., 2001. Land surface evaluation: conclusions and recommendations. In: Griffiths, J.S. (Ed.), Land surface evaluation for engineering practice. The Geological Society of London, London, pp. 241–243. , Geological Society Engineering Geology Special Publication No. 18.

Chan, R.K.S., Mak, S.H., 2007. Landslide risk management in Hong Kong. In: Ho, K., Li, V. (Eds.) Proceedings of the 2007 International Forum on Landslide Disaster Management, Hong Kong Institution of Engineers, Hong Kong, 10–12 December 2007, pp. 17–47.

Evans, N.C., Huang, S.W., King, J.P., 1997. The Natural Terrain Landslide Study Phases I & II. Special Project Report No. SPR 5/97. Geotechnical Engineering Office, Hong Kong.

Fookes, P.G. (Ed.), 1997a. Tropical Residual Soils. Geological Society Professional Handbook. A Geological Society Engineering Group Working Party Revised Report. The Geological Society of London, London.

Fookes, P.G., 1997b. Geology for engineers: the geological model, prediction and performance. (The First Glossop Lecture). Q. J. Eng. Geol. 30, 293–424.

Fookes, P.G., Lee, E.M., Griffiths, J.S., 2007. Engineering Geomorphology, Theory & Practice. Whittles Publishing, Caithness.

Fyfe, J.A., Shaw, R., Campbell, S.D.G., Lai, K.W., Kirk, P.A., 2000. The Quaternary Geology of Hong Kong. Geotechnical Engineering Office, Hong Kong.

GCO, 1987. Geoguide 2. Guide to Site Investigation. Geotechnical Control Office, Hong Kong Government, 360 pp.

GCO, 1988. Geoguide 3. Guide to Soil and Rock Descriptions. Geotechnical Control Office, Hong Kong Government, 186 pp.

Hansen, A., 1984. Engineering geomorphology: the application of an evolutionary model to Hong Kong's terrain. Z. Geomorphol. N. F. Suppl. 51, 39–50.

IAEG, 1991. A Suggested Method for a Landslide Summary. Bulletin of the International Association of Engineering Geology. No. 41 5–12.

MFJV (Maunsell Fugro Joint Venture), 2007. Final Report on the Compilation of the Enhanced Natural Terrain Landslide Inventory. Geotechnical Engineering Office, Hong Kong.

Ng, K.C., Parry, S., King, J.P., Franks, C.A.M., Shaw, R., 2003. Guidelines for Natural Terrain Hazard Studies. GEO Report No. 138. Geotechnical Engineering Office, Hong Kong. <http://www.cedd.gov.hk/eng/publications/geo_reports/geo_rpt138.htm>

Parry, S., Jonas, D.A., 2007. The use of LIDAR for landslide hazard assessments: Hong Kong case studies. Proceedings of the Conference: Engineering Geology in Geotechnical Risk Management, Hong Kong Regional Group of the Geological Society of London, London, pp. 155–161.

Parry, S., Ng, K.C., 2010. The assessment of landslide risk from natural slopes in Hong Kong: an engineering geological perspective. Quarterly Journal of Engineering Geology and Hydrogeology 43 (3), 307–320.

Parry, S., Massey, C.I., Williamson, S.J., 2002. Landslide susceptibility analysis for natural terrain hazard studies – Tsing Shan foothills area. Proceedings of the Conference: Natural Terrain – A Constraint on Development? Institute of Mining and Metallurgy, Hong Kong Branch, Hong Kong, 14 November 2002, pp. 113–123.

Parry, S., Ruse, M.E., Ng, K.C., 2006. Assessment of natural terrain landslide risk in Hong Kong: an engineering geological perspective. Proceedings of the 10th

Conference of the International Association of Engineering Geology, Nottingham, UK, 14–17 September 2006, Accepted Paper No. 299.

Parry, S., Clahan, K.B., Krug, K.. Millis, S. 2010. The importance of reading the landscape: the use of engineering geomorphology in regional landslide hazard assessments. A case study from Hong Kong. Proceedings of the 11th IAEG Conference New Zealand. Taylor & Francis Group, London, UK.

Scott Wilson (Hong Kong) Ltd., 1999. Specialist API Services for the Natural Terrain Landslide Study – Interpretative Report. Report to Geotechnical Engineering Office, Hong Kong, 32 pp. Plus 6 Appendices.

Sewell, R.J., Campbell, S.D.G., 2005. Report on the Dating of Natural Terrain Landslides in Hong Kong. GEO Report 170. Geotechnical Engineering Office, Hong Kong, 151 pp. <http://www.cedd.gov.hk/eng/publications/geo_reports/geo_rpt170.htm>

Sewell, R.J., Campbell, S.D.G., Fletcher, C.J.N., Lai, K.W., Kirk, P.A., 2000. The Pre-Quaternary Geology of Hong Kong. Geotechnical Engineering Office, Hong Kong.

Varnes, D.J., 1984. Landslide Hazard Zonation: A Review of Principles and Practice. International Association of Engineering Geology, Commission on Landslides and Mass Movement, UNESCO, France.

Wong, H.N., 2009. Rising to the challenge of natural terrain landslides. Natural Hillsides: Study and Risk Mitigation Measures. Proceedings of the 20th Annual Seminar, Geotechnical Division, Hong Kong Institution of Engineers, Hong Kong, pp. 15–54.

CHAPTER SIXTEEN

A Geomorphological Map as a Tool for Assessing Sediment Transfer Processes in Small Catchments Prone to Debris-Flows Occurrence: A Case Study in the Bruchi Torrent (Swiss Alps)

David Theler and Emmanuel Reynard
Institute of Geography, University of Lausanne, Anthropole, Lausanne, Switzerland

Contents

1. INTRODUCTION

In the Swiss alpine range (Rhone River valley), non-exhaustive surveys related to hydrological hazards indicate that ~235 localities could be reached by sediment-laden floods (Theler et al., 2007, 2008). Among these hydrological phenomena, debris flows are one of the most important

Developments in Earth Surface Processes, Volume 15
ISSN: 0928-2025, DOI: 10.1016/B978-0-444-53446-0.00016-1

processes of sediment transfer in mountainous areas (Sterling and Slaymaker, 2007) and are triggered in many geomorphological contexts. Factors acting on triggering of debris flows depend on meteorological (intensity and duration of rainfall, temperatures, snow cover), topographic (slopes), geomorphological ((in)activity of geomorphological processes, instabilities, particle size and thickness of sediment stores), thermic (permafrost evidence), geological (faults and lithological characteristics) and hydro(geo)logical (shape of the hydrological network, flow regimes, underground water circulations, glacier occurrence) parameters. All of these elements may combine so that it is difficult to define a typology of the events and to model and predict these phenomena.

Finally, a correlation between rainfall and debris-flow occurrence is not always established and a number of debris flows seem to occur when a geomorphological threshold in the channel is reached (Bovis and Jakob, 1999; Glade, 2005; Brayshaw and Hassan, 2009).

In terms of territorial planning and hydrological natural hazard management, the estimation of sediment volumes potentially mobilised in small torrential systems, as well as the identification of processes responsible of their mobilisation and transfer within the torrential system (main channel), is of great importance.

The estimation of sediment volumes potentially mobilised and transported by debris flows can be studied with the concept of 'sediment cascades' that allows the calculation of sediment budgets of several subsystems included in a geosystem (Barsch and Caine, 1984; Schrott et al., 2002). The calculation of a sediment budget necessitates the identification of processes of erosion, transportation and deposition acting within the catchment and their rates and controls (Reid and Dunne, 1996; Otto and Dikau, 2004; Otto, 2006; Beylich and Warburton, 2007). Identification and differentiation of the main sediment stores are central for conceptualising torrential systems as a succession of connected reservoirs varying in storage periods and emptying velocity. In this context, a geomorphological map is a useful tool for evidencing the processes responsible for the formation of sediment volumes in the sediment sources of alpine streams.

This chapter presents the inputs of a pragmatic geomorphological method for mapping the sediment stores in the triggering zones of debris flows. It aims at identifying, mapping and estimating, semi-quantitatively, the volumes that could reach a debris-flow channel and that could be potentially mobilised during a rain event. The method is based on both field and geographic information system (GIS) mapping, and it is particularly adapted to small and steep catchments where classical geomorphological mapping is difficult due to access difficulties and, sometimes, vegetation land cover. The mapping method and results obtained for the Bruchi torrent (Western Swiss Alps) are presented and discussed.

2. THE DEVELOPMENT OF A DYNAMIC GEOMORPHOLOGICAL MAPPING METHOD

2.1 Quick Overview of Geomorphological Mapping in Switzerland

Mapping sediment transfers in mountain watersheds involves identifying all the processes and landforms included in the erosion–deposition cycle as well as their relationships. The combination of different processes, depending on the scale, may be very complex and difficult to map.

In fact, classical geomorphological maps are generally insufficient to appreciate and represent dynamic environments like debris-flow systems, especially because their related landforms (natural levees, gullies, erosion scarps and so on) may change very quickly over a short time and space scale (Theler and Reynard, 2008). At large scales (1:10,000, 1:5000 or higher) delineating such landforms precisely and especially distinguishing erosion and accumulation landforms becomes difficult and sometimes subjective. This distinction is, however, the basis of numerous geomorphological legends that aim at designing morphogenetic maps; these are quite limited in terms of process analysis, natural hazard prevention and land planning.

This is the case for most of Swiss geomorphological mapping systems developed in the country since the 1940s and among these the ‘Swiss legend’ was the first (Figure 16.1). We present, briefly, the three most common Swiss legend systems. Developed in the 1940s by Annaheim (1944), this first legend combines two colours – red (and/or black, e.g. Baumann, 1976; Figure 16.1) for erosional landforms and green for accumulation landforms – with frames and signs in the same colours corresponding to landforms with a descriptive and morphogenetic meaning (Kienholz, 1979; Kienholz et al., 1993; Schoeneich, 1993a,b). The main advantage of these ‘morphodynamic’ maps is their simplicity by showing an instant overview of the dynamics. However, landform genesis is not clear and sometimes difficult to understand (Schoeneich, 1993a,b).

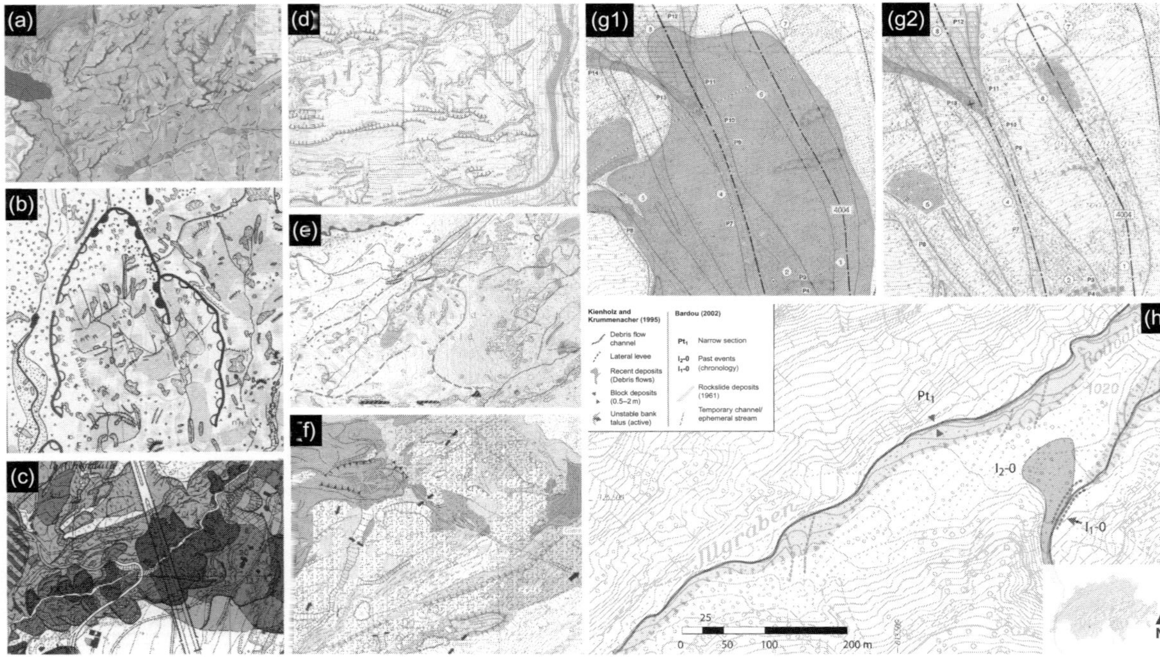

Figure 16.1 Extracts of some geomorphological maps produced in Switzerland. (a) Small-scale geomorphological map of Switzerland (Swisstopo, 2007). (b) Map of Geomorphology of Grindelwald, Switzerland: Scale 1:10,000. (c) Map of regional instabilities of Lausanne-East (Noverraz, 1985). (d) Geomorphological map of Zentralen Aargaus (Moser, 1958). (e) Geomorphological hazards map of Grindelwald (Baumann, 1976). (f) Geomorphological map of Tsanfleuron, scale 1:10,000 (Reynard, 1993). (g) Phenomena maps for gravitational processes (1) and snow avalanches (2). (h) Improvement of the phenomena legend (Kienholz and Krummenacher, 1995) in Illgraben torrent by making a distinction between punctual and potential areal alimentation of a debris-flow channel. The strict application of 'the phenomena legend' may result in a loss of information about the potential alimentation of the debris flows (Bardou, 2002).

The 'IGUL' (Institute of Geography, University of Lausanne) legend system was developed at the end of the 1980s. It is a simplification of the French R.C.P. 77 system (Tricart, 1972) for conventional signs, German GMK system (Stäblein, 1980; Leser and Portmann, 1985) for colours depicting morphogenetic domains and the Swiss 'DUTI' system (Figure 16.1), which is the name of a project on a mapping system of instable terrain in the 1980s (Noverraz, 1985) with representation of gravitational landforms. The system was simplified: morphography, slopes and lithology are not represented (Schoeneich et al., 1998). The IGUL system was designed for a 1:10,000 scale but is usable on scales from 1:5000 to 25,000 (Reynard, 1993; *cf.* Figure 16.1) and has been widely used in the Swiss Alps, Prealps and Jura (Schoeneich et al., 1998).

The Swiss 'phenomena legend' (Kienholz and Krummenacher, 1995) is used for preliminary fieldwork in hazard mapping. Despite the variety of tools available, this system leads to some inconsistencies in the field. In fact, the recommended legend for mapping the phenomena only gives a momentary vision of one single event. In theory, the phenomena map should be redrawn after each new event and all the maps should finally be superposed to have a comprehensive view of the endangered area (Theler and Reynard, 2007). The superposition of maps allows a general view of the geomorphological dynamics but without a strong interpretation of the phenomena (Bardou, 2002; Theler et al., 2007, 2008).

2.2 Identification and Delineation of Sediment Stores

The proposed approach aims at better taking into account the dynamics of processes. It is organised in to six steps (Figure 16.2). The first step consists in localising potential sediment sources within the torrential system. This phase is based on the evidence of geomorphological indices through high-accuracy digital elevation model (DEM) processing, field survey and/or topographical maps. Due to the strong heterogeneity in the sediment supply by the slopes (Johnson and Warburton, 2002), some parts of the reception basin may be completely inactive, whereas other supply zones may be identified along the transport channel and even on the upper part of the alluvial fan. A key issue is, therefore, to determine if the sediment stores are connected or not with active channels and/or the main channel (Step 2, Figure 16.2). This task is done by simulating the hydrographic network using classical ArcGIS® hydrologic tools on precise LiDAR (light detection and ranging) airborne DEM provided by Swisstopo (2007) and then on field checks. Based on the assumption that

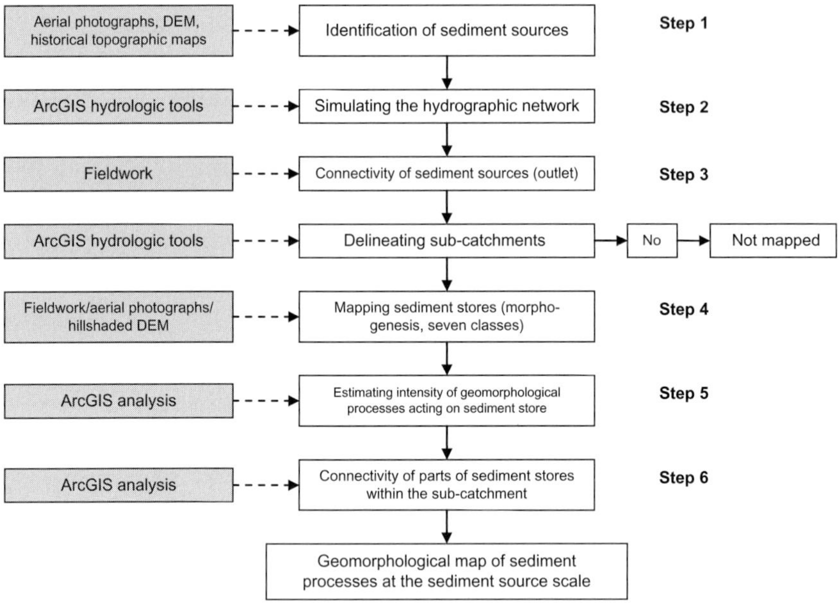

Figure 16.2 Flow chart of the procedure used in the mapping method for small alpine catchments.

sedimentary transfers correspond to hydrological fluxes, this step allows areas that drain water (and sediments) to a common outlet to be determined (Theler and Reynard, 2008; Theler et al., 2008). Drainage areas corresponding to the connected preferential channels are then delineated after verification of the connection points in the field. The approach allows, therefore, the delimitation of small geomorphological units (sub-catchments) more or less in the same way as it was proposed by Pasuto and Soldati (1999) or Bartsch et al. (2009). The properties of the sub-catchments, which may be composed of various sediment stores, are then described and mapped (Step 4).

2.3 Mapping Sediment Stores as Geomorphological Units

2.3.1 Delineating Morphogenetic Sediment Stores

Step 4 (Figure 16.2) consists in delineating sediment stores under a simple morphogenetic approach. We propose a classification of six types of stores corresponding to six processes typical of mountain contexts (Figure 16.3) such as glacial, fluvio-glacial, fluvial, gravitational, structural and periglacial landforms. Structural surfaces – i.e. free faces or rock escarpments on

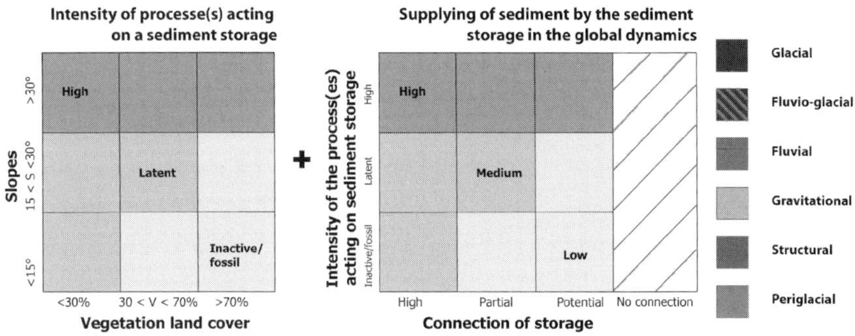

Figure 16.3 Step 5: Two matrices depicting the importance of the sediment storage in the global sediment dynamics and the main considered geomorphological processes.

fractured or much folded rocks – are also considered as storage surfaces because they have quite similar behaviour to unconsolidated materials. We use a typical panel of colours in reference to principles established by Joly (1962) and adopted in the IGUL legend that relates to different types of processes (glacial, periglacial, gravitational, fluvial and so on) responsible for the accumulation.

2.3.2 Activity of Sediment Stores

Each contributing zone (sub-catchment) is mapped by using two simple matrices (Step 5, Figure 16.2) inspired by the legend for hazard mapping in Switzerland (OFAT et al., 1997). The first matrix comprises information on *slope* and *vegetation cover*, directly derived from the DEM, and allows zones sensitive to erosion to be classified in to three classes of intensity (high, latent and inactive). Slope information is central to the creation of susceptibility maps. The minimal slope for debris-flow triggering is 25° for processes triggered on the slope and 11–12° for remobilisation processes of sediments accumulated in the channels (Bonnet-Staub, 2001). Based on different values taken from literature, three classes were defined (<15°, 15–30°, >30°), which correspond to the general slope of a fan (<15°) and angle of repose (>30°).

The role of *vegetation* in slope stability can be of a great importance (Greenway, 1987) by diminishing rain drop impact and runoff and by stabilising non-consolidated sediments through the effect of the roots. The vegetation land cover can be mapped on the basis of aerial photographs or by using the differences of two high-accuracy LiDAR airborne DEMs

(Gachet, 2009) provided by Swisstopo (2007). Three classes of land cover (>70%, 30–70%, <30%) were defined.

The second matrix takes the output of the first matrix (intensity of the process acting on the accumulation landform or sediment store) and crosses it with the connection of the stores between secondary channels and/or the main channel. Three variations in intensity of morphogenetic colours depict the morphodynamics and, more specifically, the susceptibility (low, medium, high) of the stores situated in the different sub-catchments to participate in the global sediment dynamics.

2.3.3 Connectivity of Sediment Stores

The connectivity of the sediment stores with the main channel has importance in the evaluation of volumes of sediments potentially mobilised during a debris-flow event (Zimmermann et al., 1997) and is commonly correlated to the hydrological connectivity (Croke et al., 2005). Four classes of connectivity (high, partial, potential, no connection) were defined but only the first three are used in mapping. Although the interfluves also contribute to the sediment supply, we assume that the processes involved (mass movements and superficial runoff) do not have a strong influence on the sediment dynamics and can be ignored at this scale.

2.3.4 Symbology

Specific symbols adapted to small torrential watersheds were created. They integrate generic hydrographical elements (lakes, springs), artificial objects that may disturb the flow (bridges, hydropower infrastructures), as well as dead wood that may have a high impact on flow by damming the *talweg*, increasing sedimentation upslope and increasing granulometry downslope when the dam does not occupy the whole river bed (Maridet et al., 1996; Brayshaw and Hassan, 2009). However, damming by wood deposits may also have positive impacts by limiting the sediment transfers and reducing the hydraulic energy.

3. EXAMPLE OF APPLICATION IN THE BRUCHI TORRENT

3.1 Geomorphological Settings

The method was tested in the Bruchi torrent (south-western Swiss Alps). This perennial watercourse (7.5 km length) constitutes the main tributary

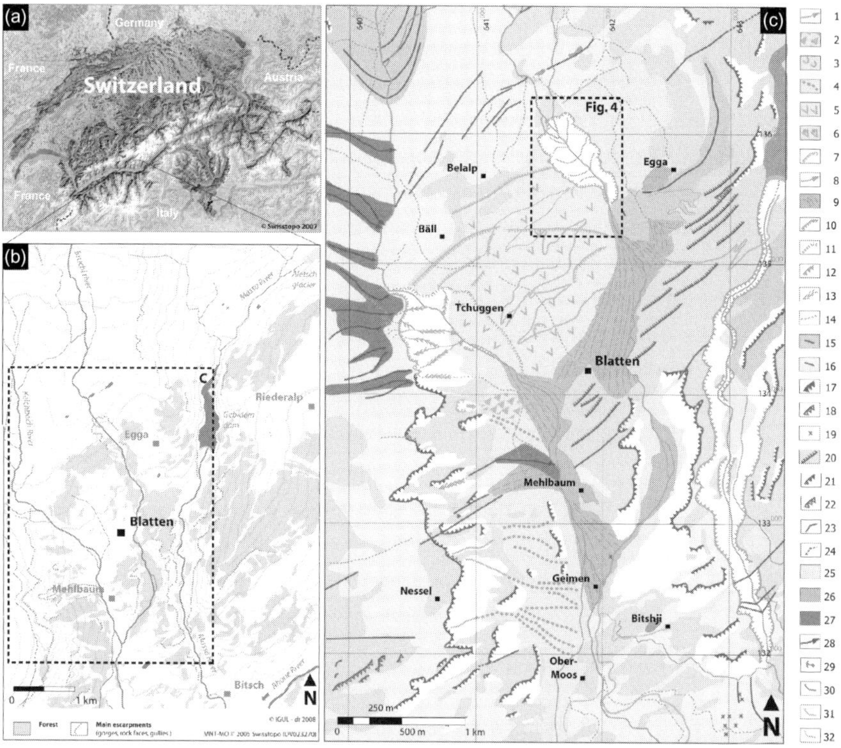

Figure 16.4 Location and geomorphological map of the study site. Geomorphological legend: (1) scree corridor; (2) rock avalanche deposits; (3) landslide; (4) vegetalised scree cone; (5) rockslide; (6) rockslide (with dislocation); (7) erosional escarpment; (8) debris-flow channel; (9) alluvial deposit; (10) active bank erosion; (11) inactive bank erosion; (12) gorge; (13) gullying; (14) gullying and gullies (1:5000); (15 and 16) Holocene and Lateglacial moraine deposit and ridge; (17 and 18) glacial escarpment (covered); (19) erratic boulder; (20) small and covered glacial escarpments; (21 and 22) rock escarpment (covered and uncovered); (23 and 24) fault (supposed); (25) bedrock covered with soil; (26) organic deposit; (27) snow avalanche deposits; (28) avalanche corridor; (29) spring; (30) hydrography; (31) dyke; (32) secondary road. Zones in white are erosional zones (e.g. free faces).

of the Kelchbach torrent, which drains the right part of the Massa River Valley (Figure 16.4). The source of the Bruchi is situated at about 2800 m asl in a depression located under the Hohstock Mountain (3226 m). The upper part of the stream is characterised by a meandering course through pastures developed on basal till with granitic and gneissic *roches moutonnées* belonging to the crystalline Aarmassif (Steck, 1966) of the Helvetic Alps.

The retreat of the Aletsch glacier system was followed by slope debuttressing, which triggered deep rockslides that particularly affected a slope located near the settlement of Tschuggen. This slope is also characterised by large gullies, inherited from geomorphological processes more active in the past by the Kelchbach and the Bruchi torrents. The whole sediment load of the Bruchi torrent is transported periodically by channelised debris flows and comes from a large active gully (4.2 km^2) situated midfield of the stream course (1600–2000 m asl). The perimeter has evolved rapidly between 1940 and 1999 (Bollschweiler, 2003), especially near the road Blatten–Tschuggen–Belalp, where regressive erosion has reached about 70 m in 30 years (Theler and Reynard, 2008).

The geomorphic activity varies spatially from one part to the other. In the upper part of the main gully, talus accumulation due to solifluction processes active in the subvertical headwalls at the top of the area is dominant. Some elements are stored temporarily at the foot of the cliffs in shallow debris accumulations, and localised rockslide deposits are also visible in the north-eastern part of the zone. On the right side of the gully, material is transported and accumulated preferentially by fluvial processes in a network of secondary gullies that are active only during pluviometric events. On the left side, the gully network, the steep slopes and the scarce vegetation cover are all indicators that these units are very active alimentation zones. This sector corresponds to the triggering zone of most debris-flow events due to the convergence of superficial runoff coming from the cliffs and flow coming from springs. Important stocks of older fluvial sediments corresponding to natural levees cover the bank and are also visible along the main channel of the Bruchi torrent. Absence of trees and lichens shows that the store is clearly unstable and is eroded during medium to large flows (Figures 16.5 and 16.6).

3.2 Results

A dynamic geomorphological map of the gullying zone of the Bruchi was established with the proposed methodology and is presented below. White and black arrows inform about the type and frequency of the main geomorphological processes acting on the sub-catchments. This information was developed using different methods to estimate sediment fluxes and denudation rates adapted for debris-flows environments (Schrott et al., 2002; Beylich and Warburton, 2007). Reference coloured lines, painted stones and wooden markers were installed during summer 2007

Figure 16.5 Different sedimentary stocks present in the studied area: (a) main channel of Bruchi torrent; (b) lateral landslide periodically drained by debris flows; (c) fractured rock escarpment at the top of the drainage basin; (d) general view downstream from the top of the basin; (e-h) rapid changes (erosion, collapses, deposit of natural levee) in different kind of stores (Pictures: April and July 2007).

and observed regularly until June 2009. These measurements highlighted a strong supply activity of two geomorphological units (that correspond to a landslide eroded at the surface by debris slides, respectively on the right and left sides of the Bruchi channel) at almost each pluviometrical event.

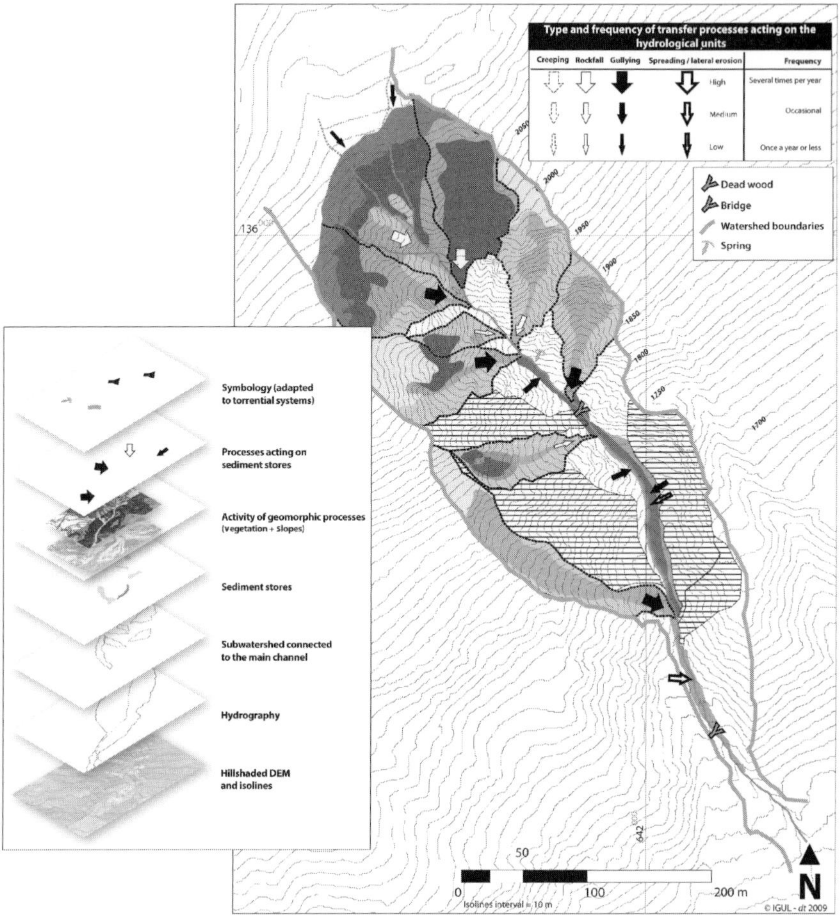

Figure 16.6 Dynamic geomorphological map of sediment transfer processes for the Bruchi torrent.

4. DISCUSSION

With respect to classical geomorphological mapping, this method focuses on specific parts at the (Bruchi) catchment scale and aims at delineating and characterising the sediment stores (potentially) supplying a watercourse. It is, therefore, a process-oriented method, whereas most geomorphological mapping systems developed in the past were much more landform-oriented legends aiming at classifying the landforms according to the main processes active in the area. Nevertheless, the proposed approach does not completely abandon the genetic approach as it classifies the stores according to the processes responsible for the deposits that are typical of alpine environments. One improvement is the use of intensities of colours aiming at representing the susceptibility of sediment stores to reach the main channel. In this sense, the proposed methodology tries to combine morphogenetic and morphodynamic information. The final aim is to give a first evaluation of the sediment volumes that can be potentially mobilised by a hydro-meteorological event (Figure 16.3).

The second item of interest with the method is the possibility of implementing the sediment cascade concept in a cartographic system. Not only are the different interconnected reservoirs defined but they are also mapped, with an indication of the type of deposits (genetic classification of deposits). This improvement gives a better comprehension of the spatial relationships of the sediment reservoirs at different scales.

Nevertheless, the method is still not sufficient for really quantifying the volumes that could be mobilised during an event. Moreover, the dynamic and frequency of the processes involved in the deposition of sediments in the main channel are also approached in a qualitative manner based on simple field measurements. This is the reason why the cartographic information must be complemented by field measurements (e.g. LiDAR) carried out in the most sensitive areas defined by the mapping method. Finally, delineation of morphogenetic stores can be problematic at this scale, and it is often not simple to delineate them.

5. CONCLUSIONS AND PERSPECTIVES

The aim of this study was to improve the morphogenetic approach traditionally used in geomorphological mapping in order to be better adapted to process-oriented methods usually used in natural hazard studies.

At the moment, the proposed method does not take into account the secondary stores and considers only the original supply process and the connection between the several stores within a hydro-geomorphological unit. Although sediment transfer starts generally from the hill slopes – where physical weathering followed by gravitational processes are predominant – the time of residence of sediments in the upper catchment is very variable depending on the topographic setting and the intensity of processes. The next step will be to assess the transfers quantitatively, estimated based on values already published and available in the geomorphological literature.

REFERENCES

Annaheim, A., 1944. Begleitwort zur Legende zur morphologischen Grundkarte der Schweizer Alpen. Basel, Manuskript (unpublished).

Bardou, E., 2002. Méthodologie de diagnostic des laves torrentielles sur un bassin versant alpin. Ph.D. Thesis (2479), EPFL, Lausanne, Switzerland.

Barsch, D., Caine, N., 1984. The nature of mountain geomorphology. Mt. Res. Dev. 4, 287–298.

Bartsch, A., Gude, M., Gurney, S.D., 2009. Quantifying sediment transport processes in periglacial mountain environments at a catchment scale using geomorphic process units. Geogr. Ann. 91A, 1–9.

Baumann, T., 1976. Geomorphologische Gefahrenkarte von Grindelwald. Blatt 2. Geographischen Institut der Universität Bern.

Beylich, A.A., Warburton, J., 2007. Analysis of Source-to-Sink-Fluxes and Sediment Budgets in Changing High-Latitude and High-Altitude Cold Environments. SEDIFLUX Manual, first ed. NGU Report 2007.053.

Bollschweiler, M. 2003. Frequenzanalyse von Murgangereignissen anhand dendrogeomorphologischer Untersuchungen. Murkegel Bruchji, Blatten b. Naters, Wallis, Schweiz. Diploma Thesis, University of Fribourg, Fribourg.

Bonnet-Staub, I., 2001. Une méthodologie d'analyse et de cartographie de l'aléa 'initiation de laves torrentielles' – Application au torrent du Bragousse (France). Int. Bull. Eng. Geol. Env. 59, 319–327.

Bovis, M.J., Jakob, M., 1999. The role of debris supply conditions in predicting debris flow activity. Earth Surf. Process. Landforms 24, 1039–1054.

Brayshaw, D., Hassan, M.A., 2009. Debris flow initiation and sediment recharge gullies. Geomorphology 109, 122–131.

Croke, J., Mockler, S., Fogarty, P., Takken, I., 2005. Sediment concentration changes in runoff pathways from a forest road network and the resultant spatial pattern of catchment connectivity. Geomorphology 68, 257–268.

Gachet, G., 2009. Analyse et exploitation des données de LIDAR aéroportés pour la caractérisation des milieux boisés de la Suisse. Thèse n°4283, EPFL, 294 pp.

Glade, T., 2005. Linking debris-flow hazard assessments with geomorphology. Geomorphology 66, 189–213.

Greenway, D.R., 1987. Vegetation and slope stability. In: Anderson, M.G., Richards, K.S. (Eds.), Slope Stability. Wiley, Chichester, pp. 187–230.

Johnson, R.M., Warburton, J., 2002. Annual sediment budget of a UK mountain torrent. Geogr. Ann. 84A (2), 73–88.

Joly, F.M., 1962. Principes pour une méthode de cartographie géomorphologique. Bull. Ass. Géogr. Franç. 309/310, 271–277.

Kienholz, H., 1979. Maps of geomorphology and natural hazards of Grindelwald, Switzerland: Scale 1:10,000. Arct. Alp. Res. 10 (2), 168–184.

Kienholz, H., Krummenacher, B., 1995. Légende modulable pour la cartographie des phénomènes. Recommandations. Dangers naturels, OFEFP et OFEE, Berne, 19 pp.

Kienholz, H., Fitze, P., Haeberli, W., Leser, H., Maisch, M., Monbaron, M., 1993. Geomorphology in Switzerland (46). In: Walker, H.J., Grabau, W.E. (Eds.), The Evolution of Geomorphology. A Nation-by-Nation Summary of Development. John Wiley & Sons, Chichester, pp. 429–439.

Leser, H., Portmann, J.-P., 1985. Die geomorphologische Kartierung in der Schweiz. (unpublished report).

Maridet, L., Piégay, H., Gilard, O., Thevenet, A., 1996. L'embâcle de bois en rivière : un bienfait écologique ? Un facteur de risques naturels ? La Houille Blanche 5, 32–38.

Moser, S., 1958. Studien zur Geomorphologische des zentralens Aargaus. Mitt. Geogr. Ethn. Ges. Basel X, 1–98.

Noverraz, F., 1985. Détection et utilisation des terrains instables. Rapport final. Ecole polytechnique fédérale de Lausanne (EPFL), 229 pp.

OFAT, OFEE, OFEFP, 1997. Prise en compte des dangers dus aux crues dans le cadre des activités de l'aménagement du territoire. Recommandations 1997, Dangers naturels, Berne, 32 pp.

Otto, J.-C., 2006. Paraglacial Sediment Storage Quantification in the Turtmann Valley, Swiss Alps. Doctoral Thesis, Rheinischen Friedrich-Wilhelms-Universität Bonn, Bonn, 195 pp.

Otto, J.-C., Dikau, R., 2004. Geomorphologic system analysis of a high mountain valley in the Swiss Alps. Z. Geomorphol. N.F. 48, 323–341.

Pasuto, A., Soldati, M., 1999. The use of landslide units in geomorphological mapping: an example in the Italian Dolomites. Geomorphology 30, 53–64.

Reid, L.M., Dunne, T., 1996. Rapid evaluation of sediment budgets. Catena Verlag, Reiskirchen 200 pp.

Reynard, E., 1993. Comparaison de cartes géomorphologiques à différentes échelles: le cas de la vallée de la Morge. In: Schoeneich, P., Reynard, E. (Eds.), Cartographie géomorphologique, cartographie des risques. Institut de Géographie, Université de Lausanne, Travaux et Recherches 9 pp. 25–30.

Schoeneich, P., 1993a. Cartographie géomorphologique en Suisse. Une bibliographie commentée et des propositions. In: Schoeneich, P., Reynard, E. (Eds.), Cartographie géomorphologique, cartographie des risques. Institut de Géographie, Université de Lausanne, Travaux et Recherches 9, pp. 1–12.

Schoeneich, P., 1993b. Comparaison des systèmes de légendes français, allemand et suisse – principes de la légende IGUL. In: Schoeneich, P., Reynard, E. (Eds.), Cartographie géomorphologique, cartographie des risques. Institut de Géographie, Université de Lausanne, Travaux et Recherches 9, pp. 15–24.

Schoeneich, P., Reynard, E., Pierrhumbert, G., 1998. Geomorphological mapping in the Swiss Alps and Prealps. Wiener Schriften zur Geographie und Kartographie Band 11, 145–153.

Schrott, L., Niederheide, A., Hankammer, M., Hurfschmidt, G., Dikau, R., 2002. Sediment storage in a mountain catchment: geomorphic coupling and temporal variability (Reintal, Bavarian Alps, Germany). Z. Geomorphol. N.F. 127, 175–196.

Stäblein, G., 1980. Die Konzeption der Geomorphologischen Karten GMK 25 und GMK 100 im DFG-Schwerpunktprogramm. Methoden und Anwendbarkeit geomorphologischer Detailkarten. Berl. Geogr. Abhan. 31, 13–30.

Steck, A., 1966. Petrographische und tektonische Untersuchungen am Zentralen Aaregranit und seinen altkristallinen Hüllgesteinen im westlichen Aarmassiv im Gebiet Belalp-Grisgighorn. Beitrage zur Geologischen Karte der Schweiz, Stämpfli, Bern, 99 pp.

Sterling, S., Slaymaker, O., 2007. Lithologic control of debris torrent occurrence. Geomorphology 86, 307–319.

Swisstopo, 2007. Géodonnées. Geodata-news 14, 1–4.

Theler, D., Reynard, E., 2007. Geomorphological mapping in high mountain watersheds: the contribution of geomorphology to the evaluation of sediment transfer processes. Landform Anal. 5/2007, 85–86.

Theler, D., Reynard, E., 2008. Mapping sediment transfer processes using GIS applications. Proceedings of Sixth ICA Mountain Cartography Workshop, Lenk, Switzerland, pp. 227–234.

Theler, D., Reynard, E., Bardou, E., 2007. From geomorphological mapping to risk assessment: a project of integrated GIS application in the western Swiss Alps. Proceedings of the Fifth Mountain Cartography, Bohinj, Slovenia, pp. 236–241.

Theler, D., Reynard, E., Bardou, E., 2008. Assessing sediment dynamics from geomorphological maps: Bruchi torrential system, Swiss Alps. J. Maps 2008, 277–289.

Tricart, J., 1972. Normes pour l'établissement de la carte géomorphologique détaillée de la France: classification codée, critères d'identification et légende pratique (1/20,000, 1/25,000, 1/50,000). CNRS 12, 37–105.

Zimmermann, M., Mani, P., Romang, H., 1997. Magnitude-frequency aspects of alpine debris flows. Eclogae Geol. Helv. 90, 415–420.

CHAPTER SEVENTEEN

Geomorphological Assessment of Complex Landslide Systems Using Field Reconnaissance and Terrestrial Laser Scanning

Malcolm Whitworth[a], Ian Anderson[b] and Graham Hunter[c]
[a]School of Earth and Environmental Sciences, University of Portsmouth, Portsmouth, UK
[b]Halcrow Group Ltd, Martlett House, Chichester, UK
[c]3D Laser Mapping Ltd, Bingham, Nottingham, UK

Contents

1. INTRODUCTION

Geomorphological mapping is a widely accepted part of conventional terrain evaluation and plays an important role in slope-stability assessments, especially where it is supported by complimentary methods such as desk study and historical analysis, aerial photographic interpretation, sub-surface investigation and monitoring (Griffiths, 2001; Lee, 2001). Standard field-based morphometric and morphological techniques are appropriate for landslide mapping, where terrain features such as slope morphology, drainage and vegetation can be used to identify slope instability (Crozier, 1984; Bromhead, 1992; Soeters and Van Western, 1996;

Developments in Earth Surface Processes, Volume 15
ISSN: 0928-2025, DOI: 10.1016/B978-0-444-53446-0.00017-3

Dikau et al., 1997). Once compiled, this information can form the basis for the compilation of detailed engineering geomorphological maps and landslide inventories for an area (Griffiths et al., 2002).

Conventional, remotely sensed data for landslide mapping have typically relied upon aerial photographic interpretation and satellite imagery; but now with the advent of terrestrial laser scanning, opportunities for detailed landslide terrain evaluation through the generation of high-resolution digital terrain models (DTMs) is possible. The technology has developed rapidly in recent years since the early applications of laser scanning for landslide morphometric studies (Rowlands et al., 2003); with recent developments in range and precision, laser scanners can now provide rapid geomorphological surveys for large areas.

This chapter discusses the application of terrestrial laser scanning to the study of a complex landslide system in the Cotswolds region of the United Kingdom by firstly discussing the field-based assessment and landslide geomorphology and then the application of terrestrial laser scanning.

2. STUDY AREA

The study was located on the Cotswolds escarpment to the east of the village of Broadway (OS Grid Reference 410000 237500). The village itself is located in the Vale of Evesham, within the Wychavon region of Worcestershire (Figure 17.1).

The Broadway study area exemplifies the important and well-established relationship between underlying geology and landslide occurrence. The geological sequence present on the Cotswolds escarpment produces extensive slope instability as a result of the presence of distinct permeability boundaries; the presence of relatively weak, clay-rich rocks underlying stronger more competent rock mass and weak surface materials created and modified under a periglacial climatic regime (Whitworth et al., 2000). The solid geology is comprised of an alternating sequence of Lower Jurassic marine clays, sands and limestones including the Charmouth Mudstone Formation, Dryham Formation, Marlstone Rock Formation and Whitby Mudstone Formation (previously termed the Lower, Middle and Upper Lias, Barron et al., 2002). The overlying superficial geology is characterised by variable thicknesses of solifluction, colluvial and landslide deposits that have formed under Quaternary periglacial conditions and more recently during the Holocene and Little Ice Age (Whitworth et al., 2000).

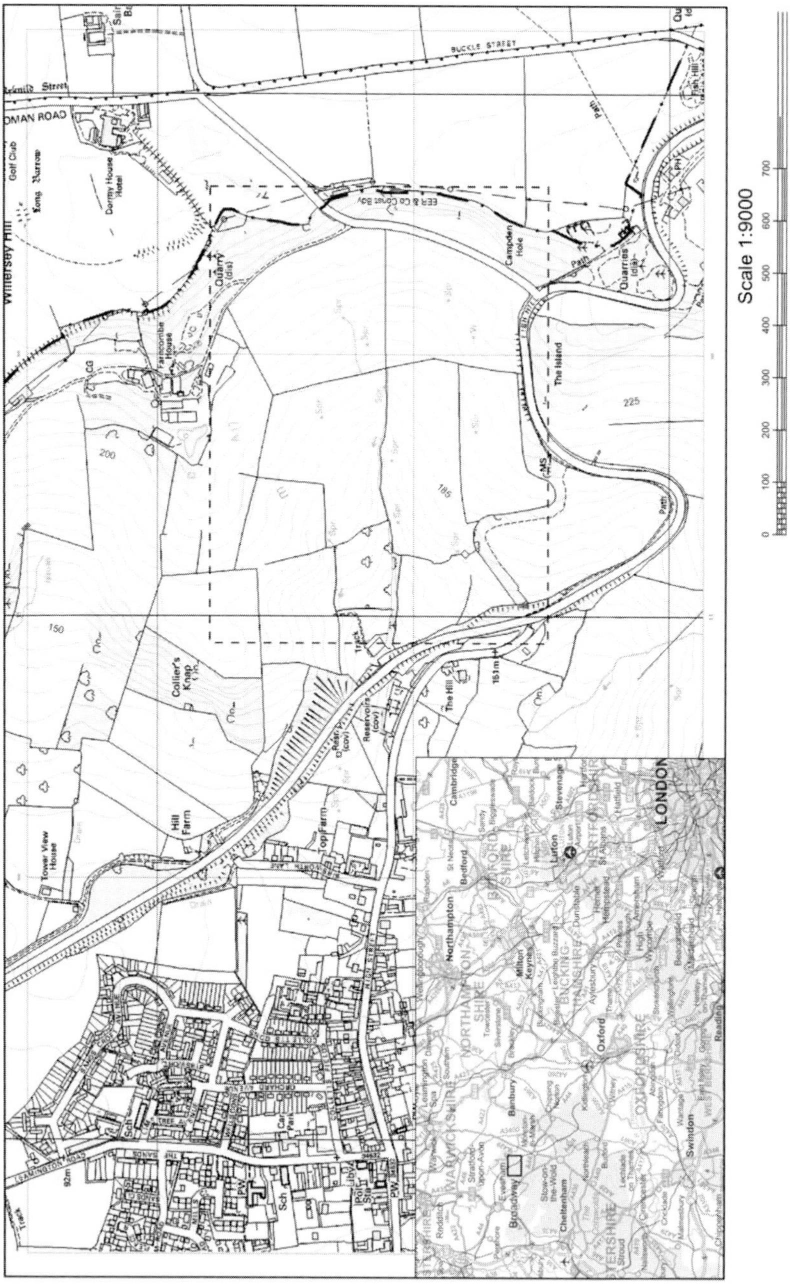

Figure 17.1 Location of the study area (dashed lines) on the Cotswolds escarpment to the west of the village of Broadway.

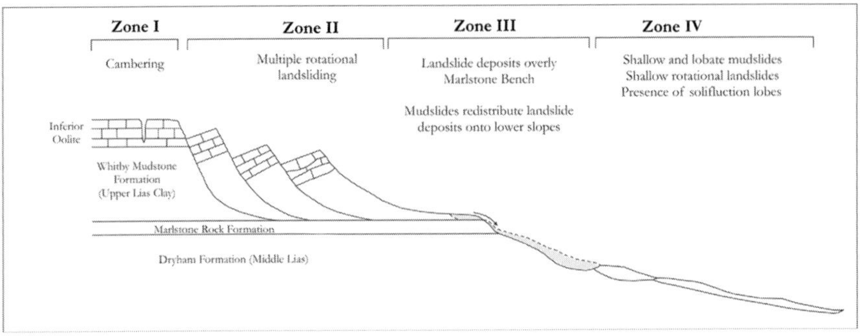

Figure 17.2 The landslide profile of valley slopes in the Cotswolds (Whitworth et al., 2005).

The landslides of the Cotswolds escarpment and similar Lias clay slopes have been widely studied (Chandler, 1970, 1971, 1974; Hawkins and Privett, 1979; Forster, 1992; Hawkins, 1996). In the Cotswolds, they described a common downslope landslide sequence that consists of the following:

1. Cambered strata in the Inferior Oolite which caps the upper part of the escarpment,
2. Zone of large-scale rotational landslides below the Inferior Oolite,
3. Zone of successive shallow rotational landslides,
4. Extensive shallow mudslides and translational landslides on the lower escarpment slopes.

This distribution involves multiple rotational failures on upper slopes, which degrade downslope through a sequence of progressively shallower successive rotational and translational failures (Figure 17.2).

3. FIELD LANDSLIDE MAPPING

The first stage of the study involved field-based geomorphological assessment of the site. This involved recording morphometric and morphographic terrain information in order to identify and map landslides in the study area. During this investigation, the landslide classification scheme described by Dikau et al. (1997) was used during field reconnaissance, assisted by other published landslide identification schemes by Bromhead (1992), Crozier (1984) and Soeters and Van Western (1996). The mapping system used was from the Engineering Geology Working Party report

(Geological Society Working Party, 1982) adapted using more recently published examples of geomorphological mapping in landslide investigations (Griffiths, 2001; Lee, 2001). The results of this field reconnaissance are shown in the engineering geomorphological map (Figure 17.3).

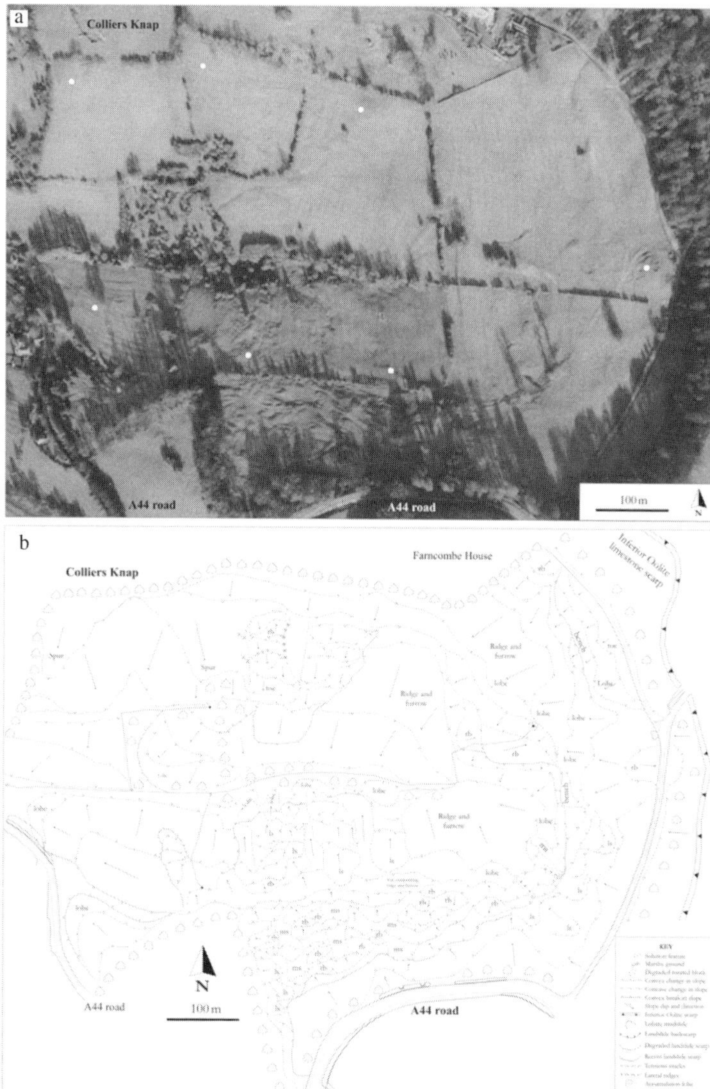

Figure 17.3 Location of the study area chosen for the laser-scanning survey: (a) aerial photograph indicating the seven laser scan locations and (b) geomorphological map of the study area.

3.1 Landslide Geomorphology

The landslides observed at Broadway are representative of the style and distribution of slope instability typically found on the Cotswolds escarpment. The morphology of the study area is controlled by a fault controlled re-entrant valley that has been eroded into the escarpment face.

The top of the escarpment is dominated by a wooded scarp face forming a distinct arcuate embayment in the escarpment, which delineates the top of the valley and the edge of the Inferior Oolite that caps the escarpment at the east (Figure 17.3). The *eastern part* of the study area is characterised by a zone of multiple rotational landslides on the flanks of the Inferior Oolite scarp. These form benches at various levels on the scarp face and consist of a series of degraded backscarps and flat or back-tilted slope units. The *northern flank* of the valley (below Farncombe House) is dominated by an extensive mudslide and complex translational landslide, whose distinct form indicates continued landslide activity. The area at Colliers Knap is dominated by a lithological bench formed by the presence of the more competent Marlstone Rock Formation at the top of the Dryham Formation.

The *southern flank* of the fault valley is dominated by a complex zone of shallow landslide activity consisting of a repeating sequence of mudslide lobes and rotational landslides (with water seepages). Below this, irregular mudslides and translational slides have disrupted the ridge and furrow cultivation system (Whitworth, 2000).

4. TERRESTRIAL LASER-SCANNING SURVEY

4.1 Background

The recent development of ground-based laser-scanning systems presents opportunities for the remote acquisition of high-density digital elevation data. The laser-scanning technology is based upon that used in traditional total stations; however, the laser scanner is an automated system that is able to acquire a dense point cloud of data from surfaces at significant distances from the scanner location. The scanner uses time-of-flight to determine the distance between the laser and a point on a reflective surface while concurrently recording the relative direction and elevation angle of each measurement. Using the known location of the scanner, it is then possible to determine the location of a reflective point to within 5–10 mm; and by repeated scanning, a laser scanner is able to record locations of millions of closely spaced points on the ground surface (termed a point cloud).

The laser system is combined with a differential GPS system and calibrated colour digital camera (although infrared cameras can also been fitted) to produce surface topographic data and photographic imagery. Typically, landslide mapping and monitoring use medium- to long-range laser scanners with above 500 m range, and the survey will often require multiple scan locations in order to image an area covered in vegetation or where the landslide is large and complex. Terrestrial laser scanning has been used to study both coastal and inland landslides (Hobbs et al., 2002; Rowlands et al., 2003; Jaboyedoff et al., 2009) and rock slope stability (Ruiz et al., 2004; Mikoš et al., 2005). Other studies have taken advantage of repeated surveys for change detection and temporal studies of landslide movement (Hsaio et al., 1994; Glenn et al., 2006) and landslide monitoring (Jones, 2006; Teza, 2007; Teza et al., 2008; Prokop and Panholzer, 2009).

4.2 Laser-Scanner Survey

The purpose of the laser-scanner survey was to generate a high-resolution topographic model of the main valley at Broadway in order to assess the use of laser scanning for landslide terrain evaluation. The study site was chosen for two reasons: firstly for the complexity of the geological terrain and associated landslides, and secondly, for the presence of ancient cultivation remains which have been disrupted by landslide activity (Whitworth et al., 2000, 2005). The aerial photograph and geomorphological map of the study area are shown in Figure 17.3.

Terrestrial laser scanning was chosen due to its high resolution, wide field of view and speed of data acquisition. Survey data was acquired at the Broadway study site using a Riegl LMS-Z420i scanner operated by 3D Laser Mapping. The scanner is a tripod-mounted system with a range of up to 800 m and a scan rate of up to 12,000 points per second (Table 17.1). The scan head is able to rotate in a 360° arc around the scanning location with a rotating mirror, which is able to scan vertically up to an 80° angle, 40° above and below the horizontal plane. A connected laptop collects recorded data on range, angles and signal amplitude for each returned laser pulse via a network connection. The scanning

Table 17.1 Summary of the Properties of the Riegl Laser Scanner Used in this Study (Riegl, 2009)

	LMS-Z420i
Vertical scan angle	80°
Horizontal scan angle	360°
Range of laser scanner	300–800 m
Measurement accuracy	10 mm
Measurement rate	Up to 12,000 points per second
Laser wavelength	Near-infrared (0.9 μm)

system also collects simultaneous photography using a digital camera, which is mounted on top of the scanner head. These can be used to generate ortho-photography when combined with the topographic data during post-processing.

Laser scanners are line-of-sight survey instruments and so when mapping complex terrain, multiple scan locations are generally required in order to provide complete coverage where the extent of the area exceeds the scanner's range or where the ground surface is obstructed by trees or buildings. The Broadway valley itself is 1000 m long and 800 m wide and consists predominantly of pastureland with isolated pockets of woodland and hedgerows (see aerial photograph in Figure 17.3). In order to provide complete coverage, the area was imaged using seven scanning sites located at vantage points on each side of the valley (Figure 17.3).

At each location, the scanner was programmed to collect data with a resolution of 0.75 m at 700 m distance resulting in a scan time of 12 min per location. To ensure the accuracy of the scanning process, ground control was provided by Leica differential GPS system, which was mounted on the scanner system. A base GPS receiver was installed at a known location for the duration of the survey. These GPS data would subsequently be used during post-processing to provide accurate location for each scanner position in order to provide the horizontal and vertical control required for the correct orientation of the surface models and georeferenced to the UK National Grid coordinate system.

The resulting data consist of a point cloud of reflected points that are heterogeneously distributed in 3D space depending on characteristics of the reflective surfaces, visibility of the surface and presence of obstacles such as tress and vegetation (Prokop and Panholzer, 2009). Consequently, post-processing is required alongside the GPS positional data. Teza et al. (2007)

provide an overview of laser-scanning processing. In this study, the processing consisted of the following three stages:

1. *Scan editing and verification* to identify and remove erroneous data and re-sample the scans if necessary (obstacles such as tress were removed manually at this stage),
2. *Registration of the scan data* to an appropriate coordinate system using differential GPS control data,
3. *Combination of individual scans* into one point cloud covering the area of interest. This involves the GPS control along with either the use of back site targets to orientate the scans or in this case, using a least-squares comparison of overlapping point cloud data.

The survey produced a large volume of data, which can present significant problems when attempting to combine multiple scans during post-processing. Each scan generated 4 million points of data, resulting in a total of 28 million points for the whole site. The raw point cloud data produced by each scan were processed using RiSCAN PRO software, transforming the coordinates of each scanned point from a position relative to the scanner to their correct grid position and elevation.

The point cloud data were subsequently combined into one data set using the GPS data and a surface matching technique in which overlapping portions of each scan are matched and orientated using a least-squares technique. During this process, duplicate data in areas of overlapping scans were included to ensure accurate representation of topography in the resulting surface model. The final point cloud data are shown in Figure 17.4a and demonstrate the high point density in the final surface model. RiSCAN PRO was used to triangulate the point cloud data and export the data as an ASCII format (x, y and z) into ER Mapper where the point cloud was gridded using a triangulation-based interpolation technique to a 0.5 m grid digital elevation model (DEM). The resulting digital surface model is shown in Figure 17.4b.

4.3 Data Analysis

The laser-scan survey resulted in the production of a high-resolution DEM of the study area, which is shown in Figure 17.4 using a relief-shaded surface model. Using this base data set, a number of topographic variables were generated including slope angle, curvature and surface roughness.

- Slope angle (degree) is defined as the angle measured between a horizontal plane at a given point on the land surface,

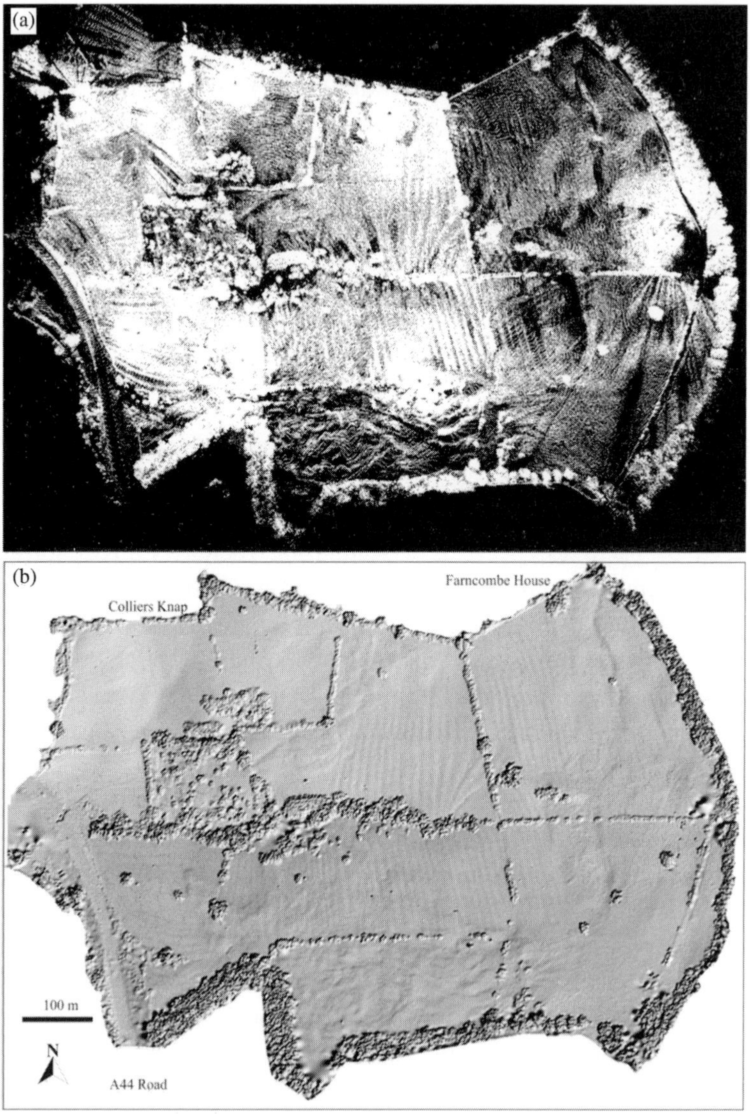

Figure 17.4 (a) Laser scan point cloud data and (b) relief-shaded image for the Broadway valley generated using terrestrial laser scanning.

- *Curvature* (or convexity) is defined as the rate of change of slope and is the second derivative of elevation. There are two components of convexity: profile convexity, which is the rate of change of slope along a slope line, and plan convexity, which is the rate of change of aspect along a contour. Positive curvature represents surface convexity, while negative curvature indicates concavity,
- *Topographic roughness* is a measure of variability in either elevation or slope angle and is typically calculated using a standard deviation filter.

These topographic derivatives are typically calculated from a digital elevation model array using a 3×3 moving kernel window to create a separate DTM of the surface. These are the most commonly generated indices for landslide and slope-stability studies (McKean and Roering, 2003; Moreno et al., 2003).

4.4 Results

The topographic variables have been used as part of a qualitative study into the application of digital topographic data for landslide identification and mapping. This has involved the interpretation of a range of topographic variables derived from the original DEM data including slope, curvature and topographic roughness, which have been analysed in conjunction with geomorphological maps of the area. The comparative images are presented and discussed.

The slope-angle image (Figure 17.5) is a very powerful tool for geomorphological assessment. This is exemplified in this example where the landsides are clearly visible in the resulting slope-angle image. This includes zones of rotational failures indicated by flat and low-angle benches alongside steep, commonly arcuate backscarps, shallow landslide activity and mudslides (Figure 17.5).

The lobate mudslide on the slopes south of Farncombe House is clearly visible along with a detailed image of the internal morphology of the landslide, which is enhanced by the slope-angle image. On the opposite side of the valley an area of disturbed ridge and furrow that has been disrupted by shallow landslide activity is visible; the zone upslope of this area contains extensive instability including rotational slides and mudslides. In this area, a sequence of successive rotational landslides is evidenced by the repeated slope changes that reflect the backscarps of these individual failures.

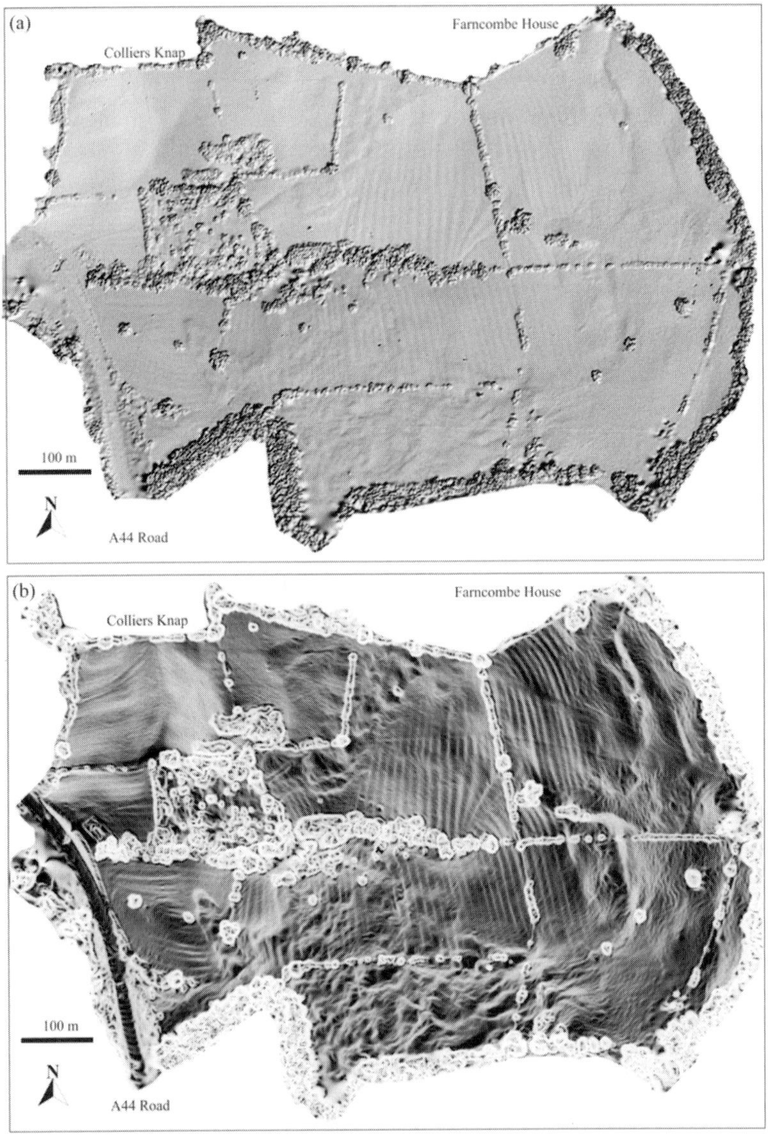

Figure 17.5 (a) Relief-shaded image and (b) slope-angle image derived from the digital elevation model of the Broadway valley generated using terrestrial laser scanning.

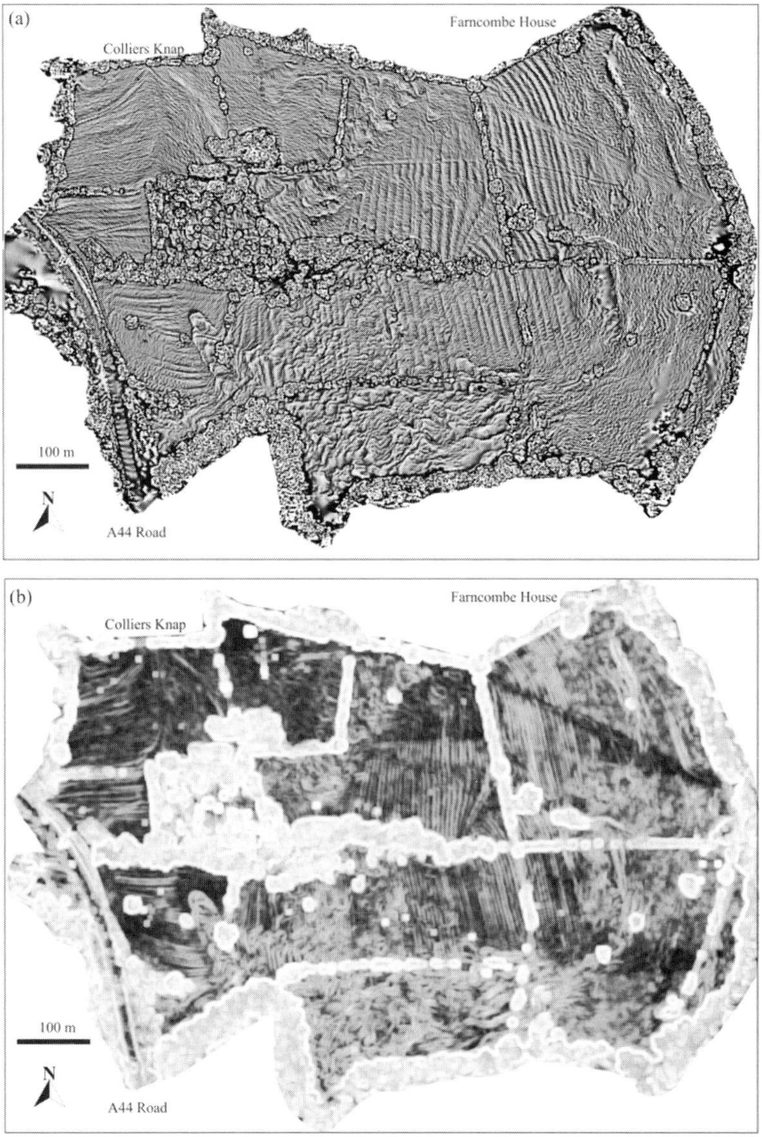

Figure 17.6 (a) Plan curvature image and (b) surface roughness image derived from the digital elevation model of the Broadway valley generated using terrestrial laser scanning.

The upper slopes below the main Inferior Oolite scarp face consist of a series of benches reflecting landslide activity at the boundary with the underlying Whitby Mudstone Formation. These have been overprinted by more recent mudslide activity and the slope-angle image highlights the frontal lobes of these landslides. At the top of the slope, the remains of a large rotated block identified during field mapping can be identified in the image (Figure 17.5).

Plan curvature was generated for the area using the digital elevation data. The results reflecting changes in slope, either convex or concave and so curvature is a useful geomorphological tool for the delineation of landslides since the measure is sensitive to changes in slope angle. In the Broadway valley, the curvature image (Figure 17.6) is particularly useful at identifying areas with pronounced slope changes such as the front of lobate mudslides and enhancing sharp slope changes associated with arcuate backscarps and displaced blocks of rotational landslides.

Finally, topographic roughness was generated by the application of a slope-angle filter and standard deviation statistical filter. The resulting image illustrates the variation in slope angle across the valley, highlighting areas where the topography is either rough or smooth. In terms of landslide mapping, this has proven useful in enhancing irregular or hummocky topography associated with areas of surface disruption from landslide movement. Consequently, areas of high topographic roughness are typically associated with areas identified as landslide, a relationship that is evident in Figure 17.6.

5. CONCLUSIONS

DTMs derived from terrestrial laser scanning provide high-resolution data sets for the geomorphological assessment of landslide terrains. The data have proven particularly useful in helping to understand the complex landslide systems present on the Cotswolds escarpment, where the use of topographic variables such as slope angle and curvature can provide important geomorphological information for landslide discrimination.

REFERENCES

Barron, A.J.M., Sumbler, M.G., Morigi, A.N., 2002. Geology of the Morton in Marsh district. Sheet description of the British Geological Survey, Sheet 217 (England and Wales). British Geological Survey, Keyworth, Nottingham.

Bromhead, E., 1992. The Stability of Slopes. second ed. Blackie, London.

Chandler, R.J., 1970. The degradation of Lias clay slopes in an area of the East Midlands. Q. J. Eng. Geol. 2, 161–181.

Chandler, R.J., 1971. Landsliding on the Jurassic escarpment near Rockingham, Northamptonshire. In: Brunsden, D. (Ed.), Slopes: Form and Process. Institute of British Geographers Special Publication 3. Institute of British Geographers, London, pp. 111–128.

Chandler, R.J., 1974. Lias clay: the long-term stability of cutting slopes. Geotechnique 24 (1), 21–38.

Crozier, M.J., 1984. Field assessment of slope instability. In: Brunsden, D., Prior, D.B. (Eds.), Slope Instability. John Wiley & Sons, Chichester, pp. 103–142.

Dikau, R., Brunsden, D., Schrott, L., Ibsen, M.L. (Eds.), 1997. Landslide Recognition: Identification, Movement and Consequences. John Wiley & Sons, Chichester, pp. 1–12.

Forster, A. (1992) The Slope Stability of the Lincolnshire Limestone Escarpment between Welbourne and Grantham. 1:50,000 Geological Map Sheet 127. British Geological Survey Technical Report WN/92/5.

Geological Society Working Party, 1982. Land surface evaluation for engineering practice: report by a Working Party under the auspices of the Geological Society. Q. J. Eng. Geol. 55, 265–316.

Glenn, N.F., Streutker, D.R., Chadwick, D.J., Thackray, G.D., Dorsch, S.J., 2006. Analysis of LiDAR-derived topographic information for characterizing and differentiating landslide morphology and activity. Geomorphology 73 (1–2), 131–148.

Griffiths, J.S., Mather, A.E., Hart, A.B., 2002. Landslide susceptibility in the Rio Aguas catchement, SE Spain. Quarterly J. Eng. Geol. Hydroge. 35, 9–17.

Griffiths, J.S., 2001. Development of a ground model for the UK Channel Tunnel portal. In: Griffiths, J.S. (Ed.), Land Surface Evaluation for Engineering Practice. Geological Society Engineering Geology Special Publications 18. Geological Society, London, pp. 129–133.

Hawkins, A.B., 1996. Observation and analysis of the ground conditions in the Jurassic landslip terrain of southern Britain. Proceedings of the Seventh International Symposium on Landslides, vol. I, Trondheim Norway, 17–21 June, pp. 3–16.

Hawkins, A.B., Privett, K.D., 1979. Engineering geomorphological mapping to elucidate areas of superficial structures with examples from the Bath area of the Cotswolds. Q. J. Eng. Geol. 12, 221–233.

Hobbs, P., Humphreys, B., Rees, J., Tragheim, D., Jones, L., Gibson, A., et al., 2002. Monitoring the role of landslides in 'soft cliff' coastal recession. In: McInnes, R., Jakeways, J. (Eds.), Instability – Planning and Management. Thomas Telford, London, pp. 589–600.

Hsiao, K. Lui, J. Yu, M. and Tseng, Y., 1994. Change detection of landslide terrains using ground based Lidar data. Proceedings of the 20th ISPRS Congress, Istanbul, Turkey, 12–23 July 2004. Commision VII, WG VII/5. <http://www.isprs.org/istanbul 2004/>

Jaboyedoff, M., Demers, D., Locat, L., Locat, A., Locat, P., Oppikofer, T., et al., 2009. Use of terrestrial laser scanning for the characterization of retrogressive landslides in sensitive clay and rotational landslides in river banks. Can. Geotech. J. 46 (12), 1379–1390.

Jones, L.D., 2006. Monitoring landslides in hazardous terrain using terrestrial LiDAR: an example from Montserrat. Q. J. Eng. Geol. Hydrogeol. 39 (4), 371–373.

Lee, E.M., 2001. Geomorphological mapping. In: Griffiths, J.S. (Ed.), Land Surface Evaluation for Engineering Practice. Geological Society Engineering Geology Special Publication 18. Geological Society, London, pp. 53–56.

McKean, J., Roering, J., 2003. Objective landslide detection and surface morphology mapping using high-resolution airborne laser altimetry. Geomorphology 57 (3–4), 331–351.

Mikoš, M., Vidmar, A., Brilly, M., 2005. Using a laser measurement system for monitoring morphological changes on the Strug rock fall, Slovenia. Nat. Hazards Earth Syst. Sci. 5, 143–153.

Moreno, M., Levachkine, S., Torres, M., Quintero, R., 2003. Geomorphometric analysis of raster image data to detect terrain ruggedness and drainage density. Lect. Notes Comput. Sci. 2905, 643–650.

Prokop, A., Panholzer, H., 2009. Assessing the capability of terrestrial laser scanning for monitoring slow moving landslides. Nat. Hazards Earth Syst. Sci. 9, 1921–1928.

Riegl, 2009. RiScan Pro User Manual: Operating and Processing Software for Riegl Laser Scanners. Version 1.5. Riegl LMS. <www.riegl.com>

Rowlands, K.A., Jones, L.D., Whitworth, M., 2003. Landslide laser scanning: a new look at an old problem. Q. J. Eng. Geol. Hydrol. 36 (2), 155–157.

Ruiz, A., Kornus, W., Talaya, J., Colomer, J., 2004. Terrain modelling in an extremely steep mountain: a combination of airborne and terrestrial lidar. Proceedings of the XXth ISPRS Congress Geo-Imagery Bridging Continents, Istanbul, Turkey, 12–23 July 2004. Commission III, WG III/3.

Soeters, R., Van Western, C.J., 1996. Slope instability recognition, analysis and zonation. In: Turner, A.K., Schuster, R.L. (Eds.), Landslides: Investigation and Mitigation. Special Report 247. Transportation Research Board, National Research Council. National Academy Press, Washington, DC, 121–314.

Teza, G., 2007. Terrestrial laser scanner to detect landslide displacement fields: a new approach. Int. J. Remote Sens. 28 (16), 3425–3446.

Teza, G., Pesci, A., Genevois, R., Galgaro, A., 2008. Characterization of landslide ground surface kinematics from terrestrial laser scanning and strain field computation. Geomorphology 97, 424–437.

Teza, G., Galgaro, A., Zaltron, N., Genevois, R., 2007. Terrestrial laser scanner to detect landslide displacement fields: a new approach. Int. J. Rem. Sens. 28 (16), 3425–3446.

Whitworth, M.C.Z., Murphy, W., Giles, D.P., Petley, D.N., 2000. Historical Constraints on Slope Movement Age: a Case Study at Broadway, United Kingdom. Geogr. J. 166 (2), 139–155.

Whitworth, M.C.Z., Giles, D.P., Murphy, W., 2005. Airborne remote sensing for landslide hazard assessment: a case study on the Jurassic escarpment slopes of Worcestershire, United Kingdom. Q. J. Eng. Geol. Hydrogeol. 38 (3), 285–300.

CHAPTER EIGHTEEN

Digital Terrain Models from Airborne Laser Scanning for the Automatic Extraction of Natural and Anthropogenic Linear Structures

Rutzinger Martin[a], Höfle Bernhard[b], Vetter Michael[c] and Pfeifer Norbert[c]

[a]ITC-Faculty of Geo-Information Science and Earth Observation, University of Twente, Enschede, The Netherlands and Institute of Geography, University of Innsbruck, Innsbruck, Austria
[b]Department of Geography, University of Heidelberg, Heidelberg, Germany
[c]Institute of Photogrammetry and Remote Sensing, Vienna University of Technology, Vienna, Austria

Contents

1. INTRODUCTION

Recently, digital terrain models (DTMs) derived from laser scanning point clouds have became an important input for automated geomorphologic analysis and mapping. Since the 1990s, the operational usage of airborne laser scanning (ALS) in many mountainous areas such as

Developments in Earth Surface Processes, Volume 15
ISSN: 0928-2025, DOI: 10.1016/B978-0-444-53446-0.00018-5

the Alps has demonstrated the feasibility of this remote sensing technique for generating high-resolution DTMs with high accuracy. In the ALS case, the laser scanner is mounted on an aeroplane or a helicopter. The position of the laser scanner is determined by a global positioning system (GPS) integrated with inertial measurement unit (IMU) measurements. The IMU additionally provides the angular attitude of the sensor platform. ALS measures the travel time of the emitted laser pulse to the hit target and back. Knowing all those components, the 3D position of the reflection (or echo) is computed with decimetre up to centimetre accuracy depending on several system and flight parameters. Depending on the system type the emitted beam is deflected on a rotating or oscillating mirror, which causes typical scanning patterns. The scan angle determines the width (swath) of the area captured within one flight strip. The reflections are stored as point coordinates (in x, y and z) with an intensity (or amplitude) value, which is related to the strength of the recorded echo. By comparison to elevation information extracted from traditional photogrammetry, laser scanning has the ability to penetrate the canopy of high vegetation and provides ground measurements even in forested areas (Kraus and Pfeifer, 1998). For further reading on the principle of laser scanning, the reader is referred to Shan and Toth (2009) and Vosselman and Maas (2010).

Since the original laser measurements represent reflections on objects (such as buildings, trees, temporal objects, fences and walls) and ground surface, it is necessary to classify them into reflections from bare Earth and from objects, which is an essential step before geomorphological mapping. The classification into terrain points also known as filtering is a very active research field where several algorithms and methods have been developed (Sithole and Vosselman, 2004, Pfeifer and Mandlburger, 2009; Briese, 2010). Finally, the 3D point cloud is interpolated and converted into a 2.5D raster model, the DTM, which is the input for many geo-related applications such as hydrological modelling, natural hazard analysis and geomorphological mapping (Geist et al., 2009).

The highly detailed representation of the Earth surface in ALS DTMs lead on the one hand to higher precision in the delineation of geomorphological units by mapping, but on the other hand it limits the application of traditional mapping approaches because the data contains disturbing sub-features, which were not apparent in DTMs with coarser resolution.

To be able to handle distortions from large-scale objects in small-scale object classification, concepts such as multi-resolution, object-based, image analysis were introduced. The basic idea is to segment homogeneous regions in the data. For these segments, features are calculated and then used to classify the scene into the classes representing the objects of interest. Finally, these objects are merged to superior objects for a specific application scale (Blaschke et al., 2008). For geomorphological mapping, this means that terrain features from a larger scale can appear on a superior geomorphological unit, which can have a disturbing effect on automated geomorphological mapping techniques at a certain scale. For example, a landslide can be segmented into several artificial sub-units by a road winding along a mountain slope. This problem did not affect DTMs with coarser resolutions or DTMs in forested areas from photogrammetry, where only low accuracy could be achieved. Hence, geomorphological phenomena are not always apparent in their original and natural shape. The Earth's surface may have been reshaped by anthropogenic activities such as the development of settlements or the construction of infrastructure, which makes it sometimes difficult to analyse the geomorphologic surface in its original appearance. Concluding, anthropogenic features disturb geomorphometric analysis or geomorphological mapping approaches. For automated geomorphological mapping, techniques are needed that are able to process the large amount of high-resolution data such as from laser scanning. Furthermore, this demands a separation of natural and anthropogenic features in the DTM.

This chapter introduces an automated method for the separate extraction of geomorphological structure lines and line features originating from roads in an ALS DTM. The chapter is structured as follows: Section 2 provides an overview of related work on line extraction and automatic geomorphological mapping using high-resolution DTMs: Section 3 introduces the method for lineament extraction and the knowledge-based classification approach for the separation of natural and anthropogenic features; Section 4 describes the data sets used and the test sites; Section 5 presents the settings of the tested workflow, discusses the results and applies the classification to recognise geomorphologic active areas. The chapter closes with a conclusion in Section 6.

2. RELATED WORK

The following state of the art focuses on line extraction from remote sensing data related to geomorphology. A comprehensive overview of geomorphologic applications and the use of ALS data from a technological perspective can be found in Höfle and Rutzinger (in press).

Several studies showed the feasibility of the derivation of geomorphological line structures from remotely sensed data using edge preserving filters (Clark and Wilson, 1994; Wladis, 1999; Jordan and Schott, 2005; Nyborg et al., 2007). Mavrantza and Argialas (2008) used five bands from Landsat ETM+ and additional derivatives such as the Normalized Difference Vegetation Index (NDVI), principal component analysis (PCA) and clustering results from automatic Iterative Self-Organising Data Analysis Technique (ISODATA) classification. Segmentation was performed using the multi-scale segmentation approach in eCognition (Benz et al. 2004) and an edge preserving filter for lineament extraction. Then a hierarchical knowledge-based classification was performed using a fuzzy-logic rule base and nearest neighbour classifiers to extract tectonic lineaments and a lithology layer of the test site. Since images of a passive sensor were used, assumptions on changing vegetation cover caused by certain geological situations had to be taken into account, which increased the uncertainty of the final classification output.

Geomorphologic studies using DTMs derived from ALS data are feasible for (i) high-resolution mapping and (ii) vegetation penetration. Glenn et al. (2006) and McKean and Goering (2004) used ALS DTMs to describe and parameterise geomorphological phenomena. They derived morphometric parameters from ALS DTMs to describe landslide morphology and activity. Anders et al. (2009) used an ALS DTM for simulating landscape evolution in an Alpine catchment. Asselen and Seijmonsbergen (2006) used a slope map derived from an ALS DTM for semi-automated geomorphological mapping in an Alpine test site. They used eCognition to segment homogeneous slope surface patches, which are then classified to the final units corresponding to a geomorphological map from field survey.

Several methods have been developed to extract structure lines from ALS data. Knowledge on terrain breaklines is not only valuable information for geomorphological interpretation but also a prerequisite for DTM enhancement in filtering (Briese, 2010). Raster-based approaches are based on first- and second-order derivatives derived from the DTM. Wood (1996) demonstrated that these terrain derivatives such as slope, aspect and curvature can be used to classify the terrain into elementary landforms such as peaks, ridges, passes, channels, pits and planes. Methods using curvature derived from ALS DTMs as basis for structure line vectorisation was presented by Brügelmann (2000) and Rutzinger et al. (2007). Brzank et al. (2008) derived structure lines using data from bathymetry and ALS to delineate channel patterns in the Wadden Sea.

Bailly et al. (2008) presented a workflow using elevation profiles from ALS point clouds to derive structure lines by wavelet transformation and watershed segmentation for determining agricultural topographies. Algorithms to derive structure lines automatically from ALS point clouds were developed by Briese (2004) to improve the DTM filtering process and by Vosselman and Liang (2009) to detect curbstones for road inventory mapping. The advantages of point cloud approaches are the higher achievable positional accuracies and maintenance of the original 3D information of the laser-scanning measurement.

3. METHOD

The workflow shown in Figure 18.1 comprises laser data processing to compute the DTM, the derivation of structure lines (Section 3.1) and the classification of geomorphologic and other line features (Section 3.2).

In the DTM processing step, the last echoes of the point cloud data have to be filtered to distinguish terrain and object points (Kraus and Pfeifer, 1998). Remaining object points are excluded manually by an operator. The terrain points are then interpolated to the DTM. This processing step was already performed by the data provider, prior to receiving the ALS data.

3.1 Structure Line Extraction

Structure lines appear as part of step edges in the DTM and can be described by the local curvature value within a certain neighbourhood. From the DTM, the maximum and minimum curvature is calculated using a moving window approach (Wood, 1996; GRASS GIS Developer Team, 2010). In the following, areas of high positive or negative curvature values are selected as they correspond to edges in the terrain. Areas of high positive curvature correspond to ridges and upper edges, whereas areas of high negative curvature correspond to valleys and lower edges. These areas are thinned and vectorised to 3D vector lines using the elevation value from the input DTM (Brügelmann, 2000; Rutzinger et al.,

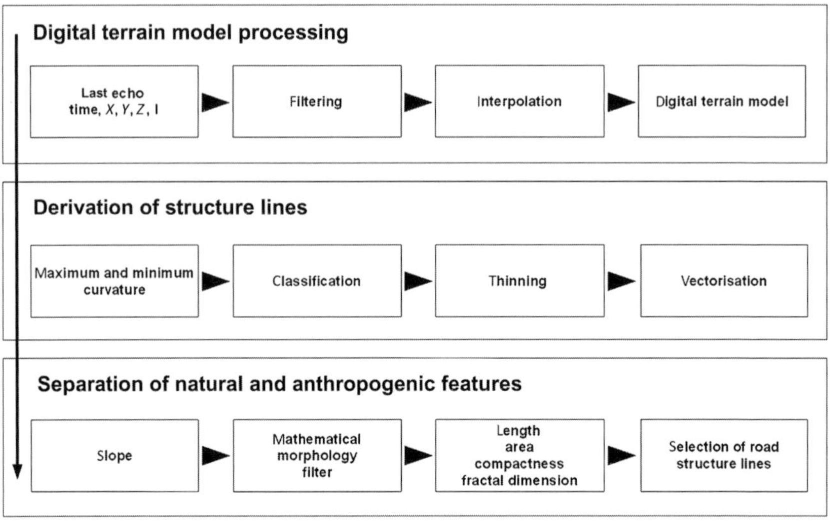

Figure 18.1 Workflow for DTM processing, structure line derivation and classification.

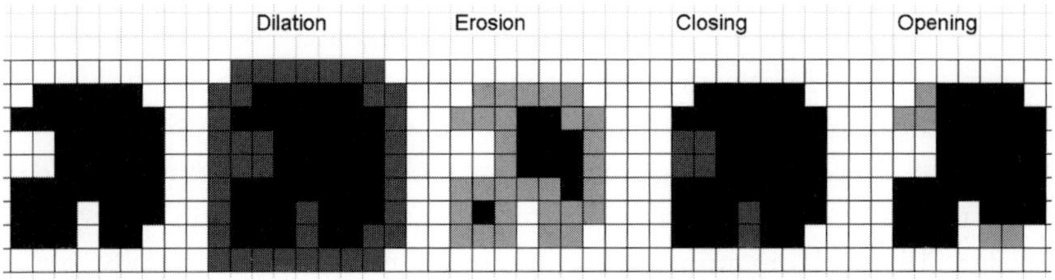

Figure 18.2 Mathematical morphology where opening is the combination of dilation followed by erosion, and closing is the combination of erosion followed by dilation. Dark grey are pixels that are added and light grey are pixels that are removed from the filtered segment.

2007). The two vector data sets on upper and lower edges are combined before being used in the classification.

3.2 Separation of Anthropogenic and Natural Structures

We assume that geomorphologic edges occur in rather steep and undulating terrain, while anthropogenic features such as roads are characterised by constant and lower slope. Initially, pixels of low slope are selected. This selection still contains several natural surface patches (pixel segments). Therefore, the selection is enhanced by mathematical filter operations such as iterative dilation and erosion (Figure 18.2). The remaining segments are selected by features describing the segment extent and shape such as length, area, compactness (Eq. (18.1)) and fractal dimension (Eq. (18.2)). High compactness and fractal dimension are significant for strongly undulating segment outlines. An ideal segment with the shape of a circle results in low compactness and fractal dimension.

$$\text{Compactness} = \frac{\text{Perimeter}}{2 \times \sqrt{\pi \times \text{area}}} \tag{18.1}$$

$$\text{Fractal dimension} = 2 \times \frac{\log(\text{perimeter})}{\log(\text{area})} \tag{18.2}$$

The separation of structure lines is enhanced by considering neighbouring lines by their length and distance to each other. Rather long line segments mainly belong to roads. Hence, all lines with a certain length are considered as seed lines to select smaller ones in the proximate neighbourhood. This procedure excludes short structure lines on non-anthropogenic-influenced terrain and reclassifies them to natural lines.

4. DATA SET AND TEST SITE

4.1 Data Set

The ALS data used was acquired in November 2005 using an Optech ALTM 3100 laser system mounted in an aircraft. The laser unit recorded x, y, z coordinates and intensity for each echo. It can record up to 4 echoes per laser shot, which mainly occurs in forested areas. The average point density in the test sites is about 4 points per square metre. The raw data point cloud was filtered using robust interpolation (see Section 3). Remaining gross errors were removed manually. The resulting DTM has a resolution of 1 m^2.

For all three test sites, reference data representing the road network was digitised by an independent operator using shaded relief maps of the ALS DTM, orthophotos (taken in 2004) and existing road vector maps (TIRIS, 2009). The road network was classified according to their width into roads (>4 m), forest roads (2–4 m) and hiking paths (<2 m).

4.2 Test Sites

The test sites are located in the Austrian Eastern Alps, south of Innsbruck, Tyrol (Figure 18.3). They represent areas with different

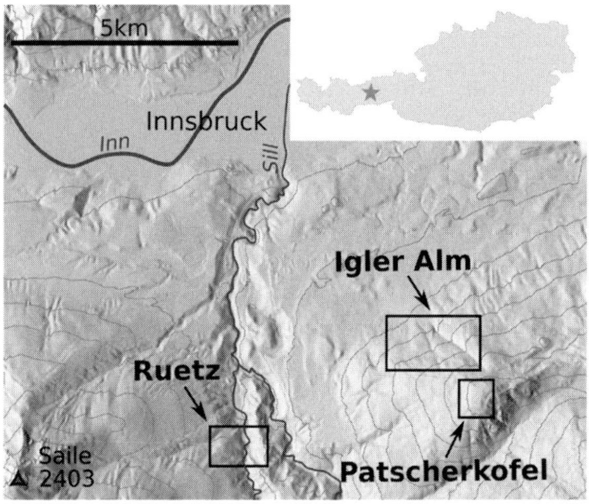

Figure 18.3 Location of the test sites.

geomorphologic situations, which were also influenced by anthropogenic activities (i.e. road construction).

The test site *Igler Alm* covers 1200 m × 2100 m of a north-west-facing slope, with two major erosion torrents. The area is completely forested and contains some logging patches. The area is connected by a dense network of forest roads and hiking paths (Figure 18.4).

The test site *Ruetz* (800 m × 1300 m) covers three cuttings leading from south to north. The valley centres are dominated by strong erosive fluvial processes, which lead to steep slopes in those areas. The slopes east and west of the river Ruetz are forested and accessed by a network of forest roads. The steep slopes are filled with glacial deposits from the last ice age. A landslide in the centre of the test site occurs as a series of corrugated and crescent-shaped structures, which reach up to the small horizontal agricultural field of the top of the slope. The landslide was triggered by the erosion of the Ruetz.

The third test site is situated above the timber line at the summit of the mountain *Patscherkofel*. It covers an area of 800 m × 750 m. This area is affected by mountain fissures, which is apparent through cracks and step edges along the crest (Gruber, 2004). Here, the stepping surface follows from east to west and is crossed by an unpaved road leading to the summit.

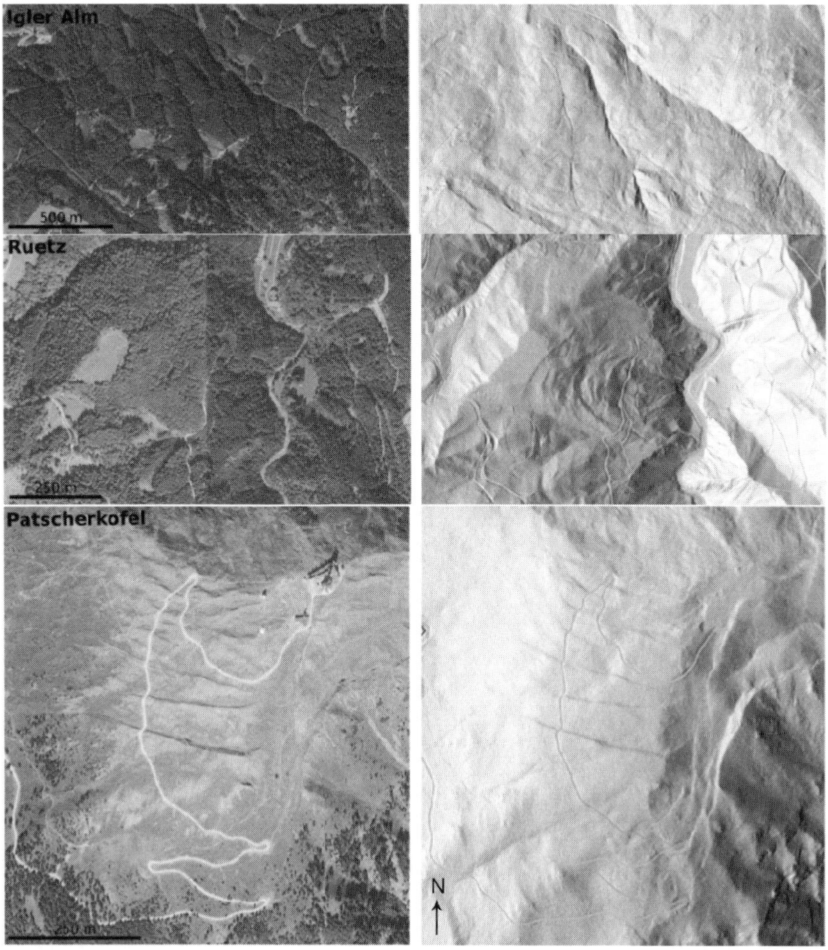

Figure 18.4 Orthophoto (left), and shaded map of the DTM (right) of the test sites Igler Alm (top), Ruetz (middle) and Patscherkofel (down).

5. RESULTS

5.1 Structure Line Classification

Minimum and maximum curvature in any direction were calculated using a 9 m × 9 m moving window. The structure lines were classified from pixel groups with curvature higher than 0.05 for upper step edges and

Table 18.1 Knowledge-Based Classification Rules

	Minimum Area (m^2)	Maximum Area (m^2)	Minimum Compactness (m^{-1})	Minimum Fractal Dimension (m^{-1})
Igler Alm	20	1,000	2.0	—
Ruetz	20	10,000	3.0	1.80
Patscherkofel	20	12,000	4.0	1.79

lower than −0.05 for the lower step edges, respectively. After vectorisation and smoothing, lines smaller than 10 m were removed to avoid noise from further processing. Road segments are selected from a slope map smaller than 15°. After iterative mathematical morphology, filtering lines shorter than 20 m are removed. The settings of the knowledge-based classification for enhancing those road segments are shown in Table 18.1. Very small and large areas are excluded. Road segments have rather high compactness and fractal dimension.

The construction of roads and paths on a sloped surface covered with sediments cuts a step into the natural surface. Material is deposited beneath the road track, whilst above the road, sediments tend to sink. This can lead to up to four parallel structure lines, which are caused by a single road.

Figure 18.5 shows the comparison of the digitised reference road network (left column) and the classified structure lines (right column). Most structure lines belonging to forest roads were detected correctly. Hiking paths with a smaller width are not evident because the 9 m × 9 m moving window for curvature computation is too large for these structures. This means that they do not meet the target scale of this analysis and can therefore not be detected.

Manual inspection shows that the method produces reliable results. In the Igler Alm test site, the separation of roads and geomorphological lines works especially well. In the Ruetz test site, the road crossing the landslide could be separated from the line structures belonging to the landslide. Forest roads may be underestimated if they have longer sections, which are steeper than 15° (e.g. test site Ruetz, centre, north). An overestimation of road structure lines occurs if geomorphological lineaments are rather straight and flat (e.g. test site Patscherkofel, centre). Ambiguity remains if roads follow natural line structures of the terrain, which, for example, is the case in the Patscherkofel test site.

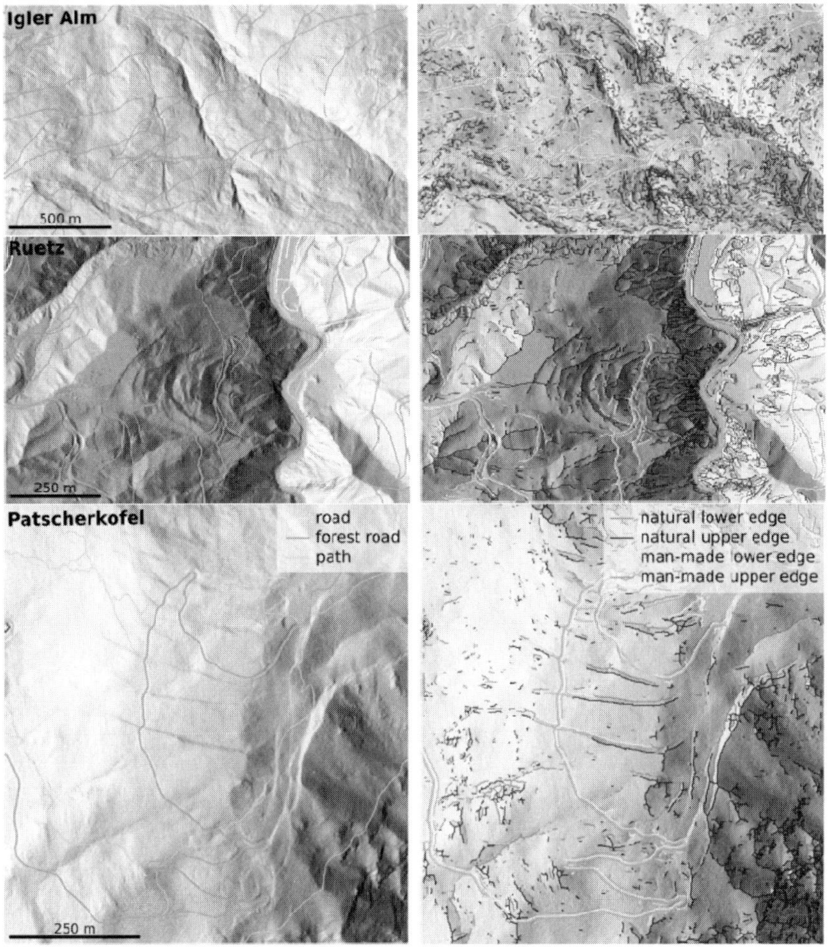

Figure 18.5 Reference road layer (left) and classified structure lines (right) divided into upper and lower edges.

5.2 Application: Geomorphological Activity Mapping

The effect of structure line derivation and separation of anthropogenic and natural features for geomorphological interpretation is tested by calculating line density images (Figure 18.6). The vector lines are converted to a raster map of 1 m resolution. For a moving window of 25 m × 25 m, the line pixels are counted to show the percentage of line coverage per area. It is evident that a geomorphological interpretation is only possible if anthropogenic features are removed. High line density indicates high

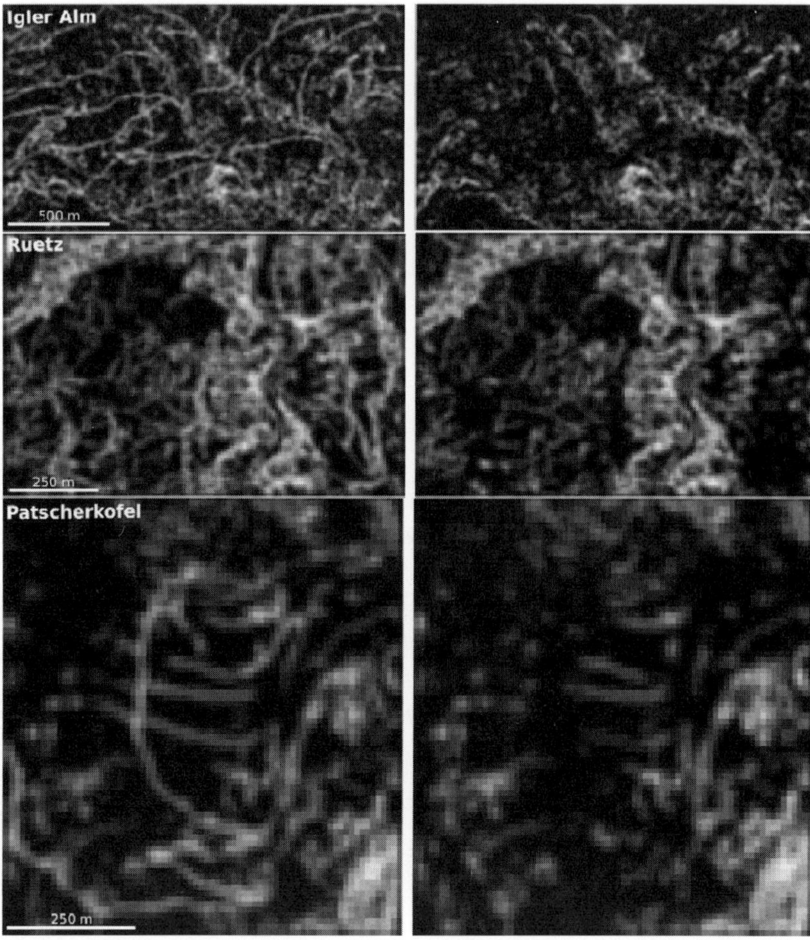

Figure 18.6 Line density map using all derived structure lines (left) and using geomorphological structure lines only (right). High line density is indicated by high brightness.

geomorphological activity, which can be caused by erosion or gravitational processes.

6. CONCLUSION

DTMs from ALS represent terrain structures in high detail even under forested areas. Nevertheless, if these models are to be used for geomorphological analysis and mapping tasks, further processing is required in order to detect anthropogenic terrain features such as roads. The presented concept to extract terrain structure lines of upper and lower edges is fully automated. The lines are then classified into roads and geomorphological structure lines. After classification, the line density maps show the enhanced interpretation possibilities if anthropogenic lines are removed.

If the aim is to study the current geomorphological processes, then all terrain features have to be considered in the analysis. If, on the other hand, the aim is to understand the overall picture and main geomorphological units as a whole, then it is necessary to exclude anthropogenic influences on the terrain surface. ALS provides reliable high-resolution terrain information for both cases of geomorphologic analysis and mapping tasks.

REFERENCES

Anders, N.S., Seijmonsbergen, A.C., Bouten, W., 2009. Modelling channel incision and alpine hillslope development using laser altimetry data. Geomorphology 113 (1–2), 35–46.

Asselen, S., Seijmonsbergen, A., 2006. Expert-driven semi-automated geomorphological mapping for a mountainous area using a laser DTM. Geomorphology 78 (3–4), 309–320.

Benz, U., Hofmann., P., Willhauck, G., Lingenfelder, I., Heynen, M., 2004. Multi-resolution, object-oriented fuzzy analysis of remote sensing data for GIS-ready information. ISPRS J. Photogramm. Rem. Sens 58 (3–4), 239–258.

Bailly, J., Lagacherie, P., Millier, C., Puech, C., Kosuth, P., 2008. Agrarian landscapes linear features detection from LiDAR: application to artificial drainage networks. Int. J. Remote Sens. 29, 3489–3508.

Blaschke, T., Lang, S., Hay, G. (Eds.), 2008. Object-Based Image Analysis, Lecture Notes in Geoinformation and Cartography. Springer, Berlin.

Briese, C., 2004. Three-dimensional modelling of break lines from airborne laser scanner data. In: IAPRS, vol. 35, Part B3, pp. 1097–1102.

Briese, C., 2010. Extraction of digital terrain models. In: Vosselman, G., Maas, H.-G. (Eds.), Airborne and Terrestrial Laser Scanning. Whittles Publishing, Boca Raton, FL.

Brügelmann, R., 2000. Automatic breakline detection from airborne laser range data. In: IAPRS, vol. 33, Part B3, pp. 109–116.

Brzank, A., Heipke, C., Goepfert, J., Soergel, U., 2008. Aspects of generating precise digital terrain models in the Wadden Sea from lidar – water classification and structure line extraction. ISPRS J. Photogramm. Remote Sens. 63 (5), 510–528.

Clark, C.D., Wilson, C., 1994. Spatial analysis of lineaments. Comput. Geosci. 20 (7–8), 1237–1258.

Geist, T., Höfle, B., Rutzinger, M., Pfeifer, N., Stötter, J., 2009. Laser scanning – a paradigm change in topographic data acquisition for natural hazard management. In: Veulliet, E., Stötter, J., Weck-Hannemann, H. (Eds.), Sustainable Natural Hazard Management in Alpine Environments. Springer, Heidelberg, pp. 309–344.

Glenn, N.F., Streutker, D.R., Chadwick, D.J., Thackray, G.D., Dorsch, S.J., 2006. Analysis of LiDAR-derived topographic information for characterizing and differentiating landslide morphology and activity. Geomorphology 73 (1–2), 131–148.

GRASS Developer Team, 2010. Geographic Resources Analysis Support System (GRASS) Software. Version 6.4.0. <http://grass.osgeo.org> (accessed 19.04.10.).

Gruber, A., 2004. Jahrbuch der Geologischen Bundesanstalt, Geological Survey of Austria, Vienna, in Bericht 2004 über geologische Aufnahmen im Quartaer der Noerdlichen Tuxer Alpen auf Blatt 148 Brenner, pp. 337–343.

Höfle, B., Rutzinger, M., in review. Topographic airborne LiDAR in geomorphology: a technological perspective. Z. Geomorphol.

Jordan, G., Schott, B., 2005. Application of wavelet analysis to the study of spatial pattern of morphotectonic lineaments in digital terrain models. A case study. Remote Sens. Environ. 94 (1), 31–38.

Kraus, K., Pfeifer, N., 1998. Determination of terrain models in wooded areas with airborne laser scanner data. ISPRS J. Photogramm. Remote Sens. 53 (4), 193–203.

Mavrantza, O., Argialas, D., 2008. An object-oriented image analysis approach for the identification of geologic lineaments in a sedimentary geotectonic environment. In: Blaschke, T., Lang, S., Geoffrey, H. (Eds.), Object-Based Image Analysis, Lecture Notes in Geoinformation and Cartography. Springer, Berlin, pp. 383–398.

McKean, J., Goering, J., 2004. Objective landslide detection and surface morphology mapping using high-resolution airborne laser altimetry. Geomorphology 57 (3–4), 331–351.

Nyborg, M., Berglund, J., Triumf, C., 2007. Detection of lineaments using airborne laser scanning technology: Laxemar-Simpevarp, Sweden. Hydrogeol. J. 15 (1), 29–32.

Pfeifer, N., Mandlburger, G., 2009. LiDAR data filtering and DTM generation. In: Shan, J., Toth, C.K. (Eds.), Topographic Laser Ranging and Scanning – Principles and Processing. CRC Press, Taylor & Francis, London, pp. 307–334.

Rutzinger, M., Maukisch, M., Petrini-Monteferri, F., Stötter, J., 2007. Development of algorithms for the extraction of linear patterns lineaments from airborne laser scanning data. In: Kellerer-Pirklbauer, A., Keiler, M., Embleton-Hamann, C., Stötter, J. (Eds.), Proceedings Geomorphology for the Future. Innsbruck University Press, Obergurgl, Austria, pp. 161–168.

Shan, J., Toth, C.K. (Eds.), 2009. Topographic Laser Ranging and Scanning – Principles and Processing. CRC Press, Taylor & Francis, London.

Sithole, G., Vosselman, G., 2004. Experimental comparision of filter algorithms for bare-Earth extraction from airborne laser scanning point clouds. ISPRS J. Photogramm. Rem. Sens 61 (1), 33–46.

TIRIS, 2009. Tiroler Rauminformationssystem. <http://tiris.tirol.gv.at> (accessed 1.08.09.).

Vosselman, G., Liang, Z., 2009. Detection of curbstones in airborne laser scanning data. In: IAPRS, vol. 38, Part 3/W8, pp. 111–116.

Vosselman, G., Maas, H.-G. (Eds.), 2010. Airborne and Terrestrial Laser Scanning. Whittles Publishing, Boca Raton, FL.

Wladis, D., 1999. Automatic lineament detection using digital elevation models with second derivative filters. Photogramm. Eng. Remote Sens. 65 (4), 453–458.

Wood, J.D., 1996. The Geomorphological Characterisation of Digital Elevation Models. Ph.D. Thesis, University of Leicester, Leicester.

CHAPTER NINETEEN

Applied Geomorphic Mapping for Land Management in the River Murray Corridor, SE Australia

Colin F. Pain, Jonathan D.A. Clarke and Vanessa N.L. Wong[1]
Geoscience Australia, Canberra, Australia

Contents

1. INTRODUCTION

Airborne electromagnetic (AEM) data are increasingly being used to map the 3D distribution of conductivity and its relationship to the distribution of materials (Worrall et al., 1999; Cresswell et al., 2007). This approach has considerable potential for enhancing studies of various geomorphic environments, for example, in the Lower Balonne area of southern Queensland (Kernich et al., 2009).

[1] Present address: Southern Cross GeoScience, Southern Cross University, Lismore, NSW, Australia.

Developments in Earth Surface Processes, Volume 15
ISSN: 0928-2025, DOI: 10.1016/B978-0-444-53446-0.00019-7

In early 2007, an AEM survey was flown along a 450 km reach of the Murray River in SE Australia. The complete study area stretches from the South Australian border eastwards to Gunbower in Victoria. A total of 24,000 line kilometre of AEM data were acquired. The overall survey area encompasses iconic wetland areas, national and state forest parks and areas of irrigation and dry-land farming. This chapter uses the Lindsay—Wallpolla area, which is a subset of the AEM survey adjacent to the South Australian border (Figure 19.1), as an example.

Salinity of groundwater and of rivers is an important issue in Australia and elsewhere. The Murray River provides water for many settlements in SE Australia, the most important being Adelaide, the capital of South Australia. It is well known that saline aquifers underlie much of the lower Murray River basin, so it is important to understand any connections between ground and surface waters. It is also important to understand the current and potential impact of land use on the salinity of both ground and surface waters. The aim of this survey, therefore, was to provide information vital for addressing salinity, land management and groundwater resource issues along the River Murray Corridor (Lawrie et al., 2009).

Within the project area, key land management questions included:

1. What is the impact of irrigation on the floodplain, river and groundwater system?
2. What is the distribution of saline groundwaters where these have the potential to impact on the floodplain and river?
3. Where are the salt stores in the unsaturated zone within the floodplain?
4. What is the potential for salt mobilisation during managed environmental inundation actions and natural flood events?
5. What is the extent of losing and gaining effects along different reaches of the river system?

Maps of salt store distribution, saline and fresh groundwater and the hydraulic properties of soil and regolith materials in the shallow sub-surface were required to address the land management questions. These maps provide a 3D perspective of the connection between salt stores and saline groundwaters to surface waterways and the land surface. Sub-surface interpretations in the survey areas are hampered by a low density of useful borehole data in the floodplain and a paucity of soil and landscape surface mapping at appropriate scales throughout the project area. Therefore, new maps of the landforms and surface materials of the area and a newly evolved geomorphic understanding of the project area were

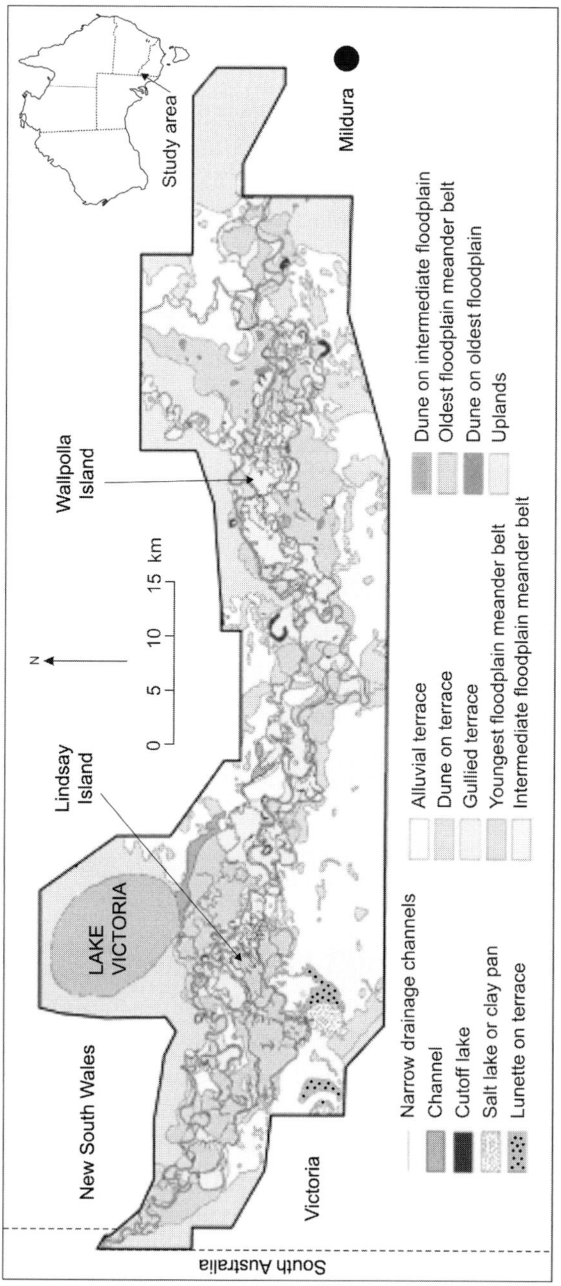

Figure 19.1 Location and landforms of the Lindsay–Wallpolla study area. The border between New South Wales and Victoria is along the southern bank of the Murray River.

required to constrain the interpretation of the AEM surveys and to address the land management questions.

The study area consists of Lindsay and Wallpolla 'islands', areas of the floodplain surrounded by secondary anastomosing channels that branch off from the southern side of the Murray River and the corresponding floodplain on the northern side of this reach of the Murray River (Figure 19.1). It includes RAMSAR-listed wetlands. Also, on the northern side is Lake Victoria, a giant oxbow system lying on an anabranch formed by Frenchman's Creek and the Rufus River (Figure 19.1). Several salt pans exist in the vicinity of Lake Victoria. The Darling River and its Anabranch also enter the Murray River in this reach at Wentworth and ~15 km west of Wentworth, respectively.

The main objective of this project was to provide information for constrained inversion of AEM data as a first step in interpreting those data to provide answers to land use questions posed for the area by the Mallee and Lower Murray-Darling Catchment Management Authorities. The studies also provide information about the materials that make up the various floodplain units to provide a framework within which to assess the utility of the AEM data to help answer the land management questions.

This chapter briefly describes the compilation of geomorphic maps of the Murray River floodplain and their use to aid interpretation of the AEM data to help answer some of the important land management questions.

2. PREVIOUS STUDIES

The regional geomorphic context of the project area was described by Bowler and Magee (1978) and Bowler et al. (2006). Studies of the Murray River floodplain include those that relate largely to the geology and its evolution in the region and soils and pedogenesis (Gill, 1973; Hills, 1975; Macumber, 1977; Brown and Stephenson, 1991). A number of studies have been undertaken in the Riverine Plain of Victoria and NSW, upstream of the present study area. These have been extrapolated to the Murray River floodplain region with respect to similarities in their evolutionary histories (Bowler and Harford, 1966; Pels, 1966; Butler et al., 1973). There have been few highly integrated studies in which geology, geophysics, soils and geomorphology have been used to address questions of land management, with the study by Rowan and Downes (1963) being a notable exception. This is particularly important, given that the floodplain of the Murray River in this region acts as an interface between the river and the regional groundwater systems, with the potential to mobilise large stores of salt under altered hydrological regimes.

Previous geomorphological studies in the region have identified a number of terraces (Kotsonis et al., 1999) in the Murray floodplain. Thoms et al. (1999) recognise that the present-day channels and rivers in this region are inset within intermediate channel systems and are therefore associated with relict floodplain surfaces that contain numerous palaeochannels and oxbow lakes. Gill (1973) named materials underlying the oldest terrace the Rufus Formation.

The Lake Victoria lunette has been emplaced and remodelled over 20,000 years of deflation (Gill, 1973). The river banks and sides of Lake Victoria are subjected to erosion with extensive blowouts and sand-falls, while the lunette associated with the lake is unusually wide and preserves horizontal bedding (Gill, 1973). The salt pans near Lake Victoria may be relicts of a former single large lake, as indicated by the presence of a shallow gypsiferous layer above the Blanchetown Formation, where the groundwater occurs within a metre of the ground surface beneath the overall area (Chen, 1995).

Stratigraphic units in the area are listed in Table 19.1. The study area is a semi-arid region with extensive aeolian deposits overlying either the Pleistocene lacustrine Blanchetown Clay or Pliocene marine sands of the Loxton-Parilla Formation. Aeolian deposits (the Woorinen Formation) occur in a regular series of linear dunes with an east-west trend stabilised

Table 19.1 Stratigraphy of the Lindsay—Wallpolla Study Area

Age		Murray Gorge		Uplands
		Floodplain	**Terrace**	
Holocene	Mid-late	Coonambidgal Formation		
	Early	Monoman Formation		
Pleistocene	Late		Rufus Formation	Woorinen Formation
	Middle			
	Early	Blanchetown Clay		
Pliocene		Loxton-Parilla Sands		

by vegetation except for the very local active sand patches. The dunes generally have calcareous B horizons with buried palaeosols. The material of which the east-west dunes are composed is pale to dark reddish-brown calcareous sand with some clay (Hills, 1975).

Reid (2007) described the hydrogeology of the Victorian side of the Murray River in the Gunbower Island to Lindsay—Wallpolla reach.

3. METHODOLOGY

The principal satellite images used to compile surface polygons were from ASTER and SPOT. Landsat images were used for comparison and infill, but they were not routinely used because SPOT and ASTER coverages were generally adequate for the Murray River Corridor project area.

Airborne gamma-ray images from Geoscience Australia's Australia Wide Geophysics Survey (AWAGS) were used to supplement areas with poor coverage of high-resolution digital elevation models (DEMs) and to provide regional scale information about lithological partitioning at the landscape surface. This is effective because the materials of the terrace composed of the Rufus Formation (Gill, 1973) are distinct from those of the active floodplain on ternary gamma-ray images. The distribution of this material as seen in the gamma-ray data was used where the absence of light detection and ranging (LiDAR) coverage precluded mapping the position of terraces by more accurate means.

Three DEM coverages were used for most of the map compilation:

1. High-resolution (2 and 5 m) LiDAR DEMs were used for the primary interpretation. Tiles at 1 m resolution were stitched together using ESRI ArcInfo and re-sampled to both 2 and 5 m,
2. Two DEMs derived from digitised contour maps at 10 m resolution and 20 m resolution were also used where necessary. The 10 m grid has a stated vertical accuracy of 2 m Australian Height Datum (AHD). These DEMs were used for infilling where LiDAR coverage was not available. A DEM acquired during the AEM survey was examined but was not used because of extreme anisotropy of the resolution along and across flight lines,
3. The Shuttle Radar Terrain Mission (SRTM) DEM was used to provide base information where other data were not available. At a scale of 1:25,000, the vertical and spatial resolutions of the SRTM DEM are too low, while the noise level is too high for mapping purposes. Therefore, the SRTM DEM was used for infill only.

Mapping at 1:25,000 ensured that sufficient detail was represented for the final output maps at 1:100,000 scale. ASTER, SPOT and LiDAR images were printed at 1:25,000 scale, and the maps were compiled by visual identification of landform features. Where LiDAR was available, individual scroll bars could be recognised. Unit boundaries were mapped onto a registered stable transparent overlay using mapping pens. The resulting line work was scanned to vector polygons and the polygons attributed using ArcGIS. The finished images were then printed for checking and field validation. This landform mapping exercise provided information on the spatial and chronological relationships between different surface units as well as an indication of depositional settings and processes for each.

Classification of Landsat, ASTER and SPOT images was tested in small representative areas but was found to give worse results than visual compilation from the DEMs. Late in the project, a technique for using LiDAR to automatically map specific landform types was under development (Halas et al., 2008) and will be reported separately.

Polygons were field checked by vehicular traverses along various tracks. Samples of surface materials and information representative of the mapping units were collected; these were used to provide information on hydrologic properties, in particular, recharge and salt loads.

Surface materials maps were compiled based on the landform maps, with gamma-ray radiometric images, surface observations, available bore holes and soil pits dug to a depth of 30 cm providing information on materials. These maps were used in conjunction with vegetation data to compile surface recharge maps and document the material products of the geomorphic processes indicated by the landform maps (Clarke et al., 2010). The shallow hand-dug soil pits were enforced by the generally extremely hard soils; anything deeper would have required mechanical excavation that time did not allow.

4. RESULTS

4.1 Regolith and Landform Units

The incised valley of the Murray River (the upstream equivalent of the Murray Gorge of Twidale et al., 1978) contains several mappable geomorphic units and their accompanying sediments (Figure 19.1). These are the alluvial terrace, raised several metres above the modern floodplain, the

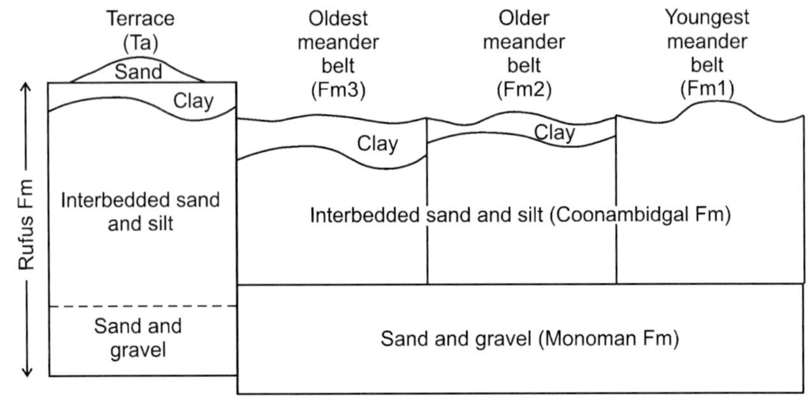

Figure 19.2 Diagrammatic representation of relationships between geomorphic and stratigraphic units. The Coonambidgal and Monoman Formations (Fm units) are inset to the Rufus Formation (Ta). The Rufus Formation varies from 5 to 12 m thick, as does the Coonambidgal Formation. The Monoman Formation is about 10-m thick at the western end and thins towards the east. The clay drape on Fm1 is absent and then increases from 0.5–1.5 m on Fm2 to 2–2.5 m on Fm3.

modern scroll plain, which is periodically inundated by floods and composed of three mappable meander tracts and a number of individual features such as dunes, lakes and lunettes.

Geomorphic differentiation of incised valley fill into terrace and floodplain deposits matches the distinct airborne gamma-ray patterns showing that the terrace units (Rufus Formation) are comparatively richer in K than in Th or U, whereas the floodplains (Coonambidgal Formation) show a strong signal in all three radioelements. The uplands, which are covered with quartz sand dune, show a very low signature in all three radiogenic elements. Despite their low resolution, these data provide a cross-check of landform mapping based on the high-resolution LiDAR.

The relationships between geomorphic and stratigraphic units are shown in Figure 19.2.

4.1.1 Uplands (U)

The uplands have sandy regolith, slightly more clayey in the swales, developed on mainly linear dunes of the Woorinen Formation. Soils are well-drained and sandy to sandy loams, with moderate amounts of carbonate in the dunes.

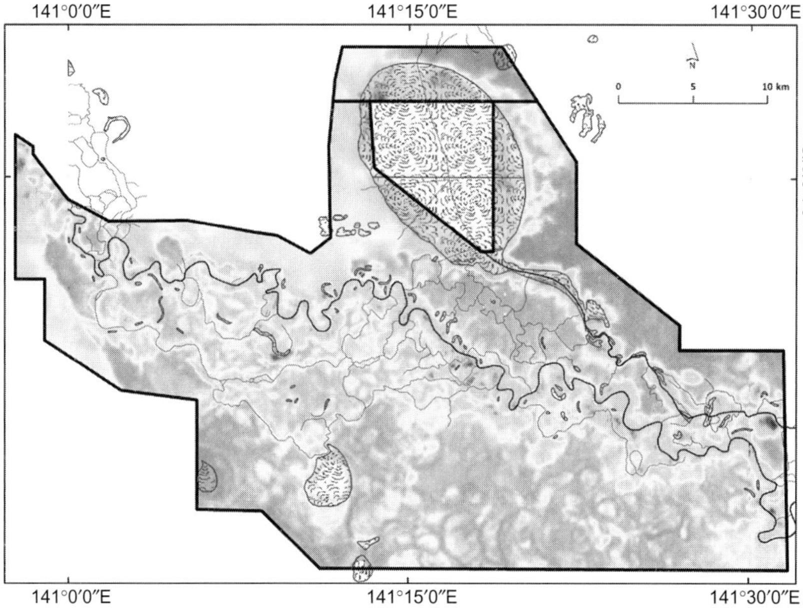

Figure 19.3 AEM slice of the western part of the study area showing conductivity from 0 to 5 m below the surface. The southern part of the image is in the terrace and clearly shows the complex of palaeo-oxbow and other palaeo-stream features that underlie the terrace landform unit. Lower conductivity values (light tones) show water-filled sandy sediments underlying young floodplains adjacent to the Murray River.

4.1.2 Alluvial Terrace (Ta)

This unit consists of clay and fine sandy alluvium and represents the Rufus Formation of Gill (1973) (Figures 19.2 and 19.3). The terrace surface has a discontinuous cover of sand dunes (Woorinen Formation) and is about 60,000 years old (Rogers and Gatehouse, 1990). The terrace is flat, with local orange sand dunes and sand sheets. Where sand is absent, the surface of the terrace consists of olive-khaki silty clays. Most soils are slightly to moderately saline. Loamy sands of relict dunes locally overlie the floodplain clays (Figure 19.2). AEM data for the 0 to 5 m slice (Figure 19.3) show that this unit is underlain by a complex of meander features.

4.1.3 Floodplain (Fm1, Fm2, Fm3)

The floodplain is formed on sediments of the Coonambidgal Formation (Butler, 1958) and consists of three distinct meander belts with well-developed scroll bars (Figures 19.2, 19.4 and 19.5).

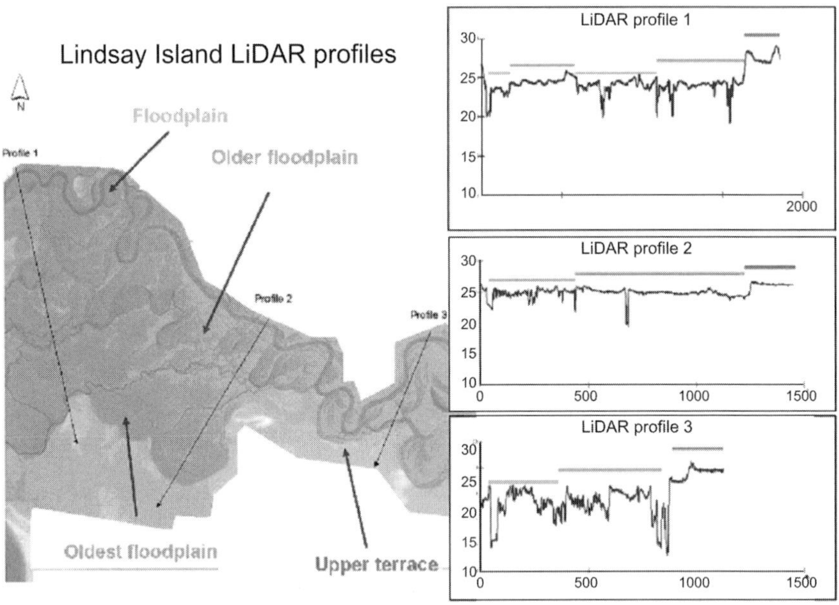

Figure 19.4 Compartmentalisation of the Murray River incised valley fill into terrace and floodplain deposits of different ages. The upper terrace is composed of Rufus Formation.

The oldest floodplain meander belt (Fm3) has a degraded scroll bar morphology. The amplitude of the scroll bars and the meander wavelength is greater for this unit than for the younger meander belts, indicating different hydrologic conditions during deposition. This unit is characterised by olive-khaki silty clay drapes over degraded scroll bars with a relief of about 2 m (Figures 19.2, 19.4 and 19.5). There are some thin (>2 m) source bordering dunes of grey sand.

The intermediate floodplain meander belt (Fm2) that has rounded morphology with scroll bars that are not as distinct as those on the modern floodplain in the LiDAR DEM (Figures 19.2, 19.4 and 19.5). Olive-khaki silty clay drapes over low-relief (~1 m) scroll bars are found in this unit. Source bordering dunes also occur on this unit.

The modern floodplain (Fm1) consists of meander belts and high-relief (2–3 m) scroll bars with crisp morphology and little or no clay draped over the surface (Figures 19.2, 19.4 and 19.5). Scroll bars are distinct in the LiDAR DEM. Surface sediments consist largely of yellow sand.

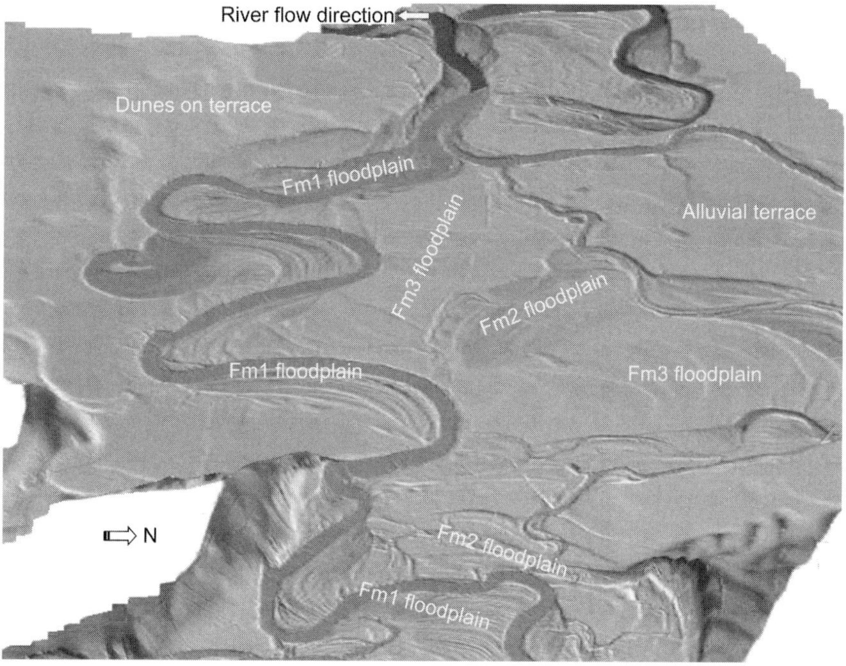

Figure 19.5 Oblique projection looking west from the Murray River at Merbein showing part of the LiDAR DEM and geomorphic elements (annotated). Width of image ~5 km.

The floodplain has a number of channels with several morphological types. The main Murray River channel sustains active although highly regulated flow and consists of a typical migrating meandering channel. Abandoned channels of the Murray River consist of broad oxbow billabongs filled with semi-permanent water. There are also sinuous anastomosing channels that have fixed banks. These channels have been superimposed on the intermediate floodplain deposits, sometimes exploiting abandoned channels, and elsewhere cutting across scroll bar and floodplain deposits. Flow in these is determined by water level in the main Murray River. During the period of observation some were flowing, while others were stagnant or dry and lined with clay, as described by Knighton and Nanson (2000).

Most soils on the floodplain are slightly to moderately saline. Loamy sands of relict dunes locally overlie the floodplain clays. Clays within the floodplain are typically smectitic and sodic. They are highly dispersive and form an impermeable seal after modest rain. Minor variations in floodplain elevation can significantly affect soil development. Scroll bars found on the oldest and intermediate floodplain units exhibit more profile development with heavier textures and are highly structured in swales compared to the corresponding crest than those on the youngest scroll bar sets. Due to the formation of a surface seal, water infiltration is limited to the upper layer of the soil profile, leading to surface ponding after rainfall and loss through evaporation.

5. APPLICATIONS

5.1 Application of Landform and Surface Materials Data to AEM Interpretation

Mathematically, constraining the AEM data from landform mapping was not possible because the landform units could not be reduced to a single mathematic value. However, the boundaries delineated by landform mapping guided the interpretation of patterns in the AEM data, especially at shallower depths. For example, there was a close correlation between the AEM patterning in the upper depth slices and distribution of landforms, such as the different generations of meander plain units, the alluvial terrace, salt and clay pans and sand dunes (Figure 19.3). This correlation allowed interpretation of the lower depth slices in terms of both sediment types and depositional environments.

5.2 Hydrogeological Issues

Compartmentalisation of the sediments has already been mention on page 498. It also appears that materials filling the incised valley of the Murray River are inset within intermediate Murray basin units. In this reach, the Blanchetown Clay is covered by dunes and slope deposits. Similarly, the Coonambidgal Formation is inset within the Rufus Formation, as suggested by Gill (1973) and shown in cross sections by Rogers and Gatehouse (1990). Modern, intermediate and oldest floodplain units and deposits compartmentalise the Coonambidgal Formation, as suggested above. Different ages of the units result in different properties. This is particularly apparent in the amount of clay at the surface which has important implications for recharge (see below). These differences occur across the axis of the floodplain because of poor interconnections between sedimentary units, and possibly down axis, within morpho-sedimentary units.

Thus the following recharge characteristics are predicted:

- All but the youngest floodplain sediments are sealed by dispersive clays with little or no recharge on intermediate and oldest floodplain and terrace units,
- Active channels have sandy bottoms and become clay-lined when active flow ceases, resulting in no recharge via abandoned channels flow. Cracking clays are of limited spatial extent and occur only in abandoned channels. There is limited recharge through surface clays via these desiccation cracks early in heavy rainfall events,
- The presence of sand dunes on the terrace leads to localised high infiltration, which results in local perching of water on the underlying floodplain clays. There may be a direct connection between source bordering dunes on the older floodplain meander belts and the underlying scroll bars, bypassing the clay drapes. However, the spatial extent of this type of dune is insignificant.

5.3 Relevance to Land Management Questions

The results of the study to date can be placed in the context of the five land management questions noted above. We make the following tentative conclusions with respect to each; considerably more work will be undertaken on these issues by the Catchment Management Authorities, using the project information:

1. What is the impact of irrigation on the floodplain, river and groundwater system?

   Irrigation has created groundwater mounds which increase the local hydraulic gradient and can cause acceleration of the flow of saline regional groundwater. Some of the irrigation mounds appear to have reached a dynamic equilibrium (Cummins and Thompson, 2005). Depending on the flow directions and depth to the water table, this may result in increased salt load into creeks and rivers or discharge of saline springs on the floodplain. Landform data and surface materials provide constraints on groundwater flow pathways and material properties of the aquifer units. They also identify zones of vertical recharge where surface salts may be flushed to the water table. This allows prediction of the impact of irrigation.

2. What is the distribution of saline groundwaters where these have the potential to impact on the floodplain and river?
   The distribution of saline groundwater can be mapped using the AEM depth slices. The landform and materials maps provide constraints on the nature of materials in the aquifers and on groundwater flow pathways. This information can then be used to assess potential impacts on the floodplain and river. Groundwater beneath the floodplain is dominantly saline, with areas of brackish water adjacent to the river flush zone. Compared to the main river channel, the creeks do not have well-developed flush zones and are essentially surrounded by saline groundwater. All these data and considerations suggest that at the time of the survey very little saline water was being discharged to the river and that it was more likely to be moving from the river to the groundwater.
3. Where are the salt stores in the unsaturated zone within the floodplain?
   This question is answered, in part, using analysis of surface soil samples to identify areas of salt storage in the floodplain. A combination of the AEM depth slices and shallow (above the water table) borehole sample analysis allows compilation of a series of products including conductive soils, surface salinity, salinity hazard and salt store maps, all included in Cullen et al. (2009). In general, the highest salt stores in the unsaturated zone are in areas remote from the Murray River and its various tributaries and anabranches.
4. What is the potential for salt mobilisation during managed environmental inundation actions and natural flood events?
   Integration of the surface data (LiDAR DEM, soil pits and satellite imagery) with AEM depth slices and borehole data was undertaken to answer this question. Products relating to salt mobilisation potential are maps of flush zones associated with the Murray River, groundwater recharge, conductive groundwater, conductive soils, surface salinity, salinity hazard and salt store maps (Cullen et al., 2009). The geomorphological interpretation of the data suggests that shallow groundwater flow may be strongly compartmentalised along the Murray River Corridor between different aged floodplain units, and possibly between different sedimentary units within each of the three mapped floodplain units. At depth, where similar sandy units of different ages may be juxtaposed, cross flow between meander belts of different ages is possible. However, preliminary interpretation suggests that environmental inundation may not move much salt within the floodplain.

5. What is the extent of losing and gaining effects along different reaches of the river system?
Similarly, integration of the surface data (LiDAR DEM, soil pits and satellite imagery) with the AEM survey and borehole data plays a large role in answering this question. These data allow identification of various potential flow pathways, particularly the channels and surface depressions along which surface and groundwater flow are most likely to occur. Integrated products relating to the losing and gaining nature of various reaches of the Murray River are found in Cullen et al. (2009), including the thickness, extent and conductivity of the flush zones, groundwater recharge, conductive groundwater, conductive soils, surface salinity, salinity hazard and salt store maps. Broad flush zones, which are also commonly deep, can be interpreted as places where the river is losing water, whereas places with no or very narrow flush zones may be stretches where the river is either not losing or may be gaining water. Deep flush zones occur along much of the Murray River in the study area suggesting that most of its length is losing water.

6. CONCLUSIONS

Compilation of the landform map of the Lindsay—Wallpolla area was greatly facilitated by the availability of a high-resolution LiDAR DEM over much of the area. The AEM data also contributed to map compilation by providing information about the conductivity variations in the materials underlying the floodplain and terrace.

The landforms and surface materials maps provided some spatial constraints on interpretation of the AEM data and were particularly useful for answering land use questions relating to the nature of materials under the floodplain, potential flow paths both on and beneath the surface and the relative rates and distribution of potential recharge across the floodplain.

ACKNOWLEDGEMENTS

We thank H. Apps, K. Cullen, D. Gibson, L. Halas, K.P. Tan and team members who gave us considerable help with this work while also busy with their own part of the larger project. K.L. Lawrie as project leader provided much needed support. The project was carried out under the auspices of the Australian Government's Community Stream Sampling and Salinity Mapping Project and managed by the Bureau of Rural Sciences. This study is published with the permission of the CEO, Geoscience Australia.

REFERENCES

Bowler, J.M., Harford, L.B., 1966. Quaternary tectonics and the evolution of the riverine plain near Echuca, Victoria. J. Geol. Soc. Aust. 13, 339–354.

Bowler, J.M., Magee, J.W., 1978. Geomorphology of the Mallee region in semi-arid northern Victoria and western New South Wales. Proc. R. Soc. Victoria 90, 5–25.

Bowler, J.M., Kotsonis, A., Lawrence, C.R., 2006. Environmental evolution of the Mallee region, western Murray Basin. Proc. R. Soc. Victoria 118, 116–210.

Brown, C.M., Stephenson, A.E., 1991. Geology of the Murray Basin, southeastern Australia. Bureau of Mineral Resources, Geology and Geophysics Australia, Bulletin 235.

Butler, B.E., 1958. Depositional systems of the Riverine Plain of south-eastern Australia in relation to soils. Commonwealth Scientific and Industrial Research Organisation, Soil Publication No. 10, 35 pp.

Butler, B.E., Blackburn, G., Bowler, J.M., Lawrence, C.R., Newell, J.W., Pels, S., 1973. A Geomorphic Map of the Riverine Plain of South-eastern Australia. Australian National University Press, Canberra.

Chen, X.Y., 1995. Geomorphology, stratigraphy and thermoluminescence dating of the lunette dune at Lake Victoria, western New South Wales. Palaeogeogr., Palaeoclimatol., Palaeoecol. 113, 69–86.

Clarke, J.D.A., Gibson, D., Apps, H., 2010. The use of LiDAR in applied interpretive landform mapping for natural resource management, Murray River alluvial plain, Australia. Int. J. Rem. Sens. 31, 6275–6296.

Cresswell, R.G., Mullen, I.C., Kingham, R., Kellett, J., Dent, D.L., Jones, G.L., 2007. Airborne electromagnetics supporting salinity and natural resource management decisions at the field scale in Australia. Int. J. Appl. Earth Obs. Geoinf. 9, 91–102.

Cullen, K., Apps, H., Halas, L., Tan, K.P., Pain, C., Lawrie, K., et al., 2009. River Murray Corridor AEM Salinity Mapping Project Atlas: Lindsay–Wallpolla and Lake Victoria–Darling Anabranch. Geoscience Australia GEOCAT # 68773.

Cummins, T., Thompson, C., 2005. Gap Analysis for the Mallee Zone Basin Salinity Management Strategy Five-Year Rolling Review. Prepared for the Murray-Darling Basin Commission. February 2005.

Gill, E.D., 1973. Geology and geomorphology of the Murray River region between Mildura and Renmark, Australia. Mem. Natl. Museum Victoria 34, 1–97.

Halas, L., Clarke, J., Tan, K.P., 2008. Integrated geomorphic techniques for the mapping of the Murray River floodplain, South Australia, using high-resolution DEM data set. Second International Salinity Forum, Adelaide.

Hills, E.S., 1975. Physiography of Victoria: An Introduction to Geomorphology. Whitcombe & Tombs Pty. Ltd., Australia, 373 pp.

Kernich, A.L., Pain, C.F., Clarke, J.D.A., Fitzpatrick, A.D., 2009. Geomorphology of a dryland fluvial system: the Lower Balonne River, southern Queensland. Aust. J. Earth Sci. 56, S139–S153.

Knighton, A.D., Nanson, G.C., 2000. Waterhole form and process in the anastomosing channel system of Cooper Creek, Australia. Geomorphology 35, 101–117.

Kotsonis, A., Cameron, K.J., Bowler, J.M., Joyce, E.B., 1999. Geomorphology of the Hattah Lakes region on the River Murray, southeastern Australia: a record of late Quaternary climate change. Proc. R. Soc. Victoria 111, 27–42.

Lawrie, K.L., Tan, K.P., Pain, C.F., Apps, H.A., Clarke, J.D.A., Cullen, K., et al., 2009. River Murray Corridor Final Report – Lindsay–Wallpolla. Geoscience Australia Record 2009/xx, GeoCat # 68771.

Macumber, P.G., 1977. The geology and palaeohydrology of the Kow Swamp fossil hominid site, Victoria, Australia. J. Geol. Soc. Aust. 24, 307–320.

Pels, S., 1966. Late Quaternary chronology of the riverine plain of southeastern Australia. J. Geol. Soc. Aust. 13, 27–40.

Reid, M., 2007. Hydrogeological Review of the Victorian Side of the Murray River Floodplain Site (Gunbower Island to Lindsay Wallpolla). CRC LEME Restricted Report 257R, 71 pp.

Rogers, P.A., Gatehouse, C.G., 1990. Late Quaternary stratigraphy of the Roonka archaeological sites. Q. Geol. Notes Geol. Surv. South Aust. 113, 6–14.

Rowan, J.N., Downes, R.G., 1963. A study of the land in North-Western Victoria. Soil Conservation Authority of Victoria Technical Communication No. 2. Government Printer, Melbourne.

Thoms, M.C., Ogden, R.W., Reid, M.A., 1999. Establishing the condition of lowland floodplain rivers: a palaeo-ecological approach. Freshwater Biol. 41, 407–425.

Twidale, C.R., Lindsay, J.M., Bourne, J.A., 1978. Age and origin of the Murray River and gorge in South Australia. Proc. R. Soc. Victoria 90, 27–42.

Worrall, L., Munday, T.J., Green, A.A., 1999. Airborne electromagnetics – providing new perspectives on geomorphic processes and landscape development in regolith-dominated terrains. Phys.Chem. Earth A 24, 855–860.

CHAPTER TWENTY

Monitoring Braided River Change Using Terrestrial Laser Scanning and Optical Bathymetric Mapping

Richard Williams[a], James Brasington[b], Damia Vericat[c], Murray Hicks[d], Fred Labrosse[e] and Mark Neal[f]

[a]Institute of Geography and Earth Sciences, Aberystwyth University, Aberystwyth, SY23 3DB, United Kingdom. rvw@aber.ac.uk

[b]Institute of Geography and Earth Sciences, Aberystwyth University, Aberystwyth, SY23 3DB, United Kingdom

[c]Forest Sciences Center of Catalonia, Crta. Sant Llorenç de Morunys, km 2 (direcció Port del Comte), 25280 Solsona, Lleida, Catalunya, Spain

[d]NIWA, 10 Kyle Street, Riccarton, Christchurch, 8011, New Zealand

[e]Department of Computer Science, Aberystwyth University, Aberystwyth, SY23 3DB, United Kingdom

[f]Department of Computer Science, Aberystwyth University, Aberystwyth, SY23 3DB, United Kingdom

Contents

Developments in Earth Surface Processes, Volume 15
ISSN: 0928-2025, DOI: 10.1016/B978-0-444-53446-0.00020-3

1. INTRODUCTION

The morphological approach (Church and Ashmore, 1998) has become established as a key tool for investigating sediment transport in large braided rivers (Carson and Griffiths, 1989; Brasington et al., 2000; Lane et al., 2003; Wheaton et al., 2010). The approach applies the sediment continuity equation to determine the bed material flux from repeated topographic surveys, and thus it links directly the spatial evolution of river morphology to areas of sediment erosion and deposition. Historically, the morphological approach has been applied in one-dimension, using cross-section data (Church and Ashmore, 1998). More recently, the availability of two-dimensional topographic data and digital elevation models (DEMs), provided by a range of new geomatics technologies and methods, has transformed this approach. These new data sources include digital tacheometry (Milne and Sear, 1997), real-time kinematic (RTK) global positioning system (GPS) (Brasington et al., 2000), digital softcopy photogrammetry (Winterbottom and Gilvear, 1997; Westaway et al., 2003), multi-beam echo sounding (Calder and Mayer, 2003), and in the last few years, airborne light detecting and ranging (LiDAR) (Thoma et al., 2005; Cavalli et al., 2008) and terrestrial laser scanning (TLS) (Milan et al., 2007). With these technological advances, new opportunities have emerged to investigate the feedbacks between flow, bed material and structure, sediment transport and vegetation, and link these to their influence on channel planform dynamics. Potentially, data can now be acquired at spatial scales that are sufficiently large to interpret reach-scale morphological change but at temporal scales commensurate with individual flood events.

In this case study, we describe a methodology that has been developed to apply TLS and empirical–optical bathymetric mapping to monitor the evolution of the braided Rees River, New Zealand, through a sequence of competent flood events. Section 2 of this chapter reviews recent developments associated with TLS and empirical–optical bathymetric mapping and considers how these technological advances can improve the geomorphological mapping of braided rivers. Section 3 introduces the Rees River study site, explains why this river was chosen for intense monitoring and describes data that were collected using TLS and optical bathymetric mapping. Section 4 presents the processing methodology used to (i) map exposed braidplain and inundated channel-bed levels and (ii) create seamless high-resolution DEMs of river bed topography. This is followed by a discussion of monitoring channel change using DEMs of difference and the management of data errors. Finally, an analysis of observed channel changes in the Rees over the summer flood season of 2009–2010 is presented.

2. TECHNOLOGICAL DEVELOPMENTS

2.1 Terrestrial Laser Scanning

Of all the technologies to emerge in recent years, TLS has transformative potential for mapping channel change, particularly in braided and ephemeral rivers, where much of the bed is sub-aerially exposed at low flow. Uniquely, TLS offers a combination of high individual point precision (2–4 mm in *xyz*), high-resolution point spacing (sub-centimetre) and fast data acquisition rates (50 kHz–1 MHz). To date, most applications of TLS in fluvial geomorphology have been restricted to small-scale investigations, generally relying upon scans acquired from a single or only a few scan locations. At this 'patch scale' ($\sim 1\ m^2$), TLS has been applied to capture the micro-scale morphology of coarse- and fine-grained gravels (Hodge et al., 2009) and has been shown to reproduce effectively the geometric properties of these complex natural surfaces (Hodge, 2010). At a larger spatial scale, Heritage and Milan (2009) surveyed a 180 m^2 gravel point bar using TLS and showed that local variations in TLS-derived elevation data were strongly related to grain size.

Although TLS can be used to map surface sedimentology at the patch and bar scale, it can also be used to map larger scale topography and through repeat survey, can be used to monitor geomorphic change. This has been aptly demonstrated by Milan et al. (2007) who monitored changes in a 6000 m^3 reach of a braided-proglacial outwash fan in the Swiss Alps. Surveys were undertaken daily, over 10 days, and achieved average point densities of 1600 points per square metre with vertical errors of ± 0.02 m in exposed braidplain topography. The resulting DEMs and DEMs of difference allowed direct quantification of erosion and deposition volumes and their spatial distribution. Although demonstrating the potential of TLS, the geomorphological insights gained from data of Milan et al. (2007) are limited by the spatial extent of braidplain surveyed. This was constrained by the rate at which the scanner could be deployed at different survey stations in-between flood events. To increase the scan frequency and thus extend the spatial scope of TLS, a number of developmental survey systems have been developed recently which deploy scanners on ground-based vehicles (Barber et al., 2008) and boats (Alho et al., 2009). These mobile or kinematic TLS systems offer the potential to acquire up to 10^9 survey points per day. Although highly attractive, the cost of such systems currently remains prohibitive, and a fully ruggedised vehicular scanning system, suitable for navigating the undulating terrain and shallow waters of braided rivers, has not yet been presented.

2.2 Bathymetric Surveying

A particular challenge associated with mapping braided rivers is surveying the inundated channel beds. Inundated areas may account for less than 10% of a braidplain at low flow, as has been observed in a range of braided river contexts (Brasington et al., 2000, 2003; Ashmore and Sauks, 2006). These inundated areas are, however, highly susceptible to topographic evolution during flood events, so it is critical to map these areas with techniques associated with minimal vertical error. Laser ranging technologies hold promise to bridge this missing link through the application of high-power, high-frequency water penetrating lasers. However, the power limitations on such systems restrict the collection of data to large footprint scans acquired from distance on airborne platforms and at significant cost. Although such 'bathymetric LiDAR' systems have been widely applied to monitor coastal topography at coarse resolution (Wedding et al., 2008; Chust et al., 2010), to date only one prototype system, the experimental advanced airborne research LiDAR (Brock et al., 2004) has been developed to map topography at the spatial resolution (<2 m spacing) required to reliably represent fluvial topography (McKean et al., 2008). As such, current strategies to survey river systems require sub-aerial data to be supplemented with externally sourced bathymetric data so that channel change in both exposed and inundated areas can be adequately quantified.

In fluvial environments, where inundated areas are typically shallow at low flows, high-resolution bathymetric maps (Gao, 2009) can be rapidly acquired using a range of active remote sensing (radar and sonar) and passive remote sensing (optical) techniques (Carbonneau et al., 2006; Flener et al., 2010). Of these approaches, empirically calibrated optical methods are particularly attractive for fluvial applications since they can be applied to readily available aerial photography, and they offer good detections of depth down to 12 m (Gao, 2009) if waters are not so turbid that the bottom of the channel is obscured. Empirical–optical methods have been demonstrated to be effective in a range of rivers (Winterbottom and Gilvear, 1997; Carbonneau et al., 2006). Westaway et al. (2003) have also successfully used empirical–optical techniques to derive inundated river topography along a 3.3 km reach of the Waimakariri River, New

Zealand. Their method used georeferenced aerial photography, an empirical model to calibrate water depths to red–green–blue (RGB) pixel values and a model of water surface elevation created by interpolating water edge elevations across inundated channels. The resulting DEM was characterised by a vertical mean error of 36 cm over inundated channels. Although this error was deemed acceptable for the typical scale of in-channel topographic changes during floods (Lane et al., 2003), it illustrates the challenge of accurately intersecting a water depth map with a water surface map to calculate channel-bed elevation. Despite this challenge, empirical–optical techniques offer considerable potential to map braided river bathymetry at a low cost and high resolution and complement ground-based surveys of exposed braidplain areas.

3. DATA COLLECTION

3.1 The Rees River, New Zealand

3.1.1 Geological and Geomorphological Setting

The Rees River is 41 km long and drains a 405 km^2 alpine basin to the east of the Southern Alps, New Zealand (Figure 20.1a). The upper

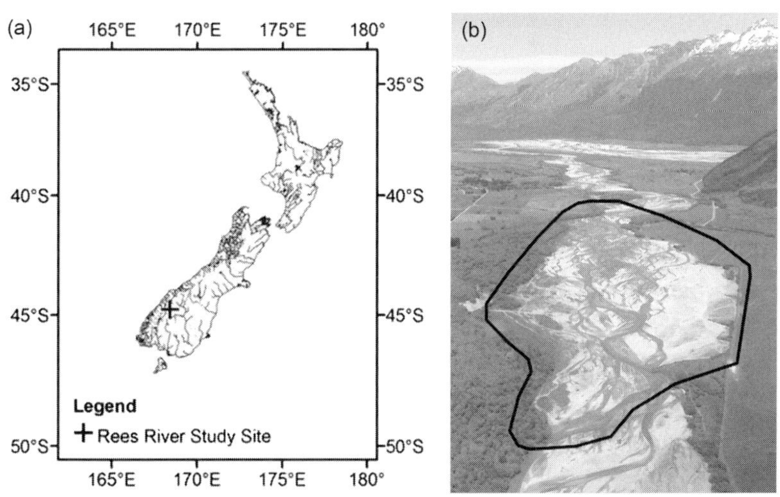

Figure 20.1 (a) Location of the Rees River in New Zealand. The Rees River Study Area, at low flow. (b) Photograph of the study area (identified by the polygon) looking downstream, towards the Rees Delta at the head of Lake Wakatipu.

catchment is dominated by schist belonging to the Mount Aspiring lithologic group (Turnbull, 2000). These schist deposits are composed of quartz–feldspar–mica, with greenshist bands (McSaveney and Glassey, 2002). The sediments are finely grained and highly susceptible to physical weathering. The upper Rees River is confined to a single channel, which has eroded into the schist bedrock. The Richardson mountains rise to the east, or true left, and the Forbes mountains rise to the west, or true right. Both mountain ranges are characterised by peaks exceeding 2000 m in elevation and glaciers sit upon the high, south-facing slopes. Present-day uplift rates across the Southern Alps, based on measurements and modelling, are up to 5 mm/year (Beavan et al., 2010). Landslides are common along the steep valley sides, and large alluvial fans extend from tributaries onto the valley floor, providing a plentiful supply of sediment.

The Rees basin was cut out by the Tyndall Glacier during the Pleistocene. This glacier is now confined to the high slopes of Snowy Creek, a former tributary to the Rees which has now been captured by the adjacent Dart River. The lower Rees slopes and valley floor are dominated by Holocene alluvial deposits derived from the erodible upper schist catchments. Downstream of the mountain front, the Rees has developed a wide, labile, braided gravel-bed. Together with the adjacent Dart River, the Rees has formed a major delta at the head of Lake Wakatipu. From an analysis of mapping and aerial photography, it is estimated that the average rate of Dart–Rees delta advanced from 1895 to 2001 was 1.4–1.6 m/year (Mabin, 2007). The incorporation of bathymetric survey data into estimates of delta advance suggests that between 1966 and 2007 the annual rate of sedimentation was approximately 2.7×10^5 m^3/year, which equates to an average catchment denudation rate of 0.3 mm/year (Wild et al., 2008). Thus, rates of sediment transport are very high in the Rees, making it an ideal catchment to study changes in river planform.

3.1.2 Study Reach

This case study focuses upon a 2.5 km × 0.7 km braided reach of the Rees River that is located approximately 7 km upstream from the Rees' mouth at Lake Wakatipu. The boundaries of the study area are shown in Figure 20.1b. Historic aerial photography indicates that this reach is highly dynamic, with the main channel *thalweg* frequently moving across the braidplain (Otago Regional Council, 2008). The true left of the braidplain is bounded by a near-continuous willow plantation. On the true right, a combination of stopbanks, cross banks and lines of willow trees defend adjacent pastoral land from erosion and flooding. The coarse gravel deposits on the braidplain are approximately 17 m deep and underlain by lacustrine silts and clays (Lalor and Purvis, 1970). Cross-section surveys of the Rees River braidplain have been undertaken on five occasions between 1984 and 2006, with the primary objective of monitoring gravel reserves (Mabin, 2007). These surveys indicate that the braidplain within the study area is slowly aggrading, with average rates of 27 mm/year and 16 mm/year upstream and downstream of the bridge, respectively. This evidence supports anecdotal evidence, reported by local residents, that the bridge's freeboard is reducing (Otago Regional Council, 2008).

The dynamic nature of Rees River study reach provides a good environment to study the interactions between flow, form and sediment transport. The braidplain can easily be accessed by vehicle, and the braidplain's width lends itself to the spatial scale that TLS and optical–empirical bathymetric mapping can be applied to. There is also a strong local interest in the dynamics of the Rees River since the position of the main *thalweg*, and levels of the braidplain, can influence flood hazards in the nearby township of Glenorchy (Whyte and Ohlbock, 2007; Otago Regional Council, 2010).

3.2 Survey Strategy

During the austral summer of 2009–2010, the evolution of the braidplain was intensively monitored to quantify flood-driven morphological change through a sequence of competent events. A gauging station was installed in a bedrock gorge approximately 8 km upstream from the study reach. The austral summer hydrograph is shown in Figure 20.2. The most common cause of precipitation in the catchment are northwesterly cold fronts moving across the main divide of the Southern Alps. Due to the steep slopes and thin soils of the catchment, the rising limb of each event is very steep. During the spring and early summer, the falling limbs are attenuated by snowmelt. By late summer, dry antecedent conditions cause very steep flood recessions; these are damped as the catchment wets up in the autumn. Winter flows tend to be low, with precipitation in the upper catchment falling as snow. Superimposed upon the hydrograph, shown in Figure 20.2, are bars that indicate when the study reach was surveyed, using a combination of TLS, aerial photography and channel depth profiling. The survey strategy is discussed below, but only the results from the

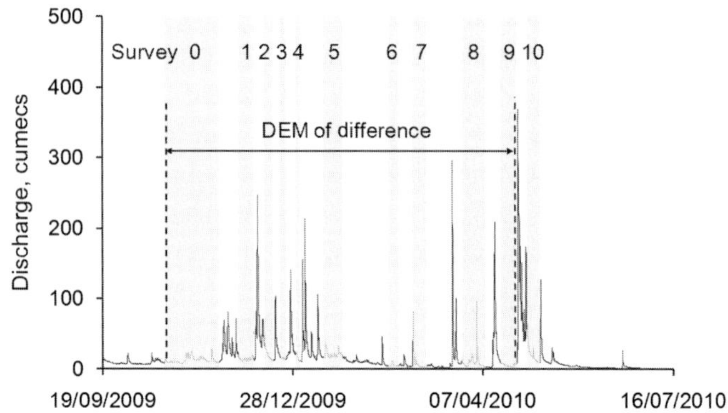

Figure 20.2 Rees River flow record at Invincible gauging station, approximately 8 km upstream of the study reach. The periods when the river was surveyed are indicated by the grey vertical bars.

first and penultimate surveys (numbered 0 and 9 on Figure 20.2) are considered in this case study.

3.2.1 Terrestrial Laser Scanning

Exposed areas of braidplain were surveyed using the ArgoScan system (Figure 20.3). This incorporates a high-frequency, gimbal-supported Leica 6100 Terrestrial Laser Scanner, an RTK GPS, a panoramic camera and an onboard systems control PC; all mounted on an Argo Amphibious All Terrain Vehicle. The Leica 6100 TLS was set to collect survey points at horizontal and vertical increments of 0.018°, corresponding to a maximum point spacing of 7.9 mm at a range of 25 m. The spatial extent of scan data acquired from a single survey location is dictated by the theoretical scanner range (79 m) and losses due to line-of-sight and scene permeability. In a fluvial setting, line-of-sight losses are typically greater across-stream due to intersecting channel banks. In practice, therefore, the distribution of scan data from a single set-up was an ellipse of 50 m × 40 m, elongated downstream and incorporated between 10 and 15 million survey points. The high-frequency phase-based laser ranging system employed by the Leica 6100 enabled this data collection within a 3 min stationary set-up. To provide seamless coverage across the study site, scans were acquired in stop—go survey mode spaced approximately every 65 m across-stream and 75 m downstream. The short scan durations enabled the collection of up to 70 locations per day, totalling one billion

Figure 20.3 The ArgoScan System.

survey points. Complete TLS coverage of the study site required approximately 300 scans (Figure 20.3).

Two reflective targets (Figure 20.3), positioned using RTK GPS in static mode, were used to provide a floating control network to co-register individual scans. The targets were positioned 10–15 m from the scanner, at an angle of approximately 120° from one another. Data were logged at each target until a minimum three-dimensional point quality of 10 mm was achieved. Blunders (Bannister et al., 1998) associated with the georectification were monitored by cross-checking the predicted scan location derived by resection from the targets, with the GPS position obtained on the ArgoScan system. Finally, the data were transformed to the New Zealand Transverse Mercator Projection, based on a network of precision GPS-located benchmarks.

3.2.2 Bathymetric Mapping

TLS surveys were undertaken at low flows. To derive channel-bed levels for the remaining areas, an empirical–optical mapping technique was used. This required aerial photographs of wetted channels and depth data, for model development and validation. Non-metric vertical aerial photographs were acquired at 1:5000 scale from a helicopter, using a Nikon D90 camera with fixed 28 mm lens and an automatic intervalometer set to acquire images every 5 s. Immediately following the acquisition of aerial photographs, depth data were obtained along two transects of the primary anabranches. These depth data were measured using the 1 MHz vertical acoustic beam of a Sontek S5 RiverSurveyor located using an integrated RTK GPS, mounted on a lightweight boat. This system enables precise geo-located depth soundings to be acquired at 10 Hz frequency as the boat was guided downstream on tethers by a single operator. Typical surveys of each anabranch took 30–40 min to acquire, resulting in 3000–4000 depth measurements along each anabranch at a point spacing of approximately 1 m.

4. PROCESSING METHODOLOGY

DEMs were derived using a data fusion of the TLS and bathymetric mapping to account for exposed braidplain and channel-bed levels.

4.1 Exposed Braidplain Mapping

Terrestrial laser scan data were processed through six main steps to generate a DEM of the exposed braidplain:

1. Individual point clouds were georeferenced to the NZTM map projection and cross-checked against the GPS-derived location of the ArgoScan system.
2. Georeferenced point clouds were merged into a single, unified point cloud containing approximately 6 billion survey points.
3. The unified point cloud was reduced to a quasi-uniform point spacing of 5 cm to limit redundancy and constrain data processing during DEM construction.
4. The decimated point cloud was manually edited to remove significant objects such as the ArgoScan system, reflective targets and their associated tripods and vegetation. Artefacts associated with laser range errors and poor-quality returns from wetted channels and atmospheric particles were also removed.
5. The cleaned cloud was spatially filtered to extract local and detrended surface statistics at a 1 m horizontal resolution. These included the maximum, minimum and mean elevations; the detrended standard deviation of elevations and higher order moments of the elevation distribution.
6. The local minimum elevation values were interpolated using Delaunlay triangulation which was then linearly resampled to a 1 m horizontal resolution raster DEM.

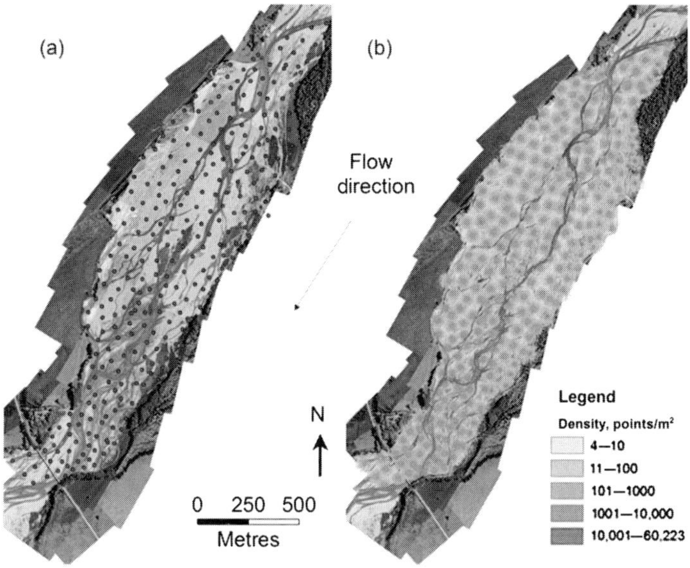

Figure 20.4 TLS survey undertaken in October 2009. (a) Location of ArgoScan stations. (b) Density of TLS points.

Figure 20.4 shows the location of the ArgoScan stations for the first survey of the Rees, which was undertaken in October 2009. The figure also reveals the data point density, shown here for the processed (reduced resolution) 5 cm point cloud. There is near-complete coverage of the exposed braidplain and 86% of the area surveyed, excluding channels, has a density greater than 10 points per square metre. Overall, the mean density is 825 points per square metre and the median point density is 221 points per square metre. The DEM constructed from these data is shown in Figure 20.5.

To assess the quality of the DEM surface, the gridded elevation values were compared to 1060 checkpoints acquired with RTK GPS. This analysis is limited by the low point precision of the RTK GPS data relative to the high point precision of the TLS data. The resulting deviations between grid elevations and point observations must, therefore, be treated cautiously and not interpreted simply as a measure of resulting DEM quality. Nonetheless, this check provides a useful test for systematic errors in the DEM, and the resulting agreement between the two data sets is encouraging with a mean error of −0.01 m, a standard deviation error of

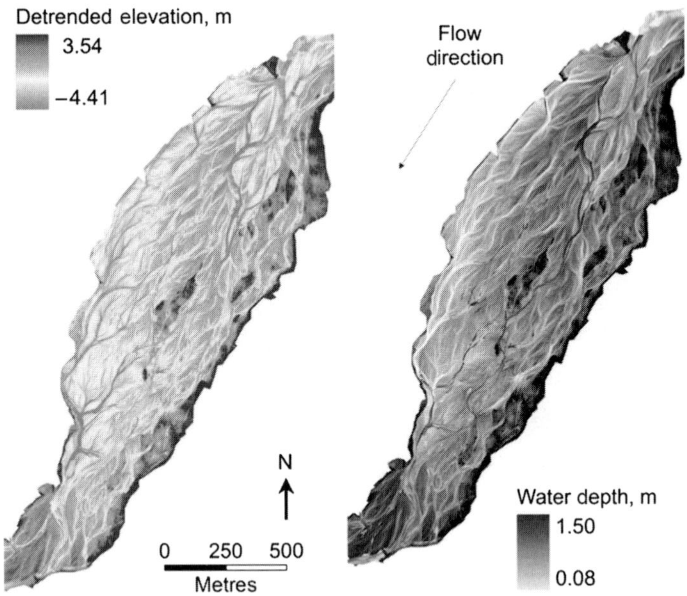

Figure 20.5 (a) Detrended DEM. The surface has been produced by calculating a mean longitudinal bed slope and subtracting this from a DEM of elevations above sea level. (b) Map of water depth derived by optical–empirical techniques.

0.09 m and a root mean square error (RMSE) of 0.05 m. The negative mean error likely reflects sampling of the particle tops by the non-invasive laser scanner, while the pole-mounted GPS measurements typically sampled elevations of the gravel interstices. The spatial pattern of errors revealed no obvious systematic error.

The quality of the DEM surface was also assessed by mapping a 1 m spatial resolution grid of standard deviations from the decimated scan data (Figure 20.6). The spatial pattern of standard deviations is closely linked to the character of the braidplain surface. A comparison between the reach aerial photograph and the standard deviation map (Figure 20.6) shows that the standard deviations across areas of exposed gravels are typically in the order of centimetre, whereas vegetated islands and banks have standard deviations >7.5 cm. The low standard deviations across exposed gravels indicate that the survey technique is capable of measuring surface topography with high precision. Thus, small magnitudes of change should be detectable from consecutive surveys.

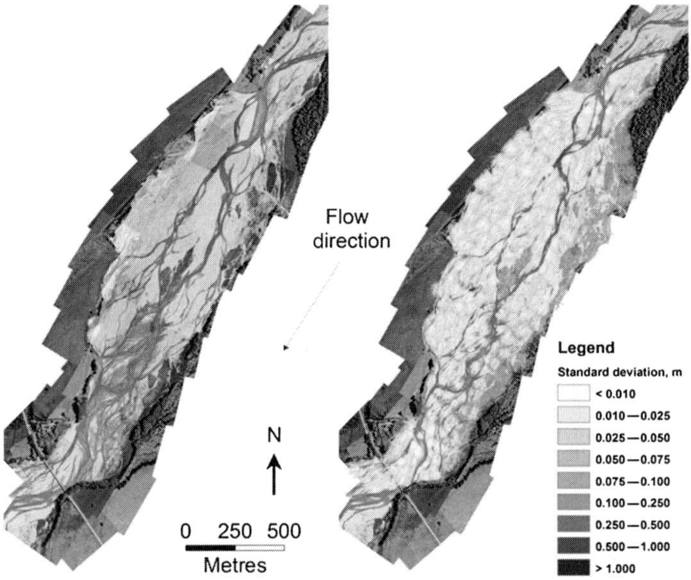

Figure 20.6 Standard deviations of 1 m gridded TLS data from the October 2009 survey. An aerial photograph is shown on the left to compare the standard deviations to surface cover.

4.2 Channel-Bed Level Mapping

Channel-bed levels were generated using an empirical—optical deep water correction model, based on non-metric aerial photographs and depth data obtained along two anabranches. First, a suite of 19 aerial photographs were georeferenced by matching objects that could be identified in the photographs and in the georeferenced TLS point cloud (from Step 2, above). Each image was georeferenced using a minimum of 15 tie-points, which were distributed across the image and then transformed by a rubber sheeting method. Subsequently, the images were resampled to a pixel resolution of 0.1 m using a nearest neighbour transformation and finally mosaicked into a 1 m resolution 3 band RGB raster. The rasters were not orthorectified because the relatively small changes in vertical relief across the study area were not considered to be a significant source of error in the georeferencing process. To complete the production of the mosaicked aerial photographs, the raster was clipped to the extent of the wet areas in the study reach. These wet areas accounted for 10% of the study reach area. The depth data, from the RiverSurveyor acoustic

Doppler current profiler (aDcp), were then used to calibrate a depth signal from the resulting RGB image. The RGB values from the mosaicked aerial photograph were extracted for each sample using a point-in-grid algorithm and the data were then randomly segmented into two classes. One class, which included two-thirds of the samples, was used to develop the empirical–optical deep water correction model. The other class was used to validate the model.

The empirical–optical model used here is based on the Lyzenga (1981) algorithm. This algorithm expresses the exponential attenuation of light with depth, using an intermediary term to account for the point where water is sufficiently deep to saturate the reflectance signal, and the channel bed no longer influences reflection:

$$X_i = \ln(\mathrm{DN}_i - \mathrm{DN}_{\mathrm{max}}) \tag{20.1}$$

where X_i is the term to fit to depth, DN_i is the pixel brightness number and $\mathrm{DN}_{\mathrm{max}}$ is the pixel brightness number in deep water. The method can be applied to any spectral band or combination of bands. The first task in developing the empirical–optical model was therefore to assess the strength of the relationships between depth and each band's reflectance. Pearson correlation coefficients for the red, green and blue bands were 0.66, 0.24 and 0.15, respectively. These results correspond to the findings of previous correlations in shallow water environments, which indicate that the red band is the most sensitive to depth, followed by the green and then blue bands (Winterbottom and Gilvear, 1997; Legleiter et al., 2004). Flener et al. (2010) found that both the red and green bands were sufficiently sensitive to be used in their empirical–optical model. For the Rees, however, only the red band was found to be sufficiently sensitive and therefore the optical–empirical model was developed using only a single-band, in the same manner as Carbonneau et al. (2006).

Least-squares regression was used to derive the optical–empirical models shown in Figure 20.7. The plot of reflectance values versus depth (Figure 20.5a) was used to select a saturation brightness value of 40, which was then used to calculate X_i , using Eq. (20.1) (Figure 20.7b). A regression quality of R^2=0.54 was calculated for the Lyzenga algorithm, which is similar to that achieved by Carbonneau et al. (2006). The scatter in the relationship (Figure 20.7b) can be attributed to differences in the global illumination of each image and variation in bed substrate, in particular the presence of algae cover. These factors could be corrected, which

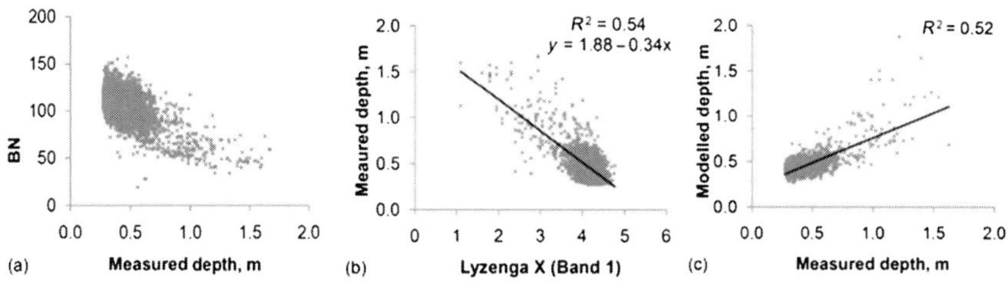

Figure 20.7 Empirical optical model used to map channel depth. (a) Pixel brightness (BN) values and measured depths for the red band of the aerial photograph. (b) Empirical optical model for the red band of the aerial photograph. (c) Modelled versus measured depths for the class of measurements used to validate the model.

may reduce the scatter. For example, the illumination of each image could be modified, a moving window could be used to sample multiple 0.1 m pixels on a fine-resolution georeferenced image and the different bed substrates could be classified and then used to develop a set of empirical–optical relationships. Here, however, a simple approach was adopted and a universal model applied.

The quality of the empirical–optical model was assessed by using the derived relationship to model channel depths for the second class of depth measurements. Figure 20.7c shows the relationship between the modelled and measured depths, giving $R^2 = 0.52$, similar to that obtained for the calibration. The mean error was −0.01 m, the standard deviation error was 0.09 m and the RMSE 0.10 m. Although the RMSE is higher than that associated with the TLS data, the error is of the same order of magnitude.

The optical–empirical model was used to estimate channel depth for each 1 m pixel in the wetted areas of the aerial photograph mosaic. The subsequent depth map is shown in Figure 20.5b. To produce a map of channel-bed elevations, these depth data must be subtracted from an estimate of the water surface elevation. This was obtained using methods applied by Brasington et al. (2003) and Westaway et al. (2003). Water edge TLS points used to construct the exposed bed DEM were used to interpolate across wetted channels and thus produce an estimated water surface elevation map. The estimated channel depths were then subtracted from the water surface elevations and then combined with the TLS-derived DEM of exposed areas to give a DEM for the whole study reach.

5. RESULTS: DEMS OF DIFFERENCE

The low errors associated with the TLS-derived DEM and channel-bed level maps suggest that the methodology is capable of delivering high-quality topographic models of the braided Rees River, which is characterised by a large spatial extent and low vertical relief. In addition to the DEM derived from surveys in October 2009, a second DEM was also constructed from survey data collected at the end of the flood season in April 2010. As shown in Figure 20.2, during the time that elapsed between these two surveys, there were a number of competent flood events. The morphological change resulting from the integration of these events can be revealed by subtracting the two (after and before) surface models to generate a DEM of difference (Figure 20.9a).

The DEM of difference derived from the fusion of TLS and empirical–optical mapping reveals spatially coherent patterns of erosion and deposition, analogous to maps of braided river evolution derived using photogrammetry and airborne LiDAR (Lane et al., 2003) and RTK GPS (Brasington et al., 2000). However, to assess whether the DEMs of difference can be used to quantify sediment budgets reliably, it is necessary to distinguish real geomorphic change from apparent changes that might arise from spurious errors or uncertainty in the individual DEMs. This theme has received substantial previous attention (Lane et al., 1994; Milne and Sear, 1997; Wheaton et al., 2010). Here, the methodology described by Brasington et al. (2003), which draws on Taylor (1997), is applied to analytically estimate a threshold level of detection (LOD) to distinguish spurious changes in DEMs of difference.

This approach attempts to predict the probability of observed changes captured by the DEM of difference, relative to an estimate of the combined uncertainty resulting from the subtraction of two error-prone surfaces. By assuming that the errors in each DEM are independent, the total, combined error in the calculation of a derived variable, U, from the addition or subtraction of two variables, Z_1 and Z_2, can be estimated as:

$$\delta u = \sqrt{\delta z_1^2 + \delta z_2^2} \quad (20.2)$$

where z_1 and z_2 are the errors associated with the variables Z_1 and Z_2, respectively. The total error can then be subjected to probabilistic thresholding by assuming that the estimates of δz are reasonably approximated by the standard deviation error. Equation (20.2) can then be modified as follows:

$$U_{\text{crit}} = t\sqrt{\text{SDE}_1^2 + \text{SDE}_2^2} \tag{20.3}$$

where U_{crit} is the critical threshold error, SDE_1 and SDE_2 are the standard deviation error in each surface, and t is the critical t-value at the chosen confidence interval:

$$t = \frac{|Z_1 - Z_2|}{\delta u} \tag{20.4}$$

where $|z_2 - z_1|$ is the absolute value of the DEM of difference. The SDE for exposed braidplain is spatially variable and can be usefully represented by a 1 m spatial resolution grid of the detrended standard deviations from the decimated scan data (described in Step 5 of the exposed braidplain mapping processing methodology, above). The SDE for wetted channels was taken to be the standard deviation error associated with the validation of the optical–empirical bathymetric model (0.09 m for the October 2009 DEM). In the absence of a spatial metric to capture uncertainty in this relationship, this was treated as constant across all the inundated channel areas. Combining errors in the hybrid DEMs and measuring the changes observed in the DEM of difference using Eq. (20.4) enables maps of the statistically significant changes to be generated for any given confidence interval. This approach to error assessment is illustrated in Figure 20.8.

Figure 20.9b shows the DEM of difference for the 84%, or the one-tailed one-sigma, confidence interval. A comparison of this DEM of difference to the raw DEM of difference (Figure 20.9a) shows that this criteria primarily removes small elevation changes identified on bar tops in the original DEM of difference. This indicates that these small elevation changes are more likely to be associated with spurious errors or uncertainty in the individual DEMs than to be associated with real morphological change. Importantly, the large magnitude morphological changes in the DEM of difference are preserved, indicating that the survey methodologies used to derive the DEM of difference are sufficient to capture the key geomorphological changes.

The total sediment budget for the study reach, between October 2009 and April 2010, can be calculated by integrating the erosion and deposition volumes (Figure 20.10). This indicates that the reach slowly aggraded

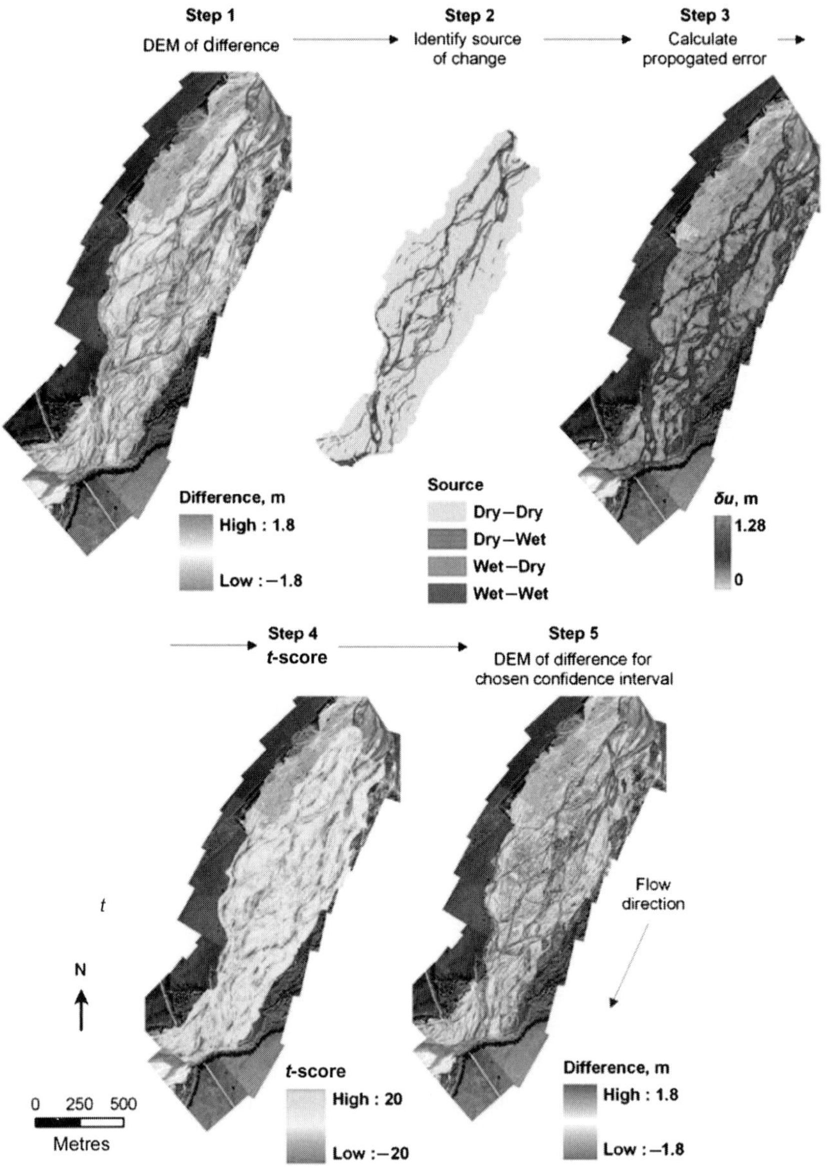

Figure 20.8 Overview of approach used to calculate DEMs of difference for particular confidence intervals. The data used to produce the DEM of difference is classified to identify the source of δu. Subsequently, δu is calculated and a *t*-score derived. The DEM of difference is then segmented for a chosen confidence interval.

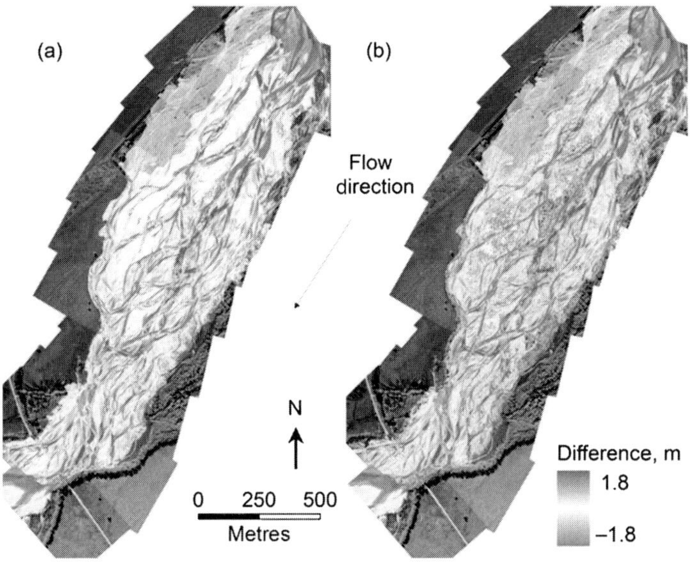

Figure 20.9 (a) DEM of difference and (b) DEM of difference for the 84% confidence interval.

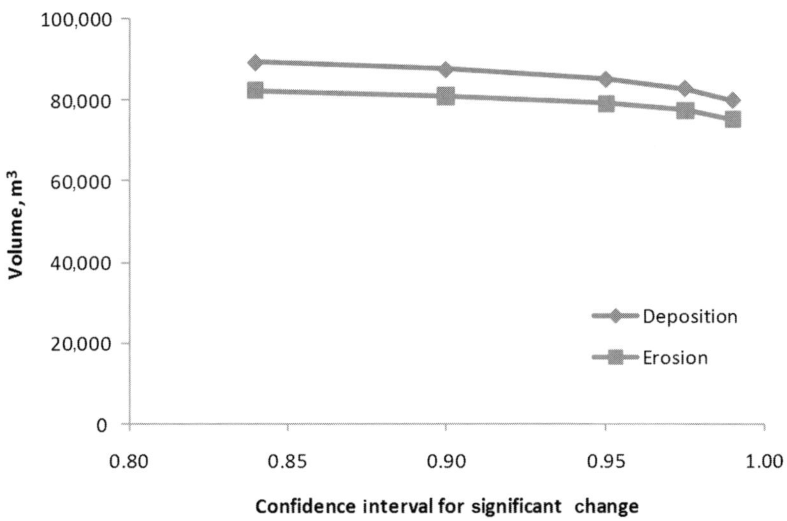

Figure 20.10 Relationship between erosion and deposition volumes and confidence interval for significant morphological change.

Table 20.1 Braidplain Area Within the Study Reach that Experienced Morphological Change During the Study Period, for a Selection of Confidence Intervals for Significant Change

Confidence Interval	Erosion, % of Study Reach Area	Deposition, % of Study Reach Area
0.84	35.5	43.3
0.9	33.0	40.7
0.95	30.4	37.7
0.975	28.4	35.4
0.99	26.2	32.8

during the study period, with deposition volumes exceeding erosion volumes by 9% for the 84% confidence interval. Overall, across the area surveyed, the reach aggraded by 11 mm. This is consistent with the trend observed from cross-section surveys during the past two decades (Otago Regional Council, 2008) and is typical of braided rivers where the braidplain is constrained between stopbanks (Davies and McSaveney, 2006). The aggradation in this reach may also, however, be a consequence of a large sediment pulse moving through the river (Hoey, 1992).

From an aerial perspective, deposition was also the dominant process. For the 84% confidence interval, 33.5% and 43.3% of the study reach experienced erosion and deposition, respectively (Table 20.1). The aerial extent of the braidplain reworking is relatively high and shows that the river is extremely active. In particular, through the central belt of the braidplain (Figure 20.9), the pattern of erosion and deposition is complex, indicating that consecutive flood events have contributed to a sequence of channel filling, avulsion and cutting. The main *thalweg* of the river has meandered extensively across the braidplain, and secondary channels have also migrated considerable distances. The true right of the braidplain, particularly in the upper half of the study area, is less dynamic with bank trimming being the dominant morphological change. Here, the channels have higher bed level, and therefore they are less frequently occupied by competent flows.

Visual inspection of the overall morphological dynamics indicates that there are striking erosional and depositional relationships within braid-bar units. Through the central belt of the braidplain, there are many places that have been reworked a number of times, as indicated by the complex mosaics of erosion and deposition. Sediment budgeting for each

Table 20.2 Contribution to Erosion and Deposition Estimates by Different Conditions of Change in the DEM of Difference, for the 84% Confidence Interval

Change Condition	Erosion		Deposition	
	Area (%)	Volume (%)	Area (%)	Volume (%)
Dry → Wet	22.7	43.9	0.4	0.3
Wet → Dry	0.3	0.3	20.3	39.7
Dry → Dry	74.8	53.5	77.4	57.8
Wet → Wet	2.1	2.3	1.9	2.2

Figure 20.8 shows the spatial extent of the conditions of change that are tabulated above. Dry indicates exposed braidplain and wet indicates channel, at the time of surveys in October 2009 and April 2010.

competent flow event would provide a more accurate assessment of morphological change because the impact of compensating erosion and deposition would be removed. Moreover, discretisation of the reach at the braid-bar or streamline unit would reveal further information on the morphological dynamics of the braidplain.

A further assessment of the DEM of difference's quality can be made by calculating the contribution to the sediment budget from different conditions of change in the DEM of difference. Since the survey methods used to derive the elevations of the exposed braidplain (referred to as dry) and channel bed (referred to as wet) are associated with different reliabilities, it is valuable to determine the relative contributions to the sediment budget from TLS data, which are more precise than the bed levels derived from the empirical–optical bathymetric mapping. The 84% confidence interval DEM of difference was classified for the four change conditions: wet–wet, wet–dry, dry–dry and dry–wet (top, middle map in Figure 20.8). Subsequently, the relative areas and volumes for both erosion and deposition were calculated (Table 20.2). For both erosion and deposition, the majority of the braidplain area that underwent morphological change, and the majority of the sediment volume, were derived from areas of the DEM of difference that were based only on TLS data (dry–dry). The remaining sediment budget was primarily calculated from areas of the DEM of difference that were based on one TLS surface (wet–dry and dry–wet). Only approximately 2% of the area and volumes, for both erosion and deposition, were based upon the least precise detection condition (wet–wet). Overall, the result that the majority of the DEM of difference is based upon the integration of two TLS-derived surfaces, with most of the remainder being derived from at least one TLS-derived surface, indicates that the sediment budget is primarily based upon precise TLS data.

The quality of the sediment budget calculations can also be assessed by evaluating how the budget changes as increasingly stringent confidence interval criteria are imposed (Figure 20.10; Table 20.1). As expected, both the volumes and areas of morphological change decrease with more stringent criteria. The rate of this change is, however, relatively slow. Thus, despite high confidence intervals, the pattern of morphological change remains consistent. Together with the other assessments of DEM quality, this indicates that the fusion of TLS and empirical–optical mapping is capable of producing high-precision and high-resolution DEMs of difference.

6. CONCLUSION

This case study describes a new methodology for monitoring the evolution of braided rivers. The development of the ArgoScan system, which incorporates TLS and RTK GPS, has enabled up to a billion high-precision survey points to be collected per day. This has facilitated the acquisition of ground-level data along a 2.5 km long braided river reach within the typically week-long low-flow conditions that occur between flood events. The surveys are characterised by high levels of precision, densely spaced scan points and a reach-scale survey extent, all of which are unparalleled in the monitoring of braided river evolution. In particular, the high density of survey points provides data for assessing the errors in the morphological budget and also has the potential for constructing sub-metre resolution DEMs. The coupling of TLS with empirical–optical bathymetric mapping has been shown to be an effective method for deriving channel-bed levels in areas inundated at the time of survey. The resulting DEMs of exposed and wet braidplain areas are rich in detail. Consequently, DEMs of difference are capable of revealing very subtle geomorphological changes which, when spatially integrated across the broad braidplain, nonetheless represent significant volumes of erosion and deposition with important consequences for the estimation of sediment budgets and transport rates by the morphological method.

The focus of this case study has been upon applying TLS to generate DEMs for use in quantifying sediment budgets using the morphological approach. In order to achieve this, scan data have been decimated and spatially filtered to produce statistical measures of surface elevations at the 1 m spatial resolution. In the subsequent analysis, the primary focus has been on constructing DEMs from the minimum elevation values and on using the detrended standard deviations for error analysis. Although this approach is appropriate for the problem being considered, it does not exploit the full richness of the scan data. For example, the spatially detrended standard deviation data are an important source of information on grain roughness and bed sedimentology. In addition, the high spatial density of the TLS points presents opportunities to undertake mapping at sub-metre spatial resolutions. Thus, the scan data have the potential to be used for micro-scale modelling as well as reach-scale analysis.

The techniques that have been introduced in this chapter are valuable for mapping landscape evolution in a range of geomorphological environments. Although data volumes from TLS can be considerable, the processing steps described here are not particularly onerous in terms of computer processing. With appropriate uncertainty analyses, the techniques are capable of differentiating real geomorphic change from likely errors. The fusion of TLS with techniques such as optical–empirical bathymetric mapping therefore provide a compelling approach for measuring topography in unprecedented detail. The interpretation of the resulting maps is likely to be an extremely valuable pursuit for explaining the feedbacks between driving processes and topographic change and for quantifying sediment budgets using the morphological approach.

ACKNOWLEDGEMENTS

This work was funded by the United Kingdom's Natural Environment Research Council Grant NE/G005427/1 with additional support from the NERC Geophysical Equipment Facility along with New Zealand's Foundation for Research Science and Technology Grant C01X0308 and Spain's Ministry of Science and Innovation's Jose Castillejo travel fund. Becky Goodsell, Michaela Cowie, Colin Rennie, Mark Smith, Jo Hoyle and Ramon Batalla assisted with fieldwork and data processing. The support of Lee Pimble and Mark Sievers of YSI Hydrodata, who assisted with setting up the Invincible Gauging Station and advising on aDcp methods, is greatly appreciated. The comments from a reviewer improved the clarity of the manuscript.

REFERENCES

Alho, P., Kukko, A., Hyyppa, H., Kaartinen, H., Hyyppa, J., Jaakkola, A., 2009. Application of boat-based laser river survey. Earth Surf. Process. Landforms 34 (13), 1831–1838.

Ashmore, P., Sauks, E., 2006. Prediction of discharge from water surface width in a braided river with implications for at-a-station hydraulic geometry. Water Resour. Res. 42 (3), 11.

Bannister, A., Raymond, S., Baker, R., 1998. Surveying. Seventh ed. Prentice Hall, Englewood Cliffs, NJ, 512 pp.

Barber, D., Mills, J., Smith-Voysey, S., 2008. Geometric validation of a ground-based mobile laser scanning system. ISPRS J. Photogramm. Remote Sens. 63 (1), 128–141.

Beavan, J., Denys, P., Denham, M., Hager, B., Herring, T., Molnar, P., 2010. Distribution of present-day vertical deformation across the Southern Alps, New Zealand, from 10 years of GPS data. Geophys. Res. Lett. 37, 5.

Brasington, J., Rumsby, B.T., McVey, R.A., 2000. Monitoring and modelling morphological change in a braided gravel-bed river using high resolution GPS-based survey. Earth Surf. Process. Landforms 25 (9), 973–990.

Brasington, J., Langham, J., Rumsby, B., 2003. Methodological sensitivity of morphometric estimates of coarse fluvial sediment transport. Geomorphology 53 (3–4), 299–316.

Brock, J.C., Wright, C.W., Clayton, T.D., Nayegandhi, A., 2004. LIDAR optical rugosity of coral reefs in Biscayne National Park, Florida. Coral Reefs 23 (1), 48–59.

Calder, B.R., Mayer, L.A., 2003. Automatic processing of high-rate, high-density multibeam echosounder data. Geochem. Geophys. Geosyst. 4, 22.

Carbonneau, P.E., Lane, S.N., Bergeron, N., 2006. Feature based image processing methods applied to bathymetric measurements from airborne remote sensing in fluvial environments. Earth Surf. Process. Landforms 31 (11), 1413–1423.

Carson, M.A., Griffiths, G.A., 1989. Gravel transport in the braided Waimakariri River – mechanisms, measurements and predictions. J. Hydrol. 109 (3–4), 201–220.

Cavalli, M., Tarolli, P., Marchi, L., Fontana, G.D., 2008. The effectiveness of airborne LiDAR data in the recognition of channel-bed morphology. Catena 73 (3), 249–260.

Church, M., Ashmore, P.E., 1998. Sediment transport and river morphology: a paradigm for study. In: Klingeman, P.C. (Ed.), Gravel-Bed Rivers in the Environment. Water Resources Center, Highlands Ranch, CO, pp. 115–139.

Chust, G., Grande, M., Galparsoro, I., Uriarte, A., Borja, A., 2010. Capabilities of the bathymetric hawk eye LiDAR for coastal habitat mapping: a case study within a Basque estuary. Estuar. Coast. Shelf Sci. 89 (3), 200–213.

Davies, T.R., McSaveney, M.J., 2006. Geomorphic constraints on the management of bedload-dominated rivers. J. Hydrol. 45 (2), 111–130.

Flener, C., Lotsari, E., Alho, P., Käyhkö, J., 2010. Comparison of empirical and theoretical remote sensing based bathymetry models in river environments. River Res. Appl. 10.1002/rra.1441.

Gao, J., 2009. Bathymetric mapping by means of remote sensing: methods, accuracy and limitations. Prog. Phys. Geogr. 33 (1), 103–116.

Heritage, G.L., Milan, D.J., 2009. Terrestrial laser scanning of grain roughness in a gravel-bed river. Geomorphology 113 (1–2), 4–11.

Hodge, R.A., 2010. Using simulated terrestrial laser scanning to analyse errors in high-resolution scan data of irregular surfaces. ISPRS J. Photogramm. Remote Sens. 65 (2), 227–240.

Hodge, R.A., Brasington, J., Richards, K., 2009. In situ characterization of grain-scale fluvial morphology using terrestrial laser scanning. Earth Surf. Process. Landforms 34 (7), 954–968.

Hoey, T., 1992. Temporal variations in bedload transport rates and sediment storage in gravel-bed rivers. Prog. Phys. Geogr. 16 (3), 319–338.

Lalor, R.M., Purvis, J.G., 1970. Report on Drilling – Rees Valley Scheelite Prospect, Glenorchy District, New Zealand. Report, Kaiser Mining & Development Ltd, 19 pp.

Lane, S.N., Chandler, J.H., Richards, K.S., 1994. Developments in monitoring and modelling small-scale river bed topography. Earth Surf. Process. Landforms 19 (4), 349–368.

Lane, S.N., Westaway, R.M., Hicks, D.M., 2003. Estimation of erosion and deposition volumes in a large, gravel-bed, braided river using synoptic remote sensing. Earth Surf. Process. Landforms 28 (3), 249–271.

Legleiter, C.J., Roberts, D.A., Marcus, W.A., Fonstad, M.A., 2004. Passive optical remote sensing of river channel morphology and in-stream habitat: physical basis and feasibility. Remote Sens. Environ. 93 (4), 493–510.

Lyzenga, D.R., 1981. Remote sensing of bottom reflectance and water attenuation parameters in shallow water using aircraft and Landsat data. Int. J. Remote Sens. 2 (1), 71–82.

Mabin, M., 2007. Glenorchy Area Geomorphology and Geo-Hazard Assessment. Report, Otago Regional Council, Dunedin, New Zealand, 83 pp.

McKean, J.A., Isaak, D.J., Wright, C.W., 2008. Geomorphic controls on salmon nesting patterns described by a new, narrow-beam terrestrial-aquatic LiDAR. Front. Ecol. Environ. 6 (3), 125–130.

McSaveney, M.J., Glassey, P.J., 2002. The Fatal Cleft Peak Debris Flow of 3 January 2002, Upper Rees Valley, West Otago. Report, Institute of Geological and Nuclear Sciences, Lower Hutt, New Zealand.

Milan, D.J., Heritage, G.L., Hetherington, D., 2007. Application of a 3D laser scanner in the assessment of erosion and deposition volumes and channel change in a proglacial river. Earth Surf. Process. Landforms 32 (11), 1657–1674.

Milne, J.A., Sear, D.A., 1997. Modelling river channel topography using GIS. Int. J. Geogr. Inf. Sci. 11 (5), 499–519.

Otago Regional Council, 2008. Channel Morphology and Sedimentation in the Rees River. Report, Otago Regional Council, Dunedin, New Zealand, 27 pp.

Otago Regional Council, 2010. Natural Hazards at Glenorchy. Report, Otago Regional Council, Dunedin, New Zealand, 50 pp.

Taylor, J.R., 1997. An Introduction to Error Analysis. second ed. University Science Books, Sausalito, CA, 327 pp.

Thoma, D.P., Gupta, S.C., Bauer, M.E., Kirchoff, C.E., 2005. Airborne laser scanning for riverbank erosion assessment. Remote Sens. Environ. 95 (4), 493–501.

Turnbull, I.M., 2000. Geology of the Wakatipu Area. Institute of Geological & Nuclear Sciences Limited, Lower Hutt, New Zealand, 72 pp.

Wedding, L.M., Friedlander, A.M., McGranaghan, M., Yost, R.S., Monaco, M.E., 2008. Using bathymetric LiDAR to define nearshore benthic habitat complexity: implications for management of reef fish assemblages in Hawaii. Remote Sens. Environ. 112 (11), 4159–4165.

Westaway, R.M., Lane, S.N., Hicks, D.M., 2003. Remote survey of large-scale braided, gravel-bed rivers using digital photogrammetry and image analysis. Int. J. Remote Sens. 24 (4), 795–815.

Wheaton, J.M., Brasington, J., Darby, S.E., Sear, D.A., 2010. Accounting for uncertainty in DEMs from repeat topographic surveys: improved sediment budgets. Earth Surf. Process. Landforms 35 (2), 136–156.

Whyte, G., Ohlbock, K., 2007. Glenorchy Floodplain Flood Hazards Study. Report, URS New Zealand Limited, Christchurch, 41 pp.

Wild, M., Cochrane, T., Davies, T., Hicks, D.M., Painter, D., Palmer, G., 2008. Recent sedimentation rates for the Rees–Dart braided river delta. Paper In J. Schmidt,

T. Cochrane, C. Phillips, S. Elliot, T. R. Davies & L. Basher (Eds.), Sediment Dynamics in Changing Environments (IAHS Red Books Publ. 325 ed., pp. 312–315): Wallingford: IAHS Press, c2008.
Winterbottom, S.J., Gilvear, D.J., 1997. Quantification of channel bed morphology in gravel-bed rivers using airborne multispectral imagery and aerial photography. Regul. Rivers Res. Manage. 13 (6), 489–499.

CHAPTER TWENTY-ONE

Uses and Limitations of Field Mapping of Lowland Glaciated Landscapes

Jasper Knight
School of Geography, Archaeology and Environmental Studies, University of the Witwatersrand, Johannesburg, South Africa

Contents

1. INTRODUCTION

In landscapes that have been recently glaciated, such as mid-latitude continental Europe and North America, glacial landforms are very significant mesoscale (10^2–10^4 m dimensions) landscape features occurring in both upland and lowland areas and on coastal plains. Key glacial landforms occurring across the glaciated mid-latitudes include drumlins, eskers, deltas and moraines, which are all upstanding, positive-relief landforms composed of glacigenic debris and associated with phases of continental ice retreat mainly in the period 20,000–13,000 years before

Developments in Earth Surface Processes, Volume 15
ISSN: 0928-2025, DOI: 10.1016/B978-0-444-53446-0.00021-5

present (BP). These landforms and the sediments contained within them are a proxy record of ice dynamics and as such are of high scientific importance. The relief that these landforms impart to the land surface, however, is also important because such landforms very strongly influence land surface patterns and properties, which include soils (soil type, thickness, geochemistry), attributes of slope drainage (depth of regolith, soil porosity, eluviation) and river patterns (locations of hills and valleys) (Collins, 1971; Mitchell and Ryan, 1997). Furthermore, glacial landforms have also strongly influenced the style and patterns of biogeography and human activity as these glacial landscapes were recolonised following ice retreat (Mitchell and Ryan, 1997; Knight, 2004) by facilitating or impeding migration routes. Today's landscapes reflect this through patterns and types of agriculture and ecosystems, and distribution of archaeological features, settlements, transport patterns, and past and present distributions of environmental resources including freshwater, peat and sediment types (Cooney and Grogan, 1994; Knight, 2001). Evaluation of all of these issues requires, as a prerequisite, accurate mapping of glacial landforms in the field (Knight, 2004).

Mapping the spatial patterns of glacial landforms has, historically, gone alongside and has emerged from standard field-based geologic mapping and surveying (Elvhage, 1980; Klimaszewski, 1990) that has been undertaken by national geological (and later by soil) surveys from the 1860s onwards. From the 1950s, remotely sensed data (air photographs and satellite imagery, including synthetic-aperture radar (SAR), Landsat MultiSpectral Scanner (MSS), Thematic Mapper (TM) and SPOT data) have been used as a common mapping tool in glacial landscapes (Derbyshire, 1958; Welch and Howarth, 1968; Ford, 1984; Rees and Squire, 1989; Szczesny, 1991). These have the advantage of being able to image much larger spatial scales that often correspond to the scale of the entire ice sheet. Digital terrain models (DTMs) have also recently been used that can visualise the nature of the land surface in three dimensions. These DTMs can be built from global digital data (e.g. SRTM), that are widely available and can fit with ground-based data mapped with the aid of differential global positioning system equipment. In many local case studies, high-resolution LiDAR and NEXTmap data have also been used (Notebaert et al., 2009). For a summary of recent mapping tools and techniques that have application to glacial landscapes, see Verbyla (1995) and Pavlopoulos et al. (2009).

1.1 Significance of Geomorphological Mapping in Glacial Landscapes

Accurate and high-resolution geomorphological mapping has several important applications in glaciated landscapes and are as follows:

1. Mapping through regional-scale remote sensing is a means by which spatial patterns of glacial landforms can be identified across an ice sheet bed. This has been performed, in particular, in Britain and Ireland (Greenwood and Clark, 2008, 2009), Scandinavia (Punkari, 1985, 1993; Ronnert and Nyborg, 1994; Hättestrand et al., 1999) and North America (Boulton and Clark, 1990). This mapping reveals both generalised patterns of subglacial lineations including drumlins and flutes (Boulton and Clark, 1990; Clark, 1993) and more detailed patterns that correspond to the relative age relationships of different types of bedforms (Knight and McCabe, 1997; McCabe et al., 1999; Greenwood and Clark, 2009). These data can also be used to reconstruct ice flow patterns and extent (Greenwood and Clark, 2009) and therefore help evaluate the controls on ice sheet evolution and dynamics as well as the geometry of the ice sheet (Fretwell et al., 2008).
2. The spatial properties and patterns can be integrated within a geographic information system (Clark, 1997) in order to examine interrelationships between different landscape elements, which may include underlying geology, topography, glacial landform patterns (including different landform types and geometries), archaeological sites and agricultural patterns. These relationships can then be quantitatively examined in order to test theories of landform development (Greenwood and Clark, 2010).
3. Mapping of the outline shapes of glacial landforms such as drumlins enables a quantitative examination of their geometric properties such as elongation (width/length ratio), density and surface area (Clark et al., 2009; Spagnolo et al., 2010). These properties are significant because they can help deduce landform genesis and ice sheet processes (Greenwood and Clark, 2009). For example, elongate landforms can be indicative of fast ice flow within ice streams (Stokes and Clark, 2002). Spatial patterns of elongation ratios and drumlin density can be related to oscillatory flow within the overlying ice sheet (Hill, 1973; Knight, 1997).
4. Accurate geomorphological mapping can help identify and quantify landscape resources of different types. For example, Knight et al. (1999) used aerial photographs and field mapping to find the distribution of glaciofluvial sand and gravel landforms, including eskers and

deltas, in Northern Ireland. The sand and gravel contained within these landforms are a source of industrial aggregate, and they form landscapes of high scenic value, which has implications for evaluating potential conflict of landscape and resource management in regions where landforms have multiple uses or where they are valued by different users (Hodges, 1995; Knight, 1998). All these varied uses of mapping analysis require accurate spatial data to be available, which yields confidence in any associated analysis derived from them.

1.2 Aims

In order to examine the application of geomorphological mapping techniques to glacial landforms found in lowland areas, this chapter presents case studies of (1) subglacial drumlins and (2) proglacial glaciolacustrine deltas found in lowland glacial landscapes and formed by the last (Late Devensian) ice sheet in north-central Ireland. These examples are then examined with respect to the complexities and challenges of geomorphological mapping in lowland glaciated landscapes, including some recommendations for their successful deployment.

2. METHODS

Geomorphological field mapping presented in this chapter is set within a wider regional mapping framework that was provided by remote sensing methods. Regional mapping was undertaken using two satellite imagery data sets: (1) from black and white Landsat MSS and false-colour composite Landsat TM imagery at a scale of 1:250,000. TM images had winter-scene coverage using bands 4, 5 and 6, which is a good combination for geological mapping purposes (Lillesand et al., 2008) and (2) using European remote sensing (ERS-1) SAR images at a scale of 1:70,000. These data were acquired on 8 April 1993 and therefore did not show a high degree of image disturbance by vegetative backscattering of the microwave radar signal. Regional geomorphological mapping from satellite data was enhanced by mapping from black and white aerial photographs at a scale of 1:10,000 and dated mainly from summer acquisition (June–September) in 1954/1955. Images were viewed stereoscopically, and major breaks of slope, demarcating individual landforms, were transferred onto 1:10,000 scale maps.

Field mapping was undertaken systematically across the entire study area by mapping upon and annotating 6 inch to 1 mile scale (approximately 1:10,000 scale) Ordnance Survey of Ireland base maps. The process of field mapping involved marking the position of breaks of slope that can be observed while field-walking along roads and other public rights of way. Breaks of slope were marked using the morphological mapping symbols of Savigear (1965) and Cooke and Doornkamp (1990). The process of morphological mapping delineates the main breaks of slope on the land surface observed systematically across the entire landscape, which in turn delineates the basal outlines of glacigenic and other landforms present therein (Figure 21.1). Breaks of slope are the most fundamental

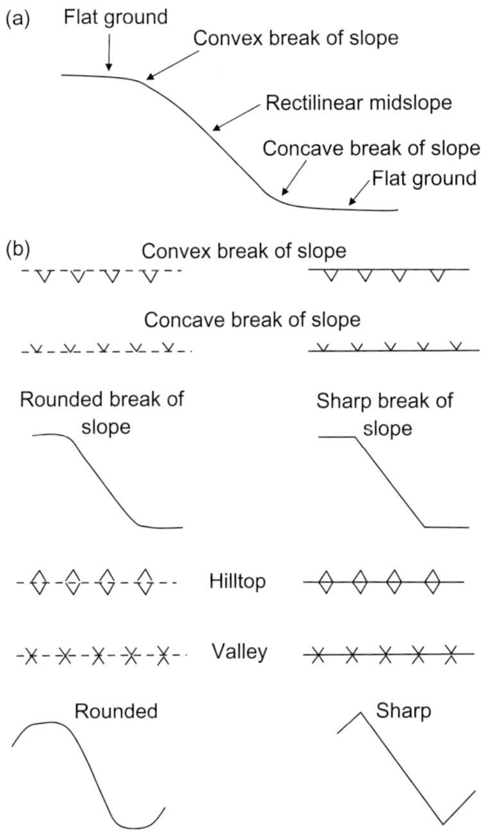

Figure 21.1 (a) Cross section of an idealised slope showing the major slope components. (b) Commonly used symbols for different breaks of slope. *After Savigear (1965) and Cooke and Doornkamp (1990).*

and objective method by which to identify landscape morphology, because they are non-genetic, and it is in this sense that they are used here. Many studies have used additional symbols in order to denote certain physical features such as surficial boulders (Sahlin and Glasser, 2008), but care must be taken to ensure that such symbols are descriptive rather than genetic. Following the completion of field mapping, the identified landforms can then be classified into different types (such as drumlins and eskers) based on both their morphology and internal sedimentology (not described here). This chapter presents two case studies of different types of glacial landforms that are representative of some of the wider issues associated with field geomorphological mapping in lowland glaciated landscapes.

3. THE CONTEXT OF GLACIAL LANDFORMS IN NORTH-CENTRAL IRELAND

The last (Late Devensian) ice sheet in Ireland was uniquely positioned to record the downstream effects of rapid climate changes in the adjacent North Atlantic (McCabe et al., 1998). The ice sheet as a whole, therefore, responded dynamically and over short time scales to climatic variability. This dynamic behaviour is imprinted on the preserved record of glacial landforms and in three main ways. First, the location of ice sheet centres can be deduced from radial patterns of subglacial landforms such as drumlins and ribbed (Rogen) moraines and striations (Smith et al., 2008; Greenwood and Clark, 2009). Changes in the location of ice sheet centres are very closely related to changes in temperature (leading to changes in ice thickness) and precipitation (leading to migration of the ice centres closer to, or further away from, the Atlantic precipitation source) (Knight, 1999). Second, patterns of subglacial landforms reflect the deformation processes by which available sediment is squeezed and moulded into landforms such as drumlins that reflect glacier direction, velocity and supply of available sediment by subglacial abrasion (Knight, 2010). Deformation can only take place under warm- and wet-based glaciers, so the presence of such landforms is an indicator of high-sediment mobility under a temperate ice sheet (Hart, 1997). However, cold/dry patches at the glacier bed, under a polythermal ice sheet, can preserve older landforms and give rise to a spatially variable pattern of sediment reworking and landform overprinting and superposition (Knight, 2002). This is seen in many areas of north-central Ireland where some drumlins are superimposed upon older ribbed moraine ridges (Knight and McCabe, 1997). Third, patterns of ice margin retreat can be reconstructed from the position and morphology of key glacial landforms such as terminal moraines, outwash and deltas. These can reveal the processes and dynamic behaviour of the ice margin, in particular, in areas adjacent to upland blocks, which can be interpreted with reference to both ice sheet processes and wider climatic controls (Knight, 2003, 2006). Glacial landform evidence, therefore, can be applied in these three ways with reference to interpretation of regional-scale climate and ice sheet properties and dynamics. Accurate geomorphological mapping is a prerequisite for these interpretations.

4. RESULTS

4.1 Drumlins

Drumlins and other subglacial bedforms are common features across lowland areas of north-central Ireland (Knight, 1997), which was a primary centre of ice dispersal during the Late Devensian glaciation (McCabe, 1987, 1993; Greenwood and Clark, 2009; Figure 21.2). The detailed distribution of drumlins in this region has been described by several workers using field mapping and remote sensing (aerial photographs and satellite imagery) but only rarely by using a combination of both methods (Charlesworth, 1924; Chapman, 1970; Knight, 1997; Knight and McCabe, 1997; McCabe et al., 1999; Greenwood and Clark, 2008). The landform patterns identified in these studies are qualitatively different to one another, which is a function of the different data sources used and their resolution, the individual interpreter and advances in understanding of glacial landforms. Different spatial patterns of landforms in these studies lead in turn to different interpretations of ice sheet dynamics, which can be resolved through better mapping procedures and outputs. In this region, drumlins of two major types occur (Knight, 1997). Diamicton (till) drumlins are composed internally of a poorly sorted admixture of glacigenic sediments that were formed and deposited subglacially. Rock-cored drumlins are those where the drumlin shape is developed almost entirely in bedrock with a minor and discontinuous till cover. These

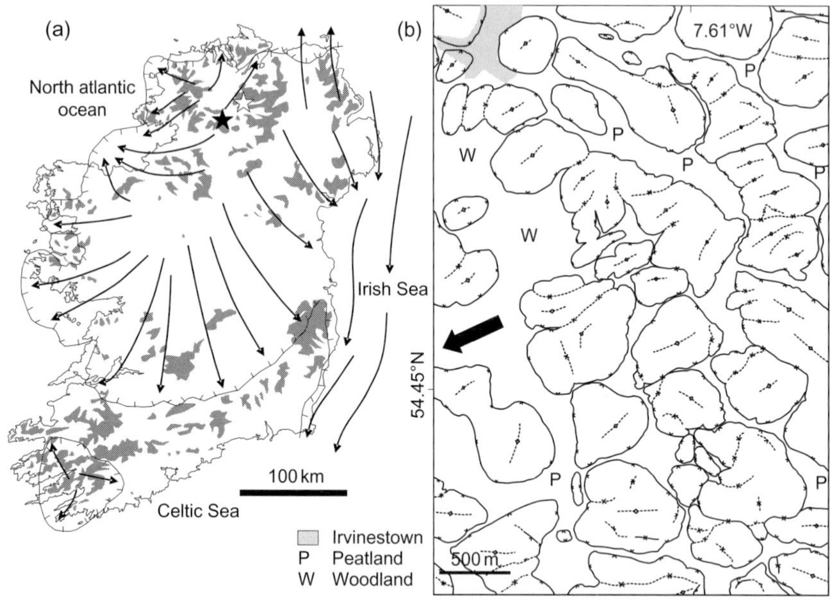

Figure 21.2 (a) Map of Ireland showing the location of the Irvinestown study region (black star). Land over 200 m is shaded. Major Late Devensian ice margins and ice flow vectors are shown. (b) Geomorphological map of drumlins around Irvinestown (shaded). The large arrow indicates regional ice flow direction. Geomorphological symbols used are shown in Figure 21.1. P, peatlands; W, woodlands. *Part (a) after Stephens et al. (1975).*

drumlin types are located in different areas, depending on till thickness, and have different morphometric properties (Knight, 1997, 2010).

Diamicton drumlins are common landforms around the town of Irvinestown (Irish grid reference H23874 58245 at 54.47°N, 7.63°W), which is located at 110 m asl in County Tyrone, north-central Ireland (Figure 21.2). Geomorphological mapping of this area highlights a number of properties of glacial landforms here. The glacial landforms can be identified and delimited by their basal enclosing break of slope with respect to the surrounding flat landscape. The landforms are oval to elongate in plan view, sometimes with more complex outline shapes, and are generally 200–780 m long, 80–680 m wide and 4–35 m in height (not shown in Figure 21.2). The landforms are sometimes isolated from one another, but are commonly touching or connected to adjacent landforms, thereby comprising compound forms of different sizes and shapes (Figure 21.3). Internally, the landforms may be partially superimposed

Figure 21.3 Annotated photograph of the drumlin landscape around Irvinestown showing a concave break of slope at the drumlin base, which demarcates a flat area of inter-drumlin peat, a convex break of slope along the drumlin crest and areas of variations in slope angle on the drumlin sides. Note that these breaks of slope are relatively smooth and are not complex over very local spatial scales. This field evidence matches with the patterns of breaks of slope identified in Figure 21.2.

and characteristically show breaks of slope that are located on the midslope area of the landforms. There are also some minor mounds located between some of the larger forms. The long-axis orientation of the landforms, identified on the basis of their longest and shortest sides, also varies across the mapped area (not all are shown in Figure 21.2). Generally, the long-axis orientation corresponds to that of the regional ice flow direction (arrowed on Figure 21.2). Based on the mapped pattern and geometric properties of the landforms, they are interpreted as flow-parallel drumlins. Also of note is that some drumlin elements form partly connected landforms (ribbed or Rogen moraine) that are elongate in a northwest-southeast direction (Figure 21.2). This suggests that overlying ice partially streamlined pre-existing ridges, forming drumlinised surfaces over their crests, consistent with regional patterns (Knight and McCabe,

1997). Drumlins here are also separated by flat areas that are presently poorly drained fen or peatland, and some drumlin margins have been oversteepened by meltwater channel incision during phases of ice retreat. In this example, geomorphological mapping successfully denotes the basal outlines of the drumlins and illustrates some of the minor breaks of slope on the drumlin surfaces and is an accurate non-genetic representation of the landforms as observed in the field.

4.2 Deltas

Glaciolacustrine deltas are common features formed where ice retreat takes place in meltwater-rich, topographically controlled environments around the margins of lowland regions where ice can become buttressed against rising bedrock slopes. In the Sperrin Mountains, located immediately to the north of the drumlin region previously described, glacial lakes were common features formed during ice retreat by a combination of high production of glacial meltwater and topographic control by the position of ice margins against bedrock slopes (Colhoun, 1970; Dardis, 1986). This interplay between ice margins, topography and meltwater/sediment generation created a pattern of dynamic lake formation and drainage that was controlled by movement of the ice margin variously blocking off or revealing drainage outlets. The glaciolacustrine deltas occurring south of the town of Gortin, County Tyrone (at Irish grid reference H49550 84504, at 54.70°N, 7.23°W) are very fine examples of such features (Figure 21.4).

The deltas are located on the south side of an east—west trending valley on the southernmost edge of the Sperrin Mountains. Morphologically, these landforms comprise a series of isolated, flat-topped hills that have steep, rectilinear slopes (Figure 21.5). The hills are not connected to each other or to the adjacent rock slopes located directly to the south (Figure 21.4). The flat tops of the landforms are not accordant with one another (not seen on the plan-view map of Figure 21.4) but are located at different elevations, at 256 and 210 m asl, with the 256 m level being the most extensive (0.3 km^2) and which extends the greatest distance northwards (0.5 km from the bedrock ridge to the south). Located in between the isolated hills is a flat to slightly undulating lower surface that is continuous across the mapped region (at 180 m asl) into which are developed nine isolated, enclosed, water-filled depressions (Figure 21.5). Incised channel features dissect the flat-topped hills and also into the 180 m surface. These

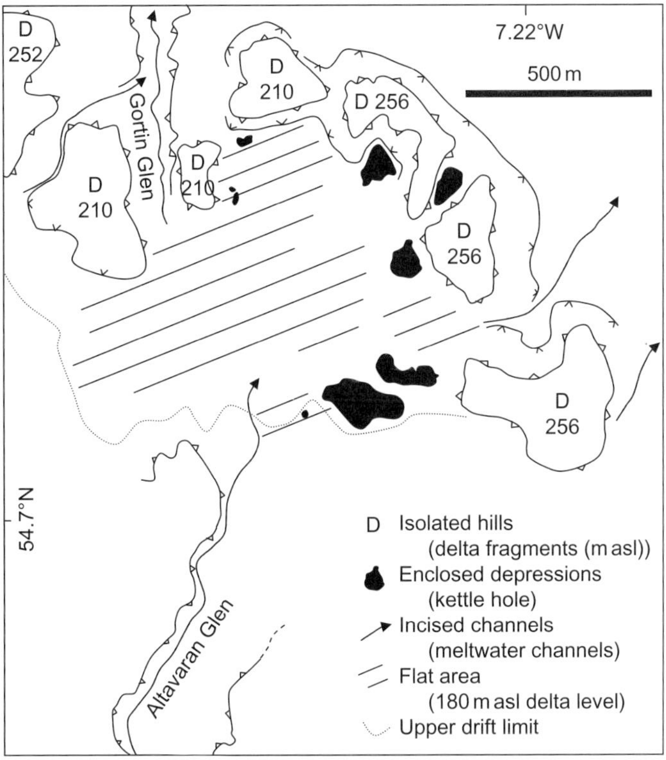

Figure 21.4 Geomorphological map of glaciolacustrine deltas south of Gortin. The regional location of Gortin is shown by the white star in Figure 21.2. Geomorphological symbols used are shown in Figure 21.1. The figure caption shows the morphological description and (in brackets) its interpretation. Delta surfaces (D) have their surface elevations shown (m asl).

incised channels are north-going, steep-sided, and they have generally poorly marked channel start and end points. A deeply incised rock-cut channel (<40 m deep) is present at Altavaran Glen, but this channel loses its identity northwards.

The non-genetic morphological description of these landforms can be interpreted with respect to ice dynamics and glacial lake formation and drainage. The flat-topped landforms are interpreted as proglacial delta tops that record phases of sediment deposition from south to north into a water body impounded by Sperrin ice (Colhoun, 1970). Meltwater and sediment from the decaying north-central Ireland ice sheet issued northwards through Altavaran Glen and into a lake with a water surface level at

Figure 21.5 Annotated photograph of the flat delta surfaces and kettle holes at Gortin.

256 m asl. Topset sediments within the deltas show that the delta surfaces corresponded to the elevation of the water surface at this level, and that this highest delta prograded northwards uniformly. Subsequently, due to lake level fall, this delta surface was partially preserved as meltwater incised downwards into the loose sand and gravel, forming the 210 m level and then the 180 m level (Figure 21.4). These three levels therefore record the three stages of glacial lake infilling and drainage. These surfaces decrease in elevation northwards, indicating that the lake level fell over time as retreat of Sperrin ice revealed progressively lower water outlets. The enclosed depressions are interpreted as kettle holes, formed by the burial by delta sediments of ice blocks detached from the ice margin, probably by iceberg calving (Benn et al., 2007). The presence of the kettle holes shows that the north-central Ireland ice margin was located right at this site, most likely near the drift limit boundary (Figure 21.4). Melting of these blocks left the kettle hole lakes that are closely associated with this ice-marginal depositional system.

5. DISCUSSION

Mapping of formerly glaciated lowland landscapes allows for the relationship between past (relict) geomorphological processes and present-day landscapes to be evaluated and the relationship between contemporary landscape patterns and human activity assessed. The landforms of these relict landscapes are no longer being affected by glacial processes, but they are however, affected by postglacial slope processes and paraglacial relaxation, resulting in a decrease in slope angle over time that leads to more topographically subdued landforms (Curry et al., 2009). This is particularly the case in clay-rich sediments, such as glacial till, that move under gravity when water-saturated. In addition, many landscapes have been strongly influenced by human modification and engineering which has also altered the external shape of many landforms. Accurate geomorphological mapping can be used as a tool to help distinguish between these factors that affect landform morphology (Minar and Evans, 2008; Smith and Pain, 2009). Few studies have considered aspects of objectivity in the procedure of field mapping, but a general problem is the role of extrapolation from limited observations. Most commonly, landforms cannot be easily observed in the round, and therefore some breaks of slope may be extended around the 'back' of the landform based upon observations that have already been made as well as assumptions of what the landform 'should' look like. For example, the basal outlines of drumlins are generally assumed to be contour-parallel but in some cases they are not.

The drumlins around Irvinestown show that, in detail, they are highly diverse in their morphology but that their basal outline can be mapped quite easily since the drumlins are surrounded by flat areas of postglacial infill (Figure 21.3). The geometry and outline shape of the drumlins are therefore clearly identifiable. Smaller, superimposed forms on top of the larger drumlins are often less easily identified or mapped. This is because these forms are located on top of a sloping surface; therefore, breaks of slope may change in direction, angle and can disappear. Long-axis orientation of drumlins is dependent on identifying the longest and shortest axes, so accurate geomorphological mapping is vital if a primary goal is to identify former ice flow direction (Clark et al., 2009). There are several other mapping issues in drumlin landscapes. Postglacial infill and slope relaxation mean that some drumlin forms have been partly buried, which acts to reduce geomorphological complexity. Postglacial soil and ecosystem development also tends to reduce the number of breaks of slope, as slopes are smoothed out over time and can also make field mapping more difficult where slope breaks are obscured. Drumlin landscapes also commonly comprise a combination of till with bedrock protrusions. Morphological mapping cannot easily convey patterns of different sediment type and thickness or identify drift limits.

The deltas near Gortin can be accurately mapped because they have sharp convex breaks of slope that define the delta surface edge and rectilinear midslopes formed by clean meltwater incision into their internal sands and gravels. In this environment, geomorphological mapping cannot, in itself, distinguish between delta surfaces of different elevations, which is critical in order to identify and correlate different delta formation stages or identify drift limits such as the termination of the rock-cut Altavaran Glen. Mapping is also unable to distinguish between relative ages of landforms which, although interpretive, are based upon relationships between delta surfaces and meltwater channels observed in the field.

5.1 The Application of Field Mapping in a Satellite Era

The advantages of applications of remote sensing, in particular satellite-derived data, to geomorphological mapping have been discussed in a number of studies, most recently by Smith and Pain (2009). In summary, the main advantages of remote sensing mapping are as follows: (1) it is cost-effective in terms of time and money, (2) numerical data of different sorts can be manipulated and analysed easily, with known data errors and with a high degree of statistical confidence, (3) these data are widely available and use now-standard computer software and visual–spatial tools such as Google Earth, (4) the data cover many spatial scales and geographical areas that would otherwise be inaccessible and (5) geomorphological mapping can go alongside and complement ecosystem and other spatial data. Increasingly, mesoscale remote sensing is also being used to identify and quantity land surface changes over time (Petit et al., 1991), which is a purpose for which field mapping has also been used (Savigear, 1962; Williams and Morgan, 1976).

The major disadvantage of remote sensing is that it cannot replace all elements of field investigations, in particular, identifying physical processes or reconstructing past environmental change which is dependent on collecting field data for sediment analysis, radiometric dating, pollen and microfossil analysis and so on. Any remotely sensed data also need to be verified and calibrated with respect to field observations which are particularly important as these data are used to drive computer models of land surface change or ice sheet dynamics (Greenwood and Clark, 2008). Finally, field geomorphological mapping is a valuable tool in students' skills training, and remote sensing cannot replicate its sensory experience. Field mapping still has a valuable place, therefore, in the range of investigative techniques used to examine Earth surface landforms and processes.

6. CONCLUSIONS

Field-based mapping in any landscape requires time and experience, but in glaciated lowland landscapes accurate and meaningful maps are particularly important because they are used as primary data to reconstruct past glacial processes and dynamics. Both drumlins and deltas, as examples of typical landforms of glaciated lowlands, pose some mapping problems, but these can be overcome where time and experience are applied, where field observations are placed in a wider, regional, context using remotely sensed imagery and backed up by observations of internal sediments.

REFERENCES

Benn, D.I., Warren, C.R., Mottram, R.H., 2007. Calving processes and the dynamics of calving glaciers. Earth Sci. Rev. 82, 143–179.

Boulton, G.S., Clark, C.D., 1990. The Laurentide ice sheet through the last glacial cycle: the topology of drift lineations as a key to the dynamic behaviour of former ice sheets. Trans. R. Soc. Edinburgh: Earth Sci. 81, 327–347.

Chapman, R.J., 1970. The Late-Weichselian glaciation of the Erne basin. Ir. Geogr. 6, 151–161.

Charlesworth, J.K., 1924. The glacial geology of the north-west of Ireland. Proc. R. Ir. Acad. 36B, 174–314.

Clark, C.D., 1993. Mega-scale glacial lineations and cross-cutting ice-flow landforms. Earth Surf. Process. Landforms 18, 1–29.

Clark, C.D., 1997. Reconstructing the evolutionary dynamics of former ice sheets using multi-temporal evidence, remote sensing and GIS. Quat. Sci. Rev. 16, 1067–1092.

Clark, C.D., Hughes, A.L.C., Greenwood, S.L., Spagnolo, M., Ng, F.S.L., 2009. Size and shape characteristics of drumlins, derived from a large sample, and associated scaling laws. Quat. Sci. Rev. 28, 677–692.

Colhoun, E.A., 1970. On the nature of the glaciations and final deglaciation of the Sperrin Mountains and adjacent areas in the North of Ireland. Ir. Geogr. 6, 162–185.

Collins, C.W., 1971. The influence of drumlin topography on field patterns in Dodge County, Wisconsin. Trans. Wisc. Acad. Sci. Arts Lett. 59, 55–66.

Cooke, R.U., Doornkamp, J.C., 1990. Geomorphology in Environmental Management: A New Introduction. second ed. Clarendon, Oxford, pp. 410.

Cooney, G., Grogan, E., 1994. Irish Prehistory: A Social Perspective. Wordwell, Dublin, 276 pp.

Curry, A.M., Sands, T.B., Porter, P.R., 2009. Geotechnical controls on a steep lateral moraine undergoing paraglacial slope adjustment. In: Knight, J., Harrison, S. (Eds.), Periglacial and Paraglacial Processes and Environments. Geological Society of London, Special Publications No. 320, pp. 181–197.

Dardis, G.F., 1986. Late Pleistocene glacial lakes in south-central Ulster, Northern Ireland. Irish J. Earth Sci. 7, 133–144.

Derbyshire, E., 1958. The identification and classification of glacial drainage channels from aerial photos. Geogr. Ann. 40, 188–195.

Elvhage, C., 1980. An experimental series of topo-geomorphological maps – with an example from a deglaciated mountain area in Jämtland, Sweden. Geogr. Ann. 62A, 105–111.

Ford, J.P., 1984. Mapping of glacial landforms from Seasat radar images. Quat. Res. 22, 314–327.

Fretwell, P.T., Smith, D.E., Harrison, S., 2008. The last glacial maximum British–Irish ice sheet: a reconstruction using digital terrain mapping. J. Quat. Sci. 23, 241–248.

Greenwood, S.L., Clark, C.D., 2008. Subglacial bedforms of the Irish ice sheet. J. Maps 2008, 332–357.

Greenwood, S.L., Clark, C.D., 2009. Reconstructing the last Irish ice sheet 1: changing flow geometries and ice flow dynamics deciphered from the glacial landform record. Quat. Sci. Rev. 28, 3085–3100.

Greenwood, S.L., Clark, C.D., 2010. The sensitivity of subglacial bedform size and distribution to substrate lithological control. Sediment. Geol. 232, 130–144.

Hart, J.K., 1997. The relationship between drumlins and other forms of subglacial glaciotectonic deformation. Quat. Sci. Rev. 16, 93–107.

Hättestrand, C., Goodwillie, D., Kleman, J., 1999. Size distribution of two cross-cutting drumlin systems in northern Sweden: a measure of selective erosion and formation time length. Ann. Glaciol. 28, 146–152.

Hill, A.R., 1973. The distribution of drumlins in County Down, Ireland. Ann. Assoc. Am. Geogr. 63, 226–240.

Hodges, C.A., 1995. Mineral resources, environmental issues, and land use. Science 268, 1305–1312.

Klimaszewski, M., 1990. Thirty years of detailed geomorphological mapping. Geogr. Pol. 56, 11–18.

Knight, J., 1997. Morphological and morphometric analyses of drumlin bedforms in the Omagh Basin, north central Ireland. Geogr. Ann. 79A, 255–266.

Knight, J., 1998. The geological conservation of glaciofluvial sand and gravel resources in Northern Ireland: an integrated approach using natural areas. In: Bobrowsky, P.T. (Ed.), Aggregate Resources: A Global Perspective. Balkema, Rotterdam, pp. 71–86.

Knight, J., 1999. Problems of Irish drumlins and Late Devensian ice sheet reconstructions. Proc. Geol. Assoc. 110, 9–16.

Knight, J., 2001. A geocultural classification of landscapes in Northern Ireland: implications for landscape management and conservation. Tearmann 1, 113–124.

Knight, J., 2002. Glacial geological evidence for stick-slip basal ice flow. Quat. Sci. Rev. 21, 975–983.

Knight, J., 2003. Geomorphic evidence for patterns of Late Midlandian ice advance and retreat in the Omagh Basin. Ir. Geogr. 36, 1–22.

Knight, J., 2004. The Ice Age inheritance of the Irish landscape. In: Parkes, M.A. (Ed.), Natural and Cultural Landscapes – the Geological Foundation. Proceedings of a Conference 9–11 September 2002, Dublin Castle, Ireland. Royal Irish Academy, Dublin, pp. 29–32.

Knight, J., 2006. Geomorphic evidence for active and inactive phases of Late Devensian ice in north-central Ireland. Geomorphology 75, 4–19.

Knight, J., 2010. Basin-scale patterns of subglacial sediment mobility: implications for glaciological inversion modelling. Sediment. Geol. 232, 145–160.

Knight, J., McCabe, A.M., 1997. Identification and significance of ice flow-transverse subglacial ridges (Rogen moraines) in northern central Ireland. J. Quat. Sci. 12, 519–524.

Knight, J., McCarron, S.G., McCabe, A.M., Sutton, B., 1999. Sand and gravel aggregate resource management and conservation in Northern Ireland. J. Environ. Manage. 56, 195–207.

Lillesand, T.M., Kiefer, R.W., Chipman, J.W., 2008. Remote Sensing and image interpretation. John Wiley and Sons, New York.

McCabe, A.M., 1987. Quaternary deposits and glacial stratigraphy in Ireland. Quat. Sci. Rev. 6, 259–299.

McCabe, A.M., 1993. The 1992 Farrington Lecture: drumlin bedforms and related ice-marginal depositional systems in Ireland. Ir. Geogr. 26, 22–44.

McCabe, A.M., Knight, J., McCarron, S.G., 1998. Evidence for Heinrich event 1 in the British Isles. J. Quat. Sci. 13, 549–568.

McCabe, A.M., Knight, J., McCarron, S.G., 1999. Ice flow stages and glacial bedforms in north central Ireland: a record of rapid environmental change during the last glacial termination. J. Geol. Soc. London 156, 63–72.

Minar, J., Evans, I.S., 2008. Elementary forms for land surface segmentation: the theoretical basis of terrain analysis and geomorphological mapping. Geomorphology 95, 236–259.

Mitchell, F., Ryan, M., 1997. Reading the Irish Landscape. Town House Press, Dublin, 392 pp.

Notebaert, B., Verstraeten, G., Govers, G., Poesen, J., 2009. Qualitative and quantitative applications of LiDAR imagery in fluvial geomorphology. Earth Surf. Process. Landforms 34, 217–231.

Pavlopoulos, K., Evelpidou, N., Vassilopoulos, A. (Eds.), 2009. Mapping Geomorphological Environments. Springer, Berlin.

Petit, C., Scudder, T., Lambin, E., 1991. Quantifying processes of land-cover change by remote sensing: resettlement and rapid land-cover change in south-east Zambia. Int. J. Remote Sens. 22, 3435–3456.

Punkari, M., 1985. Glacial geomorphology and dynamics in Soviet Karelia interpreted by means of satellite imagery. Fennia 163, 113–153.

Punkari, M., 1993. Modelling of the dynamics of the Scandinavian ice sheet using remote sensing and GIS methods. In: Aber, J.S. (Ed.), Glaciotectonics and Mapping Glacial Deposits. Canadian Plains Research Center, University of Regina, pp. 232–250.

Rees, W.G., Squire, V.A., 1989. Technological limitations to satellite glaciology. Int. J. Remote Sens. 10, 7–22.

Ronnert, L., Nyborg, M.R., 1994. The distribution of different glacial landscapes on southern Jameson Land, East Greenland, according to Landsat Thematic Mapper Data. Boreas 23, 311–319.

Sahlin, E.A.U., Glasser, N.F., 2008. Geomorphological map of Cadair Idris, Wales. J. Maps 2008, 299–314.

Savigear, R.A.G., 1962. Some observations on slope development in north Devon and north Cornwall. Trans. Inst.Br. Geogr. 31, 23–42.

Savigear, R.A.G., 1965. A technique of morphological mapping. Ann. Assoc. Am. Geogr. 55, 514–538.

Smith, M.J., Pain, C.F., 2009. Applications of remote sensing in geomorphology. Prog. Phys. Geogr. 33, 568–582.

Smith, M.J., Knight, J., Field, K.S., 2008. Glacial striae observations for Ireland compiled from historic records. J. Maps 2008, 378–398.

Spagnolo, M., Clark, C.D., Hughes, A.L.C., Dunlop, P., Stokes, C.R., 2010. The planar shape of drumlins. Sediment. Geol. 232, 119–129.

Stephens, N., Creighton, J.R., Hannon, M.A., 1975. The Late-Pleistocene period in north-eastern Ireland: an assessment 1975. Ir. Geogr. 8, 1–23.

Stokes, C.R., Clark, C.D., 2002. Are long subglacial bedforms indicative of fast ice flow? Boreas 31, 239–249.

Szczesny, R., 1991. Quaternary landforms and deposits in southern Spitsbergen on the ground of photointerpretation. Pol. Polar Res. 12, 289–343.

Verbyla, D.L., 1995. Satellite Remote Sensing of Natural Resources. CRC Press, London.

Welch, R., Howarth, P.J., 1968. Photogrammetric measurements of glacial landforms. Photogram. Rec. 6, 75–96.

Williams, A.R., Morgan, R.P.C., 1976. Geomorphological mapping applied to soil erosion evaluation. J. Soil Water Conserv. 31, 164–168.

CHAPTER TWENTY-TWO

Mapping Late Holocene Landscape Evolution and Human Impact – A Case Study from Lower Khuzestan (SW Iran)

Jan Walstra, Vanessa M.A. Heyvaert and Peter Verkinderen

Geological Survey of Belgium, Royal Belgian Institute of Natural Sciences Jennerstraat 13 B-1000 Brussels Belgium

Contents

Developments in Earth Surface Processes, Volume 15
ISSN: 0928-2025, DOI: 10.1016/B978-0-444-53446-0.00022-7

1. INTRODUCTION

The Mesopotamian alluvial plain is dominated by major rivers, most notably the Euphrates, Tigris and Karun, which throughout the Holocene have transported water and sediment from their uplands towards the Persian Gulf. The combination of a gentle gradient and high sediment supply promoted the aggradation of alluvial ridges and recurrent avulsions (channel shifts) (Morozova, 2005; Heyvaert and Baeteman, 2008). Since ancient times, societies have always depended on the position of these rivers for their economic survival: the rivers supplied irrigation water necessary for agriculture and served as principal transport routes (Adams, 1981; Cole and Gasche, 1998). There is ample evidence of human activities that influenced the natural avulsion processes in order to control the distribution of water across the plain (Alizadeh et al., 2004; Morozova, 2005; Heyvaert and Baeteman, 2008). The present-day landscape is the result of this complex interaction between natural and anthropogenic processes.

Most of the current knowledge about past river and settlement patterns is derived from ancient textual sources and archaeological surveys in the 1970s (Adams, 1981; Cole and Gasche, 1998). Although some generalised descriptions of the physical landscape are available (Buringh, 1960; Verhoeven, 1998), such a geomorphological-based context is often lacking in landscape interpretations (Hritz and Wilkinson, 2006). Recent advances in the field of remote sensing have led to a revival of landscape studies in the Mesopotamian region; in particular, following the release of declassified CORONA imagery (Pournelle, 2003; Hritz, 2005), the development of high-resolution satellite sensors (Baeteman et al., 2004; Jahjah et al., 2007; Cultraro et al., 2009), access to global imagery through online services such as Google Earth (Ur, 2006) and the availability of global elevation data through the Shuttle Radar Topographic Mission (SRTM) (Hritz and Wilkinson, 2006).

Although the prolonged presence of civilisations in the region is the basis for its unconcealed archaeological value, it also complicates a thorough understanding of the landscape. A geomorphological study can only be meaningful if it takes into account the interaction between natural and anthropogenic processes. Therefore, such an approach should not only focus on alluvial processes in the semi-arid environment but also consider humans as an important agent. Given the vast size, and limited accessibility to the region, another requirement of a suitable mapping procedure would be the use of remote sensing techniques. If possible, the source data should be inexpensive but still have sufficient and consistent quality for the entire region. The end product should also be easily interpretable, even for non-geomorphological specialists, and provide a base for further interdisciplinary analysis of landscape evolution.

The aim of this study was to develop a general procedure for the geomorphological mapping of the Lower Khuzestan plain (SW Iran), which would satisfy all aforementioned criteria. The methodology was initially developed and applied in a case study on the lower course of the Jarrahi River. The alluvial landscapes of this area can be considered representative of the rest of the Lower Khuzestan, perhaps even for the entire Mesopotamian plain, with traces of settlement and irrigation from different periods. The area has been subject to a geological field campaign (Baeteman et al., 2004; Heyvaert, 2007) and was also studied in historical sources (Ooghe, 2007; Verkinderen, 2009), thereby providing complementary data.

2. REGIONAL SETTING

The Lower Khuzestan plain is the south-eastern extension of the Mesopotamian sedimentary basin. In the north and east, the plain is bordered by the foothills of the Zagros Mountains. Subsidence of the basin and uplift of the mountains are associated with the collision of the Arabian and Eurasian tectonic plates. Orogenic uplift started during the Late Miocene and is still ongoing (Hessami et al., 2006).

The Jarrahi originates from the confluence of the rivers Marun and Ab-i Ala, both rising in the Zagros Mountains (Figure 22.1). After entering the Khuzestan plain, the river meanders for about 95 km to the southwest. The river then follows a series of rectangular segments. After another 60 km, the river enters a distributary system of numerous canals, which all end in the Shadegan Marshes. In the west, these freshwater marshes are bordered by the alluvial ridge of the Karun River, and in the south by clastic coastal *sabkhas* and salt marshes. Geological research suggests that several alluvial fans were deposited by the Jarrahi; the one farthest upstream is incised by the present-day channel (Baeteman et al., 2004). The Shadegan Marshes were formed after the latest shift of the

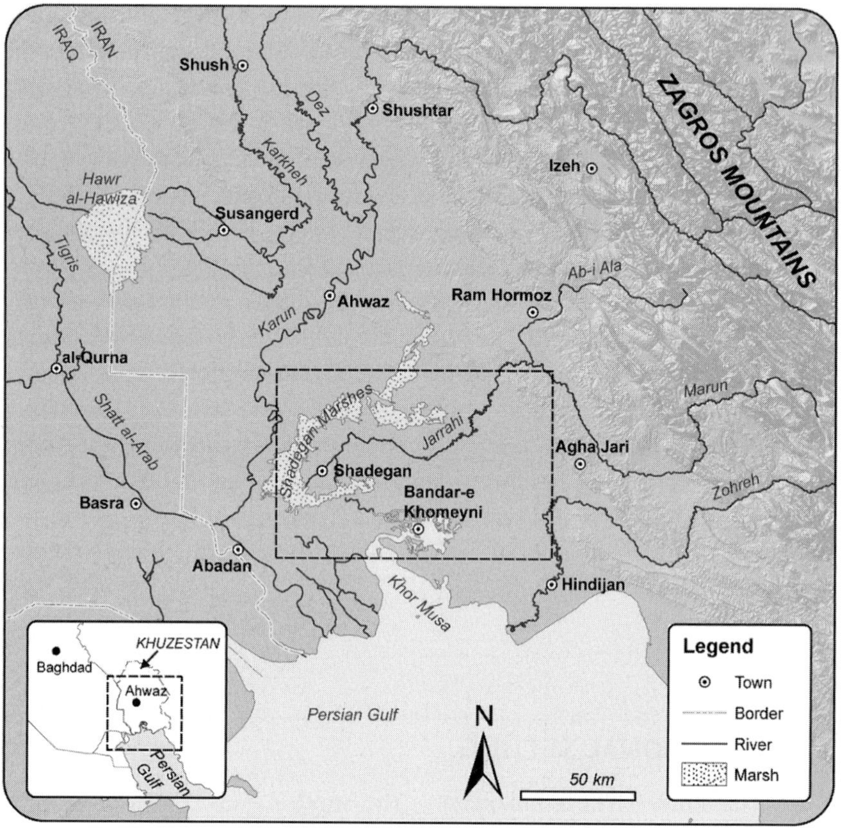

Figure 22.1 Location of the study area in Lower Khuzestan.

Karun, which caused a downstream blockage of the Jarrahi (Heyvaert, 2007).

The climate of the study area is hot and arid: in summer temperatures may rise up to 58°C, with annual rainfall is less than 200 mm (Potts, 1999). The Jarrahi receives most of its discharge from autumn and winter rains in the Zagros Mountains, causing extensive seasonal flooding of the marshes.

The capital of the district is Shadegan (Figure 22.1), which was founded in the 1740s on the banks of the Jarrahi (Layard, 1846). Limited archaeological evidence from the region proves that occupation dates back at least to Achaemenian times (539–331 BC). During the Sasanian (221–640 AD) and Early Islamic (*ca.* 640–13th century AD) periods, large tracts of land were under intensive irrigation (Hansman, 1978; Verkinderen, 2009). Up to now, no systematic surveys of the ancient irrigation systems have been carried out.

3. MATERIALS AND METHODS

The working procedure followed can be divided into three stages:

1. Acquisition and pre-processing of maps and remote sensing imagery of the area, including the scanning of hardcopy maps and images, and geo-referencing of the digital files,
2. Image interpretation of the remote sensing data and mapping of geomorphological units,
3. Presentation of the results in a geomorphological map that can be integrated with information from other disciplines (historical texts, archaeological surveys and geological data) and used for further landscape analyses.

In the following sections, a brief overview is provided of the different data sources that contributed to this study: existing cartographic material, Landsat and CORONA imagery, SRTM elevation data and Google Earth imagery. Then, the techniques used for image interpretation are treated and considerations for map design discussed. An overview of the general working procedure is illustrated in Figure 22.2.

3.1 Maps

The study area is covered by various large-scale map series (Table 22.1). The British and Iranian topographical maps helped in identifying modern infrastructural features (such as roads, railways, canals and settlements), which were ambiguous from satellite imagery alone. The geological maps provided information on the distribution of structural–geological elements and geological formations, which were incorporated in the geomorphological legend. The maps were digitally scanned and geo-referenced in ESRI ArcGIS.

3.2 Landsat

The Landsat Program was designed to provide repetitive global coverage of the Earth's landmasses, in particular for geological and land cover mapping. Its multi-spectral capabilities are suited to distinguishing between

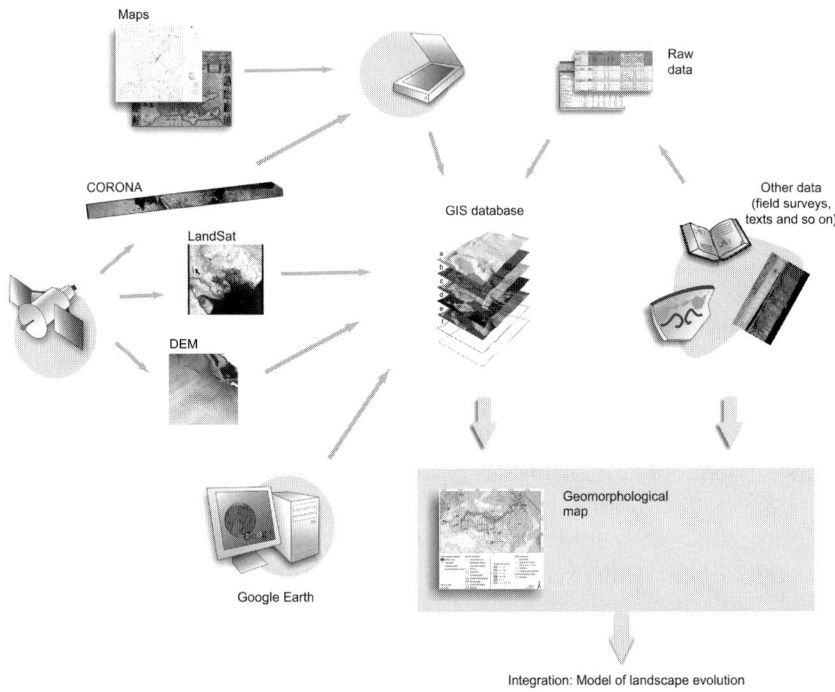

Figure 22.2 Flowchart illustrating the working procedure followed in this study.

Table 22.1 Map Sources Used in This Study

Publisher	Publication Date	Survey Date	Scale	Comments
British War Office	1962–1963	1941	1/50,000	K701 series, reprints of Indian Survey maps
National Cartographic Centre of Iran	1998–2001	1991–1995	1/50,000	Text in Farsi
Iranian Oil Operating Companies	1966–1972	1928–1971	1/100,000	Geological maps

aspects of lithology and vegetation (USGS, 2003). Landsat imagery has proven its value for regional mapping of geomorphological units, especially in remote areas (Pain, 1985; Alwash et al., 1986). Even today, Landsat remains a significant resource for geomorphologists due to its repeat coverage, large scene size and low cost (Smith and Pain, 2009). Since the end of 2008, the entire Landsat archive held by the US Geological Survey is available free of charge (http://earthexplorer.usgs.gov). Because the data are standard-processed at a high level of geometric accuracy (better than 30 m; Gutman et al., 2008), they are also suitable as ground control for geo-referencing other materials.

Drainage and vegetation patterns were found to be important indicators for landform features in the plain. Since subtle variations in vegetation vigour and soil moisture are best accentuated in near-infrared images (Lillesand and Kiefer, 2008), pan-sharpened false-colour composites were created, using spectral bands 4, 3, 2 and panchromatic band 8. This image-processing task was performed with standard functions available in ERDAS IMAGINE. A set of two Landsat scenes from different dates (Table 22.2) were acquired to account for the seasonal variability of certain landscape units.

3.3 CORONA

CORONA images were acquired by the first generation of US photo-reconnaissance satellites between 1959 and 1972. After their declassification in 1995, they have found particularly good use in geoarchaeological studies of the Near East (Philip et al., 2002; Ur, 2003; Hritz, 2005), as they provide a unique record of the landscape just before many features were destroyed by modern, large-scale cultivation. A drawback of the images is the large image distortions due to the oblique and panoramic viewing angle. Inexpensive, high-resolution digital scans of CORONA imagery can be obtained from the USGS (http://earthexplorer.usgs.gov).

The relevant CORONA images for this study area are from KH-4A missions, with a best ground resolution of 3 m (see Table 22.2). Image patches of 5.5 cm × 10.5 cm (corresponding to *ca.* 18 km × 36 km on the ground) were individually geo-referenced in ArcGIS, using about 20 control points measured from a pan-sharpened Landsat image. A second-order polynomial function was applied to allow for the large panoramic image distortions. Independent checkpoints revealed a satisfying accuracy of 30 m, which is comparable to the quality of the control data.

3.4 Google Earth

Google Earth provides worldwide, free-to-access imagery, partly at high resolution (http://earth.google.com). At the time of writing (October 2009), recent SPOT imagery had just become available for the entire

Table 22.2 Remote Sensing Data Used in This Study

Sensor	Acquisition Date	Scene IDs	Number of Bands	Resolution	Comments
Landsat ETM+	28 July 2001	Path 165, row 39	8	15/30/60 m	Wet season
Landsat ETM+	20 January 2002	Path 165, row 39	8	15/30/60 m	Dry season
CORONA KH-4A: mission 1035-1	23 September 1966	Revolution 040D, frames 16–19	1	*ca.* 3 m	Acquired from USGS; *ca.* 30% cover
CORONA KH-4A: mission 1045-2	5 February 1968	Revolution 182D, frames 78–82	1	*ca.* 3 m	Provided by CAMEL; full cover
Google Earth/ QuickBird	2003–2007		3	0.6 m	*ca.* 34% cover
Google Earth/ SPOT	2008–2009		3	2.5 m	
SRTM	11–22 February 2000	Tile 46_06		90 m	DEM

study area, dating from 2008 to 2009. Smaller parts of the area are currently covered by QuickBird imagery, mostly acquired between 2003 and 2007. The downside of using this imagery is the lack of control over image quality and acquisition time. Nevertheless, the superior resolution proved helpful for identifying features that otherwise remained ambiguous from Landsat imagery alone.

3.5 Shuttle Radar Topographic Mission

The SRTM produced the most complete, high-resolution digital elevation model (DEM) of the entire Earth (Farr et al., 2007). The DEM has a ground resolution of 90 m and its accuracy was estimated at 1.16±2.57 m (based on 37 spot heights from topographical maps). However, due to the extremely low relief, small systematic and random errors resulted in unfavourable striping and salt-and-pepper patterns. The presence of such noise prevented a quantitative analysis of DEM derivatives, other than average slopes along surface profiles.

In this study, free post-processed SRTM data from the Consortium for Spatial Information of the Consultative Group for International Agricultural Research (CGIAR-CSI) were used (http://srtm.csi.cgiar.org). Visual interpretation of the DEM was performed in ArcGIS, while analysis of slope profile was carried out in ERDAS IMAGINE.

3.6 Image Interpretation

Like traditional aerial photo-interpretation, the visual analysis of satellite imagery relies on several basic characteristics of the surface. These are tone, texture, pattern, shape, context and scale, all qualitative attributes, and their use is very much a matter of experience and personal bias (Drury, 1987).

- *Tone* refers to the colour or relative brightness; it is related to reflectance properties of the surface material, but also to illumination conditions, image processing and printing or display; therefore, absolute tone is of less use than relative tonal differences between objects,
- *Texture* is a combination of the magnitude and frequency of tonal change in an image; scale and resolution determine which features dominate texture,
- *Pattern* is the result of the spatial arrangement of different tones and textures which make up the image scene; it is related to the arrangement of vegetation, topographic features, drainage channels or geological structures,
- Interpretation of particular tones, textures, patterns and shapes always depends on their *context and scale.*

The image attributes may change depending on the time of day and year of image acquisition, due to changes in illumination conditions, vegetation cover and soil moisture content. Also, the attributes vary within the radiometric spectrum; subtle differences in soil and rock can be more readily detected using multi-spectral combinations instead of panchromatic images.

Image interpretation techniques have found extensive use in certain fields of landscape research, such as landslide inventories (Soeters and Van Westen, 1996) and archaeological prospection (Scollar et al., 2009). In these fields, standardised mapping procedures have been developed in order to improve their objectivity. So far, this has not been the case for flat alluvial landscapes in semi-arid environments such as Lower Khuzestan.

3.7 Mapping Procedure

All above-mentioned source data were combined in an ArcGIS project, allowing management and interpretation of data in separate layers. First, modern infrastructural elements displayed on contemporary topographical maps were identified and traced from the Landsat imagery. This not only provided a useful reference layer but also avoided confusion of these elements with other linear features, as ground checks were not readily possible.

The hydrological network was mapped using both Landsat and CORONA imagery. Natural watercourses (including meandering, straight and incised river sections, crevasse channels and ephemeral streams) and anthropogenic elements (irrigation and drainage canals) were all combined in a single layer. In the case of migrating meanders, their position was recorded from both image sources. Small alluvial landforms were digitised simultaneously, including meander cut-offs (oxbows), scroll bars and crevasse splays.

Palaeohydrological features were mapped in another layer. Superimpositions and connections revealed the relative chronology of palaeoriver channels, irrigation patterns and their associated alluvial landforms. Their identification was based principally on the interpretation of CORONA images. On a few occasions, traces of palaeochannels were detected from Google Earth, which were apparently not noticed from the CORONA images. Although these instances indicate the limitations, CORONA imagery remained the most effective data source, as large tracts of land are subject to modern land use, thereby obscuring the traces of palaeofeatures.

The final stage in the mapping process was the identification of the broad landscape units, based mainly on the interpretation of land use, vegetation and drainage patterns from Landsat imagery. The combination of two scenes from different seasons helped the distinction of seasonal marshes. The SRTM DEM was helpful for identifying alluvial ridges, which rise several metres above the surrounding floodplains, and the anticlines that border the study area. In addition, surface profiles extracted from the DEM were used to estimate average slope angles of the alluvial fans.

3.8 Map Design

Geomorphological maps are transmitters of information about the form, origin, age and distribution of landforms together with their formative processes, rock type and surface materials (Brunsden et al., 1975). They are not only a way of presenting data but also the result of a method of research, revealing associations of landforms, which is essential for the understanding of both individual landforms and landscapes (De Graaff et al., 1987).

A general geomorphological map can emphasise different aspects of landforms (Van Zuidam, 1985): morphology (describing the relief), morphogenesis (describing origin and process), morphochronology (relative dating) and morphoarrangement (describing spatial arrangements and relationships). These different aspects can be depicted in the map by coloured area symbols, patterns and line symbols, depending on the importance that is assigned to each aspect. Also an important quality of a map is the readability of relevant information (Gustavsson et al., 2006), even more so when it is to be used by non-specialists.

The map legend developed in this study was designed in line with the rationale of the project, that is reconstructing the evolution of alluvial and human-induced Holocene landscapes. It was therefore decided that the geomorphological map should distinguish at the highest levels on the basis of genesis and chronology, both depicted by colour. Further subdivision of alluvial landforms was represented by symbol patterns, whereas their associated hydrological elements were coloured consistently, according to their (relative) chronology. Since the alluvial plains in Lower Khuzestan are extremely flat, descriptors of relief were mostly irrelevant; only escarpments of alluvial terraces and incised channels were depicted by line symbols.

Although the image interpretation and digitisation of geomorphological features was performed at a scale of 1/50,000, the final map sheets were produced at a scale of 1/100,000. Bearing in mind the dimensions of mapped features and the vast size of the study area (9.6×10^3 km^2), this was considered an appropriate scale. Printed map sheets were fitted with a topographical background to aid orientation at a desk or in the field.

4. RESULTS

4.1 Interpretation Key and Map Legend

The image interpretation key in Table 22.3 summarises the diagnostic characteristics of all map units, as observed from the satellite imagery. The

Table 22.3 Image interpretation key and legend designed for geomorphological mapping of the alluvial plains in Lower Khuzestan

Map unit	Symbol	Definition	Characteristics Visible on Images	Source[a]
* *Anthropogenic features*				
Paved road		Tarmac surface	Dark, distinct straight line	T, L
Unpaved road		Car or cattle track	Bright straight line (easily confused with abandoned canals or embankments)	T, L
Railway			Dark, distinct straight line	T, L
Embankment			Bright-coloured line (contrasting with darker floodplain), usually on riverbank	T, C, GE
Urban area		Mixture of buildings, roads and gardens	Grey, mottled texture	T, L
Archaeological site (small/large)		Abandoned settlement or ruins	Soil contrasting with surroundings; no or disturbed vegetation; often circular shape; poor drainage – ponding in local depressions, erosion gullies on mounds	C, GE
Irrigation/drainage canal		Artificial watercourse for irrigation/drainage of cultivated land, currently active	Dark, usually straight line (but sinuous when reuse of a former natural channel); dense vegetation in/along canal; clearly associated with modern land use	T, L
Ancient irrigation/drainage canal[b]		Abandoned artificial watercourse	Bright-coloured line (contrasting with darker floodplain); ancient field systems may be visible as angular patterns	L, C

★ *Alluvial/fluvial landforms*				
Alluvial chronology[b]	III IIa II Ia I Unknown			
River channel with migrating meanders		Natural, meandering watercourse; evidence of migrating meanders after 1960s	Dark water surface (but bright in visible colour when high sediment load); adjacent lands usually intensively cultivated; migration of meanders judged from multi-temporal images	L, C
River channel with stable meanders		Natural, meandering watercourse; stable meanders after 1960s		L, C
Straight river channel		Natural watercourse; low sinuosity, possibly human-controlled		L, C
Incised river channel		Natural watercourse; incised in older floodplain		L, C
Palaeoriver channel		Abandoned river channel	Dark, upsilted channel (poorly drained) and/or scroll bars; may remain visible in cultivated land as soil/vegetation marks or in field pattern	L, C
Ephemeral stream		Natural watercourse; only carries water periodically	In dry season bright, unvegetated; in wet season dark	L, C
Scroll bars		Ridges and swales on inside of river bends, resulting from lateral accretion	Curved lines due to contrast in soil and vegetation between ridges (bright, sparse vegetation) and swales (dark, poorly drained, dense vegetation)	C

(continued)

Table 22.3 (continued)

Map unit	Symbol	Definition	Characteristics Visible on Images	Source[a]
Oxbow		Former meander, cut off from main channel and filled with water or marsh	U-shaped, dark soil, poorly drained; may remain visible in cultivated land as soil/vegetation marks or in field pattern	C
Crevasse channel		Ephemeral channel branching out on a fan-shaped body, formed after breach through levee	Distributary channel system, usually leaving from a cut in the outside bend of the main river	L, C, GE
Crevasse splay		Fan-shaped body, formed after breach through levee	Bright soils contrast with dark-coloured flow paths; patchy vegetation	L, C, GE
Human-induced splay		Fan-shaped body, induced by human activity (e.g. diversion of floodwater or dike breach)	Associated with canal and/or embankment	L, C, GE
Floodplain, incised or terraced		Flat terrain adjacent to river, subject to flooding; terrace levels result from lateral erosion after river incision; higher floodplain levels may be out of reach of present river floods	Dark, poorly drained soils adjacent to river channel; traces of oxbows and scroll bars may be present; higher terrace levels have brighter, well-drained soils and sparse vegetation	S, L, C, GE
Terrace scarp		Sharp edge between floodplain levels	Evidence of gully erosion and/or irregular vegetation cover	S, L, C, GE
Alluvial ridge		Ridge formed by river, raised decimetres/meters above floodplain as result of overbank deposition	Bright soils, contrasting with darker floodplain; good internal drainage; alluvial ridges of present rivers usually have intensive land use; abandoned ridges often non-vegetated and scoured by wind	S, L, C

Alluvial fan		Fan-shaped body, deposited by avulsive channels, characterised by a (seemingly) distributary channel system	Bright soils; drainage via meandering or distributary channels; patches of dense vegetation and land use	L, C
Bajada		Apron of alluvial fans at the foot of steep hills, resulting from flash floods	Bright soils; drainage via multiple braided, ephemeral streams; sparse vegetation	S, L, C
Flood basin, marsh		Low-lying area, periodically flooded	Open water or dense vegetation in wet season; bright soils and mottled, dead vegetation in dry season	L
Continental *sabkha*	× ×	Low-lying area, periodically flooded; formation of salt crusts in dry season	Open water in wet season; extremely bright soils in dry season; poorly drained; no vegetation	L
Lake		Permanent water body	Open water year-round	L
⋆ *Aeolian landforms*				
Dune field		Area of windblown deposits	Bright soils; well drained; no vegetation	L, C
⋆ *Coastal landforms*				
Tidal channel	——	Channel in tidal flat, subject to tidal cycle	Curved line, contrasting colour with surrounding flats	L
Tidal flat		Mudflats and sandbanks along the coast; intersected by network of tidal channels	Dark uniform soils; no vegetation; drained by tidal channels	L

(*continued*)

Table 22.3 (continued)

Map unit	Symbol	Definition	Characteristics Visible on Images	Source[a]
Clastic coastal *sabkha*	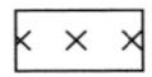	Supratidal zone of the tidal area, covered with salt crust; only inundated during spring tides	Extremely bright soils; no vegetation	L
⋆ *Geological-structural features and landforms*				
Geological stratum		Outcropping solid rock	Colour depending on rock type; usually rugged relief, densely drained by ephemeral streams; sparse vegetation	GM, L
Anticline	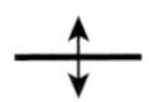	Central axe of upward-curving fold		GM
Syncline		Central axe of downward-curving fold		GM
Fault, joint	—	Fracture in solid rocks	May be visible due to linear disturbance of vegetation, or drainage anomalies	GM

[a]C, CORONA; GE, Google Earth; GM, geological maps; L, Landsat; S, SRTM DEM; T, topographical maps.
[b]Chronological order (I-III) of alluvial landforms and irrigation systems is depicted by colour or grey shades.

table also presents the map symbols used in the legend. At the highest level, landforms are subdivided into anthropogenic, alluvial, aeolian, coastal and geological-structural units, each represented by a different colour or grey shade. The relative chronology of alluvial and anthropogenic elements is also depicted by colours or grey shades. A common colour/grey shade scheme is applied to both anthropogenic and alluvial features in order to emphasise their relationships.

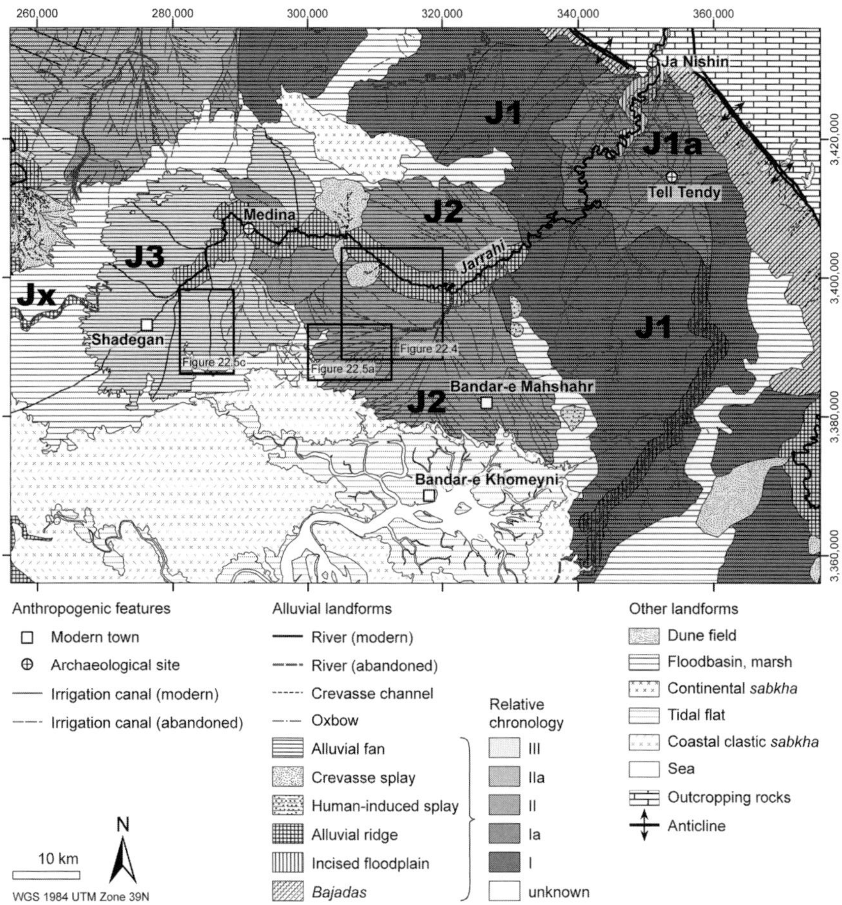

Figure 22.3 Geomorphological map of the study area.

4.2 The Geomorphological Map

Figure 22.3 shows a generalised version of the geomorphological map of the study area. A sample of the map at full scale (1/:100,000) is displayed in Figure 22.4.

The map reveals three alluvial fans successively deposited by the Jarrahi River (units J1, J2 and J3). The fans were distinguished based on their characteristic irrigation patterns, which clearly indicate that human control played an important role in their development.

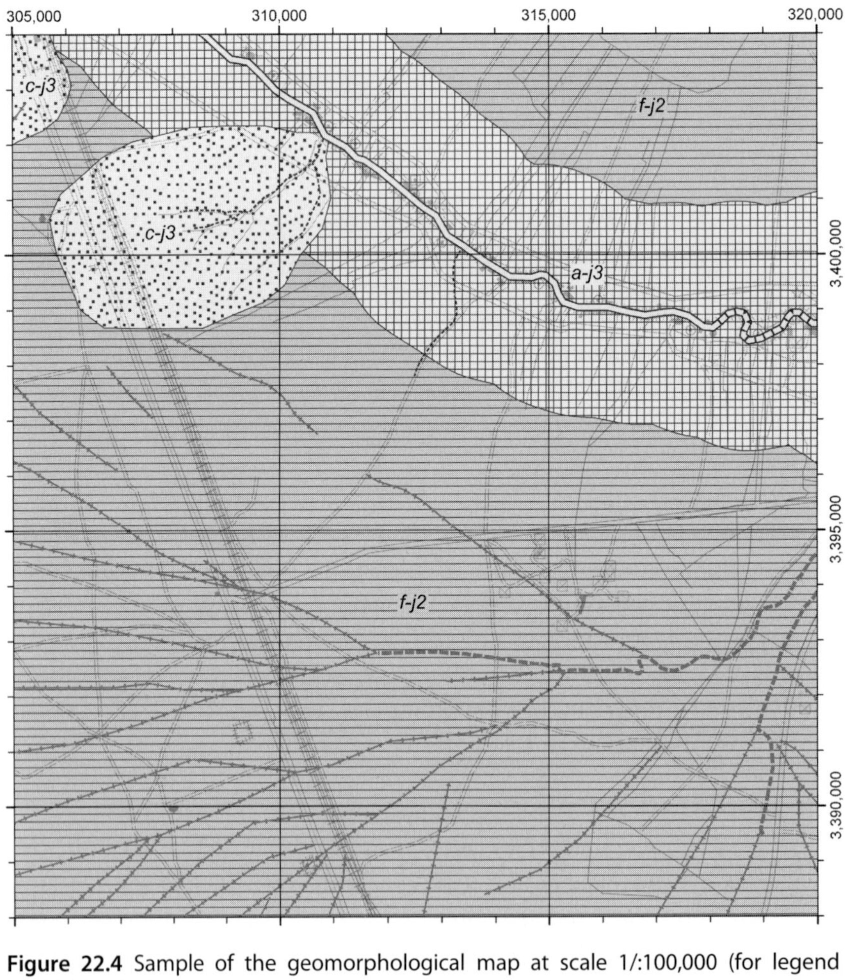

Figure 22.4 Sample of the geomorphological map at scale 1/:100,000 (for legend see Table 22.3).

4.2.1 Fan J1

The first, and largest, fan stretches from the location where the Jarrahi enters the plain to the tidal flats of Khor Musa. In the northwest and southeast, the fan gradually merges into adjoining fans. SRTM elevation drops from 32 m at the head to 2 m at the toe, with an average slope of about 0.05%. At the fan's apex is a 1.5 km wide river valley, cut 8 m deep into the original surface. Within this river section, the Jarrahi flows through anastomosing channels, and the valley bottom is dotted with numerous meander cut-offs. After 8 km the valley suddenly narrows, and the river continues via a series of entrenched meanders. Further downstream, the river gradually emerges from its entrenchment and continues over an elevated alluvial ridge.

Dense patterns of ancient irrigation canals were identified across the fan. A network of distinct canals diverges from the fan's apex and spreads across the fan over a distance of more than 20 km on both sides of the incised river valley (indicated as unit J1a). Less distinct linear features 'emerge' next to the valley, their upper sections apparently eroded by the incising river. Further away from the Jarrahi, the surface of the fan becomes mottled and featureless, indicative of long-term inactivity and weathering. Still, some faint, curvilinear traces were observed and mapped. These lines coincide with the position of slightly elevated lobes visible on the SRTM DEM and were interpreted as old alluvial ridges of the Jarrahi.

4.2.2 Fan J2

The second fan is located immediately downstream of the first one. Its surface is intersected by a large, diverging network of ancient canals, all ending in typical 'herringbone' field patterns (Figure 22.5). SRTM elevation of the fan ranges from 12 to 2 m above sea level, with an average slope of about 0.04%. The alluvial ridge of the present-day Jarrahi is raised 2–3 m above the fan surface; its deposits and modern land use patterns have obscured most of the earlier traces alongside the river.

4.2.3 Fan J3

The third fan comprises the currently active system, extending westwards into the Shadegan Marshes. Numerous canals branch off from the Jarrahi and distribute water and sediment over countless ditches. The extremely narrow, elongated fields between the closely spaced ditches are used for intensive rice and date cultivation (Figure 22.5). The irrigation canals are

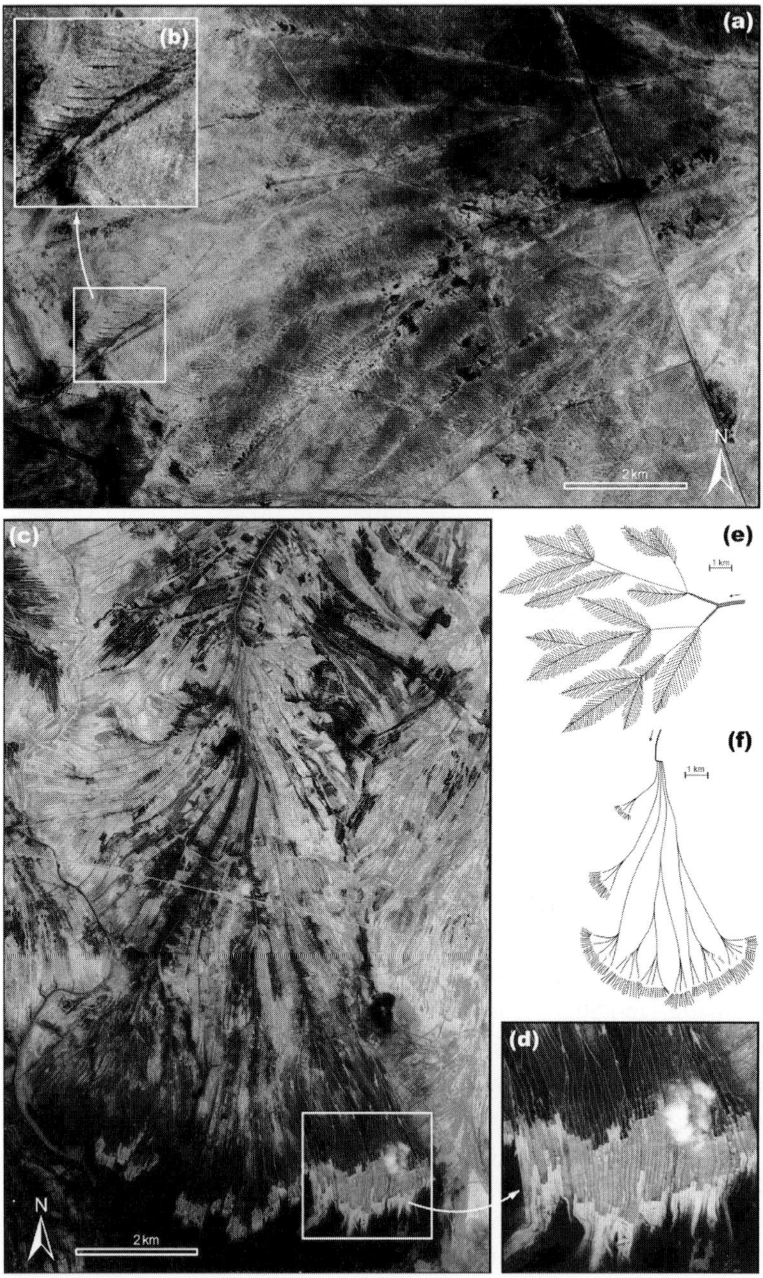

Figure 22.5 Typical examples of irrigation patterns visible on CORONA imagery (extracts from frame DS1045-2182DF080): abandoned 'herringbone' patterns of fan J2 (a and b) and active distributary system of fan J3 (c and d). The line drawings are a schematic representation of both irrigation networks (e and f).

typically curvy and irregular in contrast to the straight plan forms of the second fan. Some main branches on the eastside of the fan are abandoned, leaving bare patches of land behind. SRTM elevation of this fan varies between 10 and 2 m, with an average slope of only 0.02%–0.04%.

4.2.4 Palaeochannel Jx

To the west of the present-day fan, a meandering palaeoriver channel crosses the Shadegan Marshes. Its levees are clearly visible, raised above the water surface even at high water. At both ends the channel is covered by recent alluvial deposits: in the east by the Jarrahi alluvial fan and in the west by the alluvial ridge of the Karun. The size and orientation of the palaeochannel suggest it represents a former course of the Jarrahi.

5. DISCUSSION

5.1 Chronology

The chronological framework of the landscape was further developed with the help of information from other disciplines, taking into account the close association between ancient settlements, irrigation patterns and the alluvial landforms. This multidisciplinary approach led to the following reconstruction.

- Initially, the Jarrahi built up the large fan (J1) at the foot of the Marun and Agha Jari anticlines. An earlier course of the Jarrahi deposited a south-directed alluvial ridge, along the archaeological site of Tell Tendy. The site was populated at least between the Achaemenian and Parthian periods, but probably had a long occupation history before (Hansman, 1978),
- Incision of the river into the fan cuts off the water supply and leads to abandonment of the irrigation canals and Tell Tendy. A dam across the river and large intake canals near the site of Ja Nishin represent an ambitious effort to revive irrigation of the old fan. Dated material alongside the canals indicates that the system was active in Sasanian times (Hansman, 1978), but an earlier date cannot be ruled out,
- The channel incision could be explained by several external factors, such as tectonic activity, a change of base level or a change in river discharge. However, fan-head entrenchment is a common process

(Bridge, 2003) and simply can be the result of an avulsion and internal adjustment of the river to its new (downstream lower) gradient,

- While incising into the first fan, the Jarrahi started building up a new fan (J2). The sheer size of the irrigation patterns suggests a Sasanian or Early Islamic origin (Verkinderen, 2009), although there is no direct archaeological evidence available,
- Eventually, the locus of sedimentation shifted again further westwards, building up the present-day fan (J3). The earliest historical evidence is the founding of Shadegan in the 1740s (Layard, 1846), whereas archaeological remains of the previous capital town Medina/Dawraq were dated to the seventeenth to eighteenth centuries AD (Hansman, 1978). A radiocarbon-dated sample of organic material (350–430 cal BP; Baeteman et al., 2004) underlying the alluvial deposits confirmed the recent age of this fan,
- The position of palaeochannel Jx within the chronological framework is unclear, except it predates both the present-day Jarrahi and Karun.

5.2 Wider Implications

In this study, a variety of irrigation patterns were detected from satellite imagery. Their close association with alluvial landforms demonstrates the importance of human-induced processes in the evolution of the landscape in this region. On the other hand, the irrigation patterns also reflect the organisation of the societies that built them. Their detailed recording is therefore not only essential for the understanding of Mesopotamian landscapes but also of the past societies that formed an integral part of it.

Massive, technically advanced works, like the dam of Ja Nishin and the canals of fan J2, would have been possible only through strong central administration. They fit perfectly well in the large colonisation programmes of the Khuzestan plain, of which remnants have been attested throughout Khuzestan (Graadt van Roggen, 1905). Traditionally, these weirs are attributed to Sasanian times, but it should be stressed there is no material proof for this date (Verkinderen, 2009).

It is known that after the beginning of the twelfth century, population and agriculture declined throughout Khuzestan; the great irrigation systems decayed and large areas reverted to extensive cultivation (Christensen, 1993; Verkinderen, 2009). The relatively primitive system of the present-day fan should be seen in this light. This irrigation network simply uses gravity flow from the naturally elevated position of the alluvial ridge, without the need for technological infrastructure. Similar practices are well known from ancient times when natural crevasse-cuts formed ideal loci for irrigation and settlement (Wilkinson, 2003; Morozova, 2005).

6. CONCLUSIONS

In this study, a working procedure was developed that can be used for the consistent geomorphological mapping of the Lower Khuzestan alluvial plain, based on easily accessible and inexpensive remote sensing data. The map legend is specifically suited to the landscapes of this region, dominated by alluvial processes and with significant human-induced influences. The presented image interpretation key provides guidance on the recognition of relevant landscape elements from satellite imagery.

The working procedure was developed within the context of a case study on landscape evolution in Lower Khuzestan. The produced geomorphological map provided insights into the Late Holocene evolution of the Jarrahi River. Extensive irrigation patterns cover the different alluvial fans and clearly illustrate the important role of human impact on the landscape. It is therefore recommended to use a multidisciplinary approach in such landscape studies.

ACKNOWLEDGEMENTS

We would like to thank Cecile Baeteman and Anne Mather for comments on an earlier draft of this paper. We thank Olivier Wambacq for the drawing of Figures 22.2 and 22.5. The Center for Ancient Middle Eastern Landscapes, University of Chicago, kindly provided the digital scans of CORONA mission 1045-2 (including the imagery displayed in Figure 22.5). The research was undertaken within the framework of the Interuniversity Attraction Pole 'Greater Mesopotamia: Reconstruction of its Environment and History' (IAP 6/34), funded by the Belgian Science Policy.

REFERENCES

Adams, R.M., 1981. Heartland of Cities: Surveys of Ancient Settlement and Land Use on the Central Floodplain of the Euphrates. University of Chicago Press, Chicago, IL.

Alizadeh, A., Kouchoukos, N., Wilkinson, T.J., Bauer, A.M., Mashkour, M., 2004. Human−environment interactions on the Upper Khuzestan plains, Southwest Iran. Recent investigations. Paléorient 30 (1), 69−88.

Alwash, M.A., Zaidi, S.M.S., Terhalle, U., 1986. Description of arid geomorphic features using Landsat-TM data and ground truth information (Wadi Fatima, Kingdom of Saudi Arabia). Catena 13 (3), 277−293.

Baeteman, C., Dupin, L., Heyvaert, V.M.A., 2004. The Persian Gulf Shorelines and the Karkheh, Karun, and Jarrahi Rivers: a geo-archaeological approach. First Progress Report. 1. Geo-environmental investigation. Akkadica 125, 155–215.

Bridge, J.S., 2003. Rivers and Floodplains: Forms, Processes, and Sedimentary Record. Blackwell Publishing, Oxford.

Brunsden, B., Doornkamp, J.C., Fookes, P.G., Jones, D.K.C., Kelly, J.M.H., 1975. Large scale geomorphological mapping and highway engineering design. Q. J. Eng. Geol. 8 (4), 227–253.

Buringh, P., 1960. Soils and Soil Conditions in Iraq. Ministry of Agriculture, Baghdad.

Christensen, P., 1993. The Decline of Iranshahr: Irrigation and Environments in the History of the Middle East 500 B.C. to A.D. 1500. Museum Tusculanum Press, Copenhagen.

Cole, S.W., Gasche, H., 1998. Second and first millennium B.C. rivers in northern Babylonia. In: Gasche, H., Tanret, M. (Eds.), Changing Watercourses in Babylonia. Towards a Reconstruction of the Ancient Environment in Lower Mesopotamia. Ghent University, Ghent, pp. 1–64.

Cultraro, M., Gabellone, F., Scardozzi, G., 2009. The virtual musealization of archaeological sites: between documentation and communication. Third ISPRS International Workshop 3D-ARCH 2009, Trento, Italy.

De Graaff, L.W.S., De Jong, M.G.G., Rupke, J., Verhofstad, J., 1987. A geomorphological mapping system at scale 1:10,000 for mountainous areas. Z. Geomorphol. 31, 229–242.

Drury, S.A., 1987. Image Interpretation in Geology. Allen & Unwin, London.

Farr, T.G., Rosen, P.A., Caro, E., Crippen, R., Duren, R., Hensley, S., et al., 2007. The Shuttle Radar Topography Mission. Rev. Geophys. 45. <http://dx.doi.org/10.1029/2005RG000183>

Graadt van Roggen, D., 1905. Notice sur les anciens travaux hydrauliques en Susiane. Mémoires de la Délégation en Perse VII, 167–207.

Gustavsson, M., Kolstrup, E., Seijmonsbergen, A.C., 2006. A new symbol-and-GIS based detailed geomorphological mapping system: renewal of a scientific discipline for understanding landscape development. Geomorphology 77, 90–111.

Gutman, G., Byrnes, R., Masek, J., Covington, S., Justice, C., Franks, S., et al., 2008. Towards monitoring land-cover and land-use changes at a global scale: the global land survey 2005. Photogramm. Eng. Remote Sens. 74 (1), 6–10.

Hansman, J., 1978. Seleucia and the Three Dauraks. Iran 16, 154–161.

Hessami, K., Nilforoushan, F., Talbot, C.J., 2006. Active deformation within the Zagros Mountains deduced from GPS measurements. J. Geol. Soc. 163, 143–148.

Heyvaert, V.M.A., 2007. Fluvial Sedimentation, Sea-Level History and Anthropogenic Impact in the Great Mesopotamian Plain: A New Holocene Record. Ph.D. Thesis, Vrije Universiteit Brussel, Brussels, Belgium.

Heyvaert, V.M.A., Baeteman, C., 2008. A Middle to Late Holocene avulsion history of the Euphrates river: a case study from Tell ed-Dēr, Iraq, Lower Mesopotamia. Quat. Sci. Rev. 27 (25–26), 2401–2410.

Hritz, C., 2005. Landscape and Settlement in Southern Mesopotamia: A Geo-Archaeological Analysis. Ph.D. Thesis, University of Chicago, Chicago, IL.

Hritz, C., Wilkinson, T.J., 2006. Using Shuttle Radar Topography to map ancient water channels in Mesopotamia. Antiquity 80 (308), 415–424.

Jahjah, M., Ulivieri, C., Invernizzi, A., Parapetti, R., 2007. Archaeological remote sensing application pre-post war situation of Babylon archaeological site – Iraq. Acta Astronaut. 61, 121–130.

Layard, A.H., 1846. A description of the province of Khúzistán. J. R. Geogr. Soc. Lond. 16, 1–105.

Lillesand, T.M., Kiefer, R.W., Chipman, J.W., 2008. Remote sensing and image interpretation. John Wiley & Sons, New York.

Morozova, G.S., 2005. A review of Holocene avulsions of the Tigris and Euphrates rivers and possible effects on the evolution of civilizations in lower Mesopotamia. Geoarchaeology 20 (4), 401–423.

Ooghe, B., 2007. Off the Beaten Track: Travellers, Maps and the Landscapes of Ottoman Mesopotamia. Ph.D. Thesis, Ghent University, Ghent, Belgium.

Pain, C.F., 1985. Mapping of landforms from Landsat imagery: an example from eastern New South Wales, Australia. Remote Sens. Environ. 17 (1), 55–65.

Philip, G., Donoghue, D., Beck, A., Galiatsatos, N., 2002. CORONA satellite photography: an archaeological application from the Middle East. Antiquity 76 (291), 109–118.

Potts, D.T., 1999. The Archaeology of Elam: Formation and Transformation of an Ancient Iranian State. Cambridge University Press, Cambridge.

Pournelle, J.R., 2003. Marshland of Cities: Deltaic Landscapes and the Evolution of Early Mesopotamian Civilization. Ph.D. Thesis, University of California, San Diego, CA.

Scollar, I., Tabbagh, A., Hesse, A., Herzog, I., 2009. Archaeological Prospecting and Remote Sensing (Topics in Remote Sensing). Cambridge University Press, Cambridge.

Smith, M.J., Pain, C.F., 2009. Applications of remote sensing in geomorphology. Prog. Phys. Geogr. 33 (4), 568–582.

Soeters, R., Van Westen, C.J., 1996. Slope instability recognition, analysis, and zonation. In: Turner, A.K., Schuster, R.L. (Eds.), Landslides. Investigation and Mitigation (Special report (National Research Council (U.S.) Transportation Research Board). National Academy Press, Washington, DC, pp. 129–177.

Ur, J., 2003. CORONA satellite photography and ancient road networks: a northern Mesopotamian case study. Antiquity 77 (296), 102–115.

Ur, J., 2006. Google Earth and archaeology. SAA Archaeol. Rec. 6 (3), 35–38.

USGS, 2003. Landsat: a global land-observing program (fact sheet 023-03, March 2003). <http://landsat.gsfc.nasa.gov/pdf_archive/USGS_landsat_factsheet.pdf> (accessed 25.10.09).

Van Zuidam, R.A., 1985. Aerial Photo-Interpretation in Terrain Analysis and Geomorphological Mapping. Smits Publishers, The Hague.

Verhoeven, K., 1998. Geomorphological research in the Mesopotamian flood plain. In: Gasche, H., Tanret, M. (Eds.), Changing Watercourses in Babylonia. Towards a Reconstruction of the Ancient Environment in Lower Mesopotamia. Ghent University, Ghent, pp. 159–245.

Verkinderen, P., 2009. Tigris, Euphrates, Karun, Karhe, Jarrahi: Tracking the Traces of Five Rivers in Lower Iraq and Khuzistan in the Early Islamic Period. Ph.D. Thesis, Ghent University, Ghent, Belgium.

Wilkinson, T.J., 2003. Archaeological Landscapes of the Near East. University of Arizona Press, Tucson, AZ.

CHAPTER TWENTY-THREE

Military Applied Geomorphological Mapping: Normandy Case Study

Peter L. Guth
Department of Oceanography, United States Naval Academy, Annapolis, MD

Contents

1. INTRODUCTION

Geomorphology affects military operations, even when soldiers have no technical understanding of the Earth sciences. The landscape determines the speed and ease of travel, produces obstacles and makes some locations key terrain for both sides in a conflict. Borders often follow natural routes or barriers, and landforms dictate invasion routes. Armies differ in their employment of geoscientists and in the training in geoscience they provide to officers, but they have always recognised that the terrain can provide a significant force multiplier.

Geomorphological mapping can be applied for current operations or retrospectively for historical operations to understand what happened and apply those lessons to current operations. Tate (2006) described terrain analysis for decision-making in military geography, whereas Fleming et al. (2009) described the role of geographical information systems in military operations. The US Army doctrine on terrain analysis (Department of

Developments in Earth Surface Processes, Volume 15
ISSN: 0928-2025, DOI: 10.1016/B978-0-444-53446-0.00023-9

the Army, 1990, 2000) has not kept up with advances in computer technology and is not written from an Earth science perspective.

Military operations exist at a number of scales, and the geomorphological mapping required and the data resolution change dramatically. As spatial scales become larger (generally considered better), the volume of data and processing time increase exponentially, and the area that can be examined in detail decreases. At the smallest scale (largest area of operations), digital elevation models (DEMs) with a point spacing of 30″ or 1 km allow visualisation of an entire theatre of operations. Satellite sensors like MODIS, with a 250 m spatial resolution and near-daily temporal resolution, can show weather systems, snow cover and vegetation condition. At intermediate scales, DEMs like the 3″ (90 m) SRTM provide detail for brigade or battalion level planning, and satellites like the 30 m resolution Landsat Thematic Mapper provide regional coverage but with only a 2–3 week revisit capability. The largest scale analysis can use LiDAR DEMs with 1 m or better spatial resolution, and imagery with comparable scales, and will show every building, boulder and shrub. While vital to a platoon or squad in combat operations, the level of detail may be detrimental to higher level military units that can view the detailed coverage as noise. A division commander does not plan where each tank goes but only needs to know which terrain allows reasonable tank mobility.

2. THE NORMANDY LANDINGS IN WORLD WAR II

The Allied landings in Normandy in World War II represent perhaps the largest combined arms operation in military history, involving air, land and sea forces from a number of nations (Cole, 1951; Hammond, 1994). After landing on 6 June 1944, the Allies spent the next 6 weeks fighting through the bocage terrain with small fields and pastures surrounded by dense shrubbery and flooded marshes along the coast. On 25 July, American units launched an attack to break out of Normandy, and within days armored and mechanised units had broken the German lines and begun a mad dash across France.

The Allies employed geoscientists to prepare maps (Rose et al., 2006; Rose, 2008) to plan the Normandy landings and subsequent operations. The maps in this chapter show the locations of the front lines digitised from the daily situation maps of the First United States Army Group (FUSAG) and later the Twelfth Army Group (Library of Congress, 2009). These maps have a scale of 1:500,000 and show the situation depicted by the Allied headquarters – the fog of war and generalisation could affect the accuracy, and the roughly 80 m pixels on the maps means that when zooming in on Google Earth, the locations for the front lines will be only approximate.

For clarity, in this chapter, the front lines for only six dates are shown on the maps; daily location for all dates is available at www.appgema.net for display in Google Earth.

1. 7 June 1944: One day after the landings, Allied forces were split with a western region around Utah Beach and inland drop zones for airborne forces, and a long, narrow belt along the other beaches.
2. 12 June 1944: Allied forces had a single front line well inland from the beaches, with the largest penetration inland moving from Omaha Beach towards the town of Saint-Lô.
3. 25 July 1944: The Cherbourg Peninsula had been cleared but otherwise the front lines had not advanced significantly. On this day, the COBRA Operation began.
4. 1 August 1944: In 1 week, Allied forces had advanced over 40 km on the west side of the Normandy Peninsula, with little change on the east side.
5. 3 August 1944: Elements of six divisions had broken out, passing Rennes, 80 km from the front lines on 1 August. From this point on, the German army would be unable to establish a continuous Western front until the middle of October along the border of Germany.
6. 8 August 1944: Allied divisions had reached Saint-Malo, Rennes, Laval, and, up to 120 km from the front lines on 1 August.

3. TERRAIN ANALYSIS

This study uses readily available digital data (Table 23.1) to show the kinds of analysis possible anywhere in the globe. The scale of these data ranges from 15 to 450 m, appropriate for looking at strategic considerations of the Normandy campaign. These data postdate World War II by 50–60 years, but the general features of the landscape have probably not changed significantly at this scale.

Table 23.1 Data Sets Used

Data Set	**References**
SRTM 3″ (90 m) digital elevation model	Farr et al. (2007)
15″ (450 m) drainage network and drainage basins	Lehner et al. (2008)
Water bodies	USGS (2009)
Coastline	NGA (2009)
Landsat ETM+ imagery (15 m panchromatic band and 30 m for five multispectral bands)	GLCF (2009)
Corine land cover 2000, 100 m resolution (CLC2000)	EEA (2007), Copenhagen

Military success depends on the friendly and enemy troops, the commanders and the weather in addition to the terrain, so that geomorphic mapping cannot explain all past operations or predict success in the future. Nevertheless, understanding the landscape may be one of the most important aspects to understanding military operations.

4. GEOMORPHIC MAPS OF NORMANDY

The simplest but probably most effective application of geomorphic mapping uses digital topography to provide base maps to display the military situation and the area of operations. The most effective displays use shaded relief (also known as hillshading or reflectance maps) to clearly show the nature of the terrain (Kraak and Ormeling, 2010). This can be combined with colour tinting from the elevation (Figure 23.1) or created in greyscale when additional overlays require the use of colour. Four regions in Figure 23.1 illustrate how geomorphology affected the campaign. Region A has flat, low terrain along the coast, and the front lines here changed little during the first 2 months. After 7 weeks of slow going to expand the beachhead and clear the Cherbourg Peninsula, the initial drive that opened up the battle started near Saint-Lô along moderately higher ground. The final breakout occurred through region D, again with moderately higher ground, and not in the highest terrain around C.

Slope maps can be rapidly computed from DEMs and supply predictions of cross-country mobility. Where soils, vegetation and the weather

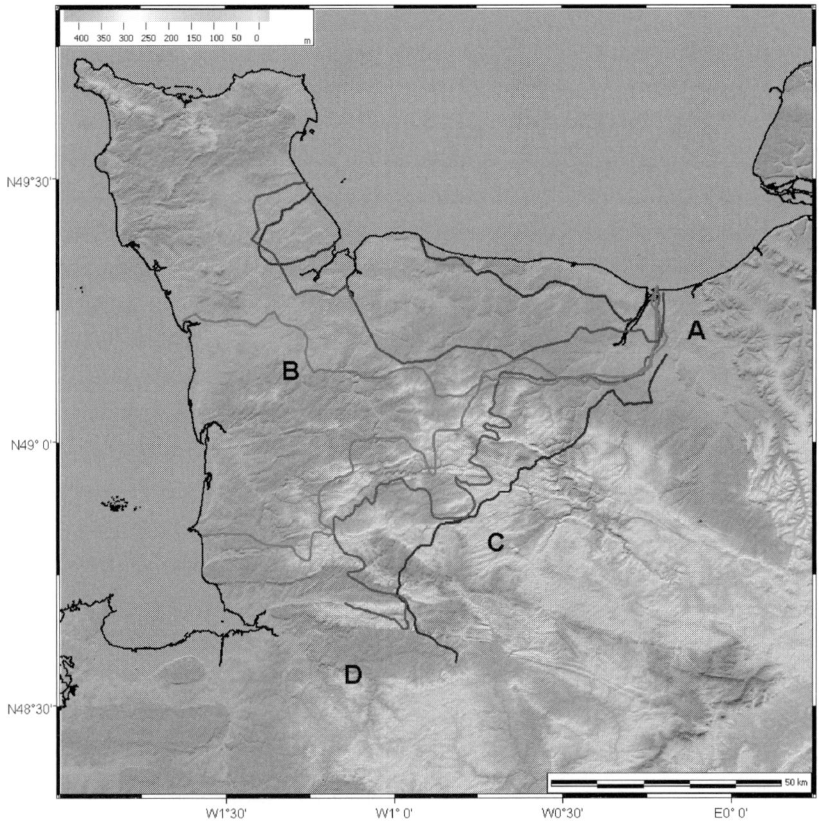

Figure 23.1 Shaded relief map of the Normandy Peninsula. The front lines at key dates during the first 2 months of the campaign are indicated and four regions with different landforms are labelled from 'A' to 'D'.

also affect mobility, getting these into digital format greatly complicates the analysis. Figure 23.2 shows a slope map for Normandy. Although isolated slope values up to 85% occur, the colour scale stops at 45%, the limit for tank mobility (Department of the Army, 1990). At this scale, large slopes stand out along some of the river valleys and especially in the 'Normand Switzerland'. The smallest slopes occur in the region taken before 25 July, and the breakout occurred in regions with mostly moderate slopes. Figure 23.3 shows the slopes along the landing beaches and clearly shows the steep cliffs that border Omaha Beach. This unfavourable terrain undoubtedly contributed to significantly higher casualties at Omaha Beach.

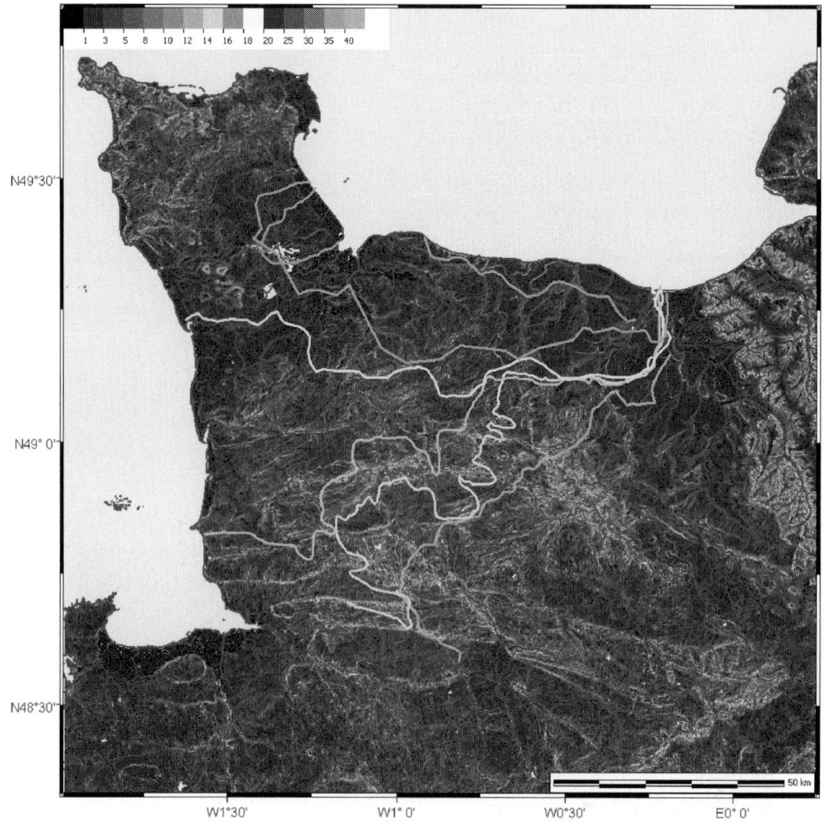

Figure 23.2 Slope map of the Normandy Peninsula, with slopes in per cent.

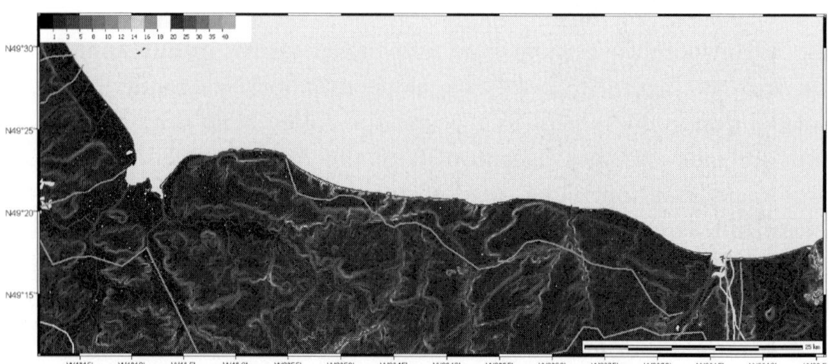

Figure 23.3 Slope map of the invasion beaches, at the full resolution of the 3″ DEM.

Geomorphic maps can combine terrain categories and reveal difficult terrain. Figure 23.4 shows the terrain in Normandy under 10 m in elevation and with slopes less than 3% which could be easily flooded. Two regions stand out: the blocked eastern flank of the allied beachhead and the region between the Omaha and Utah beaches.

Guth (2003) developed a method to quantify terrain organisation, the degree to which ridges and valleys align in the same direction. Figure 23.5 shows an organisation overlay for Normandy, computed for regions 2500 m on a side; this parameter depends on the size of the analysis region. The orientation of the lines depicts the orientations of the ridges and valleys, and the length of the lines depicts the strength of the organisation. Cross-country mobility will be greatest parallel to the lines and most restricted perpendicular to the lines.

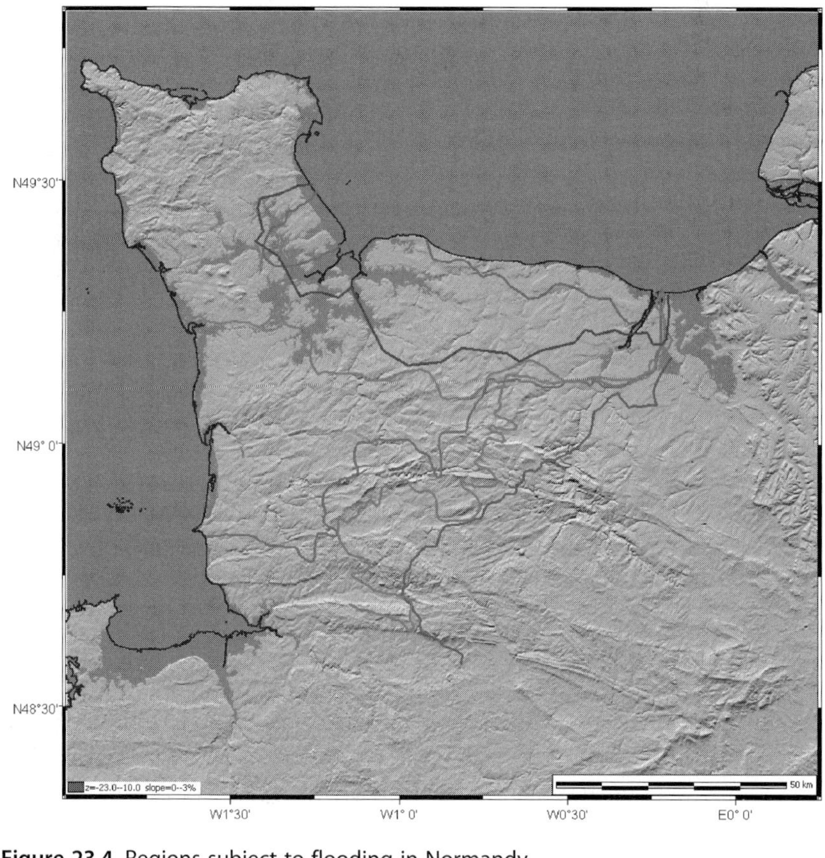

Figure 23.4 Regions subject to flooding in Normandy.

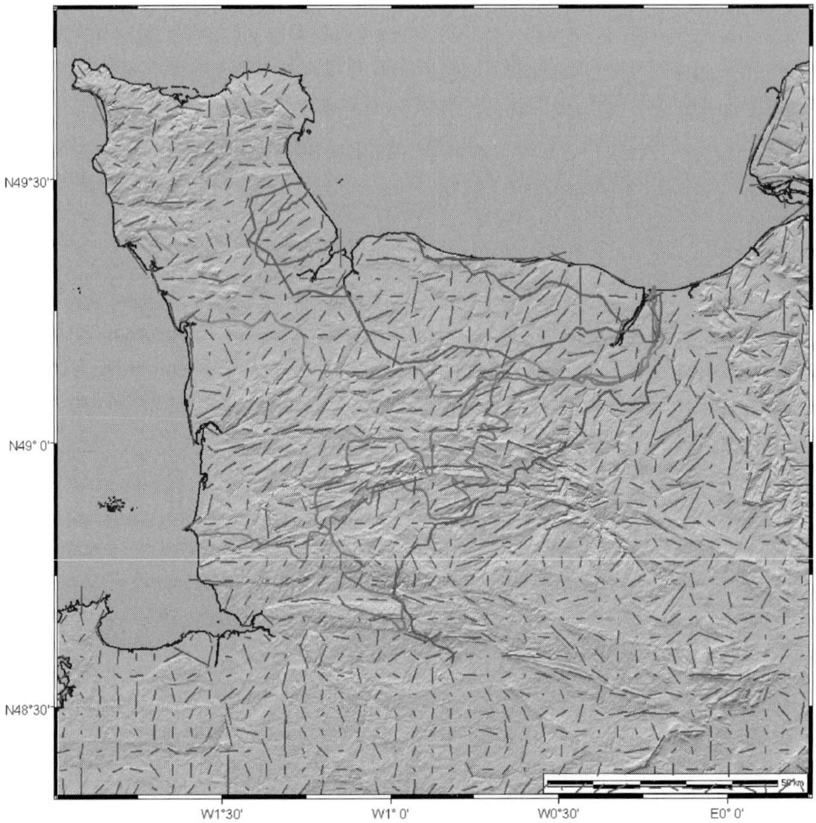

Figure 23.5 Terrain organisation (Guth, 2003) map of Normandy. The length of the lines show the degree to which valleys and ridges share similar orientations, and the orientation of the lines shows the direction in which cross-country mobility will be maximised.

Curvature and convexity measures can also help to show topography. Schmidt et al. (2003) discussed the challenges in mapping curvature as a second derivative compared to slope as a first derivative. Wood (1996) developed landform classifications that can identify ridges and valleys, although his algorithm requires three to four parameters and must be tuned to the characteristics of the DEM and the study region. MICRODEM (Guth, 2008) provides two alternative and simpler methods as shown in Figure 23.6 (with ridges in green and valleys in blue on the color monitor). At this scale, the contrast between the fine texture of the bocage country in Normandy and the more open terrain where the Allies broke out appears obvious.

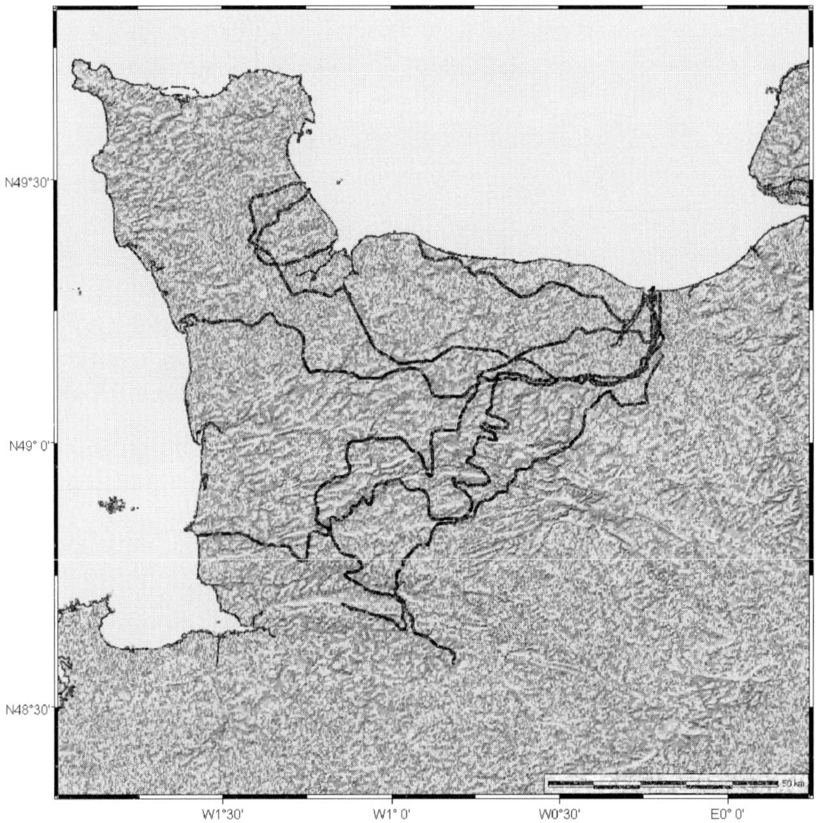

Figure 23.6 Ridge and valley classification of Normandy. Note that the areas of the original landings have very complex, fine scale patterns, and that the ridges and valleys have a much larger scale pattern in the regions where the breakout took place.

In addition to traditional geomorphological mapping, satellite imagery provides useful information for military operations. Figure 23.7 shows an edge filter using a Laplace or Gaussian filter for the 15 m resolution Landsat ETM+ Band 8. This highlights roads, streams, coastlines and the hedgerows along the field margins that created many of the obstacles that slowed military operations in Normandy. This overlay can be rapidly and automatically generated and provides an estimate of the obstacle density. Satellite imagery from the Landsat series of satellites can also be used for land classification. This can be performed in image processing software using supervised or unsupervised classification, whilst a number of national and international organisations have published land cover data sets (Figure 23.8).

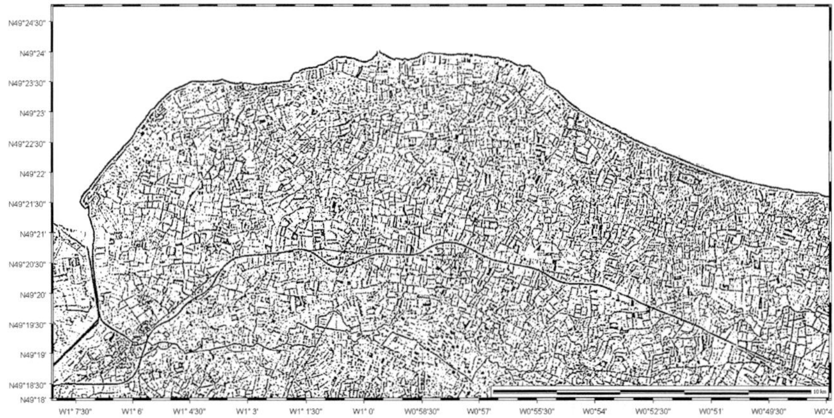

Figure 23.7 Edge map from ETM+ Band 8, using Laplace or Gaussian filter for the region around Omaha Beach.

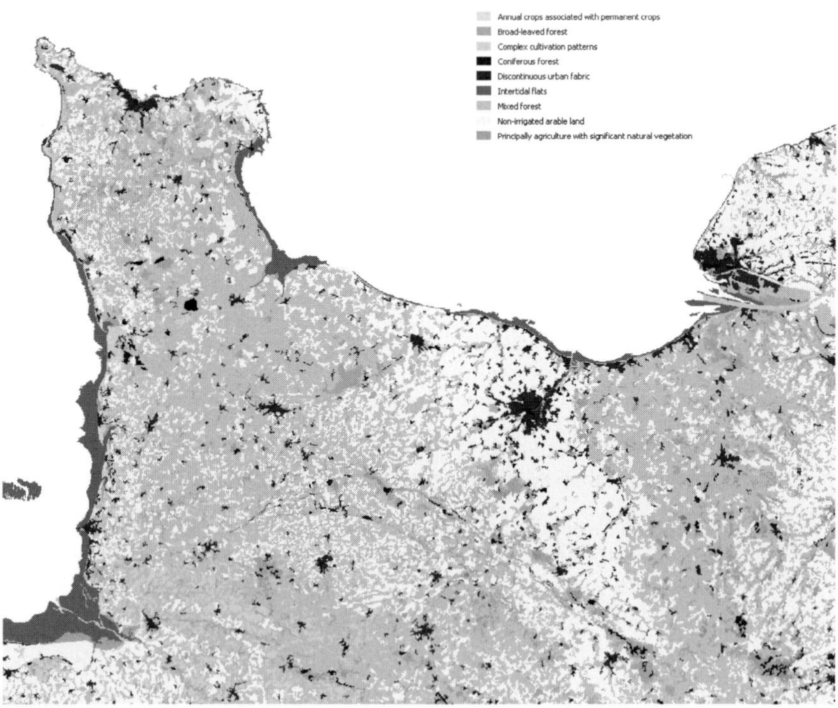

Figure 23.8 Corine land cover 2000 (CLC2000) for Normandy. © *EEA (2007), Copenhagen.*

These typically show two to three dozen categories (often far fewer in any given area), which will depict human impacts on the landscape and the type of vegetation, both of which will impact military operations.

5. CONCLUSION

The physical landscape affects military operations. Whether planning current or future operations, or trying to understand historical events, geomorphic mapping provides powerful tools. Shaded relief maps provide the most useful and intuitive displays, but slope, curvature and organisation maps allow for additional exploration of the battlefield. In the Normandy campaign of 1944, Allied forces took almost 2 months to slowly build up forces and fight though restrictive terrain near the landing beaches. While other factors of troops, equipment and leadership played their roles, the breakout coincided with reaching more favourable terrain for rapid mechanised operations.

ACKNOWLEDGEMENT

KMZ files for the maps shown in this chapter are available online at http://www.usna.edu/Users/oceano/pguth/website/kml/normandy_geomorph/normandy_geomorph_mapping.htm and can be overlaid on current high-resolution imagery in Google Earth.

REFERENCES

Cole, H.M., 1951. Cross-Channel Attack. U.S. Army in World War II. Government Printing Office, Washington, DC. Available at <http://www.history.army.mil/books/wwii/7-4/7-4_Contents.htm#toc>

Department of the Army, 1990. Terrain Analysis: Field Manual 5-33. Available at <http://www.globalsecurity.org/military/library/policy/army/fm/5-33/index.html>

Department of the Army, 2000. Topographic Operations: Field Manual 3-34.230. Available at <http://www.globalsecurity.org/military/library/policy/army/fm/3-34-230/index.html>

EEA [European Environment Agency], 2007. Corine Land Cover 2000 (CLC2000) 100 m – Version 9/2007. Available at <http://dataservice.eea.europa.eu/dataservice/metadetails.asp?id=1007>

Farr, T.G., Rosen, P.A., Caro, E., Crippen, R., Duren, R., Hensley, S., 2007. The shuttle radar topography mission. Rev. Geophys. 45 (2), RG2004, doi:10.1029/2005RG000183.

Fleming, S.D., Hendricks, M.D., Brockhaus, J.A., 2009. The role of GIS in military strategy, operations and tactics. In: Madden, M. (Ed.), Manual of Geographic Information Systems. American Society for Photogrammetry and Remote Sensing, pp. 967–985.

GLCF [Global Land Cover Facility], 2009. Available at <http://glcf.umiacs.umd.edu/index.shtml>

Guth, P.L., 2003. Eigenvector analysis of digital elevation models in a GIS: geomorphometry and quality control. In: Evans, I.S., Dikau, R., Tokunaga, E., Ohmori, H., Hirano, M. (Eds.), Concepts and Modelling in Geomorphology: International Perspectives. Terrapub Publishers, Tokyo, pp. 199–220.

Guth, P.L., 2008. Geomorphometry in MICRODEM. In: Hengl, T., Reuter, H.I. (Eds.), Geomorphometry: Concepts, Software, Applications. Developments in Soil Science Series. Elsevier, Amsterdam, pp. 351–366.

Hammond, W.M., 1994. Normandy: US Army Center of Military History. Available at <http://www.history.army.mil/brochures/normandy/nor-pam.htm>

Kraak, M.-J., Ormeling, F., 2010. Cartography: Visualisation of Spatial Data. Third Edition. Prentice Hall, Harlow, Essex.

Lehner, B., Verdin, K., Jarvin, A., 2008. HydroSHEDS Technical Documentation. Available at <http://gisdata.usgs.net/HydroSHEDS/downloads/HydroSHEDS_TechDoc_v11.pdf>

Library of Congress, 2009. World War II Military Situation Maps. Available at <http://memory.loc.gov/ammem/collections/maps/wwii/date.html>

NGA [National Geospatial-Intelligence Agency], 2009. Prototype Global Shoreline Data (Satellite Derived High Water Line Data). Available at <http://msi.nga.mil/NGAPortal/DNC.portal?_nfpb=true&_pageLabel=dnc_portal_page_72

Rose, E.P.F., 2008. British military geological terrain evaluation for Operation Overlord: the Allied invasion of Normandy in June 1944. In: Nathanail, C.P., Abrahart, R.J., Bradshaw, R.P. (Eds.), Military Geography and Geology: History and Technology. Land Quality Press, Nottingham, pp. 215–233.

Rose, E.P.F., Clatworthy, J.C., Nathanail, C.P., 2006. Specialist maps prepared by British military geologists for the D-Day landings and operations in Normandy, 1944. Cartographic J. 43 (2), 117–143.

Schmidt, J., Evans, I.S., Brinkmann, J., 2003. Comparison of polynomial models for land surface curvature calculation. Int. J. Geogr. Inf. Sci. 17, 797–814.

Tate, J., 2006. Terrain analysis for decision making. In: Mang, R., Häusler, H. (Eds.), International Handbook Military Geography. Ministry of Defense, Vienna, pp. 321–333.

USGS [US Geological Survey], 2009. Shuttle Radar Topography Mission Water Body Dataset. Available at <http://edc.usgs.gov/products/elevation/swbd.html>

Wood, J.D., 1996. The Geomorphological Characterisation of Digital Elevation Models, Unpublished Ph.D. Thesis, University of Leicester, Leicester. Available at <http://www.soi.city.ac.uk/~jwo/phd/>

CHAPTER TWENTY-FOUR

Future Developments of Geomorphological Mapping

Mike J. Smith[a], James S. Griffiths[b] and Paolo Paron[c]
[a]School of Geography, Geology and the Environment, Kingston University, Surrey, UK
[b]SoGEES, University of Plymouth, Plymouth, UK
[c]UNESCO-IHE, Institute for Water Education, Delft, NL & School of Geography and the Environment, Oxford University, UK

The formalisation of geomorphological mapping (Savigear, 1965) as a central platform for recording landform data cemented its role as a key organisational framework for the study of landforms, their history, materials and the processes associated with them. However, geomorphology has a reach far beyond its academic origins and is a key integrating discipline, cross-cutting both academic and professional applications. In short, knowing *where* a landform is and *why*, as well as *what* it is made of and how it has *changed*, is an incredibly powerful tool for understanding and managing the landscape. So, although geomorphological mapping has already proved useful for a large number of applications for some considerable time, the importance and number of these is only set to increase.

Geomorphological mapping has re-emerged from a hiatus since the 1980s as the core interface between academic subjects working on commercial, governmental and research projects; it has seen the development of a pivotal role in terms of synthesising a diverse range of data. This has been partially driven by technological developments and, in particular, access to a variety of remotely sensed data sets (Smith and Pain, 2009). It is now possible to map remote regions, in greater (topographic) detail, over increasingly smaller time periods. However, it is the availability of ever higher resolution topographic data that has transformed our study of terrestrial landforms, whereas the advent of multibeam swath bathymetry and side-scan sonar has had a similar effect in the investigation of submarine landforms.

Landforms are composed of 'stuff', and remotely sensed imagery (principally satellite imagery) provides information on electromagnetic reflectance at different wavelengths. This, of itself, is useful, however, for

Developments in Earth Surface Processes, Volume 15
ISSN: 0928-2025, DOI: 10.1016/B978-0-444-53446-0.00024-0

mapping; we are often interested in 'shape'; this can be gleaned using reflectance as a proxy (through, for example, the use of shadowing), but it is preferable to acquire data on surface elevation as a direct measure of landform surface expression and thereby shape. Archive elevation data are comprised of terrestrial surveys (usually from topographic maps) and up to 15 years of synthetic aperture radar (SAR) data, but the emergence of space, aerial and terrestrial data sets has provided the opportunity for the development of compelling new applications. This is based upon the collection of space and aerial survey data using SAR and light detection and ranging (LiDAR) techniques at relatively high resolutions over increasingly larger regions. Indeed, Intermap's NextMap product leverages the capacity of SARs for operating at higher altitudes to cover large regions meaning that they are able to offer a single, consistent, digital elevation model (DEM) product for the whole of western Europe and conterminous United States – a remarkable achievement. As a result of the development of these technologies, field mapping is becoming *more* prevalent. Whereas theodolites and total stations have long been used for field mapping, they are now being augmented, and in some instances replaced, by terrestrial LiDAR surveying (Hodge et al., 2009).

The previous paragraph highlights the importance of elevation, but that is not to say that other complimentary data sets are unimportant. Far from it, geomorphological mapping is concerned with shape, materials, process and history. Therefore, techniques such as radiometrics (composition of the upper 50 cm of the surface), aeromagnetics (imaging subsurface features) and airborne electromagnetics (3D conductivity down to as deep as 100 m) are vitally important for the development of the discipline. And reflectance remains important, particularly with the increased development of spaceborne hyperspectral remote sensing. The investigation of materials is of vital importance for the understanding of past and future landform evolution: traditional field techniques are now presented with an incredible array of methodologies, such as modern radiometric dating, for even the loosest of sediments (e.g. aeolian sands) and can provide unprecedented insights into, for example, rates of sedimentation on alluvial plains. The miniaturisation of probes and data loggers augment the amount of information that can be collected on the surface and near subsurface and also increases the frequency of data collection for the contexts where repeated observations are needed (e.g. hydrology). Geomorphological mapping, and the development of thematic information, can now easily and cheaply tap into these techniques and methods.

The increasing resolution of geophysical data gathering technologies in the marine environment has revolutionised the investigation of submarine landforms. High-resolution bathymetric maps can now be compiled through the use of remotely operated vehicles (ROVs) or autonomous underwater vehicles (AUVs). This high-resolution information is used in conjunction with bathymetry, side-scan data and seismic reflection surveying collected by surface vessels to provide detailed images of the seabed that are then subject to geomorphological interpretation. Although not cheap, this work has a major role to play in the assessment of offshore resources and submarine hazards.

The technologies outlined above have allowed us to capture massive data sets at higher resolutions over shorter timescales and with increased frequency. This provides a revolutionising change to the way we can analyse and understand the Earth's surface, but it also brings with it a range of problems related to data management and manipulation. Geographic information systems (GIS) provide the ideal management solution, utilising the paradigm of data 'layers', geodetically referenced to one another, to combine and analyse. The ability to handle, manipulate and analyse large, parallel, data sets will slowly begin to be realised, but remains a significant problem, particularly with increasingly complex algorithms for data analysis. As a result, computer processing power and storage space will remain a potential constraint within current computing paradigms. However, the move towards server and software virtualisation will lead to an improvement in both stability and scalability. Although this can be conveniently conflated with 'cloud computing', this should not be confused with current consumer implementations. This is an environment focused upon server-based applications, delivered to the desktop, that are expandable and ideally suited to intensive processing tasks. As a result, it is simply a case of purchasing greater capacity, as and when required, that can then be automatically expanded.

Geomorphological mapping is now going through a period of renaissance; initial academic development during the 1960s and 1970s saw huge potential, but the fragmentation of legend systems and, more widely, a change to large-scale geomorphological field studies led to its relegation as an adjunct fringe activity. However, it was enthusiastically employed by applied fieldworkers in areas such as engineering geology, who saw its potential as an integrating practice. The dramatic increase in activity over the last decade has been driven by technology; the ability to access data over new, unexplored, regions in greater amounts of detail, all within a digital workflow, has been enthusiastically taken up by academics and practitioners alike. Of greater importance is the realisation that geomorphology really is an 'interface' discipline, providing a key link between pure and applied sciences that seek to understand and manage the Earth's surface and near surface.

Geomorphology therefore finds itself in a position of some importance, representing one of the key linkages as part of society's effort to better understand and manage its environment both currently and for future generations. So in the same way that geological mapping is a key underpinning of resource exploitation by society, geomorphology is increasingly being seen as a societal necessity. Indeed, this book should be seen as part of this general trend that emphasises the increased interest and requirement for geomorphological mapping. We have a background and history which is important for understanding the current status of geomorphological mapping, a rapidly developing set of methodologies and techniques for exploiting increasingly detailed and complex data sets and an amazingly diverse range of applications. Although this has far-reaching implications in and of itself, the training of geomorphologists to meet these new demands becomes more critical as the subject becomes increasingly recognised as a genuine applied science. However, the skills needed to develop and use the new methodologies and technologies will not reduce the importance of the practicing geomorphologist acquiring field-based expertise and experience. The best geomorphologists will still be those who have seen, mapped and measured the most landforms.

At the same time, the digital, economic and educational divide between the great and less economically developed countries may well increase and at greater rates. For this reason, there should be a stronger imperative for cooperation on research and educational programmes between these regions. The vast quantities of data available through the Internet for western countries are remarkable but remain a dream for developing countries. This is in part due to a lack of investment in data collection programmes, as well as the availability of broadband Internet access facilities, with subsequent detrimental impacts upon education, research and professional practise. It is also a matter of fostering collaboration and exposure between African, South American and Asian geomorphologists with other communities around the world.

Future trends will undoubtedly see geomorphological mapping become more ubiquitous and the products made widely available. Producers of geomorphological data and derived mapping will seek to engage stakeholders and, increasingly, the general public. Dissemination will therefore make use of distribution channels and, in particular, digital routes including virtual globes and their capacity for landscape and landform visualisation, web mapping and GeoPDFs. The sophistication with which these data sets are managed, maintained and analysed will increase. Indeed, the accelerating trend of capturing more data, in greater detail, over shorter time periods will continue, with a particular focus upon elevation point clouds and derived DEMs. The next decade should be viewed with anticipation as the discipline is significantly enhanced.

REFERENCES

Hodge, R., Brasington, J., Richards, K.S., 2009. In situ characterization of grain-scale fluvial morphology using terrestrial laser scanning. Earth Surf. Process. Landforms 34, 954–968.

Savigear, R.A.G., 1965. A technique of morphological mapping. Ann. Assoc. Am. Geogr. 55 (3), 514–538.

Smith, M.J., Pain, C.F., 2009. Applications of remote sensing in geomorphology. Prog. Phys. Geogr. 33, 568–582.

INDEX

D

E

J

K

L

M